James M. Apple is Professor in the School of Industrial and Systems Engineering at the Georgia Institute of Technology. He is a consultant to industry and Director of the annual Material Handling Management Course at Lake Placid, and was previously Professor of Industrial Engineering at Michigan State University and Vice-President in Charge of Manufacturing for the Clarke Floor Machine Company. Professor Apple is the author of *Material Handling Systems Design* and a contributing editor of the *Production Handbook*, published by The Ronald Press Company.

Plant Layout and Material Handling

THIRD EDITION

JAMES M. APPLE
Georgia Institute of Technology

JOHN WILEY & SONS
New York • Chichester • Brisbane • Toronto • Singapore

ISBN 0 471 07171-4

Library of Congress Catalog Card Number: 77–075127

Printed in the United States of America

20 19 18 17 16

Preface

Responding to the gratifying acceptance of the two previous editions, this *Third Edition* continues as a demonstration of the ordered planning necessary for efficient material flow and the preparation of effective layouts for requisite physical facilities. The advances in the field in recent years are brought in, and new chapters are devoted to process design, the use of quantitative techniques in analyzing material flow, computerized layout procedures, and the equally critical matter of facility location. As before, while layout is presented as an engineering function, the technical aspects of design and its implementation in construction and specific equipment remain secondary to the matter of coordinating layout, material handling, methods design, and production planning and control and the integration of productive facilities with the many related service and auxiliary functions.

The pedagogical features that have made it the preferred text in four-year and two-year programs in both industrial engineering and industrial engineering technology have been retained. The content has grown out of long association with numerous industries, as manufacturing executive and consulting engineer, and long experience as both college and industry teacher. Outlined in Chapter 2 and developed in the subsequent chapters is a pattern of procedure by which the many phases of layout planning can be systematically integrated in a sequence that expedites the efficient interrelation of all factors.

While the principal focus is toward the industrial establishment, recognition is given throughout to the applicability of procedures and techniques in other areas, and to the adaptability of the approaches and methods to the planning of any facility, whether library, campus, restaurant, post office, or retail store.

It would have been impossible to offer this book without the help of many, nor can this be acknowledged adequately. Teaching colleagues gave invaluable assistance in refining techniques and in reviewing and class-testing portions of the text. Many companies contributed illustrations. Engineers from industry have given willingly of their experience and have tested many of the techniques. And my students at Georgia Tech and many practitioners at seminars and short courses have contributed ideas through the interplay of questions, answers, and class discussion. My sincere appreciation goes to them all, and to users of the book, past and future, from whom comments and suggestions are most welcome.

Prevailing is a special debt of gratitude to my wife and family whose patience and cooperation were so essential to the completion of the writing.

Atlanta, Georgia
August, 1977

JAMES M. APPLE

Contents

Plant Layout and Material Handling

1

The Facilities Design Function

One of the oldest activities of the industrial engineer is plant layout and material handling. At least that is what it has been called for many, many years. It is the activity that deals with *the design of an arrangement of the physical elements of an activity*—and has always been very closely related to the manufacturing industry, where the drawing of the resulting design was known as a *plant layout.* And a good layout nearly always involves the methods of handling material as it moves through the plant; therefore, *plant layout and material handling.*

However, in recent years, as the industrial engineer broadened his outlook toward physical activity, he became aware that almost all meaningful activity required physical facilities, and most often such facilities could and should be planned and designed by following pretty much the same principles and procedures he had been using on plant layouts. So, he began to use his methodology in the design of any physical facility—therefore, *facility design,* a term just as significant for the arrangement of the physical elements of a warehouse, post office, retail store, restaurant, hospital, home, or even a factory.

In any such project, the overall objective would be to consider what the appropriate inputs might be, and design an arrangement that would move them efficiently through the facility as the required activities are performed to achieve the desired outputs (see Table 1-1). As can be seen, all of the activities involve inputs and outputs. The processing can be loosely referred to as *production* (productive activity) and in most cases requires a number of work places, machines, or other pieces of equipment, through which the inputs pass in being processed into outputs. However, lest the above be misunderstood, the facilities design engineer does not design the building—the architect or architectural engineer does this.

Definition of Facilities Design

The Facilities Design Engineer analyzes, conceptualizes, designs, and implements systems for the production of goods or services. The design is usually represented as a *floor plan,* or an arrangement of physical facilities (equipment, land, buildings, utilities), to optimize the interrelationships among operating personnel,

Table 1-1. Major Elements of Facilities Design

	Typical Inputs	Production Activities	Outputs
1. Plant	Materials and supplies	Conversion of materials to parts, assemblies, products	Products (and scrap!)
2. Warehouse	Large quantities of merchandise	Safekeeping and availability	Orders of merchandise
3. Retail Store	Orders of merchandise	Display, convenient access, transfers of ownership	Individual items for customers
4. Post Office	Letters and parcels	Sorting and accumulating	Orderly arrangement of letters and parcels
5. Restaurant	Food and supplies	Preparation of food	Meals
6. Hospital	Sick patients, medicine, supplies	Services required to "cure" patients	Cured patients
7. Home	Food, supplies, equipment, etc.	Meals and orderly activity	Happy people

material flow, information flow, and the methods required in achieving enterprise objectives efficiently, economically, and safely.

In general, the overall objective of facilities design is to get the inputs (material, supplies, etc.) into, through and out of each facility in the shortest time practicable, at acceptable cost. In industrial terms, the shorter the amount of time a piece of material spends in the plant, the less opportunity it has to collect labor and overhead charges.

Most facility design work deals with industrial facilities or plants, and this text deals primarily with that area of activity. It should be understood that an easy *translation* of the examples and terminology used will make it possible to apply the concepts, principles, and procedures to the design of any facility, for any productive enterprise.

Scope of Facilities Design

Facilities design work is frequently thought of as dealing only with the careful and detailed planning of production equipment arrangement. However, this is really only one phase of a very extensive series of interrelated activities making up a typical facility layout project.

The scope of facilities design work should include a careful study of at least the following areas of interest:

1. Transportation
2. Receiving
3. Storage
4. Production
5. Assembly
6. Packaging and packing
7. Material handling
8. Personnel services
9. Auxiliary production activities
10. Warehousing
11. Shipping
12. Offices
13. External facilities
14. Buildings
15. Grounds
16. Location
17. Safety
18. Scrap

The work of designing a facility usually starts with an analysis of the product to be made, or the service to be performed, and a consideration of the overall flow of material or activity. It progresses, step-by-step, through the detailed planning of the arrangement of equipment for each individual work area. Then the interrelationship between work areas is planned; related areas are coordinated into units, sections, or departments—which then are woven into a final layout. The detailed steps by which this work is accomplished are detailed in subsequent chapters.

Importance of Facilities Design

The importance of facilities design to the efficient operation of an enterprise cannot be overemphasized. It should be recognized that the flow of material usually represents the backbone of a productive facility, and should be very carefully planned and not allowed to grow or develop into an unwieldy octopus of confused traffic patterns. Perhaps the concept can be summarized as follows:

1. *An efficient plan for the flow of material* is a primary requisite for economical production.
2. *The material flow pattern* becomes the basis for an effective arrangement of physical facilities.
3. *Material handling* converts the static flow pattern into a dynamic reality, providing the means by which material is caused or permitted to flow.
4. *Effective arrangement of facilities* around the material flow pattern should result in efficient operation of the various related processes.
5. *Efficient operation of the processes* should result in minimum production cost.
6. *Minimum production cost* should result in maximum profit.

The material flow pattern, then, becomes the basis for the entire plant design as well as for the success of the enterprise. All too frequently, insufficient emphasis is placed on determining the most efficient plan for the flow of material through production facilities.

It should be our conclusion that the facilities design, or plant layout, process comes first. Certainly one would not build a shell for a house, and only afterward attempt to fit into it all the things necessary for a complete and comfortable home. Neither should one erect any industrial building without first completing a facilities design study and then developing a design for the facility. This will determine the desired flow of material, the most economical arrangement of physical facilities, and will serve as the basis for the building design. Of course, the architect should be consulted in the early planning stages for advice on general building construction, but his actual design work should follow that of the plant layout engineer. In this regard, one prominent architectural engineer has said:[1]

> Today's factory is no longer simply a consolidation of manufacturing activities—it is more commonly a *supermachine.* Its potential output may be enormous and its labor-reducing efficiencies may be impressive. But this super-machine is also unusually complex, sensitive, often rather inflexible, and quite frequently requires a substantial capital investment. In many instances, this supermachine will have an unforeseen impact upon sales policy, inventory levels, maintenance and control of downtime, and the work force. These considerations mean that today's industrial manager must, in effect, actively design his manufacturing and distribution facilities to meet the same economic objectives as his marketing strategy. Here, too, he must call upon the finest professional engineering and architectural team when embarking upon a facilities expansion program.

Facilities Design and Productivity

The continuing cry of businessmen is for ways of reducing costs—to offset the ever-increasing prices they must pay for labor and material. One of the prime sources of cost-improvement opportunities is in the re-designing of facilities and the methods by which they turn out their products. *Business Week*[2] has said:

> Behind the scaling up and modernization of industrial plant—and the changes in technology and organization that go with it—there is one prime motivation: to produce more goods and services with less labor. Development of an affluent society brings with it a structure of costs that almost automatically forces management to economize on labor. What was median family income a generation ago is now the official "poverty line." And what was average hourly pay for industrial workers is now the legal minimum wage.
>
> A substantial chunk of the fatter payroll is, to be sure, pure inflation, because management has tried to keep pace by boosting prices. But the practically universal expectation of a yearly increase in pay gives management no option when it comes to labor-saving. Industries that cannot achieve new standards of productivity to go along with the new standards of pay are caught in a vise.

And, following up this line of thought, one engineer has said:[3]

> A 10 per cent reduction (of labor payroll) can be anticipated under the minimum conditions of plant relocation to a modern environment, even without significant changes in methods or equipment.

[1] T. A. Faulhaber, "Planning Your Plant," *Industry,* March 1963.
[2] Oct. 17, 1970, p. 183.
[3] D. C. Stewart, "Plant Design with Labor in Mind," *Consulting Engineer,* Nov. 1968, p. 117.

Then, almost as if in conclusion, another source states:[4]

> It is obvious that if we are to achieve greater stability of unit costs and prices, productivity growth is essential. If we are to improve the "quality of life" (to be interpreted as the reader sees fit), the alternative to higher taxes and/or a reduction of other demands is higher productivity and higher output. If we are going to continue to have a high and rising standard of living, productivity is the *sine qua non.*

Increasing productivity is usually a desired result of facilities design, or re-design. It is accomplished by the design efforts necessary to carry out the several objectives of the facilities design process.

Objectives of Facilities Design

If a finished layout is to present an effective arrangement of related work areas, in which goods can be economically produced, it must be planned with the objectives of layout well fixed in mind. The major objectives are to:

1. Facilitate the manufacturing process.
2. Minimize material handling.
3. Maintain flexibility of arrangement and of operation.
4. Maintain high turnover of work-in-process.
5. Hold down investment in equipment.
6. Make economical use of building cube.
7. Promote effective utilization of manpower.
8. Provide for employee convenience, safety, and comfort in doing the work.

A brief discussion of each of these objectives will guide the layout engineer in pursuing them.

Facilitating the manufacturing process. The layout should be designed in such a way that the manufacturing process can be carried on in the most efficient manner. Some specific suggestions are:

1. *Arrange machines, equipment, and work areas* so that material is caused to move smoothly along in as straight a line as is possible.
2. *Eliminate all delays possible.* It has been said that during 80 per cent of the time a part is in the plant it is either being moved or stored—only 20 per cent of the time is productive.
3. *Plan the flow* so that the work passing through an area can be easily identified and counted, with little possibility of becoming mixed with other parts or batches in adjacent areas.
4. *Maintain quality of work* by planning for the maintenance of conditions that are conducive to quality.

These and many other suggestions will be more completely explained as the following chapters present the layout planning procedure.

[4]Machinery and Allied Products Institute, *Capital Goods Review,* March 1972.

Minimizing material handling. A good layout should be planned so that material handling is reduced to a minimum. Wherever practicable, handling should be mechanical; and all movement should be planned to move the part toward the shipping area. Where possible, the part should be "in-process" while in transit, as in painting, baking, degreasing, etc. Material handling will be dealt with in greater detail in later chapters.

Maintaining flexibility. Although a plant or department may be planned for the production of a certain quantity of a certain item, there are many occasions when it will be necessary to alter its production capabilities.

Many of the changes thus called for may be more easily made if they are anticipated in the original planning. A common way to facilitate the rearrangement of equipment is to install utility systems into which service connections can be easily tied when the building is constructed. Good examples are the electrical ducts and the cutting-compound pipe lines which are installed overhead, down the centers of bays. Such arrangements permit machines to be *plugged-out,* moved into new locations, and *plugged-in* again, almost at will.

Maintaining high turnover of work-in-process. The greatest operating efficiency is obtainable only when the material is moved through the necessary processes in the shortest possible time. Every minute a part spends in the facility adds to its cost, through the tie-up of working capital. The nearest to an ideal situation exists in the process-type industry where, by its nature, the material passes, sometimes without stopping, from the start to the finish of the process. If in-process storage of material is reduced to a minimum, the overall material turnover (manufacturing) time is reduced, the amount of work-in-process is reduced, inventory is decreased, and a lower amount of working capital is tied up therein. These savings, in turn, reduce production costs.

Reducing investment in equipment. The proper arrangement of machines and departments can aid considerably in reducing the quantity of equipment required. For example, two different parts, both requiring the part-time use of an internal grinder, may be routed through the same machine, thus eliminating the cost of a second machine. Foresight in selecting the method of processing may sometimes save purchasing a machine. If it is found that one part, as processed, calls for broaching, and will use only part of the capacity of a machine, a switch to drilling and reaming might be effected and the job done on equipment already available.

Making economical use of building cube. Each square foot of floor area in a plant costs money. One manufacturer, for example, has calculated his floor area cost to be $1.00 per square foot per month. This amount includes all overhead costs. Only if each square foot is used to best advantage can the attending overhead costs per unit of product be kept down. Floor area occupied by equipment in operation pays its own way; unoccupied, wasted, or idle floor area is a burden.

Proper layout dictates minimum spacing between machines, after the necessary allowances for the movement of men and materials have been made. With proper consideration of machine spacing in relation to other factors, much floor area can be saved. At that, many manufacturers find that only about 50 per cent of their floor area is occupied by production equipment.

Promoting effective use of manpower. A large amount of productive manpower may be wasted through poor layout practices. Proper layout, on the other hand, may increase the effective utilization of labor. Suggestions such as the following should lead to increased labor utilization:

1. *Reduce manual handling* of materials to a minimum.
2. *Minimize walking.* Twenty per cent of the time spent on one assembly line was occupied by men walking to and from material supplies and keeping up with the assembly conveyor as it moved along. This time loss was reduced considerably by bringing materials closer to the workers with specially designed racks, hoppers, and conveyors, and having the conveyor index at predetermined intervals, instead of moving continuously.
3. *Balance machine cycles,* so that, as nearly as possible, machines and workers are not unnecessarily idle. Well-balanced operation necessitates good material handling, good production control, good methods engineering, and good supervision.
4. *Provide for effective supervision.* In theory, the supervisor might stand in the midst of his group, so that he would be in immediate contact with each employee.

Although such a plan seems hardly possible, it is necessary to emphasize that a properly laid-out department is easier to supervise than one that is spread out over too large an area, is too congested, or otherwise hinders the relationship between the supervisor and his men. A well-laid-out department makes it easier for a supervisor to handle more employees, keep work moving, and conserve his time for his more important duties.

Providing for employee convenience, safety, and comfort. Satisfying this objective requires attention to such items as light, heat, ventilation, safety, removal of moisture, dirt, dust, etc.

Equipment causing excessive noise should be isolated as much as possible or enclosed in an area with sound-deadening walls and ceiling. Equipment that vibrates should be cushioned, or specially mounted, to prevent the transmission of vibration to the floor or surrounding objects. Safety must also be assured by proper planning of the layout. Machines and auxiliary manufacturing equipment must be so placed as to prevent injury to personnel and damage to material and to other equipment. Safety may be incorporated into the layout by a careful study of workplace arrangement, material handling methods, storage techniques, ventilation, lighting, fire protection, and all other factors involved in an operation.

It will be noted that often it is impossible fully to achieve these objectives. In fact, some of them are rather almost in opposition to each other. Nevertheless, each represents an important goal toward which the layout engineer must strive. When objectives in a particular situation seem to be opposed, an equitable solution must be reached that will be most effective in light of all factors considered.

Functions and Activities in Facilities Design

In carrying out the objectives stated above, the facilities design or plant layout group will perform a large number—and a wide variety—of activities. Rather than try to record an incomplete list, the *Facilities Planning and Design Activity Generator* is offered in Table 1-2. The three lists contain the *raw material* for 80, 104 theoretical potential activities of the Facility Planning and Design function. That is, any combination of the words (one from each column) will yield a possible activity or project. Actually, of course, some will be *nulls,* and in others a change in word sequence may be required for proper sense.

It will be noted that the center column pretty well identifies the entire range of facilities design interests. The two outer columns serve to delineate the tremendous number of potential activities the designer might become involved in.

The Continuing Need for Facilities Design Work

One might wonder, "What does the facilities designer do, once a facility is designed?" Well—he is not among the unemployed! In almost every plant, there are always changes and improvements to be worked out, new equipment to be integrated, future plans to be developed, and many other related tasks. In fact, one large manufacturer, occupying 1,200,000 sq ft of plant, has said that their layout turns over every five years—at a cost of $4.00/sq ft to re-arrange. This certainly generates a large amount of facilities design work.

Then, too, there should always be a long range plan, or layout—possibly as follows:

10 years ahead—Estimate of total space requirement, by function.
5 years ahead—Area Allocation Diagram, by department (see Chapter 12).
2 years ahead—Rough layout of area under consideration.
1 year ahead—Detailed layout of area.
6 to 12 months ahead—Appropriation approved.

Such a long-range plan provides for the orderly development of both improvement and expansion, and assures that all moves will be made in the direction of the master plan, which of course is being continuously up-dated.

Table 1-2. Facilities Planning and Design Delineation of Activities and Project Generator

Actions	Areas of Interest	Nature of Interest
1. Advise (on)	1. Assembly	1. Activities
2. Allocate	2. Auxiliary services	2. Analysis
3. Analyze	3. Building	3. Appropriations
4. Approve	4. Construction	4. Automation
5. Classify	5. Containers	5. Availability
6. Coordinate	6. Cost	6. Benefits
7. Design	7. Distribution	7. Budgets
8. Determine	8. Equipment	8. Capability
9. Develop	9. Fabrication	9. Capacity
10. Establish	10. Facilities	10. Changes
11. Estimate	11. Flow	11. Contracts
12. Evaluate	12. Grounds	12. Controls
13. Forecast	13. Handling	13. Costs
14. Improve	14. Location	14. Criteria
15. Investigate	15. Maintenance	15. Data
16. Measure	16. Manpower	16. Efficiency
17. Minimize	17. Manufacturing	17. Equipment
18. Model	18. Material	18. Expansion
19. Modify	19. Office	19. Facilities
20. Monitor	20. Packaging	20. Feasibility
21. Organize	21. Packing	21. Flexibility
22. Plan	22. Plant layout	22. Flow
23. Predict	23. Pollution	23. Functions
24. Promote	24. Process	24. Hazards
25. Provide	25. Production	25. Incentives
26. Qualify	26. Productivity	26. Installation
27. Reduce	27. Repair	27. Interrelationships
28. Review	28. Safety	28. Layout
29. Select	29. Salvage	29. Liason
30. Specify	30. Scrap	30. Location
31. Study	31. Security	31. Long range planning
32. Supervise	32. Space	32. Measurements
33. Test	33. Storage	33. Mechanization
34. Up-date	34. Traffic	34. Methods
	35. Transportation	35. Modifications
	36. Utilities	36. Needs
	37. Warehousing	37. Objectives
	38. Waste	38. Operations
		39. Organization
		40. Performance
		41. Plans
		42. Policies
		43. Problems
		44. Procedures
		45. Processes
		46. Programs
		47. Progress
		48. Records
		49. Regulations
		50. Replacements
		51. Reports
		52. Requirements
		53. Resources
		54. Schedules
		55. Services
		56. Sources
		57. Specifications
		58. Standards
		59. Systems
		60. Tests
		61. Trends
		62. Utilization

The three lists contain the raw material for 80, 104 theoretical potential activities of the Facilities Planning and Design functions. That is, any combination of the words (one from each category) will yield a possible activity or project. Actually, of course, some will be *nulls*, or may require a change in word sequence to be properly meaningful.

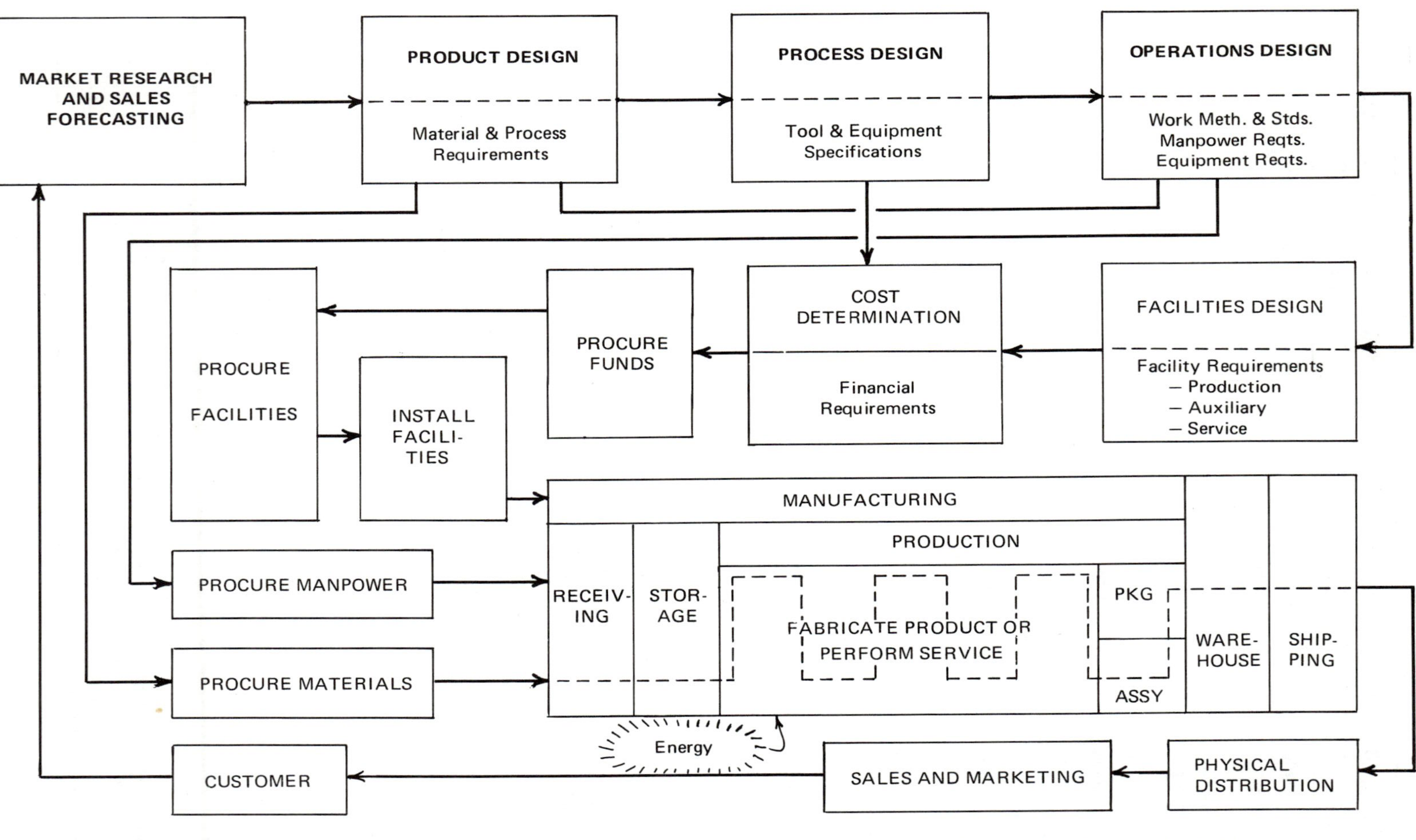

Figure 1–1—The enterprise design process—general interrelationships among major system elements, functions, and activities.

The Enterprise Design Process

Almost without exception, any enterprise would be designed somewhat following the sequence of activities shown in Figure 1-1. The enterprise aims to satisfy *customer needs* by way of:

1. *Market research*—Determining what the customer wants.
2. *Sales forecasting*—Determining how many, or how much.
3. *Product design*—Delineating product or service details.
4. *Process design*—Determining how to make the product, or provide the service.
5. *Operation design*—Working out methods for effecting the processes, and from that, the number of machines and amount of manpower required.
6. *Facilities design*—Determining the material flow paths, and designing the arrangement of activities to provide an orderly and efficient movement of material through the overall process.

It is with this step that most of the balance of this book is concerned, although it will inevitably become involved in other, interrelated, activities. Subsequent steps in the enterprise design process are:

7. *Equipment design* (as necessary).
8. *Building design*—By the architect and his associates.
9. *Financing* the facility.
10. *Procurement*—(a) building, (b) equipment, (c) manpower.
11. *Installation of facilities.*
12. *The actual manufacturing, or productive process.* For this text, let us distinguish between: (a) *manufacturing*—the conversion of materials and supplies into products, such as typewriters, lamps, foods, automobiles, etc.; and (b) *the productive process*—the organization and use of human effort to accomplish a desired result, in the form of a product or a service, (dry cleaning, retailing, banking, food service, etc.).
13. *Warehousing* of the finished goods.
14. *Distribution* of goods, *via*
15. *Marketing and sales,* to the
16. *Customer*—whose use, evaluation, complaints, suggestions, etc., come back to the enterprise by way of Market Research, beginning the whole cycle over again.

The reader can easily visualize many examples of this cycle with reference to products he is familiar with, such as automobiles, appliances, bicycles, and many others (as well as services) that are subject to re-design or model changeovers, based—at least partially—on customer demands. And, of course, it is this cycle that keeps the facilities designer on his toes, to keep the facility capable of meeting the ever-changing demands on its capabilities.

The Facilities Design Process

In order to carry out the design work implied in the foregoing discussion, the facilities designer should follow an orderly procedure, to insure that he has done a complete and accurate job in producing the facilities design, or plant layout. Almost regardless of type of facility, the design process will closely follow the sequence of steps listed (to be detailed in the later chapters indicated):

1. Procure *basic data* (3).
2. Analyze *basic data* (3).
3. Design *productive process* (4).
4. Plan *material flow pattern* (5, 6, 7).
5. Consider general *material handling plan* (14, 15, 16).
6. Calculate *equipment requirements* (11).
7. Plan individual *work stations* (11).
8. Select specific *material handling equipment* (15, 16).
9. Coordinate groups of *related operations* (11).
10. Design *activity interrelationships* (8).
11. Determine *storage requirements* (9, 10).
12. Plan *service and auxiliary activities* (9, 10).
13. Determine *space requirements* (11).
14. Allocate activities to *total space* (12, 13).
15. Consider *building types.*
16. Construct *master layout* (17).
17. *Evaluate, adjust, and check layout* with appropriate persons (18).
18. *Obtain approvals* (18).
19. *Install layout* (18).
20. *Follow-up on implementation of the layout* (18).

It will be noted that the chapter numbers do not follow in sequence. For that matter, neither—in many cases—will the steps themselves. Since no two layout design projects are the same, neither are the procedures for designing them. And, there will always be a considerable amount of jumping around among the steps, before it is possible to complete an earlier one under consideration. Likewise, there will be some backtracking, going back to a step already done—to re-check, or possibly re-do a portion, because of a development not foreseen.

However, the steps do have a general sequence, as do the chapters, although the consideration of some topics will not always be compatible with the planning sequence. Figure 1-2 shows the relationships between the several steps of the process, in a flow chart format.

Types of Layout Problems

Although the discussion to this point may have given the impression that all facilities design or layout projects are for brand new facilities, this is not at all the case. More frequently the problem involves the re-layout of an existing process or an alteration of some sort in the arrangement of certain equipment. Layout problems are of several types:

Design change—Frequently a change in the design of a part calls for changes in the process or operation to be performed. This change may require only minor alterations of the existing layout, or it may result in an extensive re-layout program, depending on the nature of the change.

Enlarged department—If, for one reason or another, it becomes necessary to increase the production of a certain part or product, a change in the layout may be called for. This type of problem may involve only the addition of a few machines for which room can easily be made, or it may call for an entirely new layout if the increased production calls for a process different from the one used before. For example, if compressors were being made in hundreds, ordinary toolroom equipment might be used. However, if the schedule were changed to thousands it might be expedient to install a related group of special-purpose machines.

Reduced department—This problem is nearly the reverse of what was stated above. If production quotas were reduced drastically and permanently, it would be necessary to consider using a process different from the one previously used for high production. Such a change would probably require the removal of present equipment and planning for the installation of other types of equipment.

Adding a new product—If a new product is added to a line, and it is similar to the products already being made, the problem is primarily one of enlarging a department. If, however, the new product differs considerably from those in production, a different problem presents itself. The present equipment may be used by adding a few new machines here and there in the existing layout, with a minimum of rearrangement; or it may be found necessary to set up a completely new department or section—possibly a new plant.

Moving a department—Moving a department may, or may not, present a major layout problem. If the present layout is satisfactory, it is necessary only to shift to another location. If, however, the present layout has not been satisfactory, an opportunity presents itself for the correction of past mistakes. This may amount to a complete re-layout of the area in question.

Adding a new department—This problem may arise from a desire to consolidate, let us say, the drill press work from all departments into one central department; or it may result from the need for establishing a department to do work never before performed. Such a case would arise if it were decided to make a part which had previously been purchased from an outside firm.

Replacing obsolete equipment—This may require movement of adjacent equipment to provide additional space.

Change in production methods—Anything more than a small change in a single workplace is very likely to have an effect on adjacent workplaces or areas. This will require a re-working of the area involved.

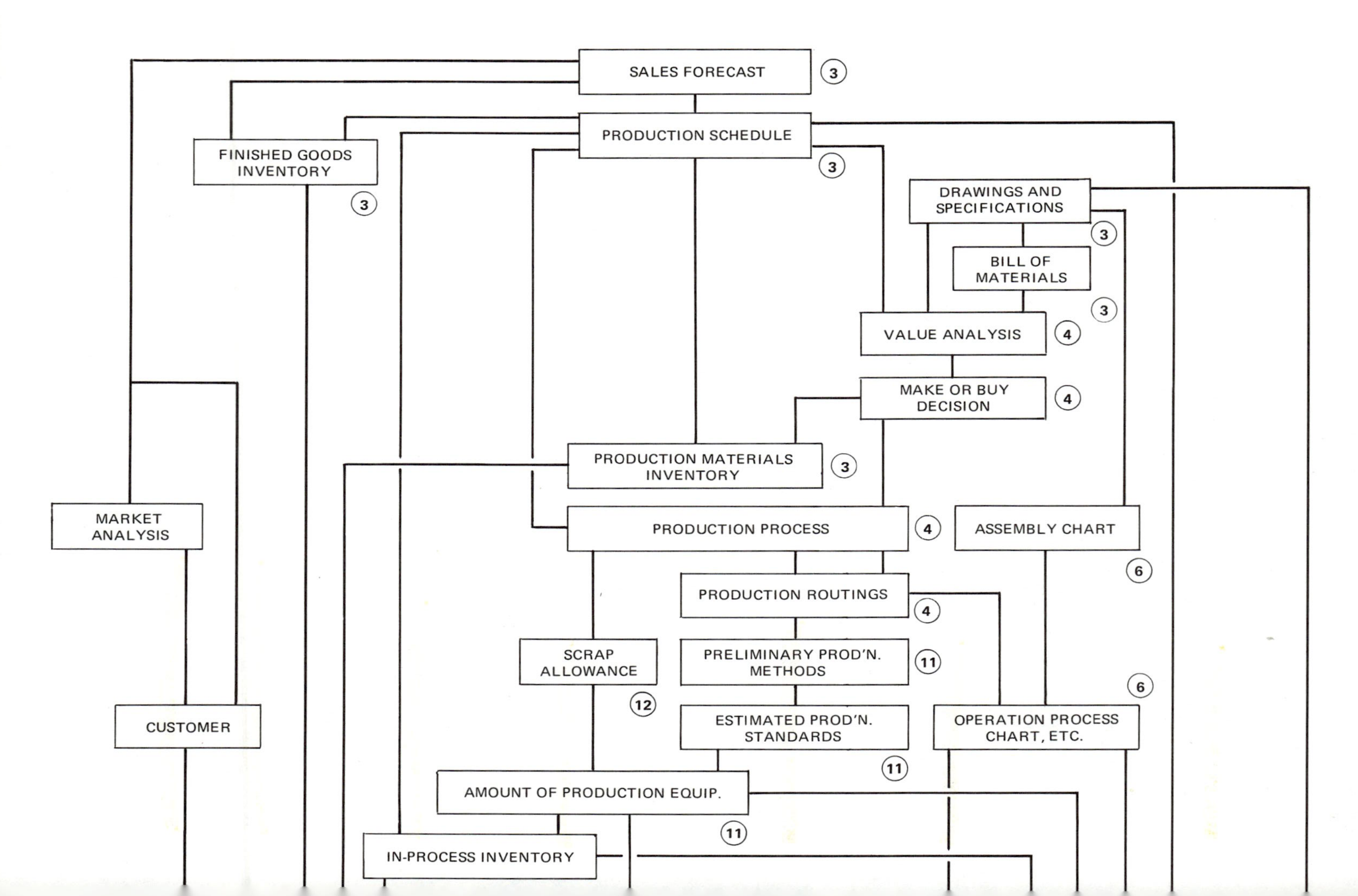
SALES FORECAST
3
PRODUCTION SCHEDULE
3
FINISHED GOODS INVENTORY
3
DRAWINGS AND SPECIFICATIONS
3
BILL OF MATERIALS
3
VALUE ANALYSIS
4
MAKE OR BUY DECISION
4
PRODUCTION MATERIALS INVENTORY
3
MARKET ANALYSIS
PRODUCTION PROCESS
4
ASSEMBLY CHART
6
PRODUCTION ROUTINGS
4
SCRAP ALLOWANCE
12
PRELIMINARY PROD'N. METHODS
11
CUSTOMER
ESTIMATED PROD'N. STANDARDS
11
OPERATION PROCESS CHART, ETC.
6
AMOUNT OF PRODUCTION EQUIP.
11
IN-PROCESS INVENTORY

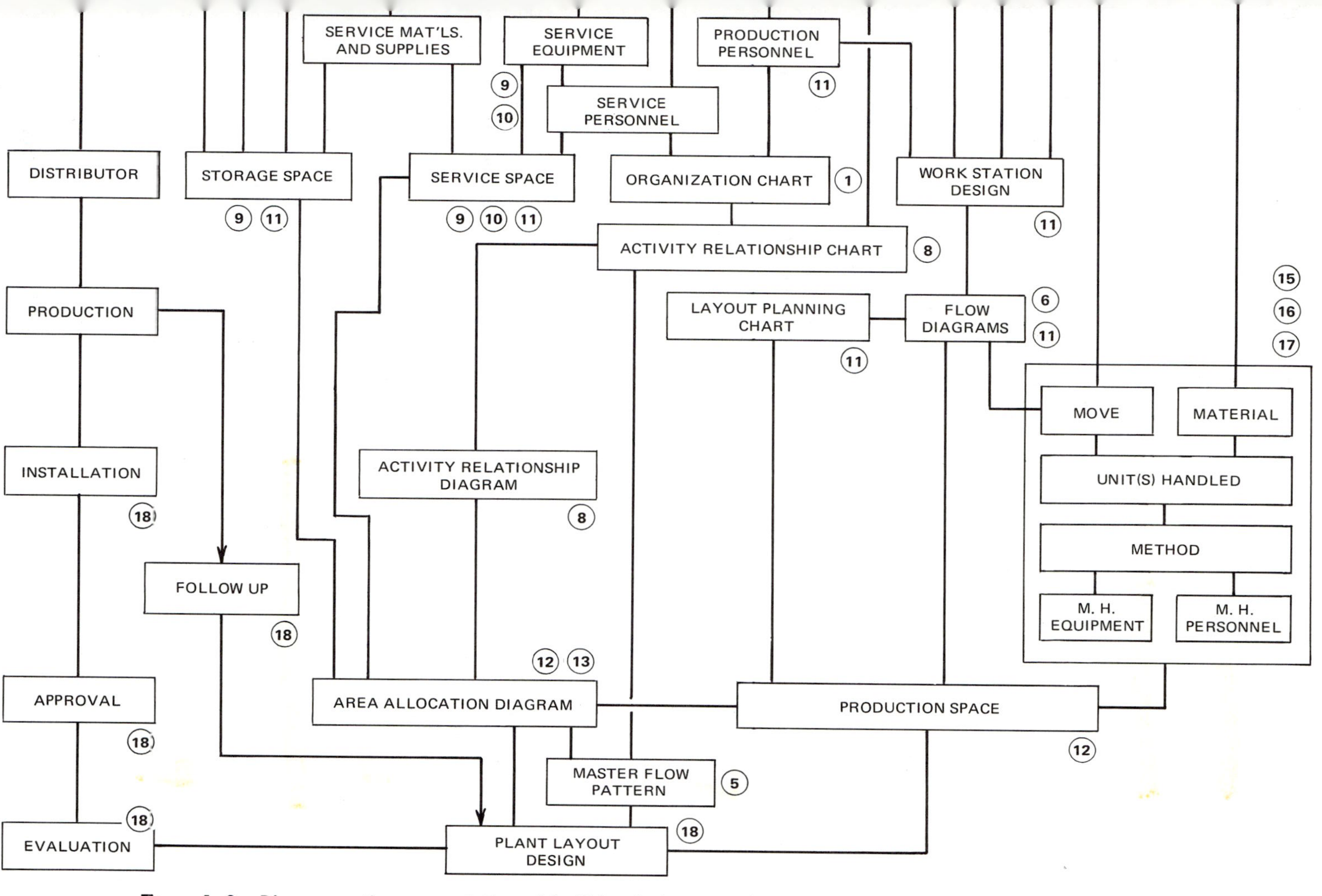

Figure 1–2—Diagrammatic representation of facilities design procedure (numbers indicate related chapter).

Cost reduction—This, of course, could be the cause or result of any of the above situations.

Planning a new facility—This presents the biggest problem in layout. Here the engineer generally is not limited by restrictions of existing facilities. He is free to plan the most effective layout he can devise. Buildings can then be designed to house the layout after it has been completed. This is where the ideal in layout can be attained. The facility can be completely laid out for the most efficient manufacturing. Then walls can be planned around the layout and the proper form of physical structure decided upon.

Each of these situations may present itself to the layout engineer. Each is as fascinating as another. Each presents its own peculiar problems to be solved; and though the engineer may have done his best on a completed layout, he always feels that a better way might possibly have been discovered.

And in addition to the above normal reasons for layout problems or projects, there are many abnormal situations or difficulties that may indicate the need for a study of an existing layout. Some of these indicators are:

1. Building not suited to requirements.
2. Failure to apply line production techniques when applicable.
3. Product design or process changes made without making necessary changes in the layout.
4. Installation of additional equipment without considering relationship to existing flow pattern.
5. Unexplainable delays and idle time.
6. Stock control difficulties.
7. Decreased production in an area.
8. Crowded conditions.
9. Many men moving material.
10. Bottlenecks in production.
11. Backtracking.
12. Excessive temporary storage.
13. Obstacles in material flow.
14. Scheduling difficulties.
15. Wasted "cube."
16. Idle people and equipment.
17. Excessive time in process.
18. Poor housekeeping.

The Marks of a Good Layout

In view of the preceding discussion, it would appear that efficient layout has certain desirable characteristics which should be evident from even a casual survey or observation. Among the most important of these are:

1. Planned activity interrelationships
2. Planned material flow pattern
3. Straight-line flow
4. Minimum back-tracking
5. Auxiliary flow lines
6. Straight aisles
7. Minimum handling between operations
8. Planned material handling methods
9. Minimum handling distances
10. Processing combined with material handling

11. Movement progresses from receiving towards shipping
12. First operations near receiving
13. Last operations near shipping
14. Point-of-use storage where appropriate
15. Layout adaptable to changing conditions
16. Planned for orderly expansion
17. Minimum goods in process
18. Minimum material in process
19. Maximum use of all plant levels
20. Adequate storage space
21. Adequate spacing between facilities
22. Building constructed around planned layout
23. Material delivered to employees and removed from work areas
24. Minimum walking by production operators
25. Proper location of production and employee service facilities
26. Mechanical handling installed where practicable
27. Adequate employee service functions
28. Planned control of noise, dirt, fumes, dust, humidity, etc.
29. Maximum processing time to overall production time
30. Minimum manual handling
31. Minimum re-handling
32. Partitions don't impede material flow
33. Minimum handling by direct labor
34. Planned scrap removal
35. Receiving and shipping in logical locations

The Layout Function

The layout function is a staff service, usually associated with the manufacturing or production activity. However, there are several organization levels at which layout work is performed, depending upon the relative size of the company and the relative importance of the layout function to the operation of the enterprise.

In the small plant, there is usually no formal Plant Layout Department. The layout work which must be done will be the result of the combined efforts of one or more foremen, the general manager, sometimes the company president, and the engineer and draftsman, who will all be in on the planning.

In a large plant, where there is much layout work to be done, a staff, perhaps of a dozen or more people, will spend full time working on problems in plant layout. Each person will be skilled in certain areas of layout work. Tasks will be subdivided, the individual parts being performed by trained experts; and a coordinated layout will finally be built up from their combined contributions.

There are certain large facilities which require little in the way of repeated layout work because of the nature of their processes. Installations of this kind are those in the process industries field—steel, rubber, petroleum, glass, etc.—where the plant, once satisfactorily laid out, remains relatively unchanged for a long period of time.

The Position of the Layout Department in Different Organizations

As was indicated above, the small plant will probably not have a Plant Layout Department, the responsibility for the function resting with the president or general manager, engineer, or foreman. When special problems arise, the man responsible for the project will solicit aid from whichever persons in the plant seem able to contribute to their solution.

Where a regular layout group has been organized, this group will be assigned a definite place in the organization plan.

The position of the layout department in the organization varies, as do the duties, with the size of the facility, the nature of the product, the importance of the layout function, and the organizational plan.

In a survey of 70 plants in the United States, from 600 to 102,000 employees (average, 2,500), the author found that 35 or more different titles were given to the person actually in charge of plant layout. With slight alterations in the specific titles as reported, the following list indicates typical titles assigned, from most common to least common.

1. Plant Engineer
2. Plant Layout Engineer
3. Supervisor of Plant Layout
4. Methods and Equipment Engineer
5. Master Mechanic
6. Industrial Engineer
7. General Superintendent
8. Foreman
9. Superintendent of Maintenance
10. Manager of Production Engineering
11. Engineer
12. Technical Assistant
13. Manufacturing Analysis Engineer
14. Manager of Engineering Design and Construction
15. Superintendent of Planning
16. Chief Engineer
17. Process Engineer

Likewise, the title of the person to whom the above individual reports varied widely. Again with slight changes in titles reported, that person is:

1. Works, Plant, Factory, or General Manager
2. Vice-President
3. Works or Plant Engineer
4. Plant or General Superintendent
5. Chief Engineer
6. Master Mechanic
7. Methods Engineer
8. Manager of Industrial Engineering
9. Production Engineer
10. President
11. Manufacturing Engineer
12. Tool Engineer
13. Supervisor
14. Standards Head
15. Planning Engineer
16. Department Manager

In two other surveys of the industrial engineering function, results showed that about 62 per cent of the time, the layout group is a part of the Industrial Engineering Department. It would appear that the increased emphasis on layout has tended to place the activity in closer proximity to other related functions.

Assuming the organization chart of a typical industrial enterprise of 2,500 employees to be as in Figure 1-3, the layout group is most commonly located as

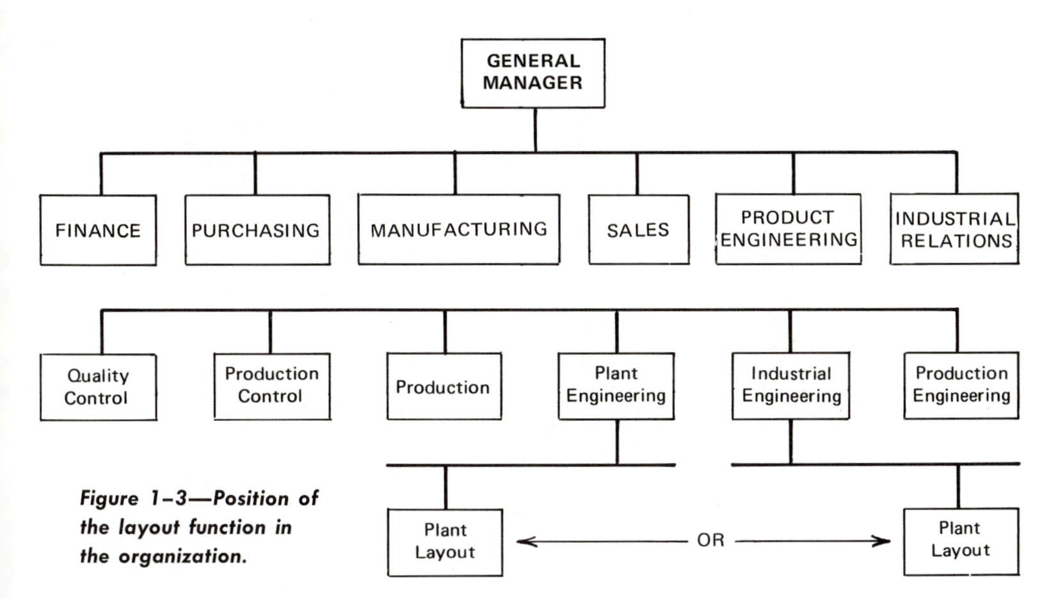

Figure 1-3—Position of the layout function in the organization.

shown. As an aid to interpreting Figure 1-3, other major activities commonly performed by the two functions, in which facility design or plant layout work is usually done, are:

A. *Plant Engineering*
1. Construction
2. Housekeeping and custodial services
3. Maintenance and repair
4. Material handling
5. FACILITIES DESIGN
6. Pollution control
7. Safety and security
8. Transportation
9. Utilities
10. Shops (crafts)

B. *Industrial Engineering*
1. Cost and economy studies
2. FACILITIES DESIGN
3. Information and control systems design
4. Organization design
5. Operations research
6. Production engineering
7. Systems engineering
8. Wage administration
9. Work methods
10. Work standards
11. Special studies and projects

Organization of the Layout Department

Where a Plant Layout Department exists, it will probably handle such activities as those indicated throughout this chapter. Its organization could be somewhat as shown in Figure 1-4.

Functions will vary, of course, from plant to plant. In the smaller plant, activities indicated above would be grouped together, while in larger plants they might be subdivided, depending upon the amount of work to be done, the number of people involved, the functions performed by other departments or divisions.

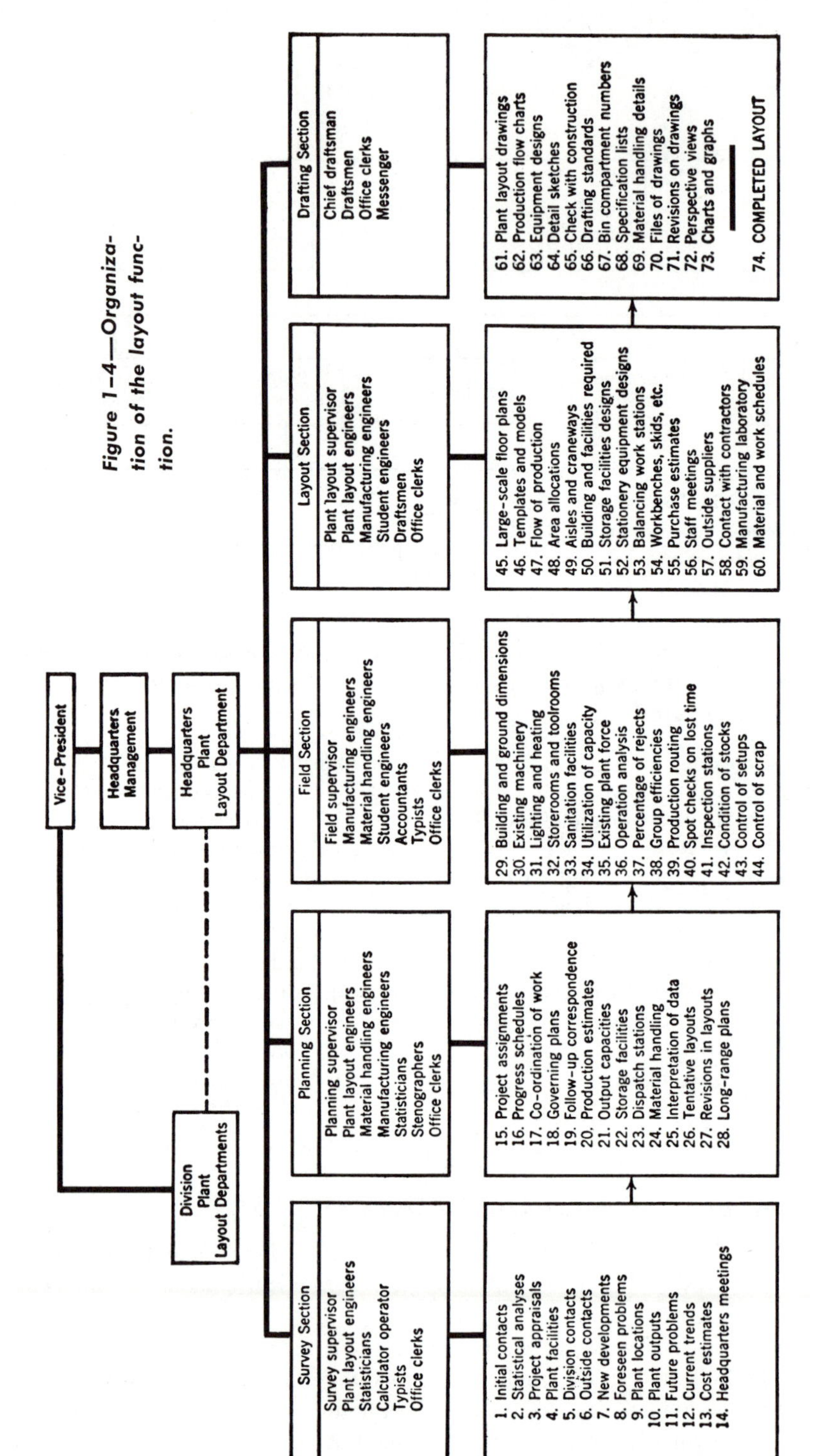

Figure 1-4—Organization of the layout function.

Nevertheless, all these activities, as well as those implied in Table 1-2, exist in every plant. It is the degree to which they are either necessary or done that varies. In a small manufacturing plant of 50 to 100 employees, one person may be responsible for all the activities indicated. In a plant of 8,000 to 10,000 employees, making automobiles, household appliances, etc., 30 to 40 men might be required to handle the same range of activities.

Table 1-3. Inter-Departmental Help in Layout Design

A. Sales Department

1. Determining quantities for production runs
2. Determining quantities to manufacture for replacement parts

B. Purchasing Department

1. Finding necessary factory equipment
2. Procuring equipment at lowest practicable cost

C. Product Engineering Department

1. Providing blueprints and parts lists
2. Offering information on manufacture, as gained in research and development work

D. Industrial Relations Department

1. Helping to design safety into the layout
2. Looking after employee comfort and services in layout
3. Training personnel for new jobs necessitated by new layouts

E. Finance Department

1. Aiding in determining cost of layout
2. Aiding in keeping equipment records

F. Production Engineering Division

1. Design of processing and production methods
2. Design of special tools and equipment
3. Planning operation sequence
4. Specifying machines and equipment to be used
5. Trying out tools, etc.
6. Mechanization and automation
7. Production capability analyses
8. New process and equipment development

G. Industrial Engineering Division

1. Determining work methods for each work area
2. Determining production standards for all operations
3. Determining machine capacities and number of machines needed
4. Aid in comparing effectiveness of methods between alternate layouts
5. Suggestions on methods

H. Production Control Division

1. Supply routings or operation lists
2. Determine production schedules
3. Aid in planning materials flow
4. Offer suggestions on materials handling methods
5. Plan storage methods and space requirements

I. Production Division

1. Suggestions on machine arrangement
2. Suggestions on human relations problems involved in layout
3. Ideas on materials handling

J. Plant Engineering Division (other than plant layout group)

1. Aid in planning for utilities
2. Plan for building changes or construction
3. Move machinery and equipment
4. Install machinery and equipment

K. Quality Control Division

1. Help plan layout to maintain quality in processes
2. Assure proper handling to protect product from damage
3. Plan scrap disposal

Relationship Between Layout and Other Departments

If the finished layout is to represent the best possible arrangement, the wholehearted cooperation of many people is necessary. Included in this group of people who can contribute information or offer suggestions will be not only members of the Plant Layout Department, but also persons from practically every other department and division in the plant. Some of the departments and divisions thus concerned, and the areas in which their cooperation is needed are listed in Table 1-3. These are only a few of the ways in which other departments and divisions can be of invaluable aid to the layout group. The utmost in cooperation is needed among all concerned if a project as large as a layout is to be successfully completed.

Conclusion

This chapter has attempted to define the facilities design and plant layout function, to briefly outline its importance and its activities and responsibilities, and to indicate its place and role in the organization.

Questions

1. What is meant by *facilities design?*
2. Explain how it may apply for an enterprise other than manufacturing.
3. What is the overall objective of facilities design?
4. Explain the scope of facilities design.
5. Why is the facilities design function important?
6. What role does facilities design play in improving productivity?
7. What are the objectives of facilities design?
8. What are some of the activities of the facilities design function?
9. Why isn't the facilities design work over with when the layout work is done and the plant is operating?
10. Briefly discuss the enterprise design process. Use a sketch, if desired.
11. Distinguish between manufacturing and production.
12. Develop a list of the steps in the facilities design process.
13. What other types of facilities design or layout problems are there—besides new facilities?
14. What are some of the marks of a good layout?
15. Discuss the position of the layout function in the organization.
16. In what ways does the layout function rely on other organizational functions? Give some specific examples related to several functions.

2

Overview of Facilities Design Procedure

As was pointed out in Chapter 1, the facilities design procedure involves the planning and design of an arrangement of:

1. Production equipment
2. Handling equipment
3. Auxiliary equipment
4. Space
5. Land
6. Buildings

in such a way that the work being done in the facility flows through in an efficient and economical manner.

Such an objective is not accomplished without carefully following a plan or procedure that will insure that as many factors as possible are properly taken into consideration and designed into the layout. A quick and careless layout job cannot result in an efficient enterprise. Nor, as some people seem to think, can it be done just as well by a shop foreman as by a trained and experienced engineer or technician.

Factors for Consideration in Facilities Design

Some idea of the complexity of the task facing the facilities designer can be obtained by studying Table 2-1, which lists the categories of factors to be considered in the design of a proposed facility. Because *factors* are involved in nearly every topic covered in this book, it may be well to review the meaning of the term. *Webster's New Intercollegiate Dictionary* defines it as follows: *one of the elements that contributes to producing a result; a component, part, element.* For our purposes, in a slightly amplified definition, *factors* are: *elements of the operating system* that, *singly or in groups, influence—positively or negatively—the attainment of the system objective.* Although it may be felt that some of the categories in Table 2-1 are factors, it should be recognized that there can be a great many factors within each of the categories listed—and their classification must be determined by the analyst, within the scope of the problem he may have in hand. The author has developed literally hundreds of factors, each within a specific category. Definitive lists will be found throughout this book, as individual facets of the overall problem are considered. Some idea of the complexity of the entire process can be obtained by looking in the index, under "Factors"! But, the categories alone should impress the designer with the extreme importance of not overlooking or ignoring any factor of significance.

Table 2-1. Principal Categories of Factors for Consideration in Facilities Design

1. Building	16. Flexibility	31. Offices	46. Shipping
2. Business trends	17. Flow	32. Organization	47. Site
3. Communications	18. Government/Legal	33. Packaging	48. Storage
4. Community	19. Grounds	34. Packing	49. Supervision
5. Competition	20. Health	35. Personnel	50. Throughput
6. Costs	21. Inspection	36. Pollution	51. Time frame
7. Customer	22. Intangibles	37. Processes	52. Transportation
8. Distribution	23. Location	38. Product	53. Unions
9. Ecology	24. Long range planning	39. Production control	54. Utilities
10. Economic	25. Maintenance	40. Quality control	55. Warehousing
11. E.D.P.	26. Management policy	41. Receiving	56. Waste
12. Equipment	27. Manufacturing methods	42. Refuse	57. Work methods
13. Expansion	28. Market	43. Safety	58. Work standards
14. Financial	29. Materials	44. Security	59. Yards
15. Fire protection	30. Material handling	45. Services	60. Zoning

Facilities Design as a Coordinating Function

This chapter presents a preview of the entire facilities planning and design procedure as a means of orienting and introducing the detailed discussion of later chapters. The design of an efficient layout can best be achieved if the problem is attacked in a logical and orderly manner. The following step-by-step procedure is presented as a guide or basic thought pattern to assure proper consideration of all aspects of the overall problem. The steps may not follow in a specific predetermined order, for much interaction of suggested activities will occur because of the complex functional interrelationship existing between layout design personnel and other members of an organization. For example, planning the material flow pattern (*Step 4* below) cannot be properly completed without giving consideration to proposed individual work stations (*Step 7*) and to the direction of flow through them, which is not considered in detail until later in the planning process.

It should be pointed out also, that conclusions reached at any step in the procedure are subject to revision when conditions are changed as a result of later findings or after more detailed consideration in subsequent stages of the planning process.

The Facilities Design Procedure

With this brief note of caution, the design of an effective layout might proceed somewhat in the following manner.

1. *Procure basic data*—The facilities designer must rely on the several staff activities for the basic data necessary to the design of the layout. He must procure, or develop, such data as:

a. Sales forecast
b. Quantity to be produced
c. Production schedule
d. Inventory policy
e. Drawings (Figure 3-7, page 44)
f. Parts list (Figure 3-8, page 45)
g. Production routing (Figure 4-13, page 81)
h. Operations to be performed
i. Preliminary methods
j. Production time standards
k. Scrap percentages
l. Existing layouts
m. Building drawings
n. Floor and ceiling load limits

These and many other factors, such as fall in the categories of Table 2-1, will be taken up and related to the overall design process. Figure 1-2 (pages 16–17) indicates some of the interrelationships of such data.

2. *Analyze basic data*—The work of the facilities designer begins with analysis of the data to determine the desired interrelationships, and then preparing it for subsequent planning steps. One of the analytical techniques useful at this early stage is the *Assembly Chart* (see Figure 6-2, page 127). This device gives a quick and early glimpse of the possible flow of materials.

3. *Design production process*—The next step is to determine the processes by which material is to be converted into the parts and products desired, or, for a non-manufacturing enterprise, the procedures by which services are to be performed. In the manufacturing situation, the process engineer examines the blueprints and other data and determines the processes and operations (Chapter 4). The results of his efforts are shown in the *Production Routing* of Figure 4-13 (page 81). With this additional data, the analyst can construct an *Operation Process Chart* (see Figure 6-5, page 130). This chart adds the operations from the *Production Routings* to the data shown on the *Assembly Chart,* extends the usefulness of the basic data, and provides a better impression of the potential material flow pattern.

4. *Plan material flow pattern*—In achieving a layout that facilitates the production process (objective No. 1!), the general overall material flow pattern must be carefully designed to assure minimum movement and expeditious interrelation of the several component-part flow paths. A good flow pattern will guide the materials. The *Assembly Chart* and *Operation Process Chart* will aid in visualizing a general material flow pattern. Based upon study of these two charts and related data on quantity and frequency of material movement, the preliminary material flow pattern should be developed. Then, after a preliminary idea of the flow of materials has been obtained, it becomes necessary to consider the other factors having an important bearing on the final flow pattern. A carefully planned flow pattern follows certain principles and general methods of material flow in the integration of material movement with the related factors. (Some of these factors are covered in Chapter 5.)

After a careful consideration of the various factors affecting the flow pattern, along with the analysis of the product itself, it is necessary to correlate the activities involved in the processing of the various components into one overall master flow pattern. The flow planning commonly results in a preliminary flow pattern, such as in Figure 6-12 (page 138). It is inherent in such a plan that changes will be necessary as more details are developed in later stages of the layout design. Flexibility is a highly desirable characteristic of a good layout, and a compromise must be worked out so as to balance the objectives of the layout and still permit future changes without too much interference with the existing layout.

5. Consider general material handling plan—The material handling system converts the *static* flow pattern into a *dynamic* flow of material through the plant. The ideal system will consist of an integrated combination of methods and equipment designed to implement the flow of material. It will consist of the most effective methods of performing every handling task—from the unloading of material delivered by suppliers, to the loading into carriers of finished goods on their way to customers. In between it will handle material into, through, and out of each production or related activity. The system will likely involve both manual and mechanical methods, and such equipment as conveyors, cranes, and industrial trucks. On the basis of the flow pattern, some decisions should be made as to the general methods that could be used in handling material and, in some instances, the type of equipment that may be used. However, detailed handling methods between operations and specific equipment may not be decided upon until after individual work stations have been planned (*Step* 7).

6. Calculate equipment requirements—Before the layout process can continue much farther, it will be necessary to determine how many pieces of each type of equipment (manufacturing, service, and auxiliary) will be required. Preliminary consideration of the number of machines was undoubtedly made in earlier steps. Here, however, final decisions must be made as to the quantity of equipment as a basis for planning individual work stations and calculating space requirements for each activity area. Also, the number of operators must be determined. If these decisions have not already been made (Figure 4-13, page 81) they must now be at least estimated.

7. Plan individual work areas—At this point, each operation, work station, area, process, etc., must be planned in detail. The interrelationships between machines, operators, and auxiliary equipment must be worked out. Consideration must also be given to operator cycling, multiple machine operation, principles of motion economy, and material handling to and from the workplace. Although the flow pattern has established the OVERALL path for the movement of material, each workplace is a stopping point in the planned flow of a specific item. Therefore, each must be tied into the overall flow pattern, and the flow through each workplace must be planned as an integral part of the overall plan. Ideally, each workplace will be designed in detail, as in Figure 11-10 (page 264). Most commonly the methods engineer will assist in this portion of the layout design, or will do it as a part of his function. (The details involved in operation and work area planning are presented in Chapter 11.)

8. Select specific material handling equipment—Next to the method of processing, material handling is probably the most important factor to be considered in layout planning. Because of the need for discussing the handling problems in some detail, Chapters 14 to 16[1] are devoted to this subject. Eventually, specific methods of material handling must be decided upon for each move of material or item. Many factors must be considered in the selection of handling methods. (These factors and methods of relating them to other phases of the layout planning process are covered in Chapter 16.)

9. Coordinate groups of related operations—When individual operations or work areas are designed, the interrelationships between work areas should also be planned. Likewise the interrelationships between related groups of operations or activities should be worked out. This step may have been started, or even completed, during consideration of individual work places (*Step 7*) or previously under flow patterns (*Step 4*). If not complete, then individual work places, production centers, departments or processes should now be integrated and tied together. A start can be made at this time toward coordinating the flow diagrams of individual processes, into the master flow pattern as originally developed (*Step 4*) and as subsequently modified. Another technique useful at this point, is the *Layout Planning Chart* (see Figure 11-12, page 267). This chart records the major steps in each operation, and forces a consideration of the steps (particularly material handling) between the operations. It serves a very useful function in pointing out omissions in planning and guiding the further development of an effective integration of operations.

The net result of the coordination might also be shown in a *Flow Diagram* (see Figure 11-13, page 270) for each area. It can be seen that the several *Flow Diagrams* represent, in reality, the initial overall layout—in several segments—but still need coordination and adjustment. It should be pointed out again, that the procedure outlined in this chapter, is not a 1, 2, 3 affair. The experienced planner has most likely been jumping back and forth between the steps, as the need for coordination between the steps becomes apparent. For this reason, some of the work described here may already have been done—or at least have received some consideration—during the planning of other phases of the layout.

10. Design activity relationships—Most of the preceding steps have dealt primarily with production activities. It now becomes necessary to interrelate these with auxiliary and service activities with respect to degrees of closeness required by material, personnel, and information flow. Also to be considered are such negative relationships as noise, odor, and dirt. These interrelationships, positive

[1]Further details on material handling are contained in J. M. Apple, *Material Handling Systems Design* (New York: The Ronald Press Company, 1972).

and negative, are analyzed with the aid of the *Activity Relationship Chart* and the *Activity Relationship Diagram* (discussed in Chapter 8; see Figures 8-1 and 8-4, pages 204 and 209).

11. *Determine storage requirements*—Again, this topic has probably been considered in previous steps, but plans should now be crystallized for the storage of raw material, goods-in-process, and finished products. Square-foot and cubic-foot requirements should be calculated, with thought given to the location of storage areas on the layout. This can be implemented with the aid of a form such as in Figure 11-5 (page 258) which helps to analyze the *Parts List* (Figure 3-8, page 45) and convert it to space needs for a given inventory. (Further details will be found in Chapter 11.)

12. *Plan service and auxiliary activities*—Service and auxiliary activities probably have already been given some thought, but primary considerations were for production-oriented activities. Before further work on the details of the layout it is necessary to consider the many service activities and auxiliary functions that assist the production activity in getting its work done, for each of these must be properly laid out and effectively integrated into the master plan. An indication of the complexity of this step can be gained by looking at the *Plant Service Area Planning Sheet* (Figure 11-19, page 284).

Depending upon the size of the plant, some or all of the service activities must be carefully studied, in order to determine which are needed. Later, during the space planning and final design aspects of the planning, the details of many of these service activities must be worked out. For instance, the contents of such as the locker room, tool room, and first aid and food service areas must be determined, to permit an estimate of the space requirements for each. (Details are covered in Chapters 9 and 10.)

13. *Determine space requirements*—After designing the flow pattern and giving some thought to the auxiliary and service activities, it is necessary to make a preliminary estimate of the total space required for each activity in the facility, and a first estimate of the total area of the proposed facility. Storage space was estimated in *Step 11,* and some of the service activities in *Step 12.* Production space needs are estimated with the aid of a *Production Space Requirement Sheet* (see Figure 11-11, page 266). Or they may be arrived at by estimating an appropriate number of square feet per machine (possibly different for different machines) and then multiplying by the number of machines. After estimating space needs for each activity or function, the *Total Space Requirement Work Sheet* (Figure 11-20, page 286) re-caps the needs for all activity areas.

It should be emphasized that space determinations made at this stage are estimates. They should very likely be optimistic enough to be sure there is sufficient area. Later steps in the planning process will result in closer figures, but only the final layout will show accurately the total space needs. (Space determination is covered in Chapter 11.)

14. Allocate activity areas to total space—The *Total Space Requirement Work Sheet* provides for an *area template* for each activity listed. These area templates are cut out to an appropriate scale and arranged to satisfy the desired or required relationships between them. The *Activity Relationship Chart* and *Diagram* will be found helpful in determining these relationships.

Then an *Area Allocation Diagram* (see Figure 12-21, page 315) can be made, based on an arrangement of the area templates to meet the requirements of the *Activity Relationship Diagram.* Very likely, the *Area Allocation Diagram* will involve developing a *Plot Plan,* which will show the interrelationships between the internal flow of materials and the external flow—by means of various transportation modes. It will also detail relationships with surrounding facilities, such as the power plant, parking areas, storage yards, and adjacent buildings. It will be found that these have a considerable bearing on the orientation of the proposed building, and for that matter, on the selection of the site, if that has not already been done.

A *preliminary* layout has now been established, showing relative areas for the required activities, and the relationships appropriate to established needs. (Area allocation is covered in Chapter 12.)

15. Consider building types—If some consideration has not already been given to buildings, it should be done at this point. In some instances the building is only a shell around and over a layout. In others the building may be an integral part of the production facility. In any case, building type, construction, shape, and number of floors, should be considered, to arrive at some tentative conclusions. However—let it be emphasized again—the building usually comes after the layout. As important as a building may seem to be, it is the layout that forms the basis for the efficient operation of an enterprise. The layout should never be squeezed into, or altered to fit a building, if it can be avoided by designing the layout first.

16. Construct the master layout—As indicated previously, the several work area plans and flow diagrams must be integrated. This means that a lot of adjustments will have to be made to accommodate the melding of the various work area plans, flow diagrams, and service and auxiliary activities, with the flow pattern—into a working composite.

This step is the culmination of the detailed work and planning done in the preceding *steps.* It is here that the final layout is prepared with the aid of templates, tapes, etc.—commonly to a scale of $\frac{1}{4}'' = 1$ ft. Three-dimensional scale models of physical facilities are frequently used in place of the more traditional two-dimensional templates. With the completion of the master layout, the plant layout engineer should have accomplished the design of an efficient production facility, as envisioned at the start of the project. The finished plant layout might appear as in Figure 17-9 (page 408). (The various techniques and procedures used in making the final layout are detailed in Chapter 17.)

17. *Evaluate, adjust, and check layout with appropriate personnel*—No matter how carefully or scientifically the previous steps have been carried out, there are always personal and judgement factors to be considered. Often they are of greater importance than the engineering aspects—when the final decisions are made. Therefore, it is only proper that the facility designer and his associates evaluate and check over their work at this stage, prior to submitting it for examination and approval. Preliminary checks might also be made with others who have contributed to the layout planning, such as those concerned with methods, production, and personnel safety. Many of these specialists will have worthwhile suggestions to offer on matters the layout engineer may have overlooked because he was "so close to the forest he couldn't see the trees." (Evaluation of the layout is covered in Chapter 18.)

18. *Obtain approvals*—In the final stages, the layout must be formally approved by certain plant officials, depending on plant policies and procedures, who may have either special knowledge of certain phases of the proposed operation, or a broad understanding of the overall relationship between various phases of the operation. (Approval of the layout is covered in Chapter 18.)

19. *Install layout*—Because the layout designer has designed the layout, it is logical that he should closely supervise the necessary work involved in the installation of the layout. He should be sure all work is done according to the plans set forth in the approved layout. Any changes found to be desirable as construction work progresses should be thoroughly investigated and, if made, approved by proper persons. The layout designer should cooperate with the architect and construction engineers to see that the planned layout is properly incorporated into the building itself. (Some of the details to be considered in installing the layout are discussed in Chapter 18.)

20. *Follow up on implementation of layout*—Just because the layout has been installed as planned, is no guarantee that it will work as planned. Equipment does not always work the way it is supposed to. Men do not always follow the methods they are supposed to. Material does not always move the way it is supposed to. And supervision is not always capable of molding together the many facets of the layout to make it operate as planned. As a plant manager once said to the author, "Something is wrong in this plant—it doesn't *hum* the way it should!" And then, no plant layout is ever perfect—the layout designer must continually take note how the layout affects the production operations. When opportunities for improvement are observed, they should be properly evaluated, and changes made if they are found to be desirable. (Implementing and following up on layout performance are dealt with in Chapter 18.)

Conclusion

It should be emphasized again that the procedure outlined above is intended only as an overview, or as the basis for a thought pattern to be followed in setting out to design a facility. Much skipping back and forth between the steps is to be expected. Many changes will be made as work progresses, and the procedure should not be considered a rigid plan for work. Also, in the many types of layout problems, the project may not require certain portions of the procedure. In such cases, it will be necessary for the layout designer to decide which steps must be performed in order to satisfactorily carry out the project. No layout project is routine, no two are alike, and each calls for a careful and thorough consideration of all its aspects. Subsequent chapters will present detailed considerations and instructions for carrying out the above procedure.

Questions

1. What are some of the categories of factors for consideration in facilities design?
2. In what respects is facilities design a coordinating function?
3. What is the necessity for following a predetermined facility layout procedure?
4. What are some examples of situations in which the procedure might not be followed?
5. What are some of the basic data required for production planning? Why?
6. What is a material flow pattern?
7. Show by a sketch the relationship between the overall flow pattern and an individual workplace.
8. How are service and auxiliary activities interrelated with the production function?
9. Who in an organization should be consulted prior to completion of the layout? What does each contribute to the final plan?
10. What are some of the relationships between the layout and the building that will house the layout?
11. Why should the layout engineer be interested in the actual construction of the plant building?

3

Preliminary Enterprise Design Activities

As stated in Chapter 1, the term *production* should be interpreted liberally, to encompass the *function of any enterprise that carries out the work of producing a product or service.* From this point of view, a production process takes place in any enterprise, such as those listed in Table 1-1, or a bank, a waterworks, a freight terminal, an airport, or an amusement park. In service operations, the production function would cover the part of the operation where the work is done that provides the service. In industrial operations, the production function involves the activities required to produce the parts/or products the enterprise is in business to provide. In either case, however, certain activities are necessary before the actual facility design process can or should take place.

Referring back to Figure 1-1, it will be seen that there are sixteen stages in the enterprise design process. One of these is facilities design, which is detailed in Figure 1-2, and also in list form on page 13. Actually, the first five stages in the enterprise design process are the source of much of the data required for the facilities design process, while the remainder of the sixteen stages are the result of the facilities design process—that is, its implementation, in the form of constructing a facility, operating it, producing the product or service and distributing it to the customer. In this chapter we consider Steps 1–3 in the enterprise design process to acquaint the reader with some of the activities preceding the facilities design process that provide the necessary data.

Basic Data Required for Facilities Design

Before we discuss these preliminary activities, let us look at the reasons for them, in terms of the data required by the facilities designer, so he can do an effective job of designing an appropriate facility. The type and amount of data will vary with the enterprise; in general, whether it is product or service oriented, required data will be what is necessary to satisfy the categories in Table 2-1. There may be more or less data as required by a specific situation, and some may have to be translated for the service enterprise. It can be seen that undertaking a facility design project is a complex matter, requiring an imposing array of information. However, there are many people who still dash out and buy or lease a building, obtain some equipment, set it up hastily, and go into production. But, they may not be in business very long—for having overlooked or ignored some of the data suggested here.

Sources of Basic Data

Reflecting back on the organization chart in Figure 1-3, and Table 1-3, it can be seen that many organizational functions are involved in providing the data. Some of these functions and their activities are discussed below. Others are taken up in subsequent chapters.

Management Policy

The most important data required by the facilities designer must come from management—in the form of their thinking and policy in relation to such matters as:

1. Sales forecasting
2. Employment
3. Growth trend
4. Investment
5. Inventory
6. Equipment replacement
7. Maintenance and repair
8. Customer service
9. Competition
10. Technology
11. Growth rate
12. Distribution
13. Overtime
14. Make/buy
15. Organization
16. Return on investment
17. Net profit
18. Staff services
19. Contemplated changes: product, process
20. Labor turnover
21. Age of work force
22. Union/non-union
23. Time constraints
24. Funds available
25. Product service
26. Warranties

While a discussion of the above items is beyond the scope of this text, it can be seen that the lack of information and data on such matters will seriously handicap the facilities designer. He would be most severely penalized by any lack of knowledge of long-range and future plans of management. And lack of such knowledge would certainly effect the flexibility and adaptability of the facility—two of management's top priority considerations.

Feasibility Studies

Two other important concerns of management, especially if the activity under consideration is a new enterprise, are the feasibility study and a related estimate of the funds required to finance the enterprise. A feasibility study will usually involve:

1. An estimate of annual costs.
2. An estimate of gross annual returns.
3. The determination of an acceptable rate of return on the proposed investment.

4. The establishment of the proposed scale of operations.
5. A final determination as to the feasibility of proceding with the project.

Again, the details of this activity are beyond the scope of this book, and the interested reader is advised to investigate the feasibility study process in other sources.

Having established the feasibility of the proposed enterprise or project, management must estimate the financial requirements. This will likely involve such activities as:

1. Establishing the preliminary organization of the enterprise or project.
2. Investigating the possible methods of obtaining land, buildings, and equipment.
3. Developing a proposed sales budget.
4. Estimating annual operating costs.
5. Estimating probable earnings.
6. Establishing a financial plan.
7. Determining sources of capital.
8. Issuing a prospectus describing the proposed enterprise.
9. Selling securities (stocks, bonds, etc.) to finance the initiation of the enterprise.

Here, too, the reader is advised to seek further details from other sources. Needless to say, the designer should make an extra effort to obtain information on these management-related matters.

Market Research

The market research activity is concerned with surveying the market to determine customer needs, (or wants). Then, it accumulates and analyzes data and information on all problems concerned with the marketing of products or services. It involves evaluation of expected costs and potential profits—usually in gross terms, that is, not in great detail. It is also interested in supply and demand, in terms of competitors and the problems inherent in satisfying the market.

Space limitations prohibit a discussion of this procedure, but the importance of the resulting information in the facilities design process cannot be underestimated. A facility is effective only to the degree it can provide for the desires of the customer. And it can be seen that much of the market research procedure is directed towards such key items as quantity, models, variations, schedules, packaging, and acceptable costs. All of these are extremely important in planning for such as facility size, flexibility, processes, and methods.

Sales Forecasting

After market research has determined customer needs and desires, the sales forecasting activity attempts to convert the resulting data—along with information the forecast itself will contribute—into a production quantity for which the facility will be designed. This is commonly developed in terms of an estimated sales volume, most likely broken down by monthly requirements and projected

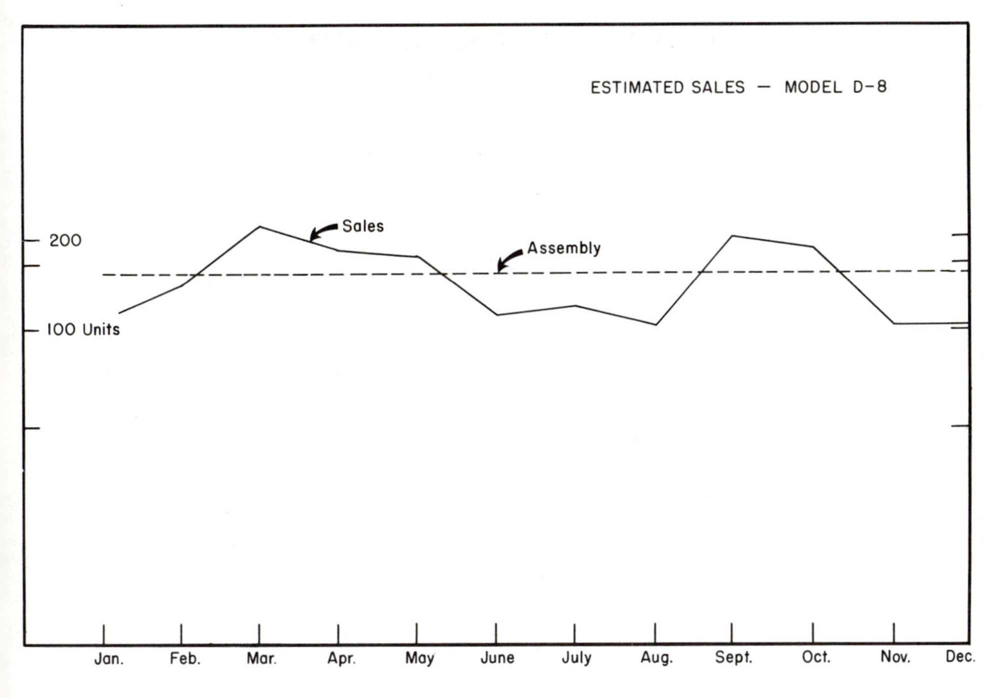

Figure 3-1—Typical annual sales curve.

ahead for several years. At the same time, it will be necessary to predict the effect of a proposed new product on any of the existing products in the line.

The sales forecast must include not only the total predicted demand for finished products, but must also indicate seasonal fluctuation, if any, along with service requirements for parts or assemblies. As suggested previously, top management should also indicate the desired inventory policy for the product—that is, how much of a stock-pile they wish to have on hand at any one time. This is especially important if sales are seasonal. Figure 3-1 shows a typical annual sales curve for a seasonal item. Figure 3-2 tabulates the necessary relationships between sales, production, and inventory. These two figures indicate the nature of the problem, although the details of such decisions and calculations are beyond the limitations of this book.

Product Engineering

The product engineering, or product design, function is charged with the development and design of the products or services that market research has determined are required by the customers. And—they must design the products not only to satisfy the customer's functional needs, but they must be economical to produce, so the customer can pay a price sufficient to result in a profit.

Product Engineering is also necessary to insure the economic growth of the

SALES, ASSEMBLY & INVENTORY – MODEL D-8

	Sales	Assembly	Balance	Cumulative inventory (100 starting inventory)
JAN	106	143	+37	+137
FEB	134	143	+9	+146
MAR	210	143	–67	+79
APR	179	143	–36	+43
MAY	170	143	–27	+16
JUNE	110	143	+33	+49
JULY	116	143	+27	+76
AUG	103	143	+40	+116
SEPT	200	143	–57	+59
OCT	180	143	–37	+22
NOV	102	143	+41	+63
DEC	104	143	+39	+102 (for carry over to next year)
TOTALS	1714	1716	–	–

Figure 3–2—Sales, assembly, and inventory relationships.

enterprise—based on a succession of new or improved products.[1] A succession of new products, or services, must be forthcoming to keep the enterprise moving ahead and in good financial condition. Figure 3-3 shows this cycle (in less detail) in terms of product life and effect on profits.

In general, the product development might proceed somewhat as follows. First, of course, comes the basic research and development work preliminary to the final design of the product to be produced. This most likely begins with an idea—from management, sales, or other source—for a new product, model, or modification. Preliminary decisions between management, sales, and product

[1] Incidentally, it is this aspect of the enterprise life cycle that keeps the facilities designer busy; he is not through with his job once a facility is designed, but is usually kept busy by the myriad changes in facilities occasioned by new models, products, and developments.

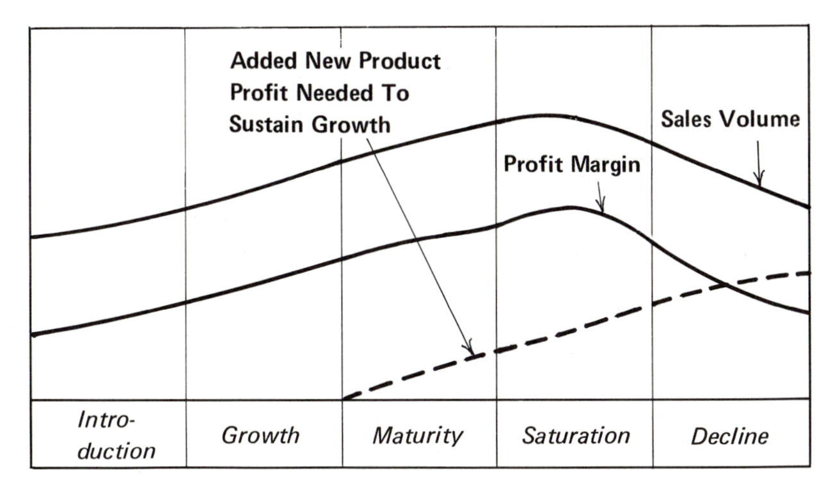

Figure 3-3—Typical product life cycle—cost relationships.

engineering most likely result in establishing a list of design characteristics or functional requirements of the proposed product.

This is followed by experimentation with basic product components. Then, after satisfying themselves that basic functions are workable, a prototype is constructed. This becomes the first working model of the new product and is built from sketches and preliminary drawings in the experimental shop or tool room.

Next comes a series of tests, trials, and modifications of the original prototype. Further testing in the laboratory or in the field, is probably followed by the construction of a small (two to six) lot of the re-designed prototypes. These, in turn, are tested, until all concerned agree that the basic product design is sound and ready for the final design stage—upon which manufacturing will be based.

At this point, however, the manufacturing people may be called in to conduct what is commonly called a *value analysis.* Involved at this stage are representatives of production (such as sheet metal, machining departments, and assembly), quality control, methods, purchasing, and possibly others. This group attacks the proposed product from a manufacturing point of view, looking for any changes that might make it easier or less costly to produce. Piece-by-piece they analyze the product, looking for such as: less expensive materials or processes, unnecessarily close tolerances, sharp corners, parts of too-high-quality for the function required, and points of difficulty in assembly. Suggestions from this value analysis are considered by product engineering and, when practicable, incorporated into the design. (For further details, refer to pages 54 and 55.)

Now—and only now—is the new product ready for the final design stage. The actual production or working drawings are made, and prints issued, along with *Bills of Materials*—complete listings of all that go into the product.

For a review of the overall product engineering procedure, see Table 3-1. In most cases, it would be desirable to relate the activities in some kind of time-table, to assure meeting the management deadline. Figure 3-4 shows an actual schedule

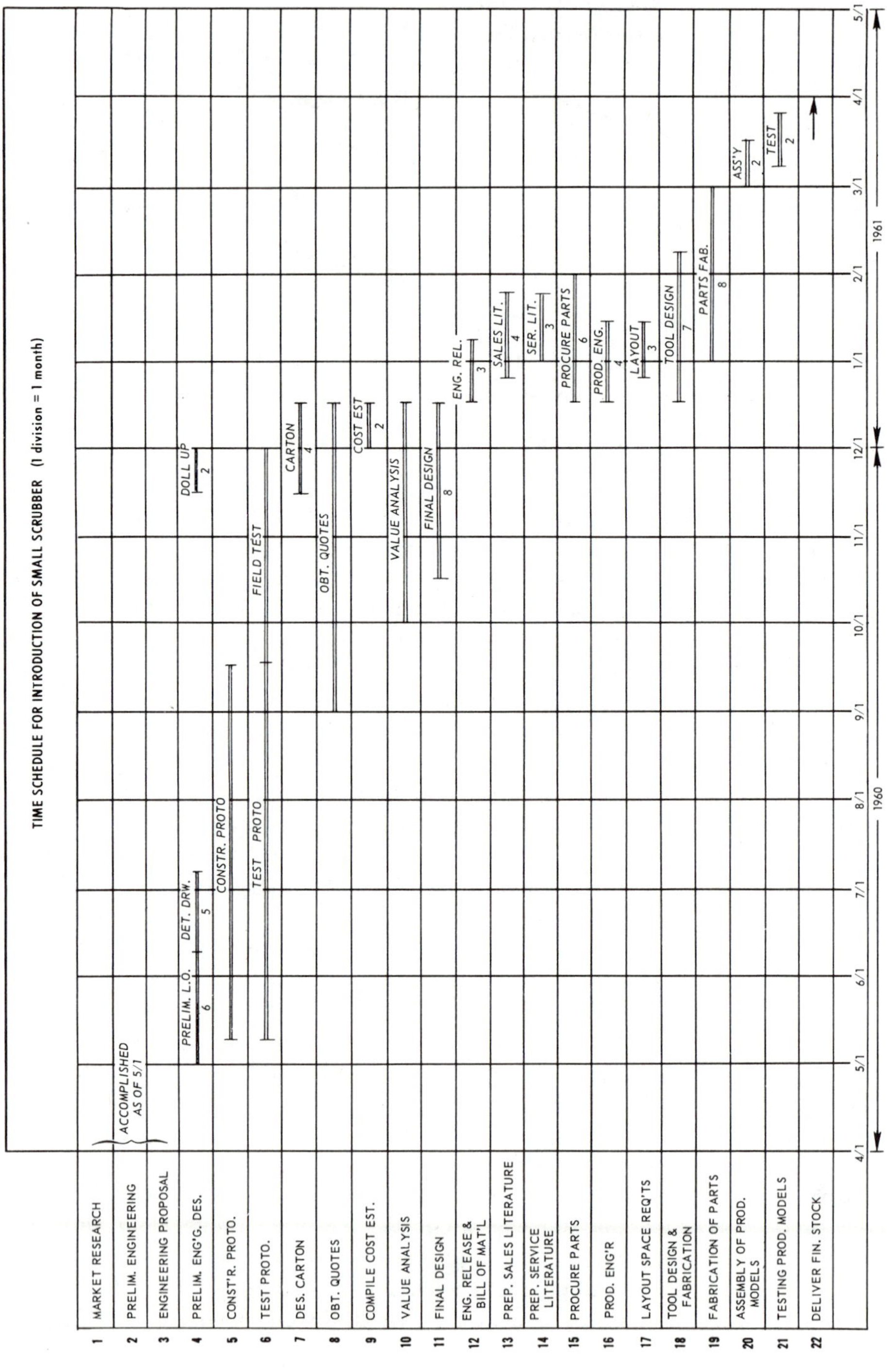

TIME SCHEDULE FOR INTRODUCTION OF SMALL SCRUBBER (1 division = 1 month)

No.	Activity	Chart annotations
1	MARKET RESEARCH	ACCOMPLISHED AS OF 5/1
2	PRELIM. ENGINEERING	
3	ENGINEERING PROPOSAL	
4	PRELIM. ENG'G. DES.	PRELIM. L.O. 6; DET. DRW. 5; DOLL UP 2
5	CONST'R. PROTO.	CONSTR. PROTO
6	TEST PROTO.	TEST PROTO; FIELD TEST
7	DES. CARTON	CARTON 4
8	OBT. QUOTES	OBT. QUOTES
9	COMPILE COST EST.	COST EST 2
10	VALUE ANALYSIS	VALUE ANALYSIS
11	FINAL DESIGN	FINAL DESIGN 8
12	ENG. RELEASE & BILL OF MAT'L	ENG. REL. 3
13	PREP. SALES LITERATURE	SALES LIT. 4
14	PREP. SERVICE LITERATURE	SER. LIT. 3
15	PROCURE PARTS	PROCURE PARTS 6
16	PROD. ENG'R	PROD. ENG. 4
17	LAYOUT SPACE REQ'TS	LAYOUT 3
18	TOOL DESIGN & FABRICATION	TOOL DESIGN 7
19	FABRICATION OF PARTS	PARTS FAB. 8
20	ASSEMBLY OF PROD. MODELS	ASS'Y 2
21	TESTING PROD. MODELS	TEST 2
22	DELIVER FIN. STOCK	→

4/1 5/1 6/1 7/1 8/1 9/1 10/1 11/1 12/1 1/1 2/1 3/1 4/1 5/1

1960 — 1961

Figure 3–4—Time (Gantt chart) schedule for a typical new product. (Courtesy of Clarke Floor Machine Co.)

Table 3-1. Product Engineering Functions and Procedure

A. Develop new product ideas

1. Review possibilities
2. Screen ideas
3. Select those with potential

B. Product conception

1. Visualize product
2. Collect data
3. Market research
   a. Customer requirements
   b. Operating specifications
   c. Sales specifications

C. Preliminary investigation and design

1. Explore alternative designs
2. Conduct feasibility studies
3. Make vendor contacts
4. Evaluate business conditions
5. Check on competition
6. Make preliminary sales forecast

D. Engineering evaluation

1. Conduct analytical investigation
2. Develop models
3. Test
4. Establish quality goals
5. Carry out value analysis
6. Evaluate production capability
7. Determine compatibility of new products with present lines and production capabilities

E. Engineering proposal

1. Preliminary design
2. Cost estimate
3. Evaluation
4. Approval

F. Prototype design

1. Preliminary drawings
2. Trial orders of components
3. Initial engineering specifications

G. Prototype construction

1. Obtain components
2. Construct prototype

H. Prototype testing and evaluation

I. Final design

1. Incorporate changes
2. Drawings, final specifications, part numbers
3. Re-construct prototype
4. Field tests—economics of manufacturing and use
5. Cost estimates
6. Package design

J. Design approvals

K. Engineering release (to Production)

L. Establish product sales and service data and policies

1. Instruction manual
2. Service manual
3. Sales data
4. Warranty data and policies
5. Service parts policies

M. Pilot production and evaluation

for a proposed new product, in the form of a Gantt chart. Frequently, this interrelationship is developed in the form of a *Critical Path Network* (see Figure 6-19).

The Powrarm

The end result of the product design activity is the design for the proposed product, which may be expressed in the form of a set of drawings (or other appropriate specifications) and a prototype (model) of the product.

At this point, we introduce the *Powrarm*—a product to be used throughout the text to illustrate many points and steps in the facilities design procedure. It is essentially a ball and socket joint, on a pedestal—not at all unlike a camera tripod. The Powrarm is used in industry to hold an object (on its ball swivel) in a desired position to facilitate work on the object—such as welding, assembly, or inspection. The Powrarm is shown in Figure 3-5, and Figure 3-6 is an *exploded-view drawing*, commonly furnished by product engineering.

The engineering drawings are usually done on tracing paper, and prints are made from them. It is these prints that are used as the basis for making or ordering the various parts.

Figure 3–5—The Powrarm completely assembled.

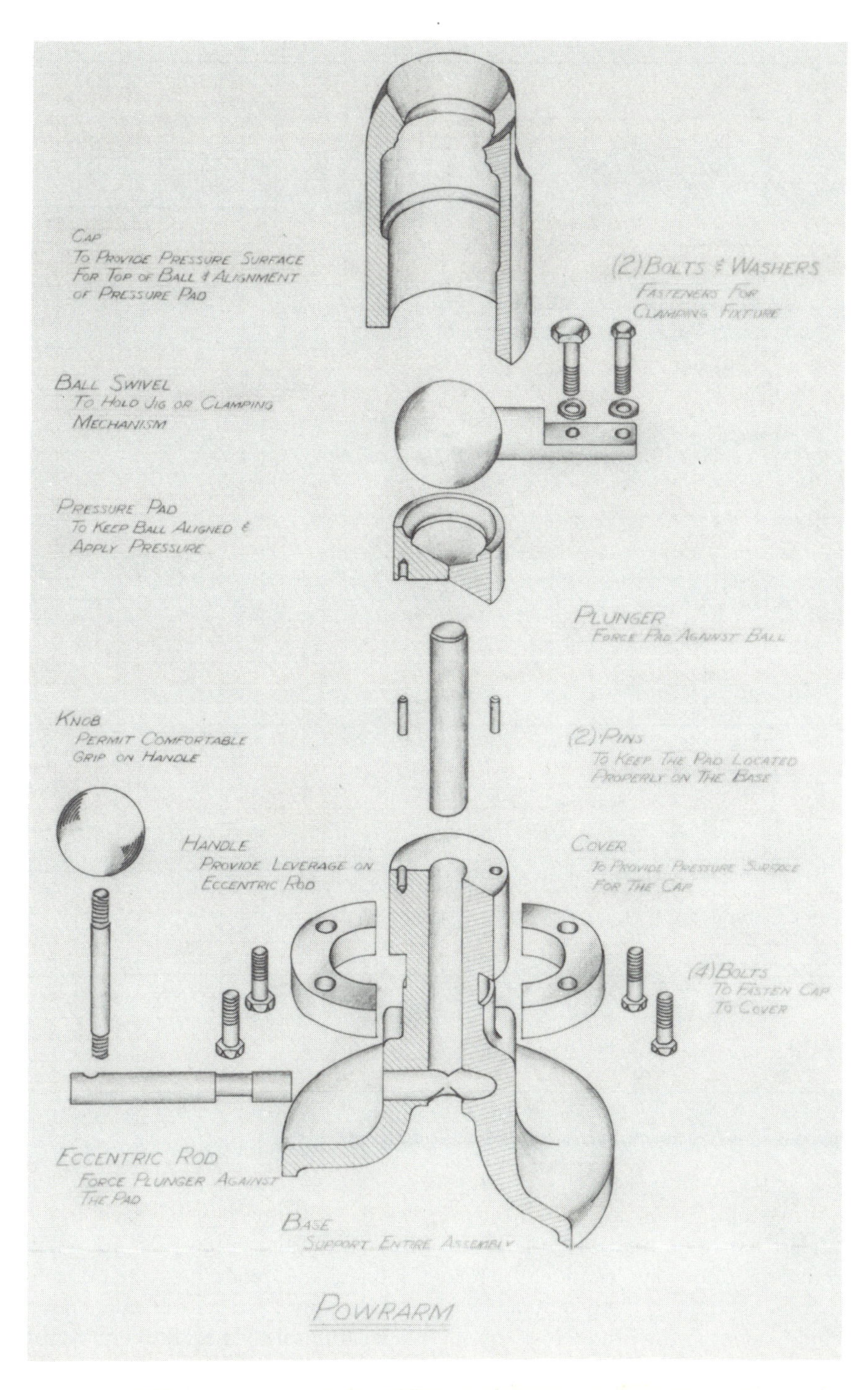

Figure 3-6—The Powrarm, exploded view, drawing.

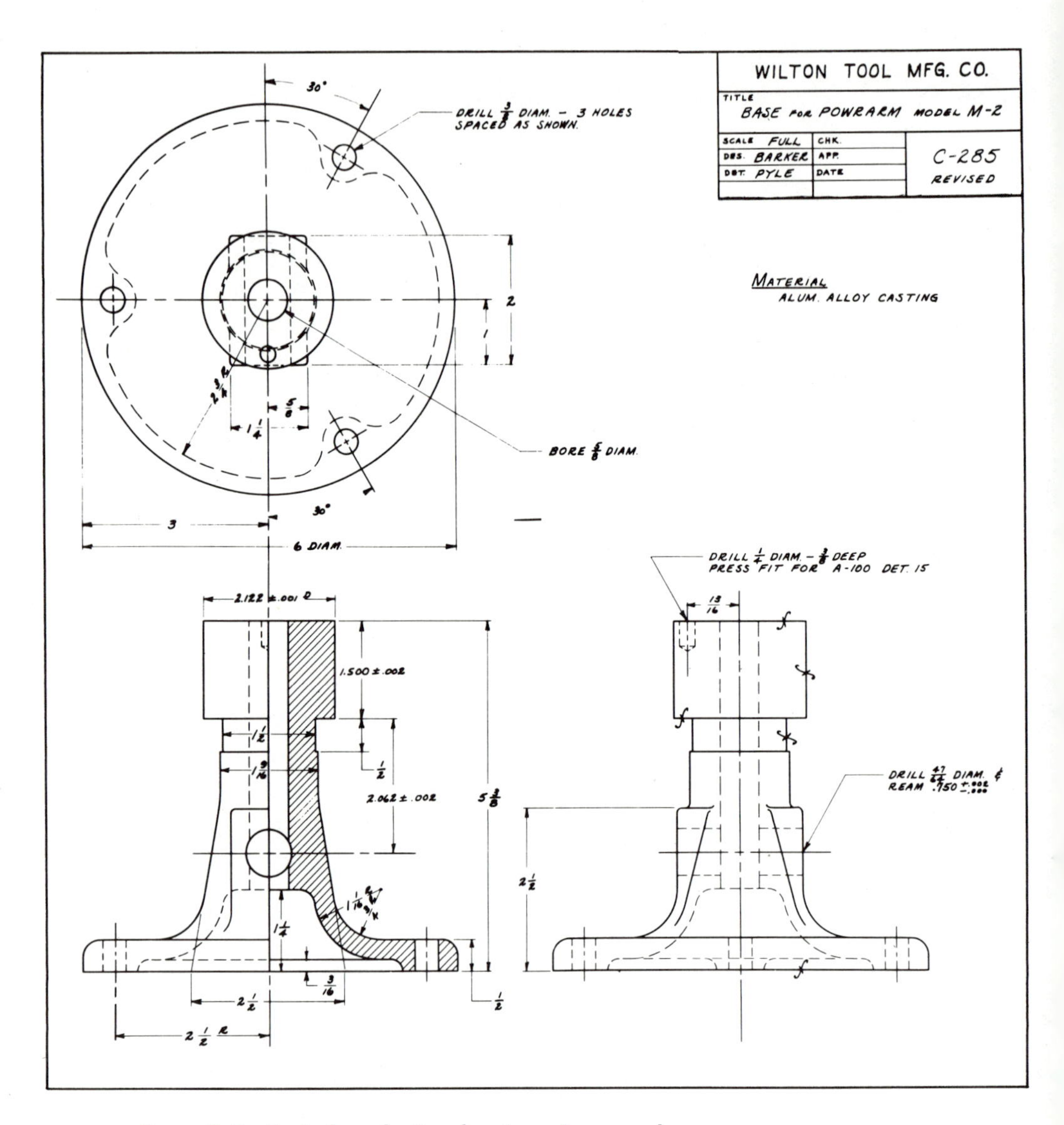

Figure 3-7—Typical production drawing—Powrarm base.

In general, prints are issued for every part—with the possible exception of standard hardware items, such as screws, nuts, and washers. Figure 3-7 is a typical production drawing. Occasionally, if a component is a standard product of some other manufacturer, the print may be general in nature, and contain only enough detail to identify the product—but with specifications listed to positively identify it as the one that is wanted.

The *Parts List* is a complete listing of all components of the product, and

A. B. C. Manufacturing Co.

PARTS LIST

For: Powrarm Model M-1 Dwg. No. D-442

Part No.	Part Name	Dwg. No.	Quant. per Unit	Material Spec.	Remarks
1	Base	C-285	1	Cast Alum.	
2	Eccentric Rod	A-143	1	C.R.S.	
3	Handle	A-143	1	C.R.S.	
4	Knob	A-143	1	Plastic	purchased
5	Plunger	A-163	1	C.R.S.	
6	Hex. Head Screw	--	4	--	$\frac{1}{4}$-20x$\frac{3}{4}$ purchased
7	Cover	A-95	1	C.A.	
8	Cap	B-111	1	C.A.	
9	Pin	A-100	1	C.R.S.	
10	Pressure Pad	A-97	1	C.I.	
11	Ball Swivel	A-98	1	C.R.S.	
12	Washer	--	2		purchased
13	Hex. Head Screw	--	2		$\frac{1}{4}$-20x1 purchased
A-1	Eccentric Assembly	A-143			pts. 2, 3, 4
A-2	Ball Swivel Assembly	D-442			pts. 11, 12, 13
A-10	Final Assembly	D-442			

Figure 3–8—Parts List, or Bill of Materials, for the Powrarm.

FORM NO. 701 REV.

CLARKE FLOOR MACHINE COMPANY
BILL OF MATERIAL

WHERE USED	CHANGE DATE	MODEL NAME	BILL OF MTRL. NO.
785450	9 15 61	HDS36 SWEEPER COMMON	785450
	EFF. DATE	LATEST E.C.N. NO.	ISSUED BY

PART CODE	QTY. REQ.		PART OR ASSEMBLY NAME	ASSY	SHEET NO.
782820	1		HDS36 EL SPECIAL 3/8 PT		
782821	1		HDS36 ELBOW MALE 8 12		
782822	1		HDS36 EL SPECIAL 1/4 PT		
782823	1		HDS36 ELBOW 90 ST 12		
782824	1		HDS36 ELBOW 90 ST 8 12		
783001	1		HDS36 FAN BLOWER		
783301	1		HDS36 FILTER HDY OIL		
783302	8		CS27 FLANGE BRG MOUNT		
783303	1		HDS36 FILTER FILLER CAP		
783304	14		HDS36 FLANGE BRG MOUNT		
783305	8		HDS36 FLANGE BRG MOUNT		
783306	2		HDS36 FLANGE BUSHING		
783310	1		HDS36 FILTER AIR		
783401	1		HDS36 GASKET FC602		
783402	1		HDS36 GASKET FC 603		
783403	1		HDS36 GASKET TANK COVER		
783615	1		HDS36 GUARD STEERING		
783623	1		HDS36 GUARD DRIVE		
783626	1		HDS36 GUARD BATTERY		
783635	1		HDS36 GUARD CHAIN		
783636	1		GUARD MUFFLER		
783701	1		HDS36 GAUGE OIL PRESSURE		
783803	1		HDS36 GAUGE AMMETER 6V		
783805	1		HDS36 GAUGE TEMPERATURE		
783806	1		HDS36 GAUGE SIGHT		
784001	1		HDS36 HOLDER REAR APRON		
784002	12		HDS36 HOLDER MAIN BROOM		
784003	2		HDS36 HOLDER SIDE APRON		
784101	2		HDS36 HOSE ASSY 12 42		
784105	1		HDS36 HOSE ASSY 6 36		
784108	1		HDS36 HOSE ASSY 6 23		
784109	1		HDS36 HOSE ASSY 6 11		
784114	2		HDS36 HOSE EXHAUST		
784117	1		HDS36 HOSE ASSY 4 34		
784118	1		HDS36 HOSE ASSY 4 38		
784119	1		HDS36 HOSE ASSY 4 17		
784120	1		HDS36 HOSE RAD TOP		
784121	1		HDS36 HOSE RAD BOT		

SHEET NO.

***Figure 3-9*—Parts List *produced from punched cards of individual components.* (Courtesy *of* Clarke Floor Machine Co.)**

should contain at least the information shown in Figure 3-8, although sometimes additional information will be desired, such as:

1. Detailed specifications
2. Source of purchased parts
3. Cost of parts
4. Part numbers of similar parts in other products

The *Parts List* is generally furnished along with the drawings, by the Product Engineering Department. Another typical *Parts List* is shown in Figure 3-9. This can be reproduced from individual punched cards or from computer files. The *Parts List* and the assembly drawings, are the first items with which the plant layout group will work in their preliminary planning activities.

Conclusion

With the presentation of the product specifications, in the form of prototypes, drawings, and parts lists, the basic data required for facilities design is at hand. Chapter 4 will discuss the work of the process engineer as he uses the above data to develop the manufacturing processes and select the equipment for producing the product.

Questions

1. Before the layout work can be started, what information must be contributed by: (a) Top management. (b) Market research. (c) Product design and engineering. (d) Industrial engineering. Briefly describe each contribution and indicate its importance in the layout planning.
2. What is the role of the feasibility study?
3. Under what circumstances would a company choose to: (a) Make components of its products? (b) Buy the components?
4. What sources might such information, as referred to in *Question 1,* come from in a smaller plant of, say, 25 to 50 total employees?
5. How does a seasonal sales pattern affect production?
6. How does one ascertain the time required to get a new product into production?
7. How might the financial policies of a company affect the introduction of a new product?
8. Why is it important to test a new product before going ahead with production. What kinds of things might one find? Discuss examples you are familiar with.
9. Briefly describe: (a) A *Production Routing.* (b) A *Parts List,* or a *Bill of Materials.*

Exercises

Select a product (not too complex) and:

A. List some significant market research questions that would be appropriate.

B. What product tests would be necessary?

C. Make a *Parts List.*

4

Designing the Process

In the manufacture of any product, the performance of any service, or the carrying out of any individual activity, the person doing it follows a certain, usually predetermined, sequence of steps. This is just as true for the production of an automobile wheel as it is for the handling of the mail, the laundering of a shirt, the selling of a bicycle, or the making of a hamburger. The steps, predetermined to make the product or service more efficiently, or more uniform, are commonly called operations. A sequence of operations and related activities is called a process. And the work involved in analyzing the product or service, and specifying the operations and equipment required, is called process design. This chapter will consider the factors, procedures and techniques involved in so designing the process (including the selection of equipment) that the operations, equipment, personnel, and material can be arranged in proper relationship to each other. The arrangement is plant layout, or facilities design.

The Production Design Procedure

Actually, the production design process involves a number of activities—and in a large enterprise, a number of people or departments. Some idea of the scope of the overall procedure, from beginning to end, is shown in Table 4-1. This lists the activities involved, and shows, by a letter *P* (Primary), the organizational functions usually charged with the activity. The letter *J* indicates a joint responsibility, while the letter *A* indicates an advisory relationship. It will be seen that the overall production design process is a joint responsibility of several organizational functions, and a serious attempt should be made to insure a smooth program of cooperation among the functions by means of organizational interrelationships and operating procedures.

The process design segment of the production design process can be defined as:

1. The analysis of the product or service.
2. The determination of what (production) operations are necessary to produce or perform it.
3. How these will be carried out.
4. What machines, equipment, tools and facilities are necessary.
5. What standards of output will govern the performance.

Table 4-1. The Production Design Team and Its Responsibilities

Production Design Procedure	Engineering Functions			
	Product	Mfg. or Prodn.	Ind'l.	Plant
Product Design				
Research and development	P			
Design	P			
Test	J	J		
Process Design				
Analyze specifications		P		
Make or buy analysis		P		
Material selection	P	A		
Process selection	P		A	
Dimensional analysis	P			
Determine manufacturing operations	A	P		
Select or specify production equipment	A	P	A	A
Specify tools and auxiliary equipment	P	A		
Establish operation sequence	A	P	A	
Prepare production routings		P		
Operation Design				
Methods analysis and design			P	
Work measurement			P	
Work standards			P	
Equipment requirements		J	J	
Manpower requirements		A	P	
Facilities Design				
Material flow design			P	
Systems design			P	
Activity relationship analysis			P	
Space allocation			P	A
Plant layout design			P	A
Storage facilities design			P	
Establish building specifications			J	J
Installation				
Building construction liason			J	J
Equipment installation liason			A	P
Try-out and Debug				
Tryout and debug		P	A	A
Evaluation studies	A	P	A	
Monitor operations		A	P	
Information feedback	A	P	A	

The process design function usually results in:

1. Operation sketches
2. Production routings or specifications
3. Tool layout drawings
4. Operation sheets
5. Detailed operation instruction sheets
6. Preliminary workplace layout drawings
7. Preliminary layout sketches

(Examples of these are shown later in this chapter—and in the subsequent chapters.)

The process design function includes such activities as:

1. Pre-production planning
2. Process feasibility studies
3. Process capability studies
4. Process capacity studies
5. Process and equipment development
6. Process design (detailed below)
7. Equipment and tool development and design
8. Production operations evaluation
9. Planning of re-work or extra operations
10. Production cost estimating
11. Manufacturing capacity analysis
12. Preparation for product, model, and design changes
13. Packaging, and packing methods and processes
14. Long-range planning as related to equipment
15. Planning procedures

Some of these are discussed below as a part of preparing basic data for use in the facilities design process. Others represent process engineering activities related to other organizational functions.

Factors for Consideration in Process Design

As will be seen throughout this book, heavy emphasis is placed on insuring that each step in the facilities design process considers as many pertinent factors as possible. This is done to avoid overlooking an item of significance in reaching an optimum decision at each step along the way.

Table 4-2 shows some of the factors that should be given consideration, if one is to design an efficient process. Many of the factors arise in carrying out the steps outlined below, while others are given consideration by or with other organizational functions throughout the design process, as well as periodically during the operation of the enterprise.

Preliminary Production Planning

Before the actual process design can begin, certain data and information necessary for making decisions must be accumulated, organized, and analyzed for use in the process engineering and subsequent production planning and related activities. Some of these activities are discussed below.

One of the first pieces of information needed by the process designer is the production volume—since so many of his decisions are based on whether he is planning to make 10, 100, 1,000 100,000, or 1,000,000—and over what period of time that quantity is to be produced. As was seen in the last chapter (and even more so in the next) the processes and methods of manufacturing depend greatly on the volume and rate of production.

The basis for the volume and rate is the sales forecast—both for the immediate period and for the longer range. Not only must the process designer plan facilities to meet the immediate needs, but he must design the processes so that they can be conveniently altered or supplemented to meet future demands of both volume and rate (future demands can be either higher or lower).

A direct result of the volume and rate analysis would be an investigation of the profitability of producing the product or service. This might well require that some of the activities discussed below be performed at this point—at least in a preliminary fashion. At some point, a decision must be reached as to whether the predicted volume will justify production: that is, the production of the required volume must be accomplished by such methods as will result in a low enough cost to permit sale of the product at a profit.

Product Analysis

After a preliminary decision has been reached as to whether to proceed with more detailed planning, the product–process relationships must be investigated in more detail. For instance, as implied above, there is the need for a searching analysis of the product design and specifications with the objective of insuring that the product will be:

1. *Functional*—and properly perform its desire function.
1. *Of the proper quality*—that is, neither better than it need be, nor worse.
3. *Acceptable to the buyer in appearance*—rather than unattractive in some respect.
4. *Producible*—at a cost compatible with a reasonable selling price.

The first task in such an analysis is the accumulation of the drawings, parts lists, bills of materials, specifications, and other information requirements necessary to accurately define the parts and the product.

Figure 4-1 is a typical drawing, which will be used in the explanation of the balance of the process design procedure. Other necessary or helpful items are a prototype part or product (Figure 3-5), or an exploded drawing (Figure 3-6). There are several important reasons for this detailed preliminary analysis of the design specifications. Among other things, the analyst is looking for:

1. Omission of necessary details or specifications.
2. Too-tight dimensions, too-close tolerances, limits, etc.
3. Errors in design or specifications.
4. Validity of implied specifications.
5. Potential processing problems.
6. Manufacturing capability.
7. Unclear dimensions, etc.
8. Difficulties or excess cost of meeting specifications.

Table 4-2. Factors for Consideration in Designing the Manufacturing Process

PRODUCT

A. Product (part)

1. Total quantity—market, trend
2. Production rate
   a. Per hour
   b. Lot size
3. Production method
   a. Continuous
   b. Intermittent
   c. Job lot
4. Life expectancy
5. Likelihood of change (stability)
6. Durability
7. Function
8. Time to get into production
9. Customer desires
10. Quality level
11. Process requirements, specifications
    a. Quality
    b. Accuracy
    c. Tolerance
    d. Appearance
    e. Finish
    f. Special
12. Estimated selling price
13. Complexity
14. Degree of standardization
15. Competition

B. Material

1. Type
2. Form
3. Size, shape
4. Properties
5. Scrap and waste
6. Finishing costs
7. Cost
8. Source
9. Estimated inventory
10. Handleability
11. Fragility
12. Availability
13. Method of receipt

PROCESS (EQUIPMENT)

C. Mechanical factors

1. Capability of performing work
2. Accuracy attainable
3. Dimensional stability
4. General vs. special purpose
5. Flexibility
6. Adaptability
7. Life expectancy
8. Potential obsolescence
9. Compatibility with present (other) equipment
10. Feeding method
11. Materials of construction
12. Durability
13. Reliability
14. Physical characteristics—size, weight, etc.
15. Foreseeable technological improvements
16. Degree of mechanization
    a. Present
    b. Potential
17. Relative complexity
18. Capacity
19. Reserve capacity
20. Need for subsequent operations

D. Operating factors

1. Interruptibility of process
2. Efficiency
3. Set-up time
4. De-bugging time
5. Frequency of use
6. Percentage of time used
7. Safety
8. Installation time
9. Manpower requirements
   a. Amount
   b. Skill
   c. Training
   d. Cost
   e. Supervision
   f. Inspection
   g. Handling
10. Human factors
11. Physical effort required

E. Cost factors

1. Investment
2. Tools, etc.

Table 4-2 (continued)

3. Installation
4. Start-up
5. Operating
6. Funds available
7. Savings
8. R.O.I.
9. Own vs. lease
10. Resale value
11. Space lost, or gained
12. Equipment cost trends

F. Building factors

1. Available space
2. Column spacing
3. Floor capacity
4. Ceiling height

G. Miscellaneous factors

1. Availability
2. Applicable standards
   a. Product
   b. Government
   c. Industry
3. Ecological consequences
4. Warranty
5. Patents, etc.
6. Intangible factors (in addition to several in above categories)
   a. Security
   b. Service availability
   c. Manufacturer's reputation
   d. Quality of service
7. Plans for expansion
8. Business trends

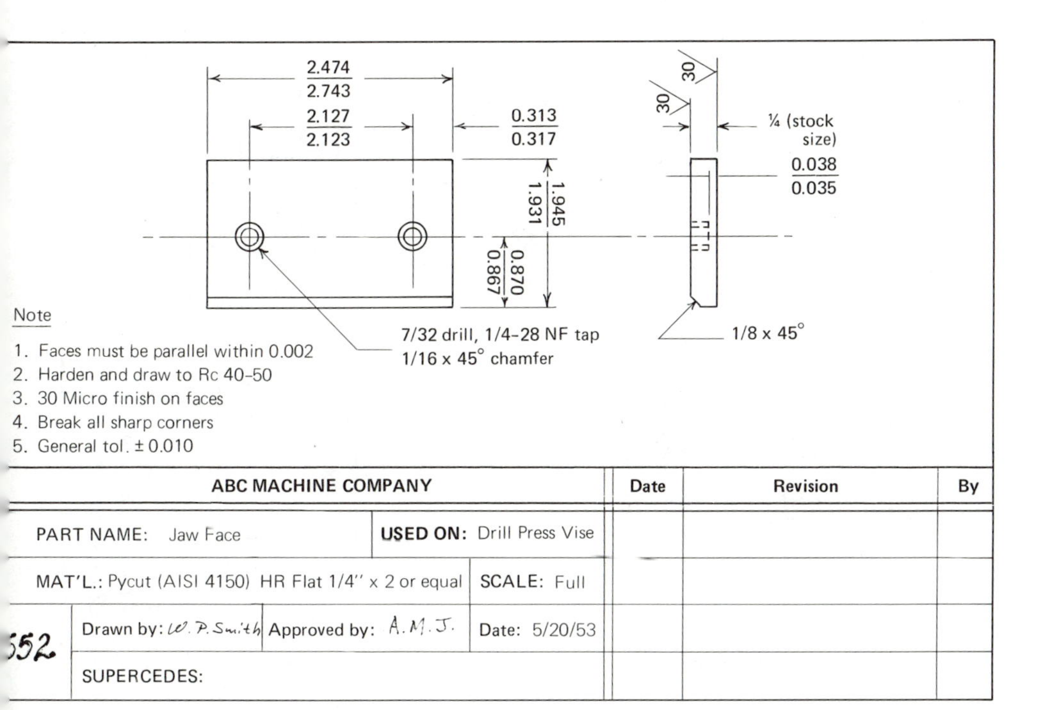

Figure 4-1—Drawing—vise jaw.

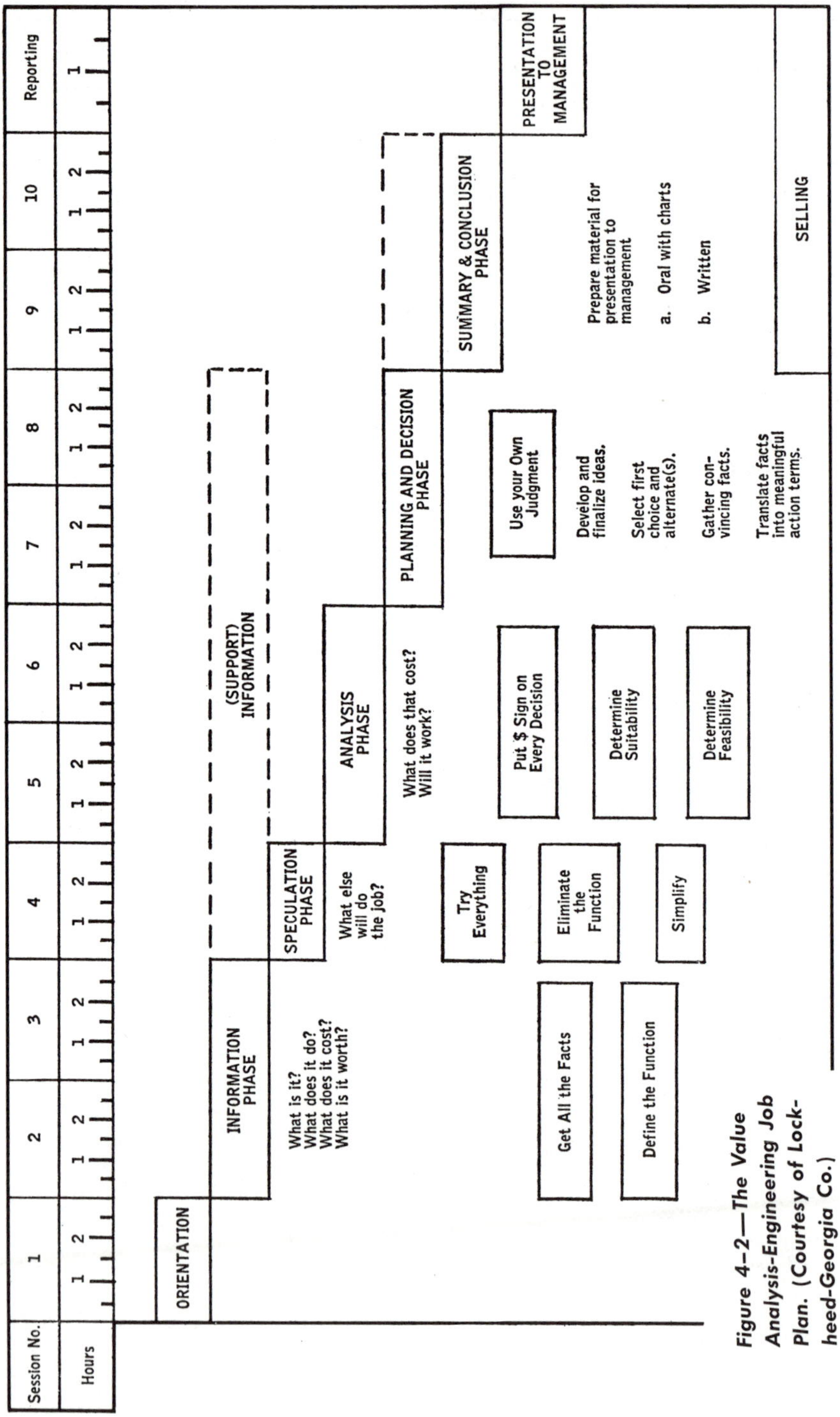

Session No.	1	2	3	4	5	6	7	8	9	10	Reporting
Hours	1 2	1 2	1 2	1 2	1 2	1 2	1 2	1 2	1 2	1 2	1

ORIENTATION

INFORMATION PHASE

What is it?
What does it do?
What does it cost?
What is it worth?

(SUPPORT) INFORMATION

SPECULATION PHASE

What else will do the job?

ANALYSIS PHASE

What does that cost?
Will it work?

PLANNING AND DECISION PHASE

SUMMARY & CONCLUSION PHASE

PRESENTATION TO MANAGEMENT

Get All the Facts

Define the Function

Try Everything

Eliminate the Function

Simplify

Put $ Sign on Every Decision

Determine Suitability

Determine Feasibility

Use your Own Judgment

Develop and finalize ideas.

Select first choice and alternate(s).

Gather convincing facts.

Translate facts into meaningful action terms.

Prepare material for presentation to management

a. Oral with charts

b. Written

SELLING

Figure 4–2—The Value Analysis-Engineering Job Plan. (Courtesy of Lockheed-Georgia Co.)

One of the techniques useful in analyzing the product is *value analysis*, or *value engineering* as it is sometimes known. The objective of value analysis is to find less expensive ways to achieve equal or better functional performance. Basically it consists of a procedure whereby each part of a product is scrutinized in detail, in an attempt to find an easier, better, cheaper way to perform a function. The procedure is somewhat as follows:

A. *Value Analysis Methodology*
1. Subject selection
2. Functional determination
3. Information collection
4. Development of alternatives
5. Design or specification of solution
6. Test and evaluation
7. Proposal and approval
8. Installation
9. Follow-up

B. *Value Analysis Process*
1. Define each function
2. Evaluate each function
3. Think creatively
4. Overcome roadblocks
5. Blast, create, refine
6. Work on specifics
7. Evaluate by comparison
8. Put dollar sign on each idea; tolerance; etc.
9. Adhere to standards

The overall value analysis–engineering plan procedure is somewhat as shown in Figure 4-2. In some companies value analysis is felt to be so important that they have established a separate department or group for it. (For additional details, see one of the several books on the subject.)

Make-or-Buy Analysis

One result of the value analysis procedure is very likely a decision as to whether to make a part or buy it from a specialized manufacturer. For example, manufacturers—even automobile plants—may buy nuts, bolts, washers, and many other parts because they do not use enough to warrant making their own, sometimes because an item requires specialized equipment that would turn out far more than their own needs.

Methods of Production

Another decision to be made, prior to the actual process design, is the determination of the method of production for a part or product. There are several alternatives, commonly referred to as:[1]

1. Continuous, mass, or line production—*product method*
2. *Similar process method*
3. Job order, job shop or intermittent production—*process method*
4. *Custom work*

See Table 4-3 for comparisons of these methods.

In continuous, mass, or line production, the work required to produce a part or product is divided into individual operations, which are usually arranged in sequence, in a "line." It has been said that no single action decreases cost and

[1]Adapted from E. D. Scheele, W. L. Westerman, and R. J. Wimmert, *Principles and Design of Production Control Systems* (Englewood Cliffs, N.J.: Prentice-Hall, Inc., 1960).

Table 4-3. Summary of Characteristics of Types of Production Processes

Characteristics	Continuous (Mass)	Similar Process	Job-Order (Intermittent)	Custom
1. Product or end result	Standard	Standard	Non-standard	Unique
2. Examples	a. Autos b. Electrical appliances c. Refining	a. Shipping and receiving b. Clerical c. Clothing manufacturing	a. General machine shop b. Machine tool manufacturing c. General engineering	a. Research b. Bridge construction c. Model making
3. Order or work unit quan.	Large	Large	Small	Usually one
4. Type of equipment	Special purpose	Special purpose	General purpose	General purpose
5. Equipment arrangement	By product	By product	By process	By process
6. Material handling equip.	Conveyor	Mobile and Conveyor	Mobile	Mobile
7. In-process inventory	Low	Relatively low	High	Relatively high
8. Worker skill level (relative)	Very low	Low	High	High
9. Supervisory difficulty	Very easy	Relatively easy	Very difficult	Highly difficult
10. Job instructions	Very few	Relatively few	Many and detailed	Many and very detailed
11. Prior-planning	Very complex, but once for all units	Relatively easy	Complex	Complex
12. Control	Very easy	Easy	Complex	Complex
13. Degree of flexibility	Very little	Some	High	High
14. Cycle time	Very short	Relatively short	Relatively short	Long
15. Balancing of work load	Very difficult	Relatively difficult	Easy	Easy
16. Unit cost	Low	Relatively low	High	Very high

Source: Scheele, Westerman, and Wimmert.

increases production more than changing from the job shop to the line or product method. In the *product method,* individual machines (or other equipment) are arranged in the order of the processes called for on the *Production Routing.* Each part travels from one machine—a drill press for instance—to the next machine—say, a lathe—for the next operation, and so on through the entire cycle of operations required. The path for each part is in effect a straight line. If the product is standardized and is made in large quantities, this product method of layout is desirable. Usually all the equipment necessary to produce a part is located in one area. On a piston, for instance, a machine is provided for each operation, from the first turning to the last inspection; and the machines are arranged in sequence so that the part can be literally passed from one operator to the next.

In determining which parts and products should be produced by the product method, it may be helpful to make use of *Pareto's law* or the *80–20 rule.* This concept holds that (approximately) 80% of sales dollars are derived from (approximately) 20% of the products. Likewise, 80% of the cost is incurred in making 20% of the components. An analysis of this type will pin-point those products and their components that should be given major attention in planning processes and operations.

In the *similar process method* of production, the work done is similar from one order to another, but not identical, for example, in shoe and clothing manufacture, mail-order packaging, and most clerical operations.[2]

> The job-order, job-shop, or intermittent method is characterized by a wide diversity of end products. Work centers are arranged around particular types of work or equipment. For example; in a plant doing machine work, the drilling machines would be in one department, the turning machines in another department, the milling machines in another department, and so on. For these reasons, the technique is often called the *process method,* in contrast to the product method previously discussed. The work then flows through these departments or work centers in batches of work called *lots.*

When the product is not or can not be standardized, or where the quantity of similar parts or products in process at any one time is low, the process method layout is more desirable because of its flexibility. But it should be noted that:[2]

> Seldom do we find a pure job-order, or job-shop type of process. Usually there is a combination of the continuous and job-shop, which we term intermittent, type of process. Obviously, wherever advantage can be taken of the continuous process, it will result in overall lower cost. By properly analyzing all of the work that goes through an activity, it might be possible to identify areas in which the sub-operation might be set up in such a manner as to simulate the continuous process. In some cases it may be cheaper to move a machine than it is to move materials from one station to the next. For example, in a garment manufacturing plant, machines can be moved around quite easily. Instead of setting up the layout on a job-shop basis for lots of garments, it might prove economical to move the machines into assembly-line setup for each product. This reduces all of the handling of the in-process inventory between operations and greatly reduces the cycle time in order to produce the finished product. This, of course, would only be done in the event that the job-orders are of sufficient size to warrant the cost of "rearranging" the machines.

Figure 4-3 compares the *product method* and the *process method* of layout.

In *custom work* the end product is unique for each job, so its processes are distinct from those of any other job, as in bridge, building, and ship construction, and in pure research and administrative activities. The only similarity would be in principle alone. Obviously, the governing factor in this process is the workers' skill level. The highest level of skills in supervisory abilities is also necessary in order to insure success.

[2]Scheele, *et at.,* p. 32.

Part	Oper. 1	Oper. 2	Oper. 3
A	lathe	drill	lathe
B	drill	mill	
C	lathe	mill	drill

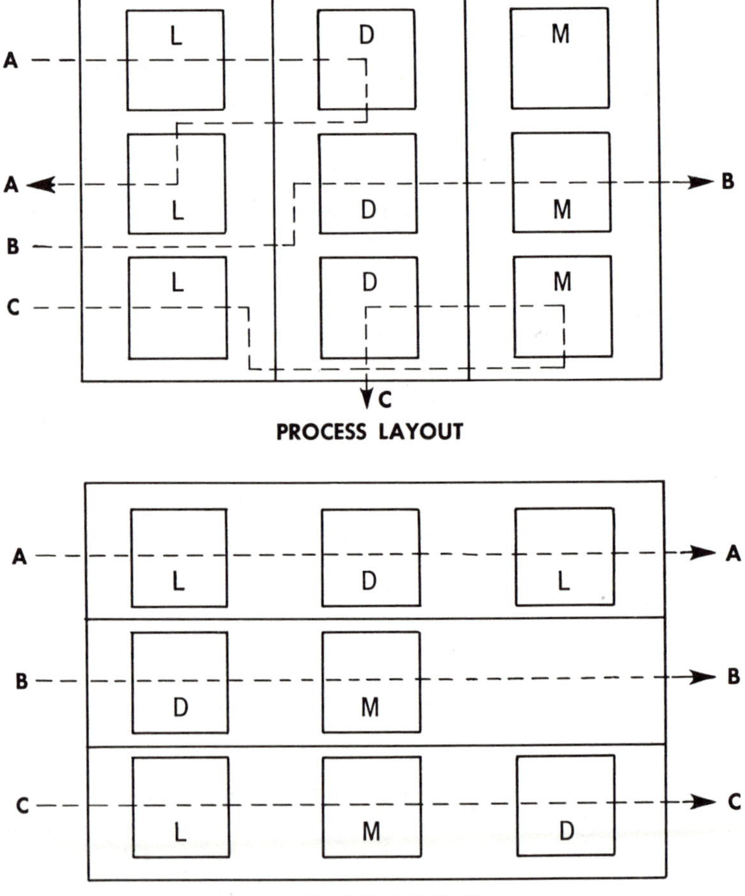

Figure 4–3—Process and product methods of machine arrangement.

Group technology. In a relatively new short-run production method often used in job shop situations, called *group technology,* essentially dissimilar parts are grouped into families based on similarities in the form or shape of the parts, rather than in the end use. This permits the use of family production lines, rather than individual machines or machining centers (or similar machine types), which

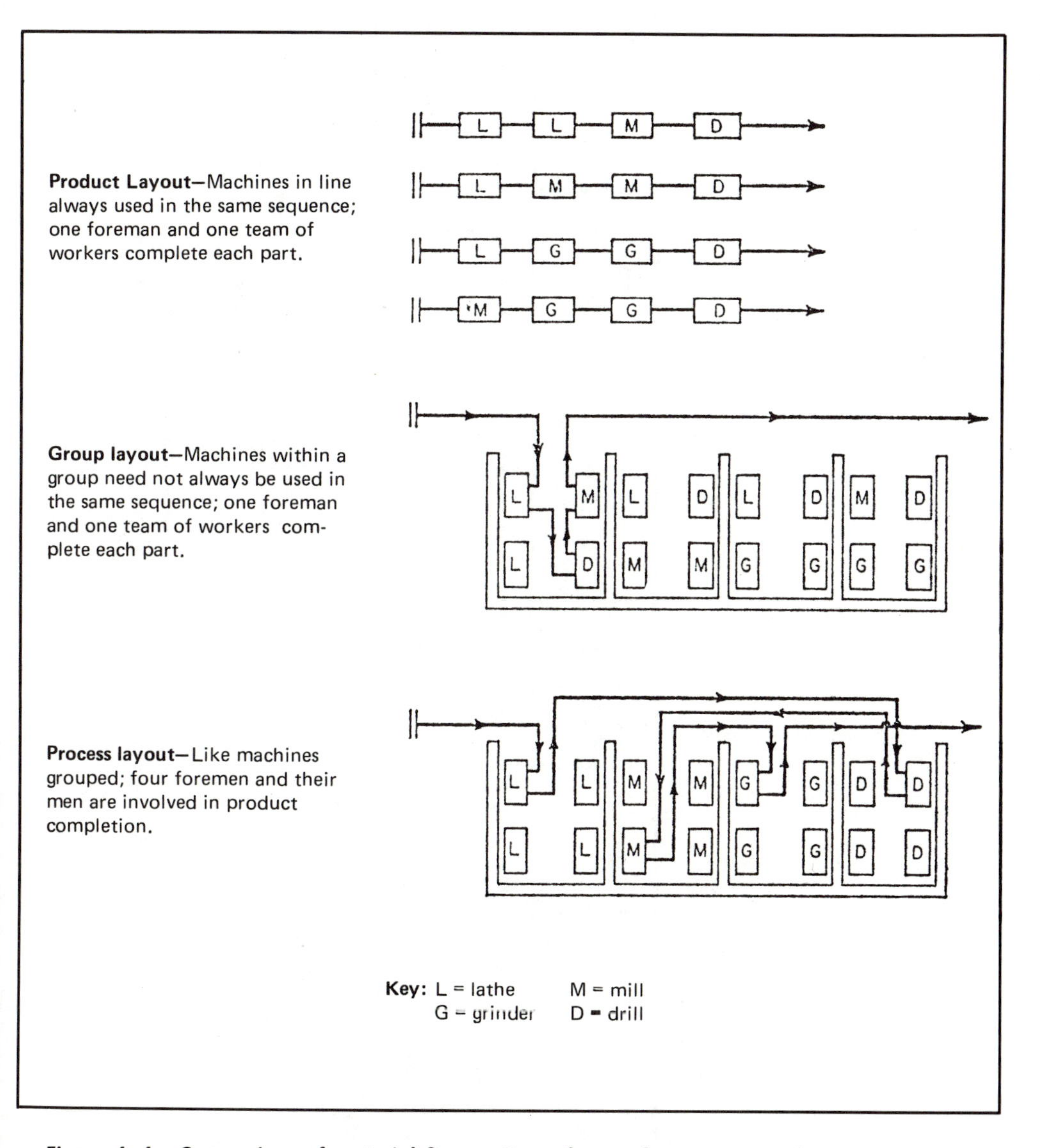

Figure 4–4—Comparison of material flow patterns for product, group, and process-type layouts (compare with Figure 4–3). (Adapted from J. L. Burbidge, The Introduction of Group Technology, John Wiley & Sons, 1975.)

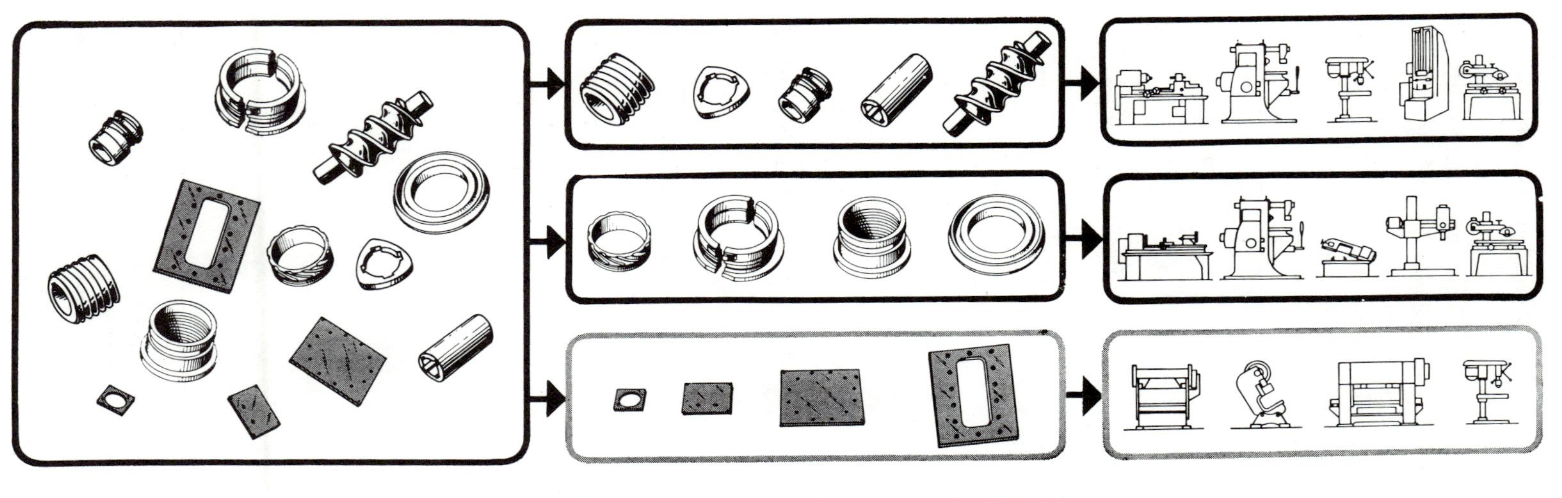

Figure 4–5—Concept of classification of parts into groups. (Adapted from "Group Technology: Best Short-Run Cost Cutter Yet," Factory, Oct. 1963: 103–4.)

permits small lots of dissimilar parts to be run through on a mass-production basis. Figure 4-4 shows the relationship between the process layout and the group layout. Figures 4-5 and 4-6 give an idea of the grouping process, and an insight into the types of machines needed to fabricate each group of parts. The details of planning for group technology are beyond the scope of this book.[3]

The Unit Process Concept[4]

Classifications of processes by type or by equipment utilized is helpful to the process designer, but such arbitrary classifications can be misleading. For example, the same machine could perform radically differing processes. A hydraulic press, for example, might:

1. *Punch* a hole.
2. *Draw* a shell.
3. *Extrude* a gear.
4. *Compact* a powder sleeve.
5. *Forge* an axle.

and a lathe might:

1. *Spin* a cup.
2. *Drill* a hole.
3. *Roll* a thread.
4. *Knurl* a handle.
5. *Form* a shank.
6. *Bore* a recess.
7. *Mill* a slot.
8. *Tap* a thread.

To think of processing strictly in terms of conventional machines is especially confusing when the volume of production becomes very large, and unusual combinations of processing of components take place.

This suggests a need to begin thinking of the simplest operations to be performed as small *building blocks* that will be synthesized in an overall planned process, and will be used in describing what is to be done to material to transform it. This building block can be called a *unit process,* defined as follows: *A unit process is a modification of material done essentially without interruption.*

Some typical unit processes are: listed in Table 4-4. Notice that the unit process does not tell which machine is to be used to modify material. It only indicates what the modification will consist of. The unit process concept has several advantages:

1. Unit processes are convenient in describing what is to be done to material to modify it.
2. Unit processes offer a convenient basis for organizing detailed process information.
3. Unit processes make it easier to synthesize a new process.
4. Unit processes make it easier to compare processes because cost characteristics can be conveniently compared.

Manufacturing a part consists of performing a group of unit processes on a piece of material until it conforms to part specifications (blueprints, material characteristics, etc.) as indicated in Figure 4-7.

[3]Interested readers might contact the Society of Manufacturing Engineers, Dearborn, Mich., for papers giving more details.

[4]Adapted from W. P. Smith and P. J. Thorson, *Basic Process Engineering*, unpublished class notes, pp. 31–33, 39.

	1ST DIGIT PART CLASS	2ND DIGIT EXTERNAL SHAPE		3RD DIGIT INTERNAL SHAPE		4TH DIGIT SURFACE MACHINING	5TH DIGIT HOLES AND TEETH
0	ROTATIONAL		Smooth without shape elements (constant diam)		Without bore or blind hole	No surface machining	No holes or teeth
1		one side increasing	No shape elements	one side increasing	No shape elements	External plane and/or circular surfaces	Axial and/or radial holes
2		one side increasing, or smooth	threads	one side increasing, or smooth	threads	External surfaces dividing each other in a given ratio	Holes in other directions
3		one side increasing, or smooth	functional grooves	one side increasing, or smooth	functional grooves	Internal plane surface and/or groove	Coupling teeth
4		several sides increasing	no shape elements	several sides increasing	no shape elements	External groove and/or slot	Spur gear teeth
5		several sides increasing	threads	several sides increasing	threads	Internal spline (polygon)	Bevel gear teeth
6		several sides increasing	functional grooves	several sides increasing	functional grooves	External spline (polygon)	Plane gear teeth (racks)
7			functional cone		functional cone	External spline & groove and/or slot, plane surface	Serrations and sprockets
8			operating threads		operating threads	External and internal splines	Wormwheels
9			spindles and others		Others	Others	Others

Component Code No.-042044120

EXAMPLE OF CODING SYSTEM

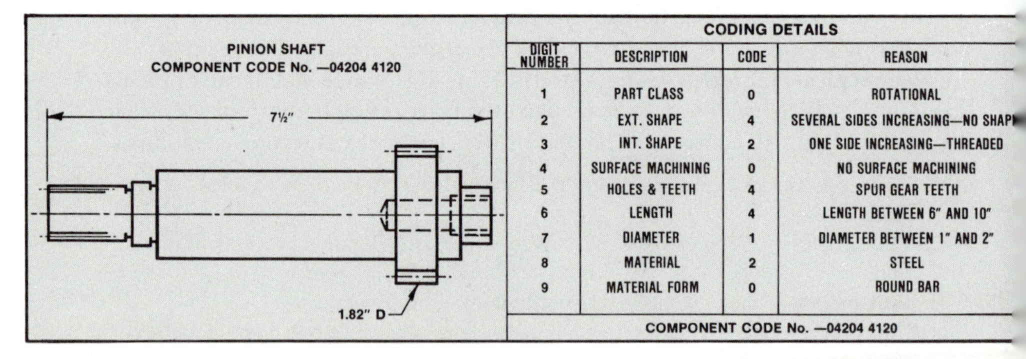

CODING DETAILS

DIGIT NUMBER	DESCRIPTION	CODE	REASON
1	PART CLASS	0	ROTATIONAL
2	EXT. SHAPE	4	SEVERAL SIDES INCREASING—NO SHAPE
3	INT. SHAPE	2	ONE SIDE INCREASING—THREADED
4	SURFACE MACHINING	0	NO SURFACE MACHINING
5	HOLES & TEETH	4	SPUR GEAR TEETH
6	LENGTH	4	LENGTH BETWEEN 6" AND 10"
7	DIAMETER	1	DIAMETER BETWEEN 1" AND 2"
8	MATERIAL	2	STEEL
9	MATERIAL FORM	0	ROUND BAR

COMPONENT CODE No. —04204 4120

The Process Design Procedure[5]

Having considered a large number and wide variety of factors—and having made decisions on many of them—it is now possible to begin the design of the process. The process design function can be defined as the analysis of the product and the

[5]This section draws heavily on Eary and Johnson, *Process Engineering*, and Smith and Thorson, *Basic Process Engineering*.

6TH OR 7TH DIGIT LENGTH 'A' AND/OR DIAMETER 'D' INCH		
0		≤1
1	>1	≤2
2	>2	≤4
3	>4	≤6
4	>6	≤10
5	>10	≤16
6	>16	≤24
7	24	≤40
8	>40	≤80
9		>80

7TH DIGIT LENGTHS INCH		
0	Smallest C≤1	B ≤1
1		1<B≤6
2		B>6
3	1<C≤6	1<B≤6
4		6<B≤15
5		B>15
6	C<6≤15	6<B≤15
7		15<B≤30
8		B>30
9	C>15	B>15

8TH DIGIT MATERIAL	
0	Cast Iron
1	Special or Malleable Cast Iron
2	Steel >25 /in.
3	Free Cutting Steel
4	Steels 2 or 3 Heat Treated
5	Alloy Steels
6	Alloy Steels Heat Treated
7	Copper Based Alloy
8	Aluminum, Zinc & Magnesium Based Alloys
9	Other Materials

9TH DIGIT RAW MATERIAL FORM	
0	Round Bar
1	Hexagon Bar
2	Bar □ Δ And Others
3	Tube and Pipe
4	Angle ⊏ T or Similar Profile
5	Sheet
6	Plate and Slab
7	Stamping, Casting or Forging
8	Welded Fabrication
9	Pre-Machined

The basic element in what has become known as "Group Technology" is a system of parts coding and classification.

With parts described by their geometries and sizes, they then can be grouped into "families" having significant similarities; and the parts in individual families then can be processed through dedicated machining work centers.

Figure 4-6—System of grouping parts into families by characteristics. (Courtesy of Production Magazine.)

determination of the manufacturing operations and facilities required to produce it to specifications as economically as possible. The procedure is usually similar to that outlined in Table 4-5 and discussed in the balance of this chapter.

1. *Obtain necessary data.* An adequate process design is affected by many external factors. It is vital to an efficient job of process design that restrictions are

Table 4-4. Example of Unit Processes

Chemical	Metal	Wood	Food	Garment
Evaporate	Drill	Rip	Blend	Cut
Humidify	Tap	Re-saw	Boil	Pink
Dehumidify	Broach	Miter	Roast	Crease
Dry	Bore	Bevel	Broil	Label
Filtrate	Form	Sand	Stir	Baste
Separate	Turn	Rabbet	Baste	Tack
Crystallize	Extrude	Dado	Whip	Sew
Distill	Shear	Groove	Knead	Hem
Absorb	Straighten	Recess	Bake	Dart
Extract	Trim	Route	Marinate	Tuck
Condense	Bend	Bore	Braise	Stitch
Mix	Stake	Drill	Fry	Quilt

known before the design is started. The process engineer who has available to him the facts that may effect the solution to his problem is in a better position than the man who does not have these facts or who finds them difficult to obtain. Much of this has been mentioned before, but some of the more important items are:

A drawing of the product and its components (Figure 3-7).

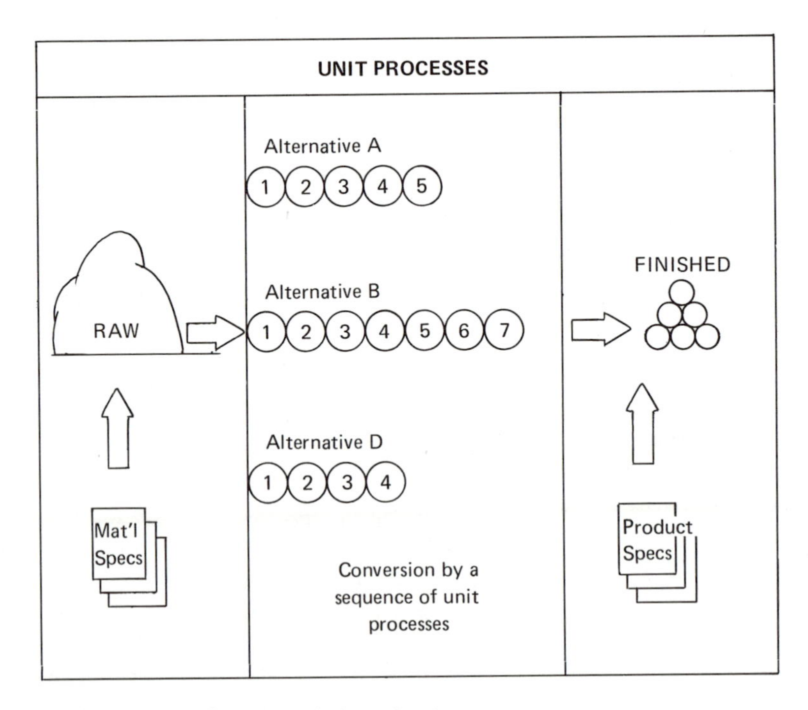

Figure 4-7—The accumulation of unit processes into a manufacturing process.

Table 4-5. The Process Design Procedure

A. Necessary data

1. Drawings (Fig. 3–7)
   a. Product
   b. Components
2. Specifications
   a. Functional
   b. Quality
   c. Appearance
3. Bill of Materials and/or Parts List (Fig. 3–8)
4. Total quantity
5. Production rate
6. Initial delivery date
7. Component availability
   a. Purchased
   b. Produced

B. Part Print Analysis

1. Characteristics of parts
   a. General description
   b. General configuration—as relates to—
      —Handling
      —Tooling
      —Machinery
      —Operation sequence
      —Production rate
      —Complexity of part
   c. Alternative materials
   d. Originating operations
   e. Recorded design changes
   f. Resistance to damage
2. Auxiliary methods of visualizing part from the print
   a. Drawings without dimensions
   b. Change in scale
   c. Cross-sectional views
3. Assemblies and sub-assemblies
   a. Drawings
   b. Exploded views (Fig. 3–6)
   c. Sample parts or models
   d. Assembly Chart (Fig. 6–2)
4. Review work
   a. Degree of symmetry
   b. Number of related surfaces to be machined
   c. Degree of relationship between surfaces
   d. Grouping of related surfaces or areas
   e. Number of surface treatments
5. Selection of materials
   a. Consider all types of material possible
      —Advantages
      —Disadvantages
      —Cost balance sheet—each part
      —Manufacturing problems
      —Procuring problems
      —Sizes required
      —Mechanical properties and qualities
      —Intangible factors
   b. Estimate waste material
      —Sale of scrap
      —Salvageable materials
      —Scrap loss
      —Waste disposal and pollution
   c. Storage problems
   d. Most economical materials
      —Raw material costs
      —Tool and die costs
      —Optimum lot sizes
6. Study specifications
   a. Explicit on print
   b. Implied
      —General knowledge
      —Convention
7. Dimensional analysis
   a. Types of dimensions
   b. Geometry of part
   c. Baselines
   d. Degree of finish
8. Tolerance analysis
   a. Cause of work piece variation
   b. Selective assembly
   c. Tolerance stacks
   d. Tolerance charts
9. Process types
   a. Raw material processes
   b. Manufacturing processes
   c. Unit processes
10. Alternative processes
    a. Information on blue print
    b. Knowledge of processes
    c. Imagination
11. Auxiliary supporting operations
    a. Receiving

Table 4-5 (continued)

b. Shipping
c. Handling
d. Inspection
e. In-process storage, packaging for

12. Construct a Critical Conditions Specification Sheet (Fig. 4–9)

13. Establish functional surfaces of work piece to be developed in process manufacture

14. Areas involved in processing
a. Locating areas
b. Supporting areas
c. Critical areas
d. Make use of degrees of symmetry

C. Construct Work List Specification Sheet (Fig. 4–10)

1. Part form and shape—all surfaces
a. Named or numbered
b. Described in detail

2. Internal and other aspects not in drawing—all additional elements of processing, such as:
a. Hardening
b. Plating
c. Stamping part number
d. Painting
e. Relieving stresses
f. Rust prevention
g. Anodizing

3. Elements of work created by processing—such as:
a. Straightening
b. Cleaning
c. Buffing

D. Combine unit processes into manufacturing operations

1. Unit process design
a. All specifications
b. Critical conditions
c. Alternatives

2. Grouping of processes—by:
a. Work place
b. Machine
c. Tool

E. Arrange operations into a logical sequence

1. Mandatory precedence
a. First
b. Last
c. Operational

2. Operational precedence—with respect to:
a. Facility layout
b. Handling
c. Equipment schedules
d. Dimensional integrity

3. Reduction of waste—with respect to:
a. Time
b. Material
c. Loading and Unloading Time
d. Handling

F. Details on equipment and tools

1. Process data
a. Product specifications
b. Manufacturing specifications
c. Review list of operations
d. List machine types for new operations
e. List required machine replacements

2. Tool data
a. Tools required
b. Gages required
c. Tool and gage drawings

G. Select and specify equipment

1. Define selection problems—such as:
a. New job
b. New mechanization
c. Machine replacement
d. Production cost
e. Production expansion
f. New technology

2. Factors in selection
a. Level of mechanization
b. Level of automation
c. Product
—Volume
—Stability
—Variety of work
—Seasonality
—Inspections
—Scrap
d. Production type
e. Potential obsolescence
f. Unit cost
g. Floor space
h. Maintenance
i. Flexibility
j. No. of operators
k. Material uniformity

Table 4-5 (continued)

l. Supervision
m. Production control
n. Life of job
o. Overhead costs
p. Complexity
q. Safety
r. Manual effort
s. Volume in process
t. Equipment cost
u. Other costs
  —Direct
  —Indirect
  —Indeterminate
  —Intangibles

3. General vs. special purpose equipment
4. Selection among alternatives—based on:
   a. Past experience
   b. Investigation
   c. Experimentation
5. Tools and auxiliary equipment

H. Estimate product costs—each alternative:

1. Annual cost
2. Return on investment
3. Other measure of feasibility
4. Cost per unit produced

I. Complete the process design

1. Manufacturing Operation Planning Sheet (Fig. 4–11)
2. Production Routings (Fig. 4–13)
3. Operation Process Chart (Fig. 6–5)
4. Number of machines and operators
5. Work place plans (Fig. 11–10)

J. Procure equipment

1. Solicit proposals
2. Decision and order

K. Installation

1. Prepare for installation
2. Supervise installation
3. Follow-up

Functional, quality, and appearance specifications—Obviously it must be known what is going to be produced in terms of size, shape, and other physical and chemical characteristics. These things have a powerful influence on processing in terms of restrictions they impose. It is important that they be expressed as clearly and as concisely as possible. It is the product designer's job to turn a complex set of functional, quality, and appearance attributes, into a usually complex set of dimensions, surface finishes, tensile strengths, carbon contents, etc., and to then apply tolerances to all of these specifications. When he does this job inaccurately, through omission of details or unwarranted tightening of tolerances, he makes the job of the process designer exceedingly difficult.

For example, the loosening of a dimensional tolerance from ± 0.001 to ± 0.005 may have a decided affect on the process design. This same thinking should be applied to all such product specifications.

Bill of Materials and/or Parts List—These were illustrated in Figures 3-8 and 3-9 and provide a complete listing of the parts and material making up the product. Obviously, no product can be profitably made unless its entire content is known at the very beginning. Eventually, each item (fabricated or purchased) must be costed out in order to determine manufacturing cost and selling price.

Total quantity required—As pointed out previously, the total quantity to be produced greatly effects or determines the processes used in manufacturing the

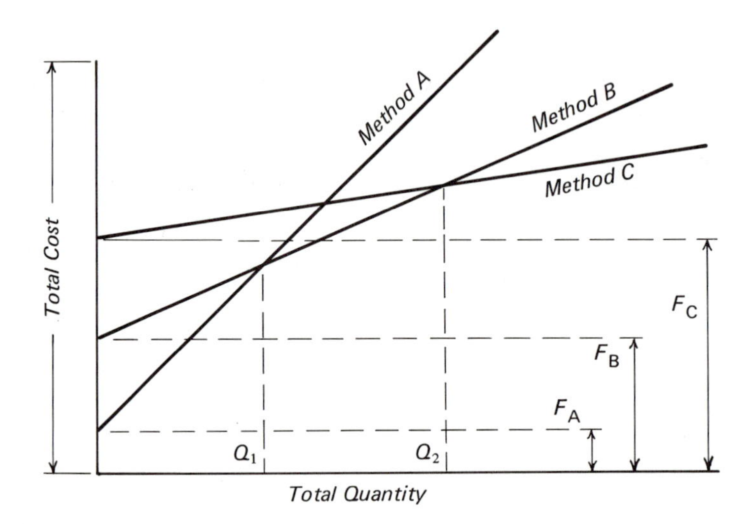

Figure 4-8—Graphical representation of alternative methods of producing a part.

product. When the quantity is low, methods should be selected that require low initial costs. Although the cost to produce each piece may be quite high, the total cost would be less than methods requiring larger initial investments. The relationship between total cost and quantity are, perhaps, best shown graphically, as in Figure 4-8. Each of the methods shown has different cost characteristics. *Method* A has a low initial cost (F_A) but the cost to make each piece after this method has been installed is relatively high. As the total quantity to be produced increases, the total cost of using *Method A* increases rapidly until at some quantity Q_1, it becomes more economical to use *Method B. Method B* has a higher initial cost (F_B), but its lower production cost after the installation makes it more economical when quantity Q_1 is required and it continues to be more economical as the quantity is increased to Q_2. At this quantity *Method C* becomes lower in total cost.

Methods A, B, and C, for example, might be the cost characteristics for an engine lathe, turret lathe, and automatic screw machine. The quantities, Q_1 and Q_2 might be 50 and 500.

However, *Methods A, B,* and *C* might refer to complex man–machine groupings. Perhaps *Method* A refers to a system for manufacturing a part with standard machine tools; *Method B* might use similar equipment with special handling devices between machines and special loading and unloading mechanisms; *Method C* might involve special machines and equipment tailored to this specific part. The quantities, Q and Q_1 might be 500,000 and 5,000,000.

It is important to remember that this cost–quantity relationship exists for any small segment of the manufacturing system as well as for a larger part of the system or even the entire system.

Rate of production required—Knowledge of the total quantity required is not enough. We must know something about the rate of production required. Knowing the total requirement to be 1,000,000 parts, processing plans might be substantially different if in one instance this meant at the rate of 1,000 per month, while in another instance this meant at the rate of 100,000 per month.

Also, the rate of consumption may not be constant. Perhaps sales are higher during certain months of the year. Processing plans should take these factors into consideration. It might be better to use different manufacturing methods for highly seasonal items than for relatively stable items even though the total quantity were identical.

Initial delivery date—Process planning must also take into account the initial delivery date required on the product as a whole as well as the individual components. High-production equipment that could efficiently produce the desired volume of production at the required rate is certainly desirable, but it must be possible to obtain, and install this equipment sufficiently ahead of the initial delivery date. This may prove to be difficult, and in some instances it may be necessary to use less efficient equipment in order to meet time demands.

Availability of Necessary Manufacturing Components—The more knowledge a process engineer has about available materials and equipment, the better position he is in to select and design the best process. This means, certainly, that records of some kind should be kept on the types of materials and processes currently used. It also means that effort must be made to learn what is available from suppliers as well as developments that are continually occurring. It would be foolish to specify a new piece of equipment if that equipment were already available in your own plant or if minor changes would make available equipment usable. As has already been pointed out it would be equally foolish to specify equipment requiring 6 months for delivery and installation when production must start in 3 months.

In the same way, money and space may be limiting factors to the process designer. Sometimes production plans must be altered because of financial considerations or space limitations. Frequently it is necessary to proceed slowly toward the optimum, in a series of stages, as money and space become available.

It is also important to remember that both long-range and short-range objective must be considered. The fact that these objectives are sometimes in conflict makes the work of the process engineer more difficult.

2. *Make preliminary Part Print Analysis.* Previous mention has been made of the need for a detailed analysis of the product—each and every individual part. The outline in Table 4-5 suggests an approach to this important preliminary step in the process design. It is sometimes a good idea to record such information as will be gathered during a part print analysis; the *Critical Conditions Specification Sheet* in Figure 4-9 is a form such as may be used for organizing some of the data, and the process designer might wish to develop more elaborate forms that will accommodate more of the details.

CRITICAL CONDITIONS AND CRITICAL AREAS SPECIFICATIONS SHEET

PART NAME: Jaw face (Drill press vise) PART NO.: 552

ANALYZED BY: W. P. Smith DATE: 6/25 SHEET: 1 OF: 1

Type I—Close Tolerances	Dimension	Finish	Geometric Relationship	Area Involved	Rank the most critical areas
1.) Faces parallel within .002, with 30 surface finish		X	X	11 Sq. In.	1
2.) 2.747/2.743 overall length	X			1 Sq. In.	3
3.) .038/.035 from bottom of tapped holes to front face	X			1/10 Sq. In.	2
4.) 2.127/2.123 c' distance between holes	X				3
5.) .870/.867 from drill to bottom	X				3
6.) .313/.317 from end to hole	X				3

Type II—Unusual Materials None

Type III—Special Treatments Required

Harden & Draw to RC 40–50 — 1

Figure 4–9—Critical conditions and critical areas specifications sheet.

3. *Construct Work List Specification Sheet.* As suggested above, the design process is made easier by recording details as they are discovered or established. The *Work List Specification Sheet*[6] (Figure 4-10), a device for recording the important details, is primarily a tabular listing of all the elements of work contained in a part. It does not attempt to tell how that work will be accomplished, but merely indicates what work must be accomplished. Subsequent steps will determine the method of accomplishing each element. The elements on a *Work List* generally fall into the following three major categories.

Elements that give shape and form to the part—These elements are readily observed by examining the drawing. Every surface on the drawing should be listed separately and if possible, given a descriptive name. Sometimes, however, surfaces are difficult to name and may be numbered for reference on the drawing.

The temptation to group surfaces together and list as one element on the *Work List* should be studiously avoided. Frequently unusual processing alternatives are overlooked because the *Work List* is made of elements containing several surfaces—especially where these surfaces have been grouped according to a preconceived processing method.

One method of listing all surfaces is to start at one end of the part and work progressively around the part. As each surface is listed it can be colored on the drawing and appropriate dimensions can be marked. This procedure is continued until all of the drawing has been colored and all dimensions marked.

Internal and surface elements of work that do not appear as dimensions or surfaces on the drawing—After all of the part surfaces have been listed, reference should be made to all notes on the drawing; these will reveal other elements of work that will be required, which should also be listed separately on the *Work List*.

Elements of work created by processing—The process itself can create additional elements of work. A hardness specification may necessitate a subsequent straightening operation. Certain machining operations create sharp edges and burrs, which must be removed. Sometimes cleaning or buffing operations are needed because of deposits made on the part during a previous operation. If these elements of work can be anticipated they should be listed separately on the *Work List*. If subsequent decisions give rise to these items, they should be listed as soon as they are realized.

Once the raw material form has been selected and a *Work List* prepared, the next step begins the actual job of process synthesis or process design. This step involves specifying for each element on the *Work List* the unit process, or in many instances, the unit processes required to complete that particular element so that it will satisfy all specifications indicated in the drawing.

Reference to the critical conditions inherent in the part should prove helpful at this step because they pinpoint areas that will generally require special attention. Frequently this means specifying several unit processes for the item on the *Work List* affected by the critical condition. For example, a surface that must

[6]Smith and Thorson.

WORK LIST SPECIFICATION SHEET			
PART NAME: Jaw face (Drill press vise)		PART NO.: 552	
ANALYZED BY: W. P. Smith		DATE: 6/27/53	SHEET: 1 OF 1

No.	View	Element Description	Unit Process Requirements (and alternatives)
1		Front face	H.T. — Surface grind — 60
2		Back face	" — Surface grind — 70
3		Left end	" — 10 Saw Mill 20
4		Right end (50)	" — Saw Mill
5		Top	" — As received
6		Bottom	" — Broach 30
7		Bottom chamfer	" — Broach
8		7/32 Blind hole (left)	" — Drill (start & bottom)
9		7/32 Blind hole (right)	" — Drill (start & bottom)
10		1/4–28 Internal thread (left)	" — Tap
11		1/4–28 Internal thread (right)	" — Tap
12		1/16 x 45° Hole chamfer (left)	" — Drill (chamfer)
13		1/16 x 45° Hole chamfer (right)	" — Drill (chamfer)
14		Harden & Draw to Rc 40–50	"
15		Sharp corners	40
		A) Top edges (2) B) Bottom edges (7)	File File 20
		C) Side edges (4)	File
16		Clean threads (?)	Wire brush 80
17		Preserve faces (?)	Dip
			NOTE: Raw material 1/4″ thick 2″ wide 20′ long

Figure 4–10—Work list specification sheet.

have an extremely good surface finish as well as high hardness might be accomplished by either of the following sequences of unit processes:

1. Mill (rough)	1. Broach
2. Harden	2. Stress relieve
3. Draw	3. Harden
4. Straighten	4. Draw
5. Grind (rough)	5. Grind
6. Hone	6. Polish

As shown in Figure 4-7, it is often necessary to list several possible groups of unit processes for a *Work List* element. Some of these alternatives can be immediately discarded because of obvious disadvantages. It is better, however, at this stage to list all possible alternatives without being too critical or selective. It is not uncommon to find that what appeared in the beginning to be a ridiculous choice may prove in the final analysis to be the best alternative. There is a time and place for subjecting all possible alternatives to critical analysis, but to attempt critical analysis during the creative process tends to choke off good ideas. *The temptation to be too critical, too soon, is perhaps the greatest retardant of creative process development and engineering.*

4. *Combine unit processes into manufacturing operations.* Once the unit processes have been established, the next task is to group these unit processes together into specific jobs to be accomplished at specific work places by specific machines and tools. These specific jobs are defined as manufacturing operations.

The theoretical manufacturing cycle already discussed would be the simultaneous performance of all of the required unit processes. By combining unit processes into manufacturing operations, this ideal situation is being approached. The advantages to be gained by doing this are fairly obvious.

In the first place, an increase in the amount of work that can be accomplished at one work station increases the percentage of manufacturing cycle time spent doing useful work and reduces the percentage of cycle time spent loading work into the machine and unloading it at the end of the cycle. Each operation that is added to the manufacture of the part adds the loading and unloading requirements for that operation. Once it becomes apparent that loading and unloading time is relatively inefficient utilization of man or a machine time, it should also be apparent that one method of reducing this waste would be to combine as many unit processes as possible into one manufacturing operation. The turret lathe, the automatic screw machine, the progressive die, the multi-station cold-roll forming machine and the transfer machine, are all examples of this concept.

The second advantage of combining unit processes stems directly from the first advantage. This additional saving is due to the elimination of the material handling, storage, and delay steps that are normally required between manufacturing operations. Whenever the work accomplished on two machines in two operations can be accomplished on one machine in one operation, the following eliminations are accomplished:

1. No provisions need be made to store parts coming out of the first operation.

2. No material handling device is needed between the two operations.
3. No paper work is required to initiate the move from the first operation to the second.
4. No separate inspection operation is necessary between the two operations.
5. No provision need be made to store parts received at the second operation.

The third major advantage of grouping unit processes together is that such grouping can very often facilitate meeting critical conditions that would be difficult to meet in any other way. For example, when two bearing diameters must be concentric with each other, it may be most advantageous to generate them at the same time. The same reasoning could be applied to other geometric relationships such as parallelism and squareness. In effect, the desired relationship is built into the set-up of one operation and the chance of compounding errors at consecutive workstations is reduced. But the advantages of combining unit processes must be sure to overcome the possible disadvantages that may be created through the combination.

One of the most serious disadvantages of combining unit processes is that special machines and tools may be required in order to accomplish the desired combinations. It is interesting to note however, that an expensive piece of special equipment is frequently less expensive than the several standard machines it might replace would be in the aggregate.

The second problem that results from combining unit processes is the increase in the amount of time required to set up the equipment, de-bug the operation, and tear down the operation at the completion of the production cycle. There is also the possibility, particularly when the combination is a radical one, that it won't work properly. It may be desirable to spend time on pilot plant studies of proposed, radical combinations of unit processes. This is particularly true in high volume industries where the savings that result from highly repetitive work become so significant.

The third disadvantage or problem is that combinations of unit processes into complex operations require people of higher intelligence, and place greater responsibility upon them. Perhaps this is a blessing, but it does present a problem in that these qualifications, or the desire to get them, may not be easy to find.

The fourth disadvantage is the high cost of downtime. Stoppage of a complex production unit immediately stops production of all unit processes being accomplished within that machine and, if it closely coupled to other complex machines, may quickly stop them too.

The final disadvantage of making combinations is probably a summary of all of the disadvantages—loss of flexibility. It is much more difficult to make changes of any kind when many combinations have been made. Potential design changes in the product, which might appear simple, may have far-reaching affects on a tightly linked production system. Changes in product volume are often difficult.

These disadvantages must be carefully balanced against the advantages of making combinations. On the following pages, methods of selecting and comparing possible alternatives will be discussed. The present problem is simply to record all possible ideas.

5. *Arrange operations into a logical sequence.* After unit processes have been grouped into manufacturing operations, the final problem of the process engineer is to determine the best sequence for the manufacturing operations. To thoroughly examine every possible arrangement would be extremely tedious. However, to fully understand the sequencing problem it is necessary to understand that, if five operations exist, then ($5 \times 4 \times 3 \times 2 \times 1 = 120$ different arrangements can be made of these operations (mathematically speaking, 5 factorial, or 5!).

It is important to note that, if an operation must be done first, in the example there are really only 24 (or $4 \times 3 \times 2 \times 1$, or 4 factorial) possibilities; and if, in addition, another operation must be performed last, only 6 possibilities remain ($3 \times 2 \times 1$, or 3 factorial). It can be seen that each restriction that is placed upon a group of operations drastically reduces the number of sequence possibilities. The practical key to the sequencing problem is to first determine all possible restrictions or limitations that must be imposed on the manufacturing operations.

There are three major limiting factors that will reduce the number of sequence possibilities. Probably the most restrictive conditions are those that are inherent in the processes that have been selected. A simple example of this type would be drilling and tapping. The hole must of course be drilled first. Frequently it is necessary to perform an operation first because succeeding operations either locate from the first operation or in some way depend upon its completion. Many of the restrictions of this type were considered when the unit processes were grouped into manufacturing operations. In many ways the problem of sequencing the unit processes within a manufacturing operation is exactly the same as the sequencing of the manufacturing operations themselves. Certain unit processes must take precedence over others. Generally speaking, these restrictions substantially reduce the number of possible arrangements of operations.

Another type of restriction involves the existing layout of physical facilities and the available material handling systems. A centralized department, such as plating, painting, or heat treating, may be located in such a way as to make certain alternative sequences more economical than others. The problem generally involves minimizing the total amount of material handling by arranging the operations in a manner that will minimize back-tracking or excess movement. The material handling problem is easy to overlook when primary attention is being focused on the development of operations. However, it must be remembered that it is necessary to get parts from one operation to the next. When certain facilities must remain fixed, one method of reducing manufacturing cost is to examine the possible sequences of operations that minimize the amount of material handling.

One final type of restriction that must not be overlooked is the existing schedule on manufacturing equipment. Where specific equipment is assigned to a particular part, this problem does not enter the picture. However, for low volume job shop activities, where many operations are performed on different parts at the one work station, it can be a critical factor in the choice of the best sequence.

It must always be remembered that the objective of the operation sequence is to guarantee the dimensional integrity of the part through every operation required to produce it. The material used represents the original condition, and the operations performed should advance it toward the part print—the desired final condition.

6. Accumulate details on equipment and tools. It now becomes necessary to collect information and data necessary to choose production and related equipment required to fabricate the parts, as outlined in Table 4-5.

7. Select and Specify equipment. The next step for the process designer is to choose the specific machinery and related production equipment. The procedure for doing this is beyond the scope of this chapter, but several aspects should be considered here.

All of the preceding effort can result in nothing worthwhile unless the process designer carefully selects the specific equipment required to carry out the unit processes and operations he has specified. Therefore, careful consideration should be given to the following.

The nature of the selection problem—Eary and Johnson[7] suggest six different ways in which the selection problem may arise:

1. For a job not previously encountered.
2. For a job previously done by hand.
3. To replace the present worn out machine.
4. To lower the cost of production.
5. For expanded production.
6. To take advantage of a technical or technological change.

In any of the above, the objective is to select the "one best" piece of equipment for the task at hand.

Factors for consideration in equipment selection—As in so many other situations, before an equipment decision can be made, a good many factors must be given consideration. Table 4-2 tabulated a number of them. Some are easy and some not so easy to work into the decision making process. After initial consideration of the levels of mechanization and automation, and various production-related factors, costs will be major concerns, beginning with equipment purchase.[8]

> *Direct costs*—commonly associated with the operation of equipment, will include the fixed costs of depreciation and other items of overhead, and the variable costs of such as personnel, power, and maintenance.
>
> *Indirect costs*—will include such equipment and method related costs as space, inventory, repair parts, and downtime, and such management related costs as re-layout, and training.
>
> *Indeterminate costs*—costs not easily pre-determined or vague, might include such frequently overlooked costs as space lost or gained, savings from inventory or production control, and changes in overhead or in product or material quality; or costs relating to installation and scheduling.

[7] Eary and Johnson, pp. 312*f*.

[8] For suggestions and procedures in applying cost considerations in equipment selection, see Eary and Johnson, pp. 325–45, and Apple, *Material Handling Systems Design*, chs. 13, 14.

Intangible factors—will include whatever applies but nevertheless defies dollar-value calculation and must therefore remain outside the usual cost comparison, such as quality and availability of equipment or parts, and the remote but vital aspects of personnel and financial policy.

General purpose vs. special purpose equipment—Yet another aspect of the equipment selection problem is the choice of general purpose vs. special purpose (manual, mechanized, automated). The analyst should be sure to identify factors relating to the advantages of each category.[9]

Approaches to selection among alternatives—Eventually the analyst must choose among the several alternative types, makes, and models of equipment available. Such decisions are made on the basis of:

1. *Past experience* of the process engineer.
2. *Investigation:* (a) discussion with other engineers; (b) contacts with manufacturer's representatives; (c) visiting machine tool shows; (d) technical society meetings and publications; (e) trade journals, catalogs, etc.
3. *Experimentation:* Many larger companies maintain manufacturing development groups that continuously search out new ways of processing materials, even to the extent of inventing equipment, which is then fabricated in the plant, or subcontracted to an appropriate vendor.

After methods, machinery, and techniques have been decided upon, specifications are established as a basis for the purchasing procedure (covered later in this chapter).

Tools and auxiliary equipment—Also to be included in the above selection process are the various tools, gages, fixtures, attachments, and other auxiliary devices necessary for the efficient operation of the machines. The procedures outlined in Table 4-5 are just as applicable to the task of selecting tools.

At this point in the chronology of the process design activities, the basic processes, manufacturing process, and unit processes have been established, and the equipment needed to carry them out has been selected. The next step is to determine their applicability by estimating the cost of performing the processes and operations by the means selected.

8. *Estimate product costs.* Inherent in the attempt to obtain a desired return on investment for the company, is the need for economically justifying the acquistion of each new piece of equipment. The cost analysis is done simultaneously with the selection of the equipment, and involves a comparison of costs that would be incurred by the alternatives under consideration. Usually, the cost calculations will result in an annual cost for each alternative, a return on investment—or other measure of feasibility—and a cost per unit produced on the equipment.

9. *Prepare Manufacturing Operation Planning Sheet.* The next task in the process design procedure is the documentation of the planning work to this point.

[9]For a more detailed discussion, see Eary and Johnson, pp. 315–25.

MANUFACTURING OPERATION PLANNING SHEET

PART NAME: Jaw face (Drill press vise) PART NO.: 552

DESIGNED BY: W. P. Smith DATE: 6/28 SHEET: 1 OF ____

No.	Operation Description	Machine Type Major Tools	Operation Sketch
10	Cut. to length, leave 1/16 stock to clean up by milling, 4 bars per load	Power hacksaw Conventional vise	*Symbols* Δ Locate □ Support → Hold Area affected *SAW* 10) (End view)
20	Straddle mill both ends, hold 2.747/2.743 overall length; mill 4 per load Operator to file (4) end burrs & (2) top edge burrs while next load is cutting & round corners on bottom edges prior to broaching	Horiz. mill (2) 6" O.D. Face milling cutters— Conventional vise	*CUTTERS* *End Location Remove Before Cutting* 20) (Side view)

Figure 4–11—Manufacturing operation planning sheet.

One of the more definitive methods of accomplishing this is with the *Manufacturing Operation Planning Sheet*,[10] Figure 4-11, using such symbols as the ones shown at the upper-right in the figure.[11] In Figure 4-11 are depicted the details of the first two operations on the drill press vise jaw shown in Figure 4-1. Although this may seem like a lot of busy work, it is extremely important for the engineer to properly document the process he has designed. Eary and Johnson[12] suggest the following, selected from a much larger list of reasons and advantages:

1. A visual aid for processing—rather than relying on memory.
2. Reduce chances of missing operations needed to manufacture the part.
3. Pictures may be checked against the part print to insure that all dimensions have been accounted for.
4. Provide processing dimensions not available on the part print.
5. Aid in visualizing a workpiece to assist in tool design.
6. Aid to production supervision, foremen, set-up men, and tool maintenance personnel in checking the tooling.
7. Aid to methods and work standards in estimating standard times for each operation.

After the *Manufacturing Operation Planning Sheets* have been completed, the necessary data are available for the *Production Routing.*

10. *Establish Production Routings.* The *Production Routing* becomes the backbone of the production activity. It is a re-cap of the data developed by the process engineer, and the primary means of communication between the product engineer and the production floor. It is often called by different names, such as the *process sheet*, or the *operation sheet.* Basically, the *Production Routing* is a tabulation of the steps involved in the production of a particular part and necessary detail on related items. Information on the routing may include:

1. Part names and numbers
2. Operation numbers and sequence
3. Operation names
4. Operation descriptions
5. Machine names and numbers
6. Tool, jig, and fixture numbers and sizes
7. Department numbers
8. Production standards (time and/or pieces)
9. Number of operators
10. Space requirements
11. Speeds and feeds
12. Effective date
13. Group numbers
14. Labor classifications required
15. Material

Typical *Production Routings* are shown in Figures 4-12 (for the *Vise Jaw Face*) and 4-13 (for the *Powrarm Base*).

An idea of the complexity and extent of the process engineer's task in a mass

[10] Smith and Thorson.

[11] For a more detailed group of symbols and an expanded method of documenting the operations in a process, see Eary and Johnson, ch. 15, on "the process picture sheet."

[12] Eary and Johnson, pp. 710*f*.

PRODUCTION ROUTING

PRODUCT: Vise PART: Jaw Face PART NO.: 552

PREPARED BY: AMJ DATE: Apr. 6 SHEET: 1 OF 1

Oper. No.	Description	Machine	Aux. Equip.	Set-up time (hr)	Hr/pc	Pcs/hr
10	Cut to length	Power hack saw	Vise	.25	.0040	250
20	Straddle mill both ends	Horizontal mill	2–6″ face milling cutters; vise	.50	.0167	60
30	Broach bottom	Broach		.30	.0083	120
40	Start drill, 2 holes (1st spindle) Bottom drill, 2 holes (2nd spindle) Chamfer, 2 holes (3rd spindle) Tap, 2 holes (4th spindle)	4 Spindle drill Press	Jig	.40	.0167	60
50	Harden and Draw	Furnace Quench tank		.50	.0050	200
60	Grind front face	Surface grinder	Magnetic chuck	.40	.0250	40
70	Grind back Face (to clean up)	(same as above)		.20	.0167	60
80	Clean out holes & apply protective coating	Hand grinder (flexible shaft)	Wire brush	.15	.0083	120

Figure 4–12—Typical Production Routing.

production industry may be gained from an example based on an automobile part, a spindle used in the transmission. It is a steel shaft about 8 inches long, slotted at one end, with two gears near the center. It takes 39 machines to make the spindle, including power hammers, centering machines, lathes, grinders, gear cutters, gear finishers, and many more. Besides the 39 machines, there must be 194 dies, jigs, fixtures, tools, and arbors. And on top of all this, 157 precision gages are needed to make sure that the work is done accurately. The spindle, moreover, is one of the simpler of the more than 10,000 parts making up a modern automobile.

11. *Make Operation Process Chart.* After *Production Routings* have definitely established the operations and their sequence, a helpful technique for visualizing the entire production process for a product is the *Operation Process Chart.* Figure

PRODUCTION ROUTING										
			DRAWING NO.: C-285							
PART NAME: Base			PART NO.: 1							
Oper. No.	Operation Description	Machine Name	Jigs, Fixtures Tools, Etc.	Dept. No.	Std. Time (hr)	Hourly Mach. Cap'y	No. Machs.	No. Opers.	Floor Space Req'd
1	Face bottom	14" LeBlond engine lathe	Special chuck Turning tool	1	.0167	60	1.46	1	
2	Face top, turn O.D., neck, drill & ream 5/8" hole	Warner & Swasey turret lathe	Special chuck Face plate Turn-neck Drill Ream	1	.042	23.8	3.58	3	
3	Drill 3 bolt holes	21" Cleereman drill press	Box jig, 3-spindle head	1	.0120	83.4	.96	1	
4	Drill pin hole	Delta drill press	Plate jig	1	.0042	238.0	.33	1	
5	Drill & ream 3/4" ecc. hole	#4 Fosdick 2-spindle drill press	Box jig	1	.0153	65.4	1.18	1	
6	Inspect			13	.018	55.5	1.34	1	
7	Degrease	Detrex		20	.0070	143.0	.52	1	

Figure 4–13—Production Routing for Powrarm base.

6-5 shows such a chart for the Powrarm (it's construction and uses will be covered in Chapter 6). As suggested in Chapter 2, the *Operation Process Chart* provides a first look at the way materials might flow in the facility under consideration.

12. *Calculate number of machines and operators.* Referring to the *Production Routing* in Figure 4-13, it will be noted that *Operation 1, Face bottom,* will require 0.0167 hr to produce one piece on the 14-in. LeBlond engine lathe.

Fundamentally the number of machines to be used is based on the required time for a unit of production. There are two major factors, however, that will combine to reduce the established production rate. One is the loss through scrap or rejects, and the other is production efficiency.

It is now necessary to refer to the basic data previously accumulated for the production volume estimated by the sales department or top management. For an example, assume that a sales forecast estimates Powrarm sales of 134,000 units per year. If there are 2,000 hr/yr, then the required production of finished good units is 134,000 ÷ 2,000 = 67/hr.

First it is necessary to allow for the scrap that will be produced at each operation on each part of the product, so that when the parts reach assembly, there will be 67 good pieces of each.

To determine this, it is necessary to start at the last operation on each part that will result in any scrap, and work back up to the first operation, figuring the scrap on each operation

$$\begin{pmatrix}\text{Good pieces desired}\\ \text{of each operation}\end{pmatrix} \div 1 - \begin{pmatrix}\text{Scrap expected}\\ \text{at that operation}\end{pmatrix}$$

and compounding it into the total number of pieces required at the first operation. (See Table 4-6 and Figure 2-1.) These calculations show that it will be necessary to purchase 78.9, or 79 pieces of material to start through *Operation 1*. If the scrap is no higher than expected, 67 good pieces will be available after *Operation 7*.

But so far we have assumed that the plant is operating at 100-percent efficiency—which is highly improbable. If it is estimated that the plant will operate at 90-per cent efficiency, then plant space, machinery, and manpower must be provided to make up for the loss in efficiency. This calculation is made by dividing the required work output of each operation by the efficiency of the plant, or operation (Table 4-6, column 6). The calculations indicate that if a plant is built, equipped and staffed to produce 74.5 units per hour at *Operation 7*, and works at 90-per cent efficiency, it will turn out 67 units. Referring to *Operation 1*, if plans are made to produce 87.7, then 90-per cent efficiency will result in 78.9 pieces, and 4 per cent scrap will reduce the total to 75.8—which is what is required for *Operation 2*, etc.; the calculations can be shown as follows:

87.7	units planned for
− 8.8	units lost due to inefficiency
78.9	units for which material must be purchased
− 3.1	units scrapped
75.8	units to go into *Operation 2*

It should be pointed out here that there is a difference between *operation efficiency* and *department* or *plant efficiency*. Even though an individual operation may be running at over 100 per cent of its assumed efficiency, other operations in the same department may be running at lower efficiencies. The efficiency of the department then is the average efficiency of all units in the department. The plant efficiency is the average efficiency of all departments in the plant. The 90 per cent used in the above calculation is a plant efficiency. In some companies, the plant figure is used in all cases; in others, the individual departmental figures are used; and in still others, the operation efficiencies are used.

Now, if 87.7 units must be planned for, equipment must be available to produce that number. The standard time for *Operation 1* (Figure 4-13) is

Table 4-6. Calculation of Machines Required

Operation No.	Machine or Function	No. of Good Pieces Desired	Expected Scrap (%)	No. of Pieces Actually Started	Basis for Planning Facilities & Manpower at 90% efficiency)	Machine Production per hr	Theoretical No. of Machines Required
1	LeBlond eng. lathe	75.8	4	78.9	87.7	60.0	1.46
2	W & S turret lathe	72.0	5	75.8	84.2	23.8	3.58
3	Cleereman d.p.	70.5	2	72.0	80.0	83.4	.96
4	Delta d.p.	68.3	3	70.5	78.3	238.0	.33
5	#4 Fosdick d.p.	67.0	2	68.3	76.0	65.4	1.18
6	Inspect	67.0	0	67.0	74.5	55.5	1.34
7	Degrease	67.0	0	67.0	74.5	143.0	.52

.0167 hr/pc, and the hourly production is 60 pc/hr. Therefore, $87.7 \div 60 = 1.46$ machines will be required to produce 75.8 good units.

The entire procedure can be recapped as in Table 4-6. As here, there are many cases where the number of machines will not come out as an integer. If, instead of 1.46 machines, the above calculation had resulted in 1.25, 3.68, or 6.42 machines, there would be a question as to the number of machines to provide.

Such decisions are largely a matter of judgment, based on such factors as:

1. How much of the machine cycle is man-controlled and might yield more pieces per hour with more favorable conditions, that is, less scrap, better overall efficiency, or a better than average operator?
2. Can the method be changed to reduce the standard time?
3. Is overtime cheaper than an additional machine?
4. Would a breakdown of a single machine shut down the line?

It would be much harder for 1 machine to turn out production for 1.25 machines than it would be for 6 to turn out production for 6.42. In the first instance the single machine must bear a 0.25 overload; while in the latter each machine must bear only $0.42 \div 6$, or a 0.07 overload. Actual decisions in such cases are made on the basis of past experience and detailed knowledge of plant conditions.

It must also be remembered that if the engine lathe used as an example here is to be used to produce other parts too, then these work loads must be added; that is, the equivalent machine requirements of the other parts must be added to the 1.46 machines determined above.

13. *Make preliminary work place plans.* Although the industrial engineer, or the methods engineer will later develop detailed work place layouts and work methods, the process designer may do some preliminary work in this direction. Work place plans are a necessary prerequisite to work methods design and the determination of work standards. Preliminary sketches of work places and estimates of work standards will be needed for determining the number of pieces of equipment and the number of operators—which are directly related to both production and storage space requirements, as well as to material handling methods. And all of this is helpful in estimating costs and justifying new equipment.

14. *Procure equipment.* After all the above activities have been carried out, there should be sufficient information on hand to proceed with the procurement of the equipment.

It should be noted that much, if not most equipment used in industry today, has been purchased after a far less rigorous analytical procedure. It can also be said, however, that the more progressive companies and the more profitable ones usually demand analyses of this type before committing themselves to substantial equipment investment.

The common procurement procedure involves soliciting proposals—based on required performance specifications—from several potential suppliers. After the quotations received are analyzed, and a decision made as to the best piece of equipment, in terms of performance, price, delivery, etc.,[13] a purchase order is issued.

15. *Prepare for installation.* The process engineer's work is not ended with the submission of technical or cost data to the purchasing agent. He should participate in any necessary follow-up activities, such as answering vendor questions, working with the vendor's engineers, and in general, seeing to it that the equipment delivered is the equipment ordered.

16. *Supervise installation.* Upon delivery, the process engineer becomes the "overseer" of the installation process, since it is his responsibility to see to it that the equipment performs the function as planned. If any variations are necessary in the installation and application of the equipment, it is he who must approve them—or when necessary, do any required re-design of any portion of the activity for which the equipment is intended.

17. *Follow up on performance.* After the equipment has been installed, the process designer is responsible for its proper operation. Therefore, he should keep an eye on the installation, study and evaluate it as necessary, and originate any changes necessary to assure that it produces as planned. Also, after a period of time, it may be advisable to revise or update the equipment. The process engineer should then originate any requests for subsequent improvements.

Computerized Process Planning

It should not be difficult for the reader to imagine that if the necessary data on various machines were put into a computer, along with the parameters of part configurations and requirements, a program could be written to print out opera-

[13]For a more detailed discussion of the selection and purchasing procedure, see Apple, pp. 423–43, and Eary and Johnson, ch. 11.

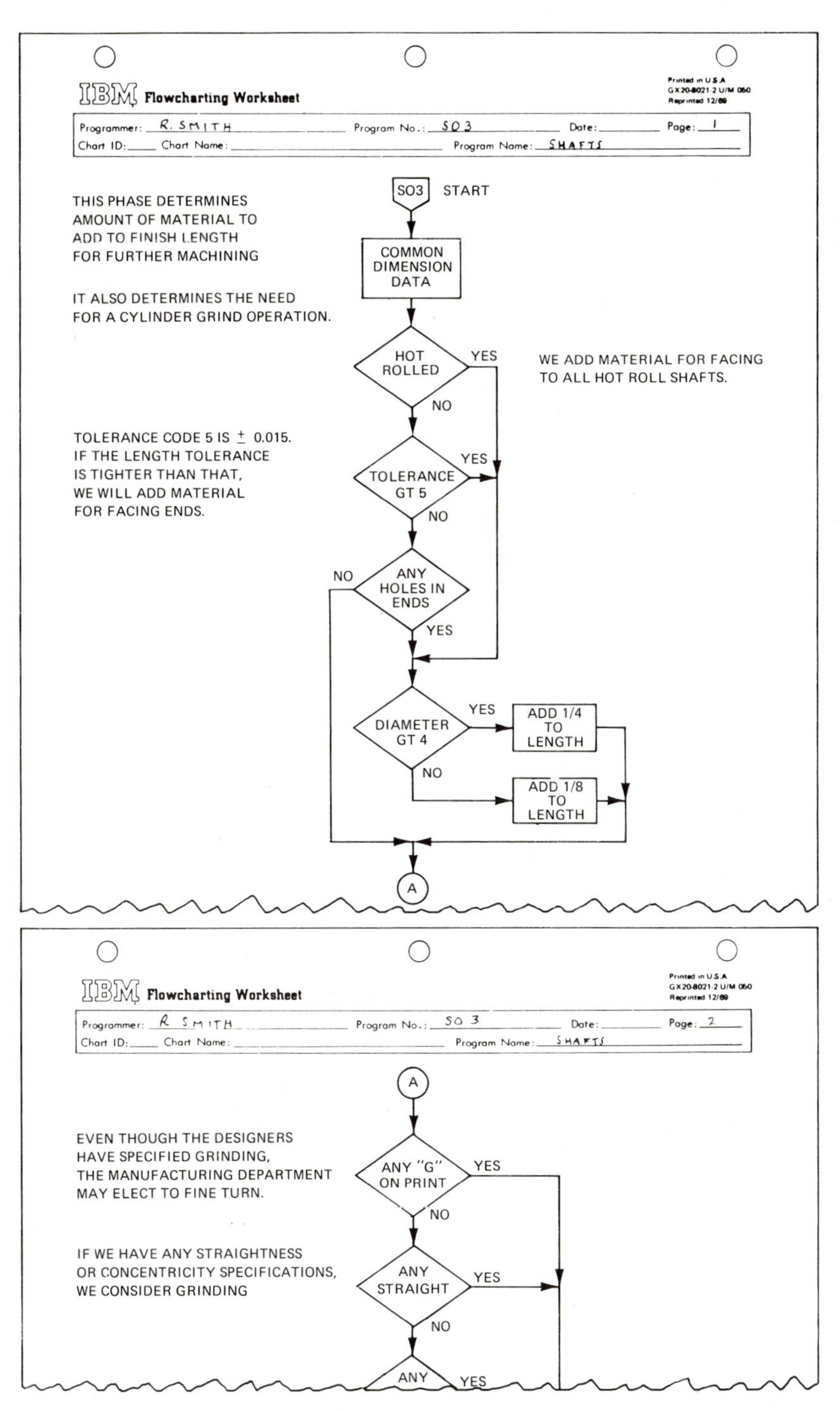

Figure 4–14—The flowcharting worksheet is a basic analytical tool for producing a methods sheet via computer. (Courtesy of Manufacturing Engineering and Management Magazine.)

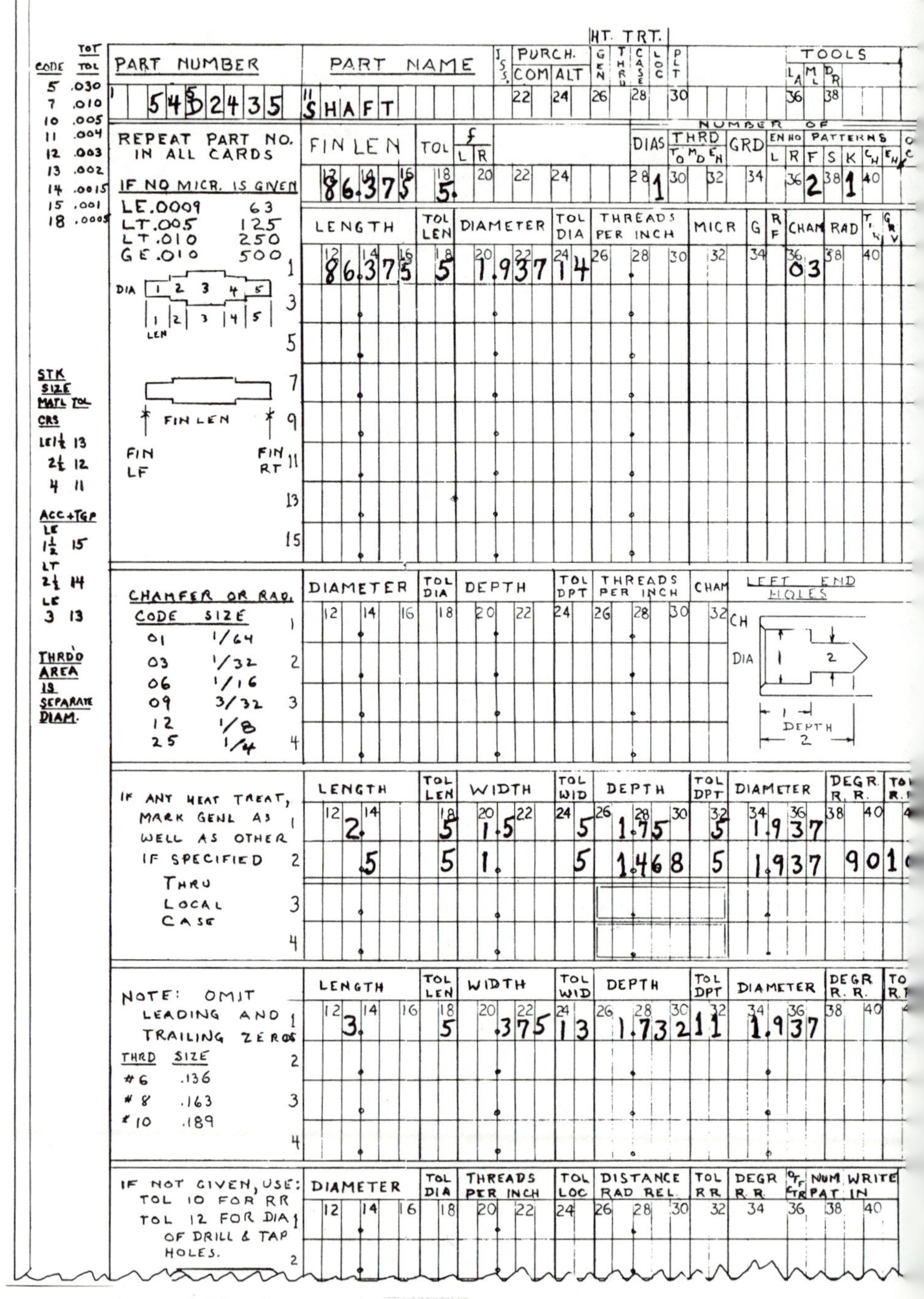

Figure 4-15—Computer input sheet.

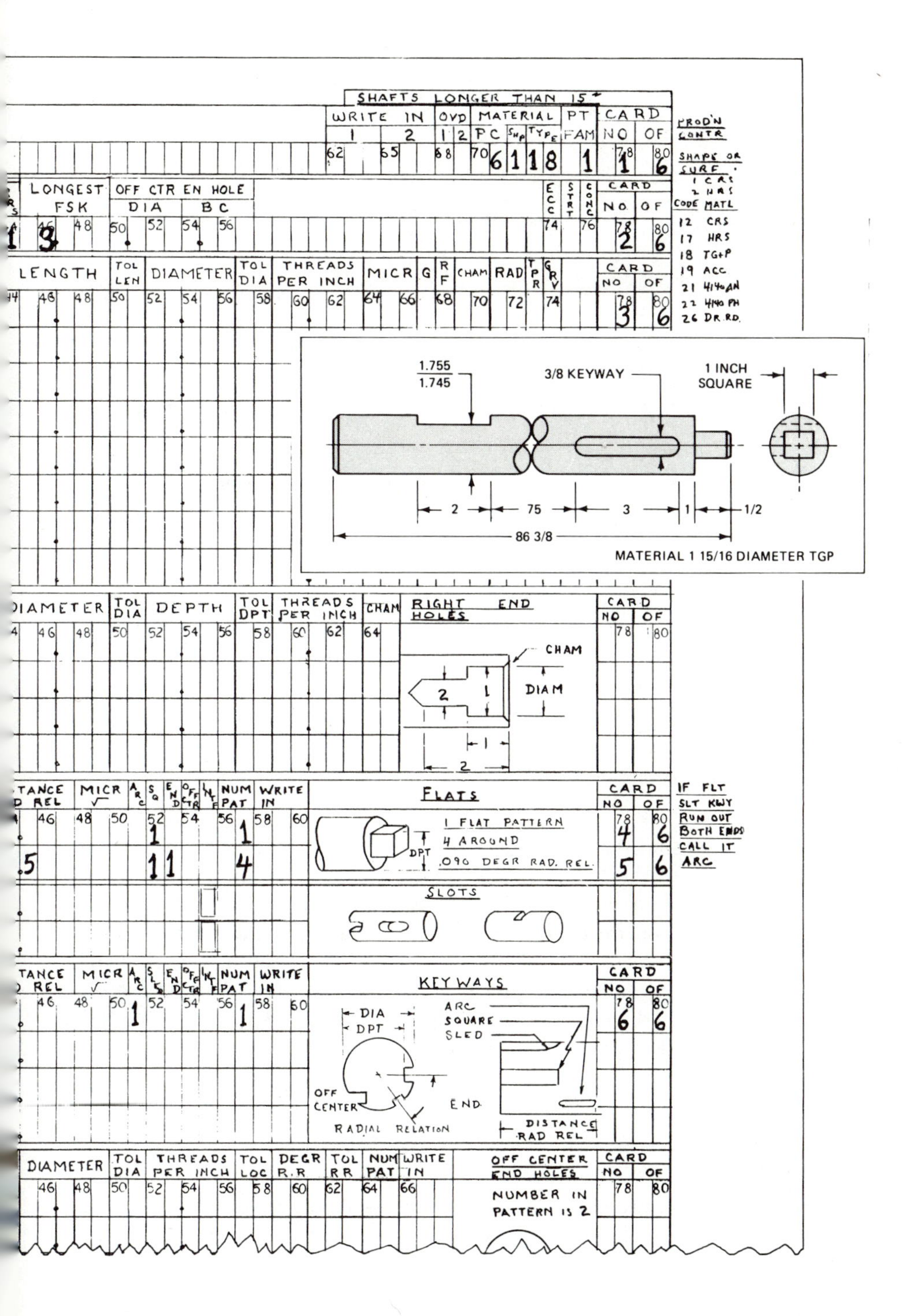

SHAFTS LONGER THAN 15"
WRITE IN
OVD
MATERIAL
PT
CARD
PC
FAM
NO
OF
PROD'N CONTR
SHAPE OR SURF.
1 CRS
2 HRS
CODE MATL
12 CRS
17 HRS
18 TG+P
19 ACC
21 4140AN
22 4140 PH
26 DR RD.
LONGEST FSK
OFF CTR EN HOLE
DIA
B C
LENGTH
TOL LEN
DIAMETER
TOL DIA
THREADS PER INCH
MICR
CHAM
RAD
1.755
1.745
3/8 KEYWAY
1 INCH SQUARE
2
75
3
1
1/2
86 3/8
MATERIAL 1 15/16 DIAMETER TGP
DEPTH
TOL DPT
RIGHT END HOLES
DIAM
FLATS
1 FLAT PATTERN
4 AROUND
.090 DEGR RAD. REL.
DPT
IF FLT
SLT KWY
RUN OUT
BOTH ENDS
CALL IT
ARC
SLOTS
MICR
NUM PAT
WRITE IN
KEYWAYS
DIA
DPT
ARC
SQUARE
SLED
OFF CENTER
END
RADIAL RELATION
DISTANCE RAD REL
DEGR R.R
TOL LOC
TOL RR
OFF CENTER END HOLES
NUMBER IN PATTERN IS 2

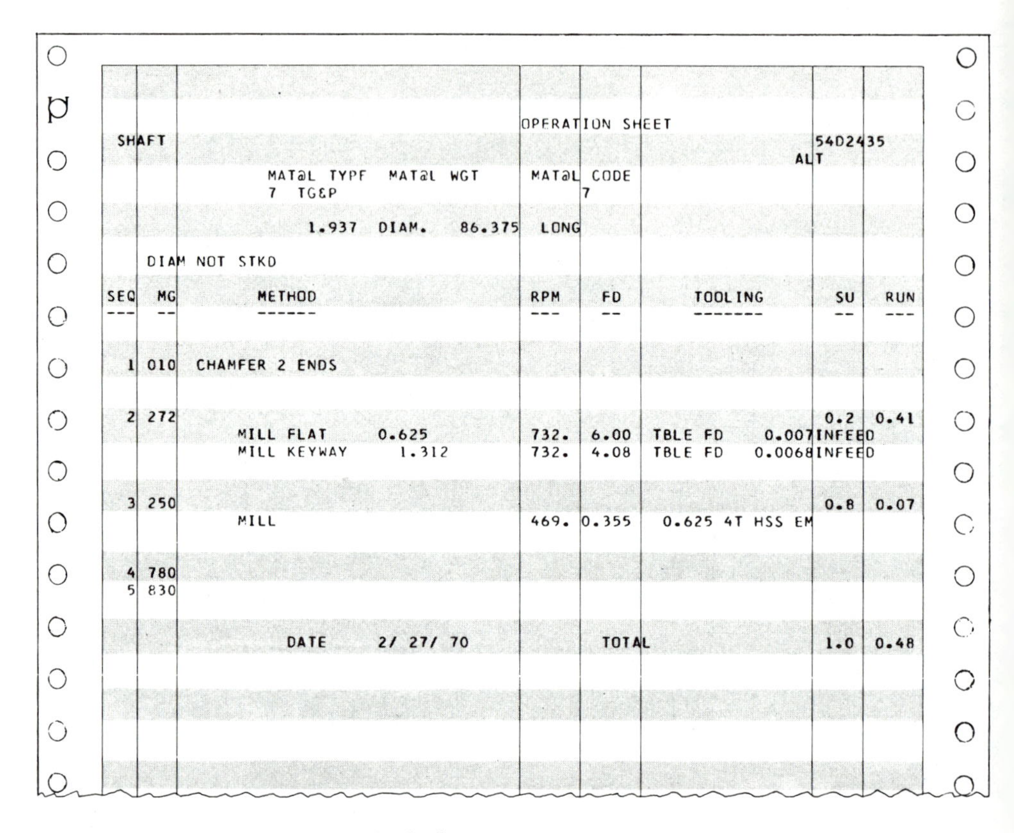

OPERATION SHEET

SHAFT 5402435

ALT

MAT@L TYPF MAT@L WGT MAT@L CODE

7 TG&P 7

1.937 DIAM. 86.375 LONG

DIAM NOT STKD

SEQ	MG	METHOD	RPM	FD	TOOLING	SU	RUN
1	010	CHAMFER 2 ENDS					
2	272					0.2	0.41
		MILL FLAT 0.625	732.	6.00	TBLE FD 0.007 INFEED		
		MILL KEYWAY 1.312	732.	4.08	TBLE FD 0.0068 INFEED		
3	250					0.8	0.07
		MILL	469.	0.355	0.625 4T HSS EM		
4	780						
5	830						
		DATE 2/ 27/ 70		TOTAL		1.0	0.48

Figure 4–16—Print-out methods sheet.

tion details, specifications, and even *Production Routings*. This, of course, has been done by some of the larger manufacturing concerns and equipment builders.

One such system, developed by IBM, is known as *Automated Manufacturing Planning.*[14] It automatically converts engineering design data into manufacturing instructions, such as sequences of operations, machine settings, work methods, tooling, and time standards. Figure 4-14 illustrates the flow chart for the procedure; Figure 4-15 the computer input sheet; and Figure 4-16, the methods sheet printed out by the computer, in the form of a *Production Routing* (Operation sheet).

Conclusion

This chapter has presented the procedure commonly followed in designing a production process, to insure that the facility designer understands and appreci-

[14]Charles Porten, "Automated Planning of Manufacturing Operations," *Automation*, May 1966, p. 85.

ates the importance of the proper prerequisites to the layout process. Only if the process design work has been done carefully, can the *Production Routing* be considered accurate enough for the layout designer to begin the conversion of paper work plans to the reality of steel and concrete.

Questions

1. Briefly outline the production design procedure.
2. Give some of the factors to be considered in designing a manufacturing process, under each of the following categories: (a) Product (part). (b) Material. (c) Equipment (process); mechanical, operating, cost.
3. What is meant by *product analysis?*
4. Briefly discuss *value analysis.*
5. Distinguish between the following production methods: (a) Continuous (or mass). (b) Similar process. (c) Job order (intermittent). (d) Group technology. (e) Custom
6. List the steps in the process design procedure.
7. What kinds of activities make up a preliminary part print analysis?
8. In what ways do product specifications affect process design?
9. What are the advantages to the process engineer of good records of equipment in the plant?
10. Why should the process engineer need to know anything about available raw materials?
11. How do processing alternatives arise?
12. Discuss the importance of considering many possible alternatives.
13. What is a *Work List Specification Sheet?* How does it differ from a list of unit processes?
14. How do critical conditions affect the choice of unit processes? Give examples.
15. Discuss the advantages of grouping as many unit processes as possible together into a manufacturing operation. Discuss the disadvantages.
16. What is the purpose of an operation sketch? What should it show?
17. Discuss the similarities between grouping unit processes into manufacturing operations and sequencing manufacturing operations.
18. Most production routings show only one possible sequence of manufacturing operations. Give examples to show why it might be better to indicate other feasible sequences.
19. Theoretically, how many different arrangements can be made of 10 manufacturing operations? If the *first, fifth,* and *tenth* operations must be performed in that order, but the others may be done in any order, how many arrangements are possible?
20. What are the three major factors that limit the number of sequence arrangements that are actually feasible? Give a specific example of each type.
21. In general, what product characteristics suggest: (a) Manually operated equipment. (b) Mechanized equipment. (c) Automated equipment.
22. List some (a) indeterminate costs, and (b) intangible factors to consider in selecting equipment.

23. What are some uses and advantages of the *Manufacturing Operation Planning Sheet?*
24. What kinds of information and data are found on a *Production Routing?*

Exercises

Select a product (or part) and:

A. Conduct a value analysis session in class.
B. Make a preliminary part print analysis.
C. Make an *Assembly Chart* (see Chapter 6 for details).
D. Construct a *Critical Conditions Specification Sheet.*
E. Construct a *Work List Specification Sheet.*
F. Construct a *Manufacturing Operation Planning Sheet.*
G. Develop a *Production Routing* from the above.

5

Designing Material Flow

In almost any enterprise one can think of, productivity is best served by an efficient flow of the elements that move through the facility. This is just as important in a library, grocery store, post office, bus station, hospital, or restaurant as it is in a manufacturing plant. In each case, elements entering the system are processed and leave the system in a changed condition. A primary objective in planning an efficient enterprise is to provide for element flows that will facilitate the efficient movement of the elements through the activities. In fact, one of the nation's largest designers and constructors of industrial buildings has said:[1]

> Smooth out materials flow and you automatically trim production costs. A plant is actually nothing more than a great collection of machinery—receiving, assembling, shipping and storage areas linked together by materials handling devices of one kind or another. The building which surrounds them is merely a protective shell that must be designed to fit their requirements. No matter how handsome a plant may be from the outside, no matter how clean and functional it may look on the inside, no matter how thoroughly it is tooled, its production efficiency will depend upon the swift, smooth flow of materials throughout the plant.

The entire flow problem arises from the need to move the elements (material, parts, people) from the beginning of the process (receiving) to the end (shipping) over the most efficient paths. In the industrial situation, each element is represented by a blueprint or an entry on a *Bill of Materials*, or *Parts List*. In other enterprises, the elements may be represented by similar analyses and listings of the items or elements that flow through the facility. The individual activities through which the elements flow while being processed, are represented in manufacturing by the *Production Routing*. Similar listings of activity sequence would guide the flow through a library, a hospital, or a post office.

The concept of flow through an enterprise may best be visualized by considering that each element entering the building flows through the building, following a prescribed path—planned or otherwise—until it arrives at the end of the process. Then, if each element has its own path through the facility, the composite of the several individual flows becomes the overall flow pattern for the entire enterprise. If one were to imagine each of these individual paths represented by a

[1] *The Comprehensive Approach to Facilities Expansion* (Detroit, Mich.: Cunningham–Limp, Inc., 1968).

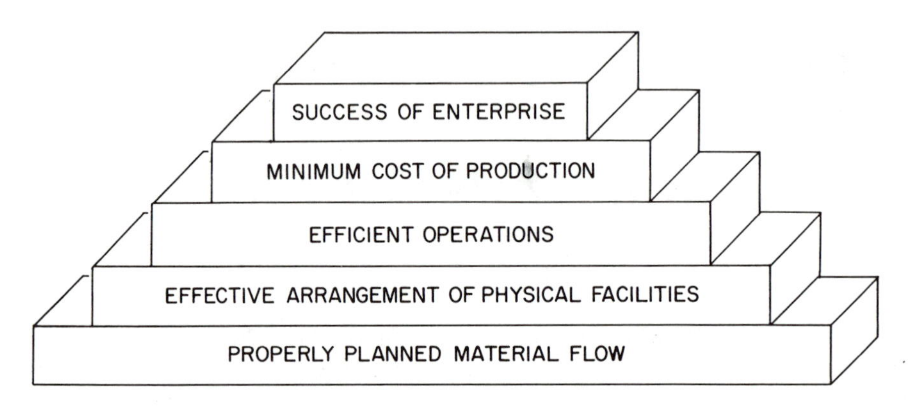

Figure 5-1—Importance of the material flow pattern.

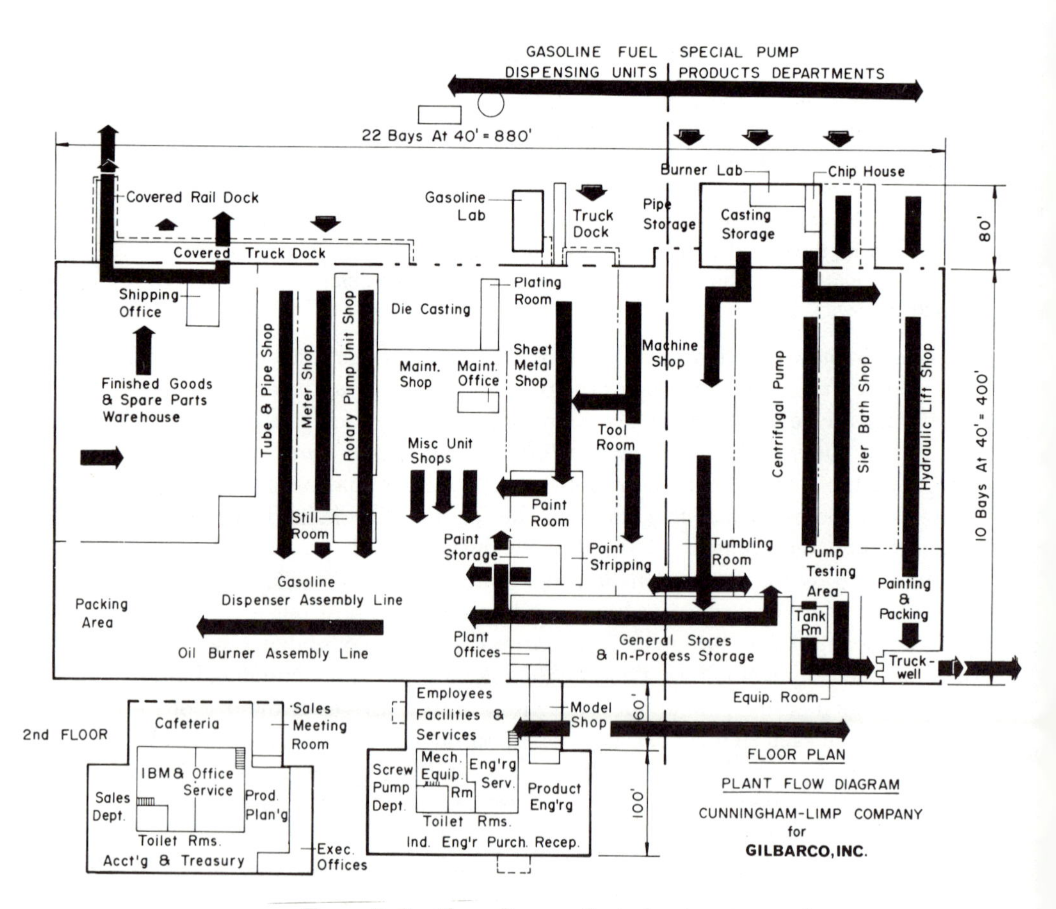

Figure 5-2—Flow diagram featuring two types of production lines. (Courtesy of Cunningham-Limp.)

piece of string on a floor plan, the resulting composite flow pattern would probably show one of two things:

1. A maze of lines going every which way, implying a serious lack of material flow planning.
2. A neat and orderly grouping of flow lines of varying densities, passing smoothly through the several areas of the facility with each component joining the main stream or assembly at the proper point, and with the overall flow pattern culminating in a single line leading into the shipping area, implying the end of the internal material flow.

Therefore, it is important that the flow pattern should be planned, rather than permitted to develop over a period of time, in a haphazard manner. Actually, the material flow pattern becomes the foundation for not only the basic design of the facility, but for the overall efficiency of the entire operation. It might be said that the overall success of an enterprise, or at least its profitability, is a direct reflection of the effort that goes into the flow planning. This can be depicted graphically as in Figure 5-1. Examples of overall material flow patterns are shown in Figures 5-2

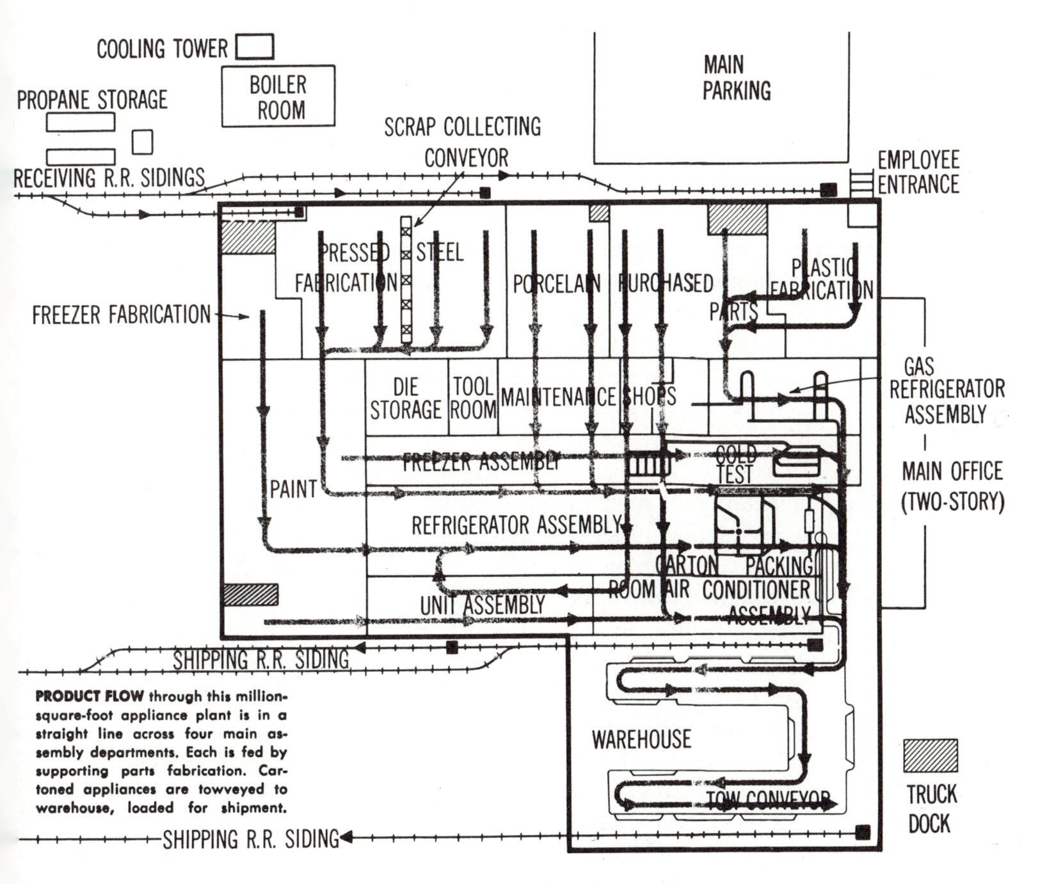

Figure 5-3—Flow diagram for Appliance Plant. (Courtesy of Factory Management Magazine.)

and 5-3. A review of these illustrations will show what is meant by—as well as the importance of—an overall flow pattern.

The Overall System Flow Cycle

While the above discussion has dealt with the internal flow of elements in an enterprise, it should be remembered that the beginning (receiving) of such a flow pattern, and the end (shipping) are also the points at which the internal flow plugs into the overall system flow cycle. This is not unlike the relationship between the individual plug-in circuits of a television set or computer, where each special-purpose circuit has its individual input and output terminals, joining it to the overall circuit, or flow pattern.

The systems approach to material flow requires visualizing each phase of the overall cycle as a segment of one all-encompassing system. This point of view requires a much broader consideration of the several interrelated flow processes involved in:

1. The movement of all elements, *from all sources of supply.*
2. All movement or handling activities *within and around the facility.*
3. The activities involved in the *distribution of the product or service* to all customers of the enterprise.

This broader, more inclusive point-of-view has the theoretical goal of conceptualizing a total solution to the overall flow problem in terms of a theoretical ideal system. It is then the task of the facilities designer to design as much of the total system flow as is currently practicable, implement the portions that are feasible, and continue to work on other portions of the theoretical system. Over a period of time, the other portions may be implemented as means become available and their installation becomes practical or economical. While the system approach may appear rather theoretical, it does serve as a worthwhile goal for the design of an overall plan. The extent to which the total system concept is carried out, depends upon the importance of the methods of handling the several elements with which the enterprise deals, as well as the practical economics of extending the overall system to the many vendor and customer locations. Obviously, the total implementation of such an approach is much simpler when the sources of supply, the productive processes, and the customers are limited in number, complexity, and intervening distances. Nevertheless, the aggressive designer should do no less than work closely with his major suppliers toward the improvement of any material flow activities in which they might have a common interest. These would include such things as packaging methods at a vendor's facility, the method of loading or shipping material to his own facility, and the effect that these two would have on the unloading, storage and use of the items at his facility.

The same philosophy applied to the finished goods portion of the handling–distribution cycle involves a similar degree of cooperation between the enterprise and at least its major customers. Here again, investigation should include a serious consideration of the packaging, packing, loading, and shipping techniques and methods. In each, detailed attention would be given to facilitating the handling and storage activities of the customers who will receive the finished products.

The interrelationships between the various activity centers within a total system are shown diagramatically in Figure 14-1 (page 340). From the system point of view, the concern of the facility designer would be for the total flow, of all elements, from all sources, to all destinations. Often the flow of materials would also include the activities necessary for efficiently handling of such items as scrap, waste, empty containers, and returned goods, back through the appropriate channels to the point at which they might re-enter the system.

Although the individual enterprise is most likely to think primarily of its own material flow cycle, it should not fail to recognize that this is actually only a segment of a larger whole. The minimum practicable consideration in implementing the material flow cycle would cover no less than the activities between receiving and shipping, including of course, all of the productive activities. However, the success of such a concept begins with a proper recognition by management of the significance of production flow and the savings potential of a serious effort directed in its behalf.

Advantages of Planned Material Flow

Too much emphasis can not be placed on the importance of determining the most efficient plan for the flow of material through facilities. It is precisely at this point, however, that many manufacturing plans fall short. Indeed, it is not at all uncommon for a management group to plan a factory building, erect it, then stand back, view it with pride, and say, "Well, let's see how we can fit the equipment in!" This, of course, is entirely the reverse of what should have been done.

Only by designing a master flow pattern, early in the planning process, can one be sure that all subsequent planning efforts are directed toward a worthwhile goal. This is not to say, however, that a flow pattern devised early in the planning stages, will not be subject to changes as planning progresses toward the final layout.

A well-conceived and carefully planned material flow pattern will have many advantages, and a good flow pattern will go a long way toward achieving several of the objectives of facilities design, as stated in Chapter 1. Some of the advantages are as follows:

1. Increased efficiency of production; productivity.
2. Better utilization of floor space.
3. Simplified handling activities.
4. Better equipment utilization; less idle time.
5. Reduced in-process time.
6. Reduced in-process inventory.
7. More efficient utilization of work force.
8. Reduced product damage.
9. Minimal accident hazards.
10. Reduced walking distances.
11. Reduced traffic congestion in aisles.
12. Basis for an efficient layout.

13. Easier supervision.
14. Simplified production control.
15. Minimal back-tracking.
16. Smooth production flow.
17. Improved scheduling process.
18. Reduced crowded conditions.
19. Better housekeeping.
20. Logical work sequence.

If such benefits as these can be achieved by planning the material flow, it is not difficult to see the trouble that could develop in an enterprise with an unplanned flow pattern. And, actually, there should be both a current flow plan, and a master flow plan or pattern for long-range planning.

The Need for a Master Flow Pattern

No production-oriented facility should exist without a master flow pattern of some sort properly recorded for use in current planning. It should represent the backbone of the entire production process, and should be consulted every time a facility change is found to be necessary. Using the master flow pattern as an overall guide will insure that no piece of equipment or department is located or moved without proper consideration being given to its relationship to the overall material flow.

Along similar lines, there should likely be a long-range master flow pattern, planned to represent expected facility expansion over 5, or 10, or more years. Again, no facility changes should be contemplated or made without reference to the long-range flow pattern, to insure that all changes are made in the general direction of the long-range overall material flow pattern. It might be pointed out, that the following of this philosophy would permit the gradual transition from a current flow pattern of questionable desirability to a future flow pattern designed to upgrade the overall production capability of the enterprise.

Factors for Consideration in Planning Material Flow

Before the actual task of designing a flow pattern can be undertaken, many factors must be considered, which, singly or in combination, will determine some of the characteristics of the flow pattern, or its relationship to other phases of the layout planning project. Not all of them can be properly considered at one time; nor can they all be adequately covered in this chapter. Some of them are listed in Table 5-1, and selected factors are treated briefly in the following overview (others are discussed or detailed in other chapters).

Material or Products

Of particular interest are material-related factors affecting volume, space, and handling (some will be covered in later chapters).

Table 5-1. Factors for Consideration in Planning Material Flow

A. Material or product (elements flowing through the facility)

1. Characteristics
   a. Receipts
   b. Shipments
2. Volume of production
3. Number of different parts
4. Number of operations
5. Storage requirements

B. Moves

1. Frequency
2. Speed
3. Rate
4. Volume
5. Scope
6. Area
7. Distances
8. Sources
9. Destinations
10. Cross-traffic
11. Required flow between work areas
12. Location of receiving & shipping

C. Handling methods

1. Unit handled
2. Possible use of gravity
3. MH principles
4. Desired flexibility
5. Equipment required
6. Possible alternatives
7. Preliminary MH plans

D. Process (activity centers)

1. Type
2. Sequence of operations
3. Possibility of performing during move
4. Specific requirements of activities
5. Product vs. process layout
6. Quantity of equipment
7. Space requirements
8. Number of sub-assemblies

E. Building

1. Size
2. Shape
3. Type
4. Number of floors
5. Location of doors
6. Location of columns
7. Aisle width, location
8. Ceiling height
9. Desired location of departments

F. Site

1. Topography
2. Transportation facilities
3. Expansion possibilities

G. Personnel

1. Number
2. Movement
3. Safety
4. Working conditions
5. Supervisory requirements

H. Miscellaneous

1. Location of auxiliary services and activities
2. Possible damage to materials
3. Cost of implementation
4. Production control
5. Flexibility
6. Expandability
7. Levels of activity

Volume of production. The desired sales and production volume were discussed in Chapter 3 as related to the sales-department and top-management contributions to the planning process. No single factor is of greater importance to layout planning than the quantity of material to be processed and the effect of this on production processes (as discussed in Chapter 4, and in Chapter 11 under space requirements).

Number of parts, products, or elements. Think for a moment of the differences in complexity of the flow pattern for such varying situations as manufacturing or providing the following products and services:

1. Yo-yos
2. Foot stools
3. Bicycles
4. Refrigerators
5. Radios
6. Typewriters
7. Automobiles
8. Aircraft
9. A campus
10. A bank
11. A hospital
12. A post office

Note that, the greater the number of components or activities, the more complex must be the flow pattern, and, likewise, the more detailed and accurate must be the plan for the facility.

Number of operations. The number of operations on each part or at each activity center is also a major factor in planning the flow pattern. For instance, a part requiring only one or two operations will probably require only one or two machines and, consequently, little space and few people to perform the work. Contrast this with the example of the automobile part that required 39 machines, 194 tools, and 157 gages. And similarly the flow between activities in a warehouse is less complex than that through a hospital. (This factor will be explored later when *Operation Process Charts* are introduced for analyzing part and operation interrelationships.)

Storage requirements. In nearly every type of enterprise, there will be a need for the storage of work awaiting processing, or awaiting movement to another area after processing. The amount—size and volume—of storage space required must be planned and will depend on a number of factors (many of which are discussed in Chapter 11).

Moves

The following are a few move characteristics worthy of preliminary thought (others will be covered in Chapter 14).

Cross-traffic. The potential interference with the orderly flow of elements through the enterprise caused by cross-traffic should be a major concern—back-tracking, crossed flow lines, etc., should be avoided.

Required flow between work areas. The flow of work from one work place to the next will be an important factor in determining the flow pattern. Therefore, interrelated work places should be close together. Complications may arise if the part must go out of line to another area in the course of its processing. This will also be true if it must backtrack to a machine a second time because to provide two identical machines would be uneconomical. (Work place design is covered in Chapter 11.)

Location of receiving and shipping activity. These points are usually the beginning and end of the material flow pattern. The receiving activity should have a high priority in orienting the flow pattern in relation to the building. Obviously it should be also closely related to the transportation serving the plant.

At the same time, the shipping activity should be located in close relation to transportation facilities, Figure 9-2 (page 218) shows some of the possible interrelationships between receiving, shipping, transportation. It will be noted that there are a number of potential configurations. Also, there are other factors to be considered which will determine whether receiving and shipping should be located in separate positions on the layout, or if they should be combined into one receiving-and-shipping area.

It should again be emphasized that it is at these two points that the internal materials flow pattern plugs into the external flow as represented by the transportation system. It is here that the overall system flow cycle (Figure 5-2) is closed, making the cycle a complete loop. (Additional details on receiving and shipping are covered in Chapter 9, planning, and Chapter 11, space needs.)

Handling Methods[2]

Material handling plans and equipment may have been given some thought, in a general way, prior to the establishment of an overall flow pattern. If this is true, it will be an important factor in the final design of the flow pattern. (Handling methods will be covered in detail in Chapters 14, 15, and 16.)

Processes

The manufacturing processes or activity centers are the reason for the flow pattern, and of particular interest in designing the flow between them are the following.

Sequence of operations. As pointed out in Chapter 4, the *Production Routing* lists the operations to be performed on each component of a product, and the order in which they are to be performed. Frequently this sequence of operations becomes the physical order in which the equipment is arranged for their performance. Obviously, the sequence of activities is just as important in other types of enterprise—such as business, government, and services.

Specific requirements of activities. Some activities, because of unusual characteristics, require special treatment in their relationships with the rest of the operations. Examples of types of process or activity with special requirements, are:

1. *Heat treating*—ventilation, fire protection.
2. *Painting*—ventilation, fire protection, heating.
3. *Plating*—ventilation, protection against acid and fumes, electrical insulation.
4. *Forging*—ventilation, heat removal, noise, and vibration damping.
5. *Foundry*—heat removal, ventilation, fire protection.
6. *Power house*—heat and dirt removal, noise damping.
7. *Heavy parts*—facilities for material handling.
8. *Final assembly*—nearness to warehouse or shipping.
9. *Inspection*—air conditioning, central location.
10. *Precision assembly*—air conditioning.
11. *Inflammable materials*—ventilation, fire protection.
12. *Executive work*—quiet.

[2]For further details on material handling, see Apple, *Material Handling Systems Design.*

Quantity of equipment. Since each piece of equipment, or activity center, will occupy a certain number of square feet, the total quantity of equipment becomes a factor in determining the gross space needs. So, not only the sequence of operations (as indicated above) but the number of machines and pieces of equipment will have an effect on the flow pattern. If several work stations are required to perform one operation—because of a relatively slow rate of production—then a preceding machine must split its output to feed it to the several slower succeeding machines, and vice versa. This not only complicates the flow pattern, but also the material handling.

Space requirements of equipment. It should always be remembered that space must be allowed not only for a piece of equipment, but also for the operator; the material to be worked on; the work completed; related auxiliary equipment such as benches, chutes, slides, stock containers, inspection devices, tool and blueprint racks, etc.; access to the machine for repairs; and access to any safety devices for emergency use.

In addition, space must be left for moving equipment with changes in layout. This is particularly a problem when columns, walls, proximity to elevator shafts, etc., are involved. (This is also covered in Chapter 11.)

Number of subassemblies. A subassembly is the result of preliminary (partial) assembly, whereby a number of parts are formed into a unit, such as a refrigerator door, an automobile instrument panel, a clock motor, a desk drawer. Subassemblies are designed:

1. To simplify the handling during final assembly.
2. To shorten the final assembly line, where overall assembly activity requires much space.
3. To reduce final assembly time, if a subassembly requires a greater amount of time than can conveniently be fitted into the operations planned for the line itself.
4. To separate from the line, equipment that for some reason would interfere with the line.
5. To reduce complications on the line, when a part must be machined after it is assembled, and where the processing equipment could not be easily fitted into the line.
6. To allow for testing a subassembled unit before introducing it into the final assembly.
7. To allow for volume production of a subassembly that is destined for several different products.

Each subassembly is, in a sense, an interruption of the overall assembly process and must be carefully worked into the flow pattern to cause as little as possible disruption of the overall flow.

Building

Since the layout more or less establishes the floor plan, it has a tremendous effect on building configuration; that is shape, size, number of columns, etc. (discussed in later chapters). Some of the other factors are the following.

Building type. The building itself, either existing or proposed, will have a close relation to the layout. If an existing building is to be used, the layout may have to be fitted into it without major building alterations; whereas if a new building is being considered, it should be planned around the optimum layout.

Number of floors. In new buildings, single stories are more common than multiple floors. Some situations, however, do call for multiple floors, and, of course, many older buildings have more than one floor. The number of floors has a considerable influence on the flow pattern, since materials may have to be moved from floor to floor.

Aisle area. The aisle area in a plant is intended for personnel traffic and material routing, but is often used for other purposes. In order to insure adequate space throughout the production area, there is a tendency, in planning new construction, to allow too much aisle area. In existing plants, on the other hand, there is a tendency to cut down on the aisles by using part of their allocated area for such things as stock, and additional equipment.

Considerable planning should go into decisions on aisle width. Aisles are non-productive areas, and each square foot used for aisles is lost to production; production pays for its floor area, whereas aisles do not. (Aisles are discussed further in Chapter 12.)

Desired location of departments. In addition to some of the factors discussed above, the location of production areas might well be concerned with the proper combination of:

1. Number of pieces to be handled per time unit.
2. Rough weight of each piece.
3. Weight of stock removed from each piece at each operation.
4. Distance over which rough pieces, semi-finished pieces, or scrap must be moved.

(This concept is developed in Chapter 7 under *From–To Charts* and in Chapter 13, under *Computerized Plant Layout.*)

Site

The plot of ground upon which a facility stands, or will stand, frequently affects the building configuration and therefore the flow pattern.

Topography. The lay of the land, in terms of contour, shape, and size, as well as sub-soil conditions, will affect the building, its configuration, its orientation, and its construction—which, in turn, affect the flow pattern.

Transportation modes desirable. Some attention should also be given to the transportation alternatives. That is, the engineer should not be overly influenced by existing transportation facilities, since it may be economically desirable to plan additional transportation facilities or alternative transportation modes. In the long run, a decision to install new transportation facilities might save many thousands of dollars over the use of existing facilities.

Transportation facilities available. Although the preceding paragraph implies that transportation facilities should be designed, rather than simply accepted, there is often no alternative but to use whatever is available. This can happen when the opportunities are limited either by volume requirements, or by the favorable proximity of suitable existing transportation.

Expansion possibilities. Possible expansion of the facility should always be kept in mind during the planning process. When designing the material flow, attention should be given to the directions in which the building might be extended. This could affect the orientation of the flow pattern within the proposed facility, or on the plant site. The flow pattern should be so designed that extension in desirable directions is both possible and logical. Potential directions of expansion should be indicated on the flow pattern. (This factor is dealt with in more detail in Chapter 12, where plot, or site, plans are discussed.)

Personnel

The people who function within the facility are of great importance in the layout design—sometimes more so than material. The personnel factor comes in for consideration in several ways.

Movement of personnel. Since all man-time can be reduced to dollars per minute, it is obvious that excess time spent in walking, costs an excess number of dollars. Consideration must be given to necessary and frequently traveled paths. For example, one large company, in the planning phase, tried to economize on plumbing by installing only one lavatory. It was calculated that this would save $4,000 in construction costs. Subsequent investigation showed, however, that it would cost $6,300 per year in extra walking time of people using the lavatory. Personnel not only must move about the plant, but they occupy space in so doing. This not only removes space that might be otherwise utilized, but it also creates a safety hazard, both to those who are walking about and to those who may be distracted by them. For these reasons, planners frequently turn to overhead walkways or to underground corridors, as illustrated in Figures 5-4 and 5-5.

Figure 5-4—*Overhead walkway for servicing and inspecting facilities. (Courtesy of The Austin Company.)*

Working conditions. The major factors to be taken into consideration in connection with working conditions are:

A. *Illumination*
   1. Lighting should be adequate and suited to the job.
   2. Use may be made of natural lighting when practicable.
   3. Artificial lighting should be immediately available for every work space and wherever there is a need for a high-intensity illumination on close work.

Figure 5-5—*Underground corridor for service areas and employee facilities. (Courtesy of Albert Kahn Associated Architects and Engineers.)*

B. *Ventilation*
   1. Ventilation must be adequate in all areas.
   2. Special precautions should be taken in locating activities causing unpleasant or unhealthy conditions, such as: painting, plating, heat treating. Fumes must be removed because of the discomfort they cause and to eliminate health and fire hazards.
   3. Lavatories, locker rooms, showers, smoking rooms, etc., should be located for convenient access and provided with good ventilation.

C. *Heating*
   1. Adequate heat must be provided for every work area.
   2. Building construction should be so planned that the layout of the heating systems provide for convenience in maintenance, expansion, and alterations.

D. *Noise and vibration.* These are two of the chief causes of unpleasantness, annoyance, and letdown in production efficiency in industrial plants. They can be overcome to a certain extent by:
   1. Proper location or isolation of equipment.
   2. Proper installation of equipment, as to both floor fastenings, and any necessary enclosures.
   3. Proper selection or design of equipment.

E. *Employee facilities.* In planning the layout, special consideration must be given to providing facilities for personal needs, rest, recreation, etc. (this subject is covered in Chapter 10).

F. *Health and safety.* The problem of healthful and safe working conditions is one of the most critical facing management. Safety records are, in the long run, constantly improving due to the increased emphasis placed on this. Factors to be considered in layout, from a health and safety point of view, are:
   1. Aisle location and widths.
   2. Equipment location.
   3. Machine and conveyor guards.
   4. Type of flooring.
   5. Floor load limits.
   6. Fire protection—extinguishers, sprinkler systems, exits, etc.
   7. First aid facilities.
   8. Light.
   9. Ventilation.
   10. Others.

The U.S. Government Occupational Safety and Health Act (OSHA) details requirements for nearly every known health or safety hazard.

Supervisory requirements. As was indicated in Chapter 1, effective supervision is one objective of good facility layout. Only if the work to be supervised by a single person is relatively centrally located can it be supervised effectively. The layout should be constructed in such a way that each supervisor can conveniently cover his area in a reasonable amount of time.

Miscellaneous

There are a number of factors that do not exactly fit into the above categories but are nevertheless important in planning the flow pattern, such as:

Desired location of service activities. These can be divided into four broad classes, serving:

1. Administration
2. Production
3. Personnel
4. Physical plant

A partial list of activities that must be included is shown in Figure 8-1. When planning the location of such service and auxiliary activities, the following factors should be considered:

1. Number of persons using service.
2. Frequency of use.
3. Equipment and floor space required.
4. Physical factors (heat, danger, ceiling height, etc.).

Cost. Although it will not be considered in detail here, it is obvious that no phase of the layout planning process can escape the close scrutiny of cost justification. Every decision made will cost more or less than some other alternative. The least costly practical alternative should be chosen in all cases. (Some details are covered in Chapter 18.)

Production and quality control. Plant design and layout have a great effect on the control of production flow, volume, and quality. Factors in flow design and layout planning affecting these controls include:

1. Number of floors
2. Shape of building
3. Flow pattern
4. Size of departments
5. Shape of departments
6. Handling operations

An ideal situation from a control viewpoint, might be to have a long, narrow, one-story building, housing a continuous process or an assembly line, with feeder lines at right angles. As we deviate from this situation to multi-story, or odd-shaped buildings and departments, the problem of keeping track of production quantity and quality becomes increasingly difficult.

Fexibility. Although flexibility on the facility site is important, as discussed above, flexibility within the facility is one of the most important characteristics of a good plant layout. In nearly every instance, the layout under consideration will be subject to the future change. Such changes are required by increases or decreases in capacity, or the addition of new products, processes, or departments. Therefore, one must keep in mind that the space may someday be occupied by activity of a greatly different nature. For example, an area might be planned for use as a light-assembly department requiring a relatively low overhead clearance of 12 feet. Future plans might develop for the use of this area as a machine shop or even a storage area—either of which might be more efficient with a higher overhead clearance of 18 to 24 feet. Similar consideration should also be given to floor construction and load limits, utility feeders, and column spacing in designing for flexibility. (Further details are covered in Chapter 12.)

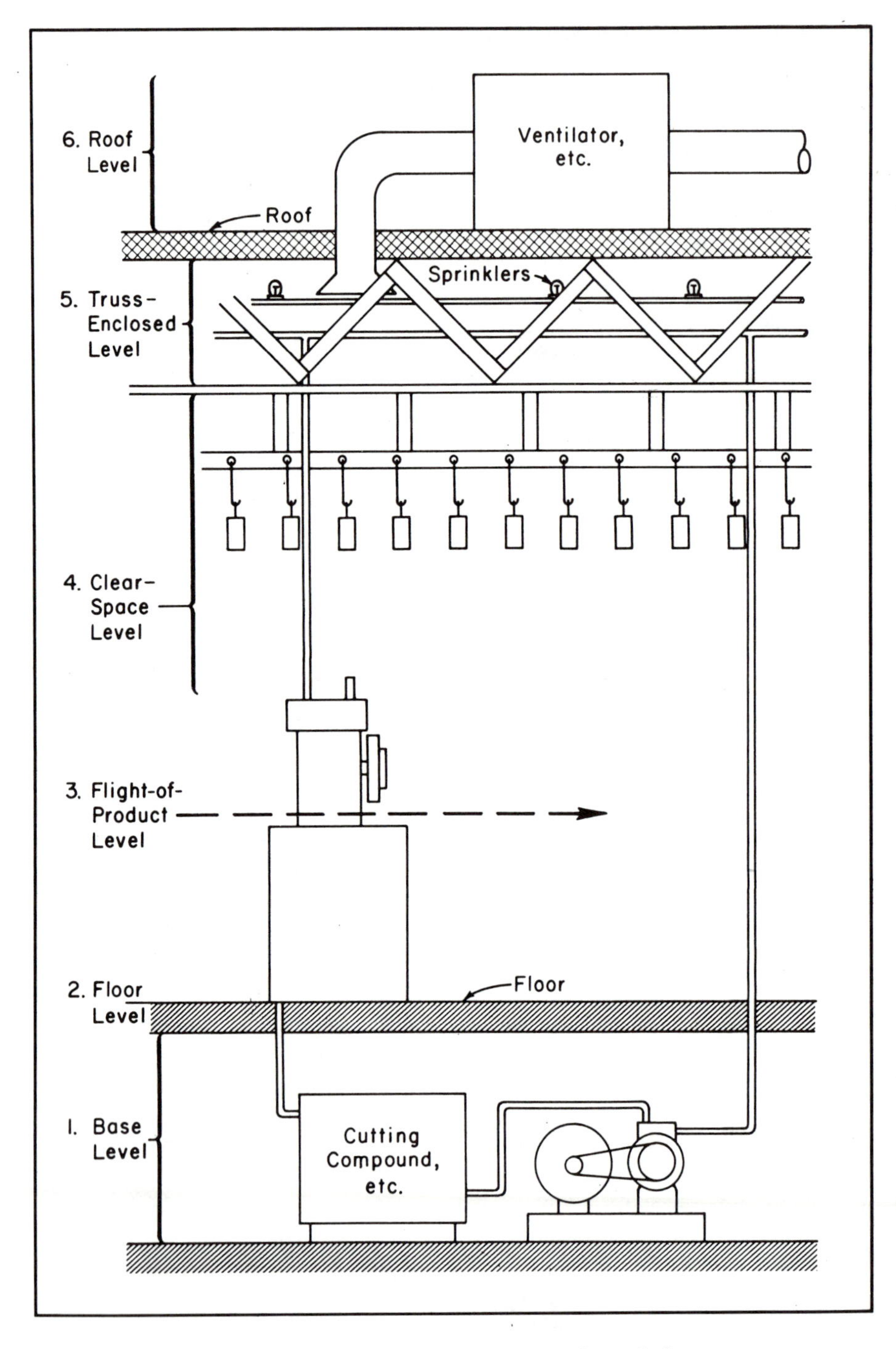

Figure 5-6—Levels of design in an industrial plant.

Levels of activity. Most of the thinking in the planning of a layout for obvious reasons revolves around activity at the floor level, that is, in the space occupied by equipment, stock, aisles, and columns. Very seldom is it sufficiently considered that plant production activity occurs also at other levels. These levels were first identified by H. H. Dasey as suggesting the great value of scale models in layout planning. The levels are:

1. *The base level.* The foundation, or under-the-floor area or space in a building, which may enclose such items as heating, ventilating, power, water, air, sewer, drain, scrap removal, and similar service facilities.
2. *The floor level.* The floor itself, which supports the equipment, stock, personnel, etc.
3. *The flight-of-product level.* This is an imaginary plane—about 36 to 46 inches above the floor—representing the line of product flow throughout the plant at work-height level.
4. *Clear-space level.* The area above the mean *operating ceiling,* that is, above the tops of the machines, etc., and below the bottom of the trusses supporting the roof. This space commonly contains conveyors, ovens, elevated stock rooms, officies, lavatories, storage, etc.
5. *Truss-enclosed level.* The area above the bottom truss chord and below the roof, which commonly encloses sprinkler lines, other utility lines, heating and ventilating equipment, etc.
6. *Roof level.* The space available on the roof, which may be used for ovens, cooling towers, water tanks, elevator shaft access, ventilators, etc.

This concept is expressed graphically in Figure 5-6. It is evident that all six levels must be kept in mind during the planning process. This will insure that the proper item will be allocated to the level at which it can make most economical use of plant space. It should never be forgotten that a facility consists of cubic feet to be efficiently utilized—not merely square feet of floor, area.

Conclusion

As indicated at the beginning of this section, the factors discussed must be considered during the process of planning any flow pattern or layout. Admittedly there are many, and all are important. The good layout engineer has bumped into or stumbled over them so many times that he has them memorized to be sure they won't be overlooked. It would be well for the beginner to start to memorize them also, to avoid the possibility of overlooking an important one.

Flow Planning Criteria

Over a period of years those who have dealt with the problems of material flow have come to a number of general conclusions about certain aspects of the flow, that might be considered criteria for material flow planning; some of them have been included in Table 5-2. It is felt that the process of designing a flow pattern will be more successful if these criteria are kept in mind throughout the planning

Table 5-2. Criteria for Designing or Evaluating Material Flow

1. Optimum material flow
2. Continuous flow—receiving towards shipping
3. Straight-line flow (as practicable)
4. Minimum flow between related activities
5. Proper consideration of process vs. product vs. group layout
6. Minimum material handling distance between operations, activities
7. Heavy material to move least distance
8. Optimum personnel flow, considering—
   a. No. of persons
   b. Frequency of travel
   c. Space required
9. Minimum of back-tracking
10. Line-production, when practicable
11. Operations combined to eliminate (minimize) handling between them
12. Minimum of re-handling
13. Processing combined with handling
14. Minimum of material in work area
15. Material at point of use
16. Material disposed at operations in convenient location for next operator to pick up
17. Minimum walking distances by operators
18. Compatible with building (present or proposed)
   a. Configuration (shape)
   b. Restrictions—strength, dimension, column location and spacing, etc.
19. Potential aisles
   a. Straight
   b. From receiving towards shipping
   c. Minimum number
   d. Optimum width
20. Related activities in proper proximity to each other
21. Provisions for expected—
   a. In-process storage
   b. Scrap flow
22. Flexibility in regard to—
   a. Increased or decreased production
   b. New products
   c. New processes
   d. Added departments
23. Amenable to expansion in pre-planned directions
24. Proper relationship to site
   a. Orientation
   b. Topography
   c. Expansion—plant, parking, auxiliary structures
25. Receiving and shipping in proper relation to—
   a. Internal flow
   b. External transportation facilities—existing, proposed
26. Activities with specific location requirements situated in proper spots
   a. Production operations
   b. Production services
   c. Personnel services
   d. Administrative services
27. Supervisory requirements given proper consideration
   a. Size of departments
   b. Shape
   c. Location
28. Production control aspects easily attainable
29. Quality control aspects easily attainable
30. Consideration given to multi-floor possibilities
   a. Present
   b. Future
31. No apparent violations of health or safety requirements

process—taking advantage of the experience of others to help prevent costly omissions or mistakes. It should be emphasized that they cannot all be followed in every case, but their consideration through the planning process will aid in arriving at a better and more effective flow pattern.

Flow Possibilities

The experience of layout engineers suggests a number of potential methods or bases for the overall material flow. Some of them pertain to the flow of material, while some refer to other characteristics of the layout problem or process. These alternatives,[3] for consideration by the flow planner, suggest that the flow might be based on the flow of material, products, personnel, or activities:

1. Requiring *similar machinery* or *equipment.*
2. Requiring *similar processes.*
3. Requiring *similar operations.*
4. Following the same *sequence of operations* or events.
5. Having *similar operation times.*
6. With *similar shape, size, purpose, or design.*
7. Requiring a *similar degree of quality.*
8. Made of the same or *similar materials.*

As the layout planner observes the preceeding, and reviews the flow documentation along with the previous lists of factors and criteria; he should attempt to draw some conclusion as to which might be appropriate to the situation at hand. It will be noted that several of the methods above relate to the *process, product,* or *group* methods of layout, which should be reviewed at this time (see pages 55–60). In all probability, the overall flow for the mass-production product will still be based on the necessary flows of the materials and components. However, for job-shop or related and auxiliary processes and operations, the flow may well be influenced by some of the above. Their individual and peculiar flow characteristics will then have to be worked into the overall flow pattern.

General Flow Patterns

The experienced layout engineer recognizes that a majority of material flow problems fit into one of a relatively small number of general flow patterns. These are illustrated in Figure 5-7, and reflect some of the basic factors in particular flow situations. A few comments will aid in understanding the reasoning or application of the patterns shown:

1. *Straight-line*—applicable where the production process is short, relatively simple, and contains few components or few pieces of production equipment.
2. *Serpentine,* or zig-zag—applicable where the line is longer than it would be

[3]Adapted from Muther, *Practical Plant Layout,* p. 175.

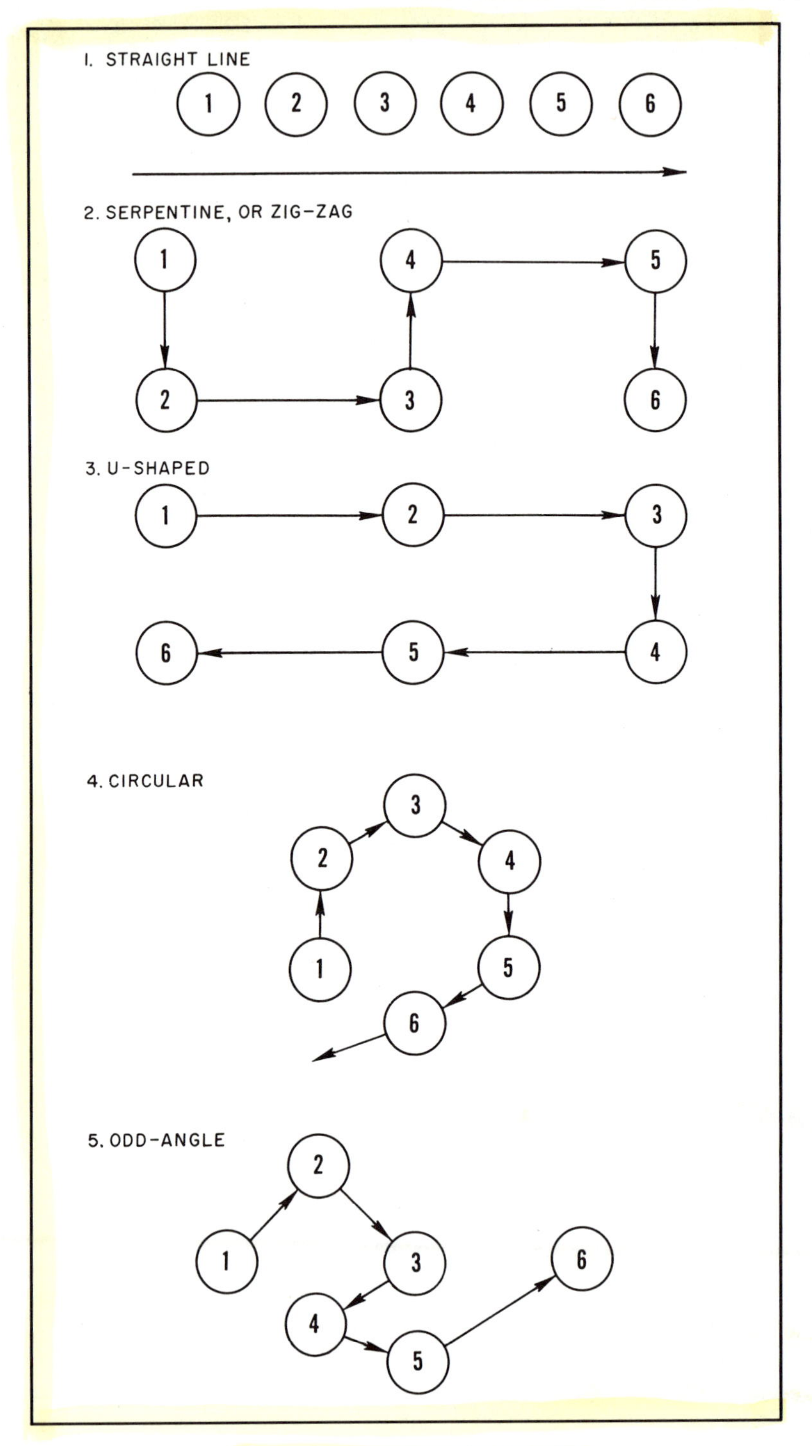

Figure 5-7—General flow patterns.

practicable to allocate space for, and therefore bends back on itself to provide a longer flow line in an economical building area, shape, and size.
3. *U-shaped*—applicable where it is desirable for the finished product to end the process in the same relative location as it begins—because of external transportation facilities, use of a common machine, etc. (also for the same reason as the *serpentine*).
4. *Circular*—applicable when it is desired to return a material or product to the exact place it started, such as: (a) for a foundry flask; (b) where shipping and receiving are at the same location; or (c) to use a machine a second time in a series of operations.
5. *Odd-angle*—no recognizable pattern, but very common (a) when the primary objective is a short flow line between a group of related areas; (b) where handling is mechanized; (c) when space limitations will not permit another pattern; or (d) where permanent location of existing facilities demands such a pattern.

If a typical facility is assumed to have receiving and shipping areas, it will be recognized that there are not too many variations of general flow patterns that can connect the two. Of course the nature of the flow pattern will reflect the number of components in the product, or the processes being performed on each. But in general, flow patterns will very likely resemble one of those in Figure 5-8. Various applications, adaptation, or combinations of the above are shown in Figure 5-8. It will be noted that the typical flow patterns represented suggest such modifications of the general flow patterns as:

1. Varying locations of receiving and shipping areas.
2. Varying numbers of product components.
3. Varying numbers of processing areas or departments.
4. Varing methods of material handling.
5. Size and configuration of the facility (if in an existing facility).

The beginning and ending points of the flow are dependent, to a certain extent, on the location of external transportation facilities, such as highways, railroad sidings, and docks or piers on navigable waterways. Regarding (*a*) and (*d*) in Figure 5-8, it is frequently assumed that such facilities are available alongside the plant. If a few long production lines are necessary, work flow would be as in (*a*). If transportation were to be available at the ends of the plant, flow might be as in (*b*); or if available at one end and one side, as in (*c*). Shown in (*e*) and (*f*) are methods of fitting a relatively long flow line into a rectangular space. In (*g*) and (*h*) are shown flow patterns involving subassembly operations.

Although it should be recognized that few if any real-world flow patterns will appear as simple as those in Figure 5-8, their characteristics will be recognized in most all designed flow patterns. For example, Figures 5-9 and 5-10, which are real-world flow patterns, will be seen to closely approximate one or a combination of the general or the typical flow patterns. Figure 5-9 shows the general flow of parts through a lift truck plant. This company designed its plant around a material handling system.

In the flow pattern in Figure 5-10, sheet metal parts come in from the press

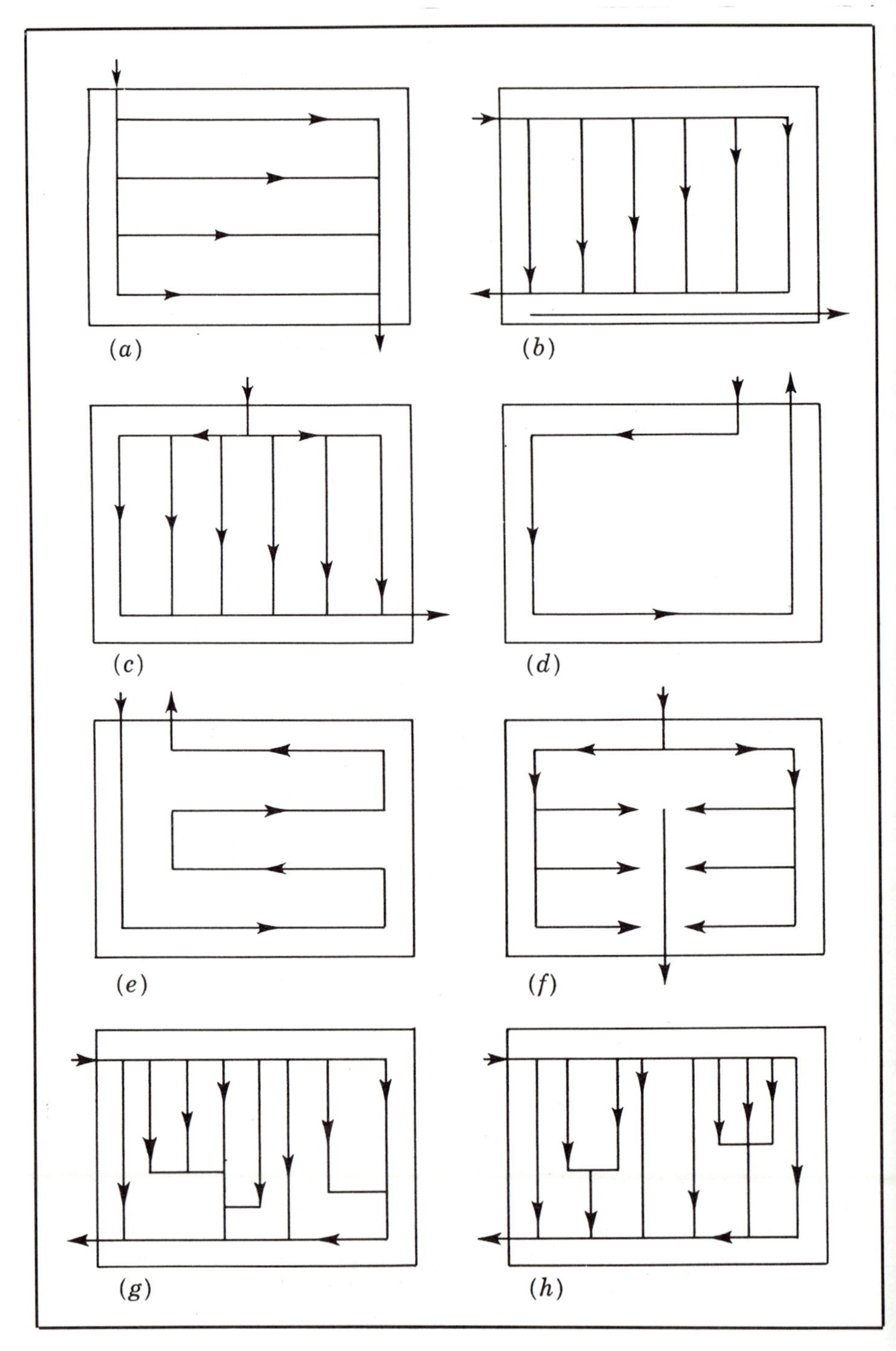

Figure 5–8—Typical material flow patterns.

and subassembly areas at the upper right. Metal finishing operations take place in the area at the upper part of the picture. Assembly begins at the left center where the outer shell is placed on the slat conveyor. At the lower left the inner shell is added. Then along the line, the refrigeration unit is added, the door is hung, fixtures are placed, inspections are made, and the unit is packed.

Many of the principles, factors, and considerations discussed have influenced the flow configurations illustrated. It is towards such a general flow pattern as these that the layout engineer is directing his efforts, using a combination of his experience, the tools and techniques appropriate, and a studied consideration of the factors related to the particular problem at hand.

No prefabricated flow pattern can be prescribed for a given situation, however. Each layout problem must be analyzed, and a flow pattern designed to best accommodate all the factors involved. At the same time, it must comply with as many of the objectives, criteria, and principles as possible. In the ideal situation, a planned material flow pattern becomes the basis for an efficient layout. The layout, in turn, becomes the basis for the design of the building—which is in reality a functional roof over an efficient material flow pattern—implemented by a proper material handling system.

Designing the Flow Pattern

Previous chapters have dealt with preliminaries, and have discussed many items of importance in planning the material flow pattern. The task now remains of actually developing the flow pattern from the information and data accumulated. Although there is no set procedure for planning the material flow pattern, it will be found helpful to proceed in an orderly manner, to avoid overlooking any of the important factors that must be taken into consideration. It will also insure the use of the applicable techniques from among the several that may be used in various planning situations. First, however, it may be useful to state an overall objective of a material flow pattern, somewhat as follows: to conceptualize, and design a composite of, the flow paths that parts, material, and personnel will follow, as required by the production operations and other requirements, from point of origin (usually receiving) to the ends of their respective paths (usually assembly, warehousing, or shipping), and to accomplish the necessary movement safely, efficiently, and economically.

Now, with all that has preceeded as background, the following is a suggested procedure, or thought pattern for carrying out the flow pattern design.

1. ***Identify and review all elements*** that will flow through the facility, such as:

   a. Materials
   b. Scrap and waste
   c. Manpower
   d. Equipment
   e. Information (including paperwork)

2. ***Collect all necessary data*** on the several elements:

   a. *Production Routings* for material
   b. Expected scrap and waste rates, volumes, locations, etc.

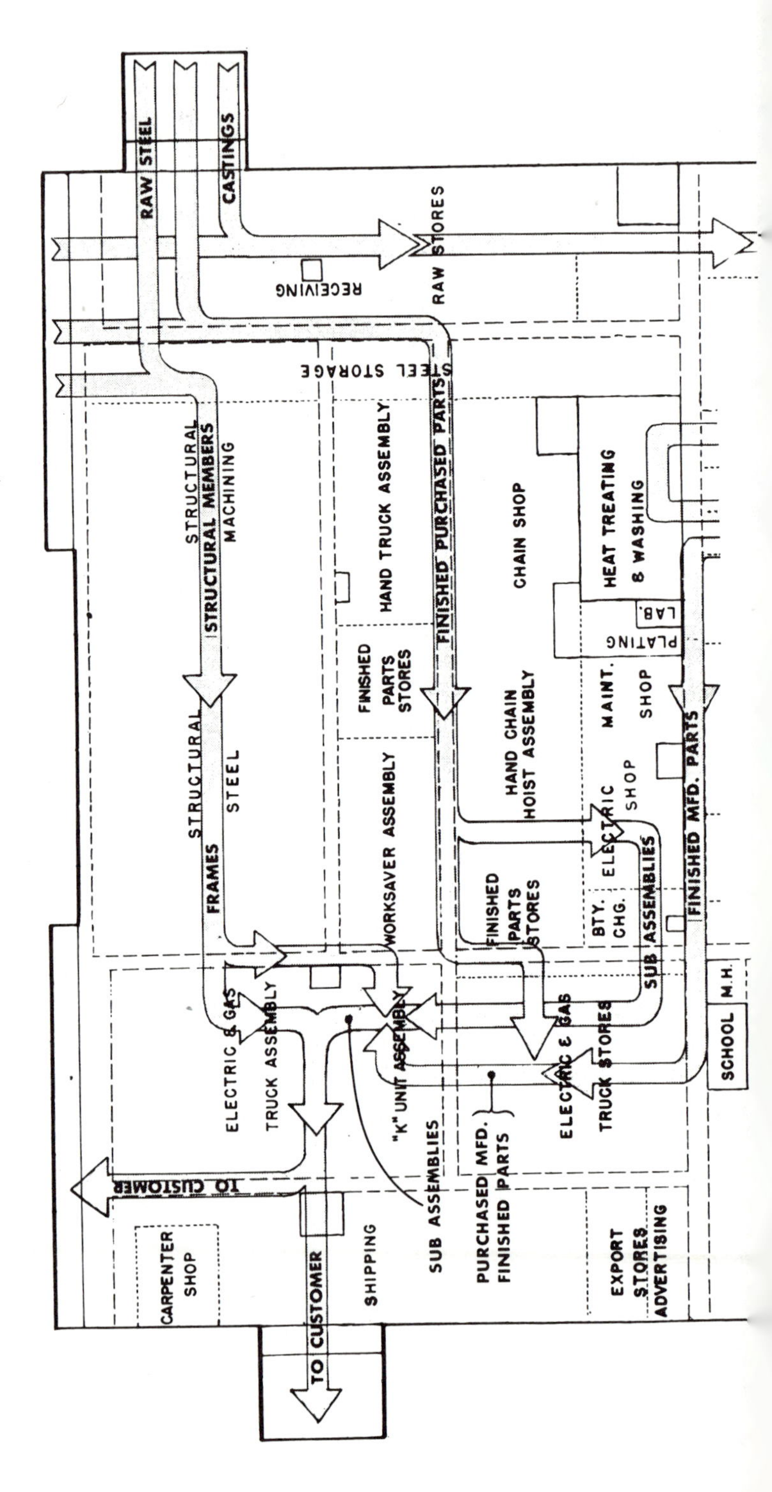
RAW STEEL
CASTINGS
RECEIVING
RAW STORES
STEEL STORAGE
STRUCTURAL MEMBERS
STRUCTURAL
MACHINING
HAND TRUCK ASSEMBLY
FINISHED PURCHASED PARTS
CHAIN SHOP
HEAT TREATING & WASHING
LAB.
PLATING
FINISHED PARTS STORES
MAINT. SHOP
STRUCTURAL STEEL
FRAMES
WORKSAVER ASSEMBLY
HAND CHAIN HOIST ASSEMBLY
ELECTRIC SHOP
FINISHED MFD. PARTS
FINISHED PARTS STORES
BTY. CHG.
SUB ASSEMBLIES
M.H.
SCHOOL
ELECTRIC & GAS TRUCK ASSEMBLY
"K" UNIT ASSEMBLY
ELECTRIC & GAS TRUCK STORES
PURCHASED MFD. FINISHED PARTS
SUB ASSEMBLIES
TO CUSTOMER
CARPENTER SHOP
SHIPPING
EXPORT STORES
ADVERTISING
TO CUSTOMER

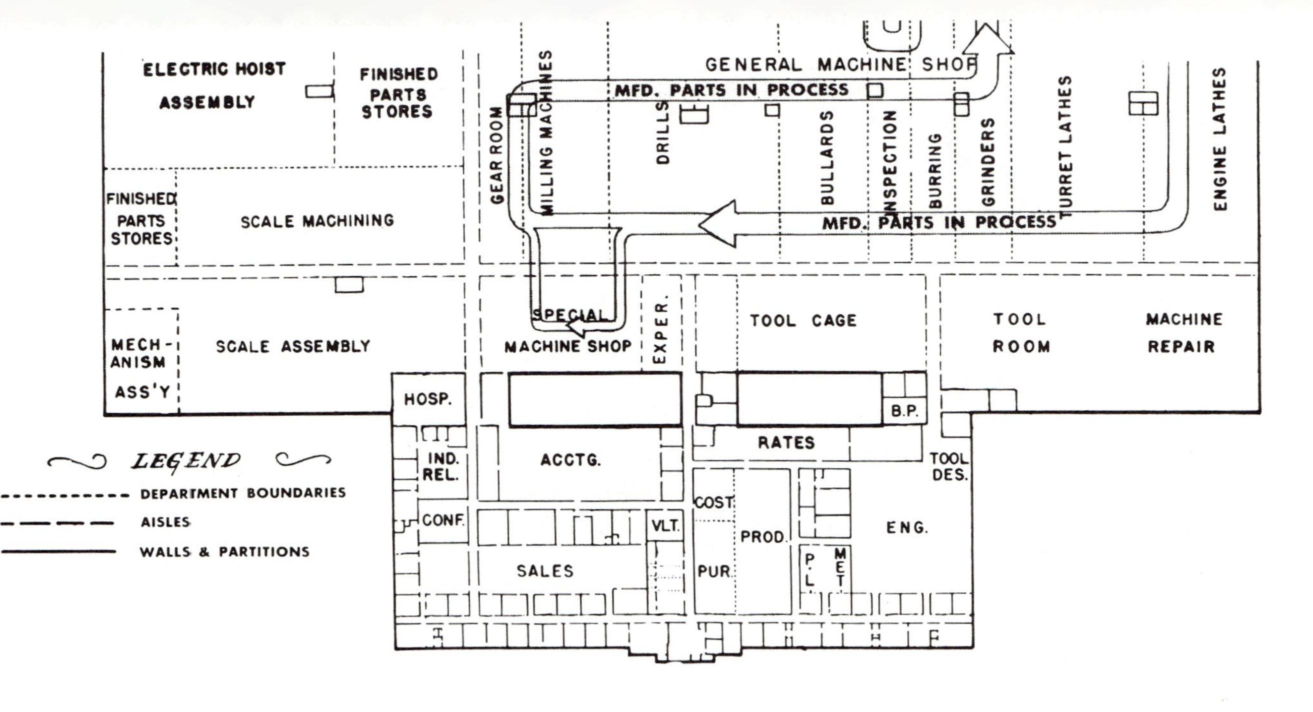

Figure 5-9—Material flow pattern for industrial truck products. (Courtesy of Eaton Corp.)

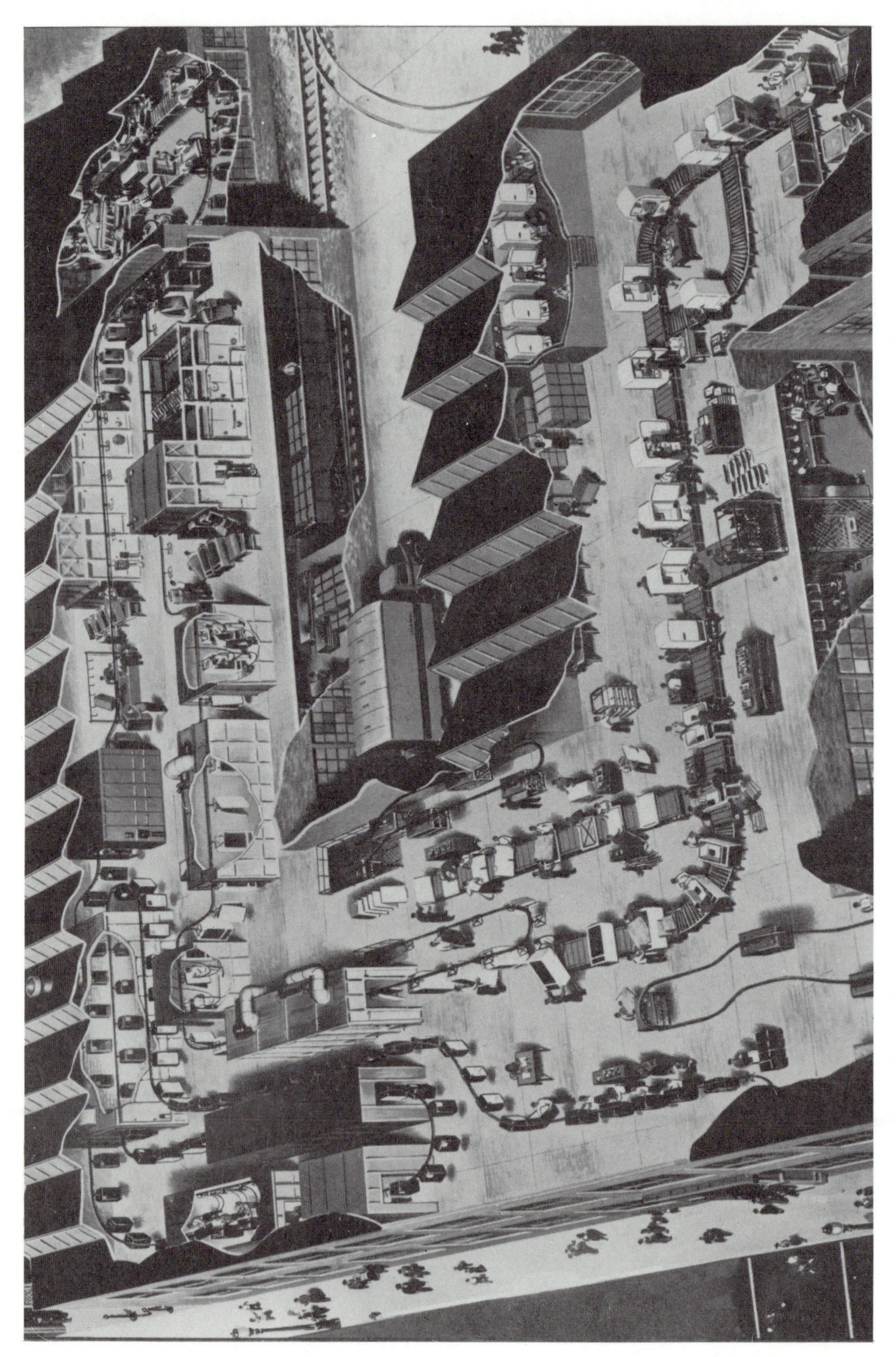

Figure 5–10—Material flow in a refrigerator plant. (Courtesy of Armstrong Cork Co.)

c. Personnel movement expected throughout the facility. This will include number of personnel, and locations of personnel in relation to facilities they utilize, such as: (1) production equipment; (2) inspection and quality control; (3) tool room, tool crib, etc.; (4) maintenance; (5) locker rooms; (6) lavatories; (7) food service; (8) other service activities.
d. Engineering data on any equipment that moves during the processing of material (this pertains primarily to portable processing equipment, not material handling equipment).
e. Information transmission and handling requirements, in terms of: (1) intercommunication systems, (2) paperwork (company mail) routes, (3) signalling devices, (4) computer remote stations, (5) production control data transmission devices (teletypewriters, etc.).

3. *Review the flow planning criteria* (Fig. 4-10)

4. *Review the factors* from among those in Table 5-1 that are closely related to the flow of the above elements, such as:

a. Material characteristics
b. Material move requirements—as visualized
c. Preliminary material handling
d. Personnel movement
e. Processing requirements and operation sequences
f. Method of production
g. Location of receiving and shipping
h. Location of raw material, in-process, and finished goods
i. Aisles—types, location, width
j. Desired, or pre-determined location, of selected activities
k. Supervisory requirements
l. Production control and quality control needs
m. Flexibility
n. Expandibility
o. Building restrictions—existing or potential
p. Topography of site

5. *Consider the several arrangement possibilities* pertaining to the product or activity and its components or divisions (Fig. 5-8).

6. *Review the analytical techniques* in the next two chapters for selection of those most appropriate for analyzing, documenting, and/or synthesizing of proposed (or existing) flows.

7. *Apply the techniques* selected above to documenting the flow of material, personnel, equipment, or information.

8. *Sketch several potential flow patterns,* based on item 7 above, as related to the present (or proposed) facility and site, and especially considering:

a. Receiving and shipping locations.
b. Transportation locations (present or proposed).

MATERIAL FLOW DESIGN EVALUATION SHEET					
		Alternative			
		1		2	
Criteria	Wt.	Eval.	Wt'd. Eval.	Eval.	Wt'd. Eval.
1. Optimum material flow					
2. Continuous flow (rec'g. to ship'g.)					
3. Straight-line flow (as practicable)					
4. Minimum flow between related activities					
5. Proper consid'n. of proc. vs. prod. vs. group layout					
6. Minimum handl. dist. between activities					
7. Heavy material move least distance					
8. Optimum personnel flow					
9. Minimum backtracking					
10. Line production, when practicable					
11. Operations combined to eliminate handling					
12. Minimum re-handling					
13. Processing combined with handling					
14. Minimum material in work area					
15. Material at point-of-use					
16. Material disposed to next pick up location					
17. Minimum walking distances by operators					
18. Compatible with building					
19. Potential aisles					

Criteria	Wt.	Alternative 1		Alternative 2	
		Eval.	Wt'd. Eval.	Eval.	Wt'd. Eval.
20. Related activities in proper proximity					
21. Provisions for in-process storage, scrap flow, etc.					
22. Flexibility					
23. Amenable to expansion					
24. Proper relationship to site					
25. Receiving and shipping properly located					
26. Activities with spec. loca. reqts. situated properly					
27. Supervisory requirements given consideration					
28. Production control aspects					
29. Quality control aspects easily attainable					
30. Consideration of multi-floor possibilities					
31. No violations of health or safety requirements					
TOTALS	/	/		/	

Figure 5-11—Material flow design evaluation sheet.

e. Expansion potential and directions.
d. Flexibility for possible future changes.
e. Others (from Tables 5-1 and 5-2), as appropriate.

9. *Review and evaluate the flow patterns.* This step can be done in several ways. One method is to evaluate it against the objectives, factors, and criteria discussed previously. These are indicated or implied in the *Material Flow Design Evaluation Sheet* shown in Figure 5-11. Each flow alternative can be checked against the items on the list, and a relative value indicated in one of the columns to the right. The evaluations may be done in terms of percentages or relative ranking, or the individual factors can be weighted to reflect their relative importance in the layout, and then evaluated.

Other evaluation techniques of a more quantitative nature could involve some of the techniques presented in Chapters 6 and 7. For instance, differences could be measured and converted to significant data for entry into a *From–To Chart.* If sufficient data are available on the amounts of material to be moved over the several segments of the flow pattern, the distances times the number of moves to be made can be converted into more realistic data, and possibly used with one of the computerized layout algorithms (Chapter 13), providing the other necessary data are also available.

10. *Revise or refine the sketches* made in item 8, on the basis of the evaluation and consideration of the factors in item 9.

11. *Resolve the best features* of each alternative into one probable flow pattern.

12. *Re-check proposed flow pattern* against criteria and factors.

13. *Draw proposed flow pattern* for guidance in the balance of the layout planning procedure.

Conclusion

This chapter has presented a consideration of the overall problem of flow through a facility and a discussion of a number of factors affecting the flow pattern. A procedure was outlined for designing a material flow pattern as a basis for subsequent steps in the layout design process. The following two chapters present a number of techniques (conventional and quantitative) useful in analyzing and planning material flow.

Questions

1. Explain the material or element flow concept, and discuss its significance in facility layout.
2. Give examples of element flow, other than material, through a facility.
3. Describe the overall system flow cycle concept.
4. What are some advantages of planned material flow?
5. Explain the significance of a master flow pattern.
6. What are some of the categories of factors to be considered in planning materials flow?
7. Discuss or explain each of the above.
8. What activities have special requirements affecting element flow, in a: (a) factory, (b) restaurant, (c) warehouse, (d) retail store, (e) residence, (f) laundry, (g) athletic stadium, (h) summer camp, (i) pick your own!
9. What are some of the factors for consideration in locating an activity in a specific location?

   Select an activity. List the factors.

10. In what ways can topography affect element flow?
11. What are some factors in locating service and auxiliary activities?
12. Discuss the importance of flexibility in facility design. Expandability?
13. Explain what is meant by the *levels of activity* concept in planning flow. Illustrate for the environments listed in *Question 8.*
14. Name some criteria for designing or evaluating material flow.
15. What are some of the bases for designing or selecting flow alternatives?
16. Distinguish between product, process, and group (technology) layouts.
17. Discuss the role of receiving and shipping activities in flow planning.

Exercises

A. Sketch a conceptual flow pattern for one of the facilities in *Question 8.*
B. Sketch a flow pattern to illustrate some of the criteria in *Question 14.*
C. Visit a local concern and sketch its material flow pattern. Comment on it. Sketch a suggested improved flow.

6

Conventional Techniques for Analyzing Material Flow

In the last chapter, some of the factors for consideration in planning material (element) flow through the enterprise were tabulated and discussed. This was followed by an outline for a procedure to be used in designing the element or material flow pattern. This and the next chapter cover some of the techniques useful in analyzing and designing the flow pattern. The techniques are divided into two categories:

1. *Conventional*—which have been in use for many years, are relatively easy to use, are primarily graphical, and are, on the whole, the best tools for their intended purposes.
2. *Quantitative*—which make use of more sophisticated mathematical and statistical methods, and are commonly classified as *operations research* techniques, frequently making use of the computer to carry out the complex calculations necessary.

This chapter will deal with conventional techniques, many of which are grossly underrated, for analyzing and improving flow and handling problems. This is possibly because they are brushed aside as too simple, in comparison with the computerized approaches—many of which fall down for lack of data that might be available if the conventional techniques were used as preliminary data collection and analysis devices.

The conventional techniques often require a lot of detail work to make an accurate record of all the moves in all the processes. They also require the gathering of many different kinds of data on the several aspects of each move, such as: the *route* over which the move is made; the *volume* moved; the *distance* travelled; the *frequency* with which the move is made; the *rate* at which the items travel, and the *cost* of the move.

Another problem, and possibly the most confusing to the analyst, is that the total flow most likely involves a large number of moves, not just one or two. These moves may take material through complex, integrated handling cycles, over a relatively long period of time. For example, a cycle may actually begin at a vendor's plant, progress through several departments of the facility under analysis, and end in a customer's storage area. In any case, it is usually a long series, with a

Much of this chapter is adapted from Chapter 8 of Apple, *Material Handling Systems Design.*

Table 6-1. Sources of Data for Material Flow Design

Characteristics of the Move	Basic Data Sources							
	Schedule	Materials	Routing	Layout	Job or Move Orders	Plot Plan	Vendors	Customers
Scope			√	√		√	√	√
Source/destination			√	√		√	√	√
Route/distance			√	√		√	√	√
Frequency	√	√			√			
Speed/rate	√	√			√			
No. of moves	√	√			√			
Area covered			√	√				
Path			√	√	√			
Location			√	√	√			
Operations in transit			√		√			
Physical restrictions				√		√		

Additional sources of data are: *carrier dimensions, maps, building drawings,* and *control procedures.*

large number of moves, interspersed by a lesser number of other activities, including production operations, inspections, and storages.

Sources of Data Required

It should be emphasized that before beginning the discussion of the techniques, all available data related to the flow should be accumulated, including data as in Table 6-1.

Appropriate data on each move in the overall flow cycle, and on every aspect of each move should be obtained and checked for accuracy. With such data at hand or available, most of the following analytical techniques can be used, supplemented by personal observation of the flow cycle, or a synthesis of a proposed flow.

Flow Planning and Analysis Techniques

There are many commonly used techniques that are helpful in the flow planning process. Some are peculiar to plant layout, some are particularly useful in the material handling phase, some are borrowed from the fields of motion economy

Table 6-2 Comparison of Planning and Analysis

	Technique	Especially Useful in—			
		Planning	Analysis	Comparison or Evaluation	Analyzing or Planning Information Flow
1	Assembly Chart	√	√		
2	Operation Process Chart	√	√		
3	Multi-Product Process Chart	√	√		
4	String Diagram		√	√	√
5	Process Chart	√	√	√	√
6	Flow Diagram	√	√	√	√
7	Flow Process Chart	√	√	√	√
8	From–To Chart	√	√	√	√
9	Procedure Chart	√	√		√
10	Critical Path Network	√			

and work simplification. Although most of the techniques were originally devised for analytical purposes, they are also useful in the planning process. The more common techniques are:

1. *Assembly Chart*
2. *Operation Process Chart*
3. *Multi-Product Process Chart*
4. *String Diagram*
5. *Process Chart*
6. *Flow Diagram*
7. *Flow Process Chart*
8. *From–To Chart*
9. *Procedure Chart*
10. *Critical Path Network*

Since it is obvious that not all the techniques are useful in every type of facility layout project, Table 6-2 will serve as a guide to the selection of the appropriate techniques for various purposes. It will be seen that most of them are equally useful in either the planning of a new project or the analysis of an existing one. Each will be commented on in this and the next chapter (some will also be covered in the chapters with which they are more closely related, and further details on some of them may be found in texts on motion study, plant layout and critical path methods, as well as in several handbooks).

Assembly Chart

The *Assembly Chart* is a graphical representation of the sequence in which parts and subassemblies flow into the assembly of a product (see Figure 6-1). It will be

Techniques Useful in Facility Layout

Primarily Applicable in Field of—								
Facility Layout						Material Handling		
Preliminary Planning	Activity Relationship	Material Flow	Evaluation of Flow Patt or. Plant Lay.	Recording Material Flow	Coordination	Planning	Analysis	Relationship Between Work Areas
√		√				√	√	√
√		√			√	√	√	√
√	√	√		√	√	√	√	√
	√	√	√	√	√		√	√
√		√		√		√	√	√
	√	√	√	√	√	√	√	√
√	√	√	√	√	√	√	√	√
√	√	√	√	√	√	√	√	√
√	√				√	√	√	√
√	√				√	√		√

seen that the *Assembly Chart* shows, in an easily understandable way:

1. What components make up the product.
2. How the parts go together.
3. What parts make up each subassembly.
4. The flow of parts into assembly.
5. The relationship between parts and subassemblies.
6. An overall picture of the assembly process.
7. The order in which the parts go together.
8. An initial impression of the overall material flow pattern.

Construction of the Assembly Chart

1. Using the *Parts List, Bill of Materials* or equivalent, and the *Routing* for the assembly process, determine what would be the last operation in the production or assembly of the product. Designate it with a $\frac{1}{2}$-inch circle in the lower right corner of a piece of paper, and briefly describe the operation, at the right alongside the circle. (It has been found most convenient to tackle the problem of making the *Assembly Chart* in reverse—that is, disassembling the product as the chart is made.)
2. Draw a horizontal line from the circle toward the left, placing a $\frac{1}{4}$-inch circle at the end, and indicating each component (its name, part number, pieces/assembly, etc.) assembled at the operation recorded. Components should be

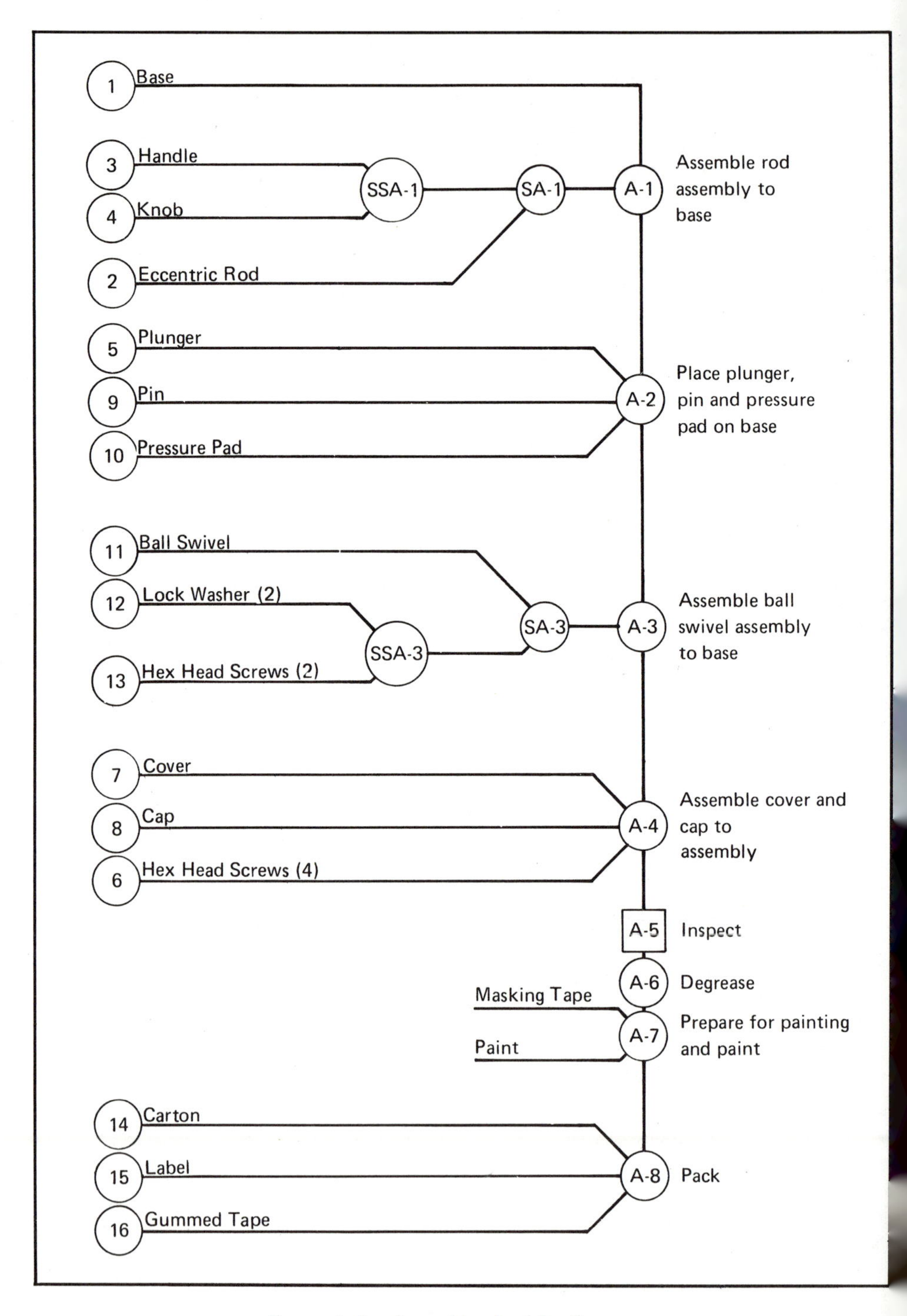

Figure 6–1—Assembly chart for Powrarm.

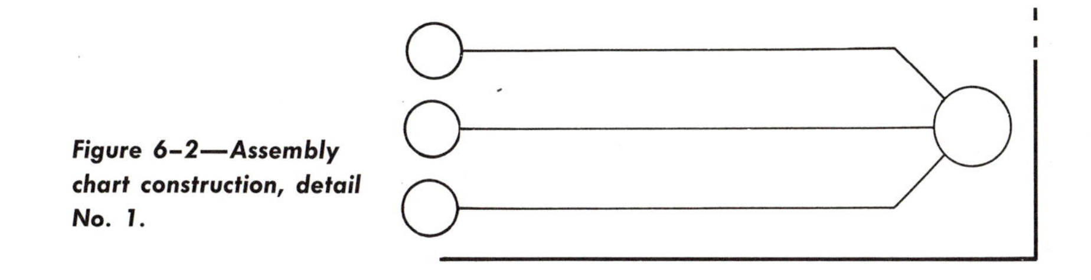

Figure 6–2—Assembly chart construction, detail No. 1.

listed in order of assembly (or disassembly), the last item at the bottom. (See Figure 6-2.)

3. When subassemblies or sub-subassemblies are encountered, run the item lines only part way to the left and terminate them with a $^3/_8$-inch circle to represent the subassembly operation. Then continue to the left until the subassembly is resolved into its component parts (see Figure 6-3). Assemblies can be numbered as indicated in Figure 6-1, after the *Assembly Chart* is complete. The line representing an individual part should again be carried to the left side of the paper and terminated with a $^1/_4$-inch circle in which the part number may be entered.
4. When the last assembly operation and its components are completely recorded, draw a short vertical line from the top of the $^1/_2$-inch circle, and enter a second $^1/_2$-inch circle to represent the next-to-last assembly operation. Indicate the components to the left, as in *Steps 2* and *3*.
5. Continue until the product has been disassembled and all components have been recorded at the left from bottom to top.
6. Check the *Chart* against the *Bill of Materials* to be sure no component has been omitted. Enter assembly and subassembly operation numbers in the circles, if desired. When completed, the items listed on the left should be in the order in which they are actually used or assembled to the product—from top to bottom.

Circles representing assemblies or subassemblies do not necessarily indicate stations on the assembly line, nor persons, but merely operations to be performed. The time required by each operation will determine what is done by each operator. The objective of the *Assembly Chart* is primarily to show the interrela-

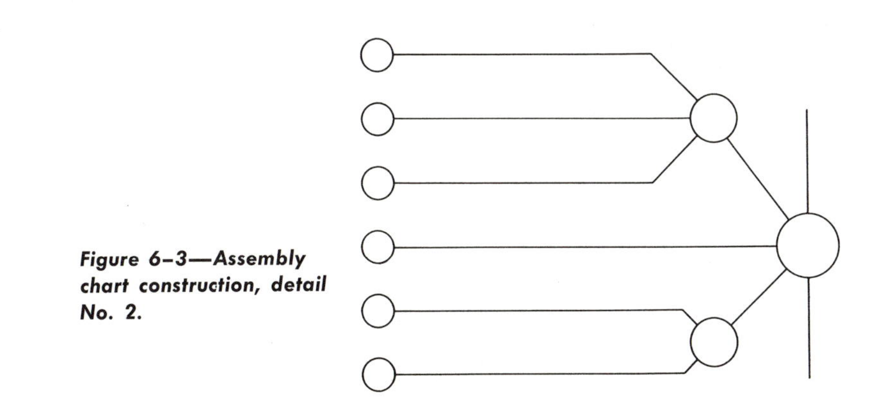

Figure 6–3—Assembly chart construction, detail No. 2.

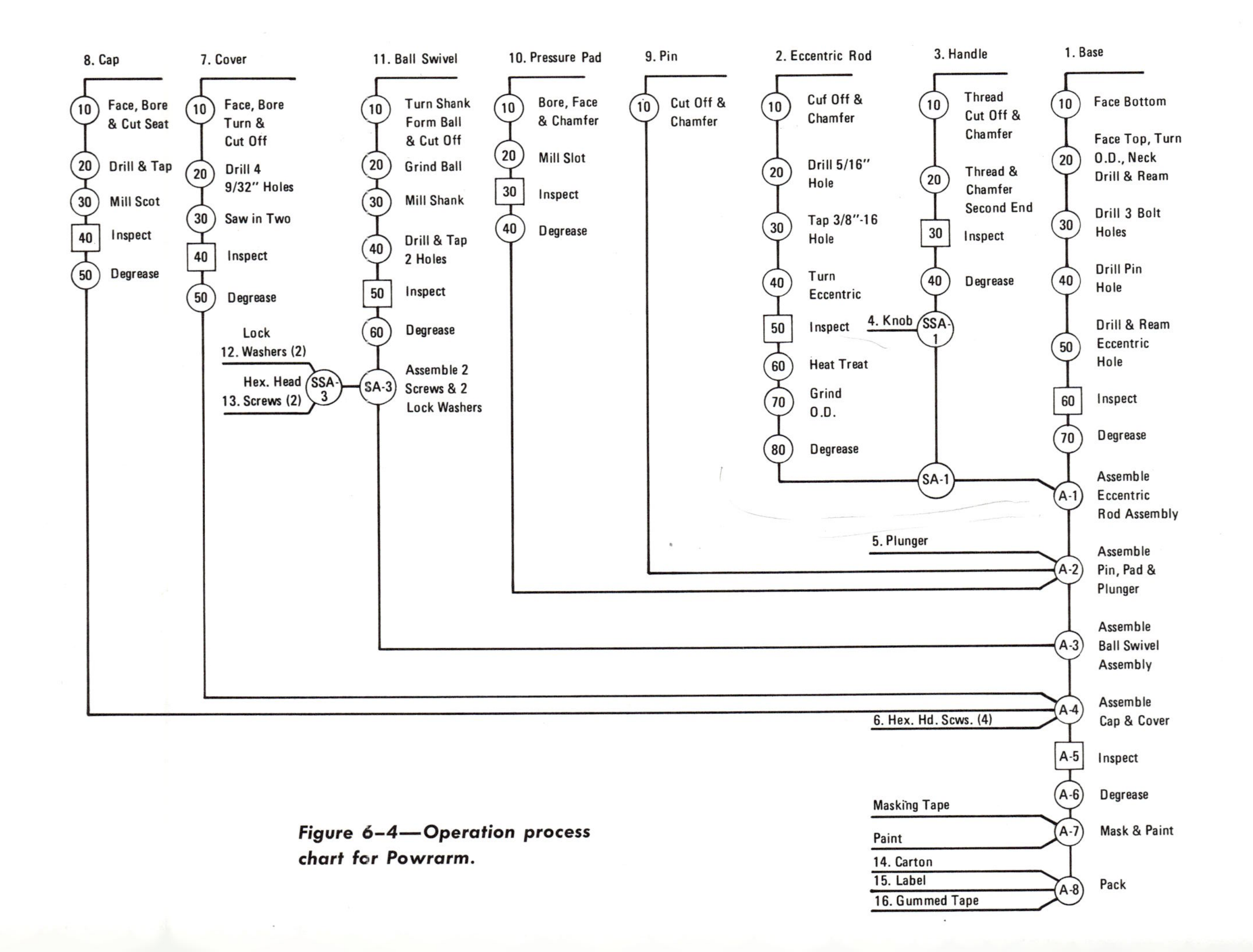

Figure 6-4—Operation process chart for Powrarm.

tionship of parts; this can also be accomplished with an exploded-view drawing, as was shown in Figure 3-6 (page 43). These techniques may also be used in teaching inexperienced workers the sequence of complex assemblies.

Operation Process Chart

This technique is the first to consider the individual operations on each part or assembly. It will provide a much more accurate picture of the production flow pattern than the *Assembly Chart,* since it adds the first quantitative data to the flow planning project. The *Operation Process Chart* (see Figure 6-4) extends the *Assembly Chart* by adding each and every operation to the graphical representation of the preliminary flow pattern previously developed.

The *Operation Process Chart* is one of the most useful techniques in production planning. Actually, it is a *diagram* of the process, and has been used in many ways as a planning and control device. With the addition of other data, it can be extremely useful as a management tool. Some of the advantages and uses of the *Operation Process Chart* are as follows:

1. Combines *Production Routings* and *Assembly Chart* for a more complete presentation of information.
2. Shows operations to be performed on each part.
3. Shows sequence of operations on each part.
4. Shows order of part fabrication and assembly.
5. Shows relative complexity of part fabrication.
6. Shows relationship between parts.
7. Indicates relative length of fabrication lines and space required.
8. Shows point at which each part enters the process.
9. Indicates desirability of subassemblies.
10. Distinguishes between purchased and manufactured parts.
11. Aids in planning individual workplaces.
12. Indicates number of employees required.
13. Indicates relative machine, equipment, and personnel concentration.
14. Indicates nature of material flow pattern.
15. Indicates nature of material handling problem.
16. Indicates possible difficulties in production flow.
17. Records manufacturing processes for presentation to others.

Construction of the Operation Process Chart[1]

1. Choose the part to be charted first. Usually a chart of the most pleasing appearance will be obtained by selecting the component on which the greatest number of operations is performed. If the chart is to be used as a basis for laying out a progressive assembly line, the part having the greatest bulk, and to which the smaller parts are assembled should be chosen.

[1]Adapted from *Operation and Flow Process Charts,* published by the American Society of Mechanical Engineers, 1947.

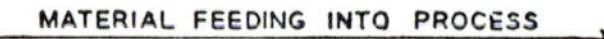

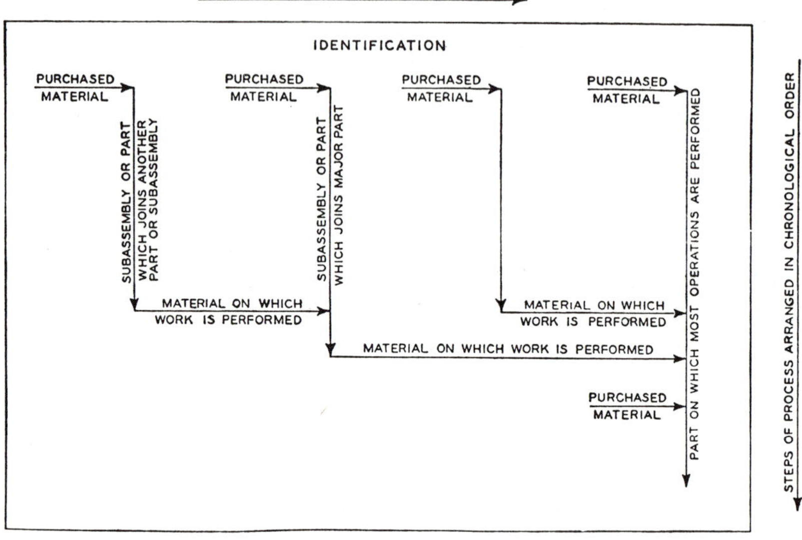

Figure 6–5—Graphic representation of principles of operation process chart construction. (From "Operation and Flow Process Charts," courtesy of the A.S.M.E.)

When the component has been chosen, start at the lower right corner of a piece of paper, and record assembly operations, as on the *Assembly Chart.* Purchased (complete) components are represented by short lines to the left, with part numbers and names shown on the lines (see Figure 6-5).

2. When all assembly and inspection operations on the major part have been entered, proceed to enter the fabrication operations, in reverse order (see Figure 6-4). After the first operation has been recorded, a short horizontal material line is drawn in the upper right-hand portion of the chart, to the right (Figure 6-4). A description of the material may be recorded directly above this line. The description may be as complete as deemed necessary. In order to identify the part itself, the name and identifying number maybe recorded directly above the material description.
3. To the right of each operation symbol, a brief description of the operation is recorded, such as *bore, turn, chamfer and cut off* or *inspect material for defects.* Other data that may be recorded are: the time allotted for performing the required work, the department in which the work is performed, operator classification, cost center, machine number, or labor cost, to the right of the symbol and below the description of the event.
4. Returning to the assembly operations already recorded, identify the last part requiring operations on it, draw a horizontal line to the far left, then up toward the top of the sheet. Indicate with $\frac{1}{2}$-inch circles the operations from the *Production Routings,* or squares for inspect, in reverse order, toward the top. At the top, identify each component. Also enter operation numbers, etc., from the *Routing.*
5. Continue in this manner, using Figure 6-4 as a guide, until all components have

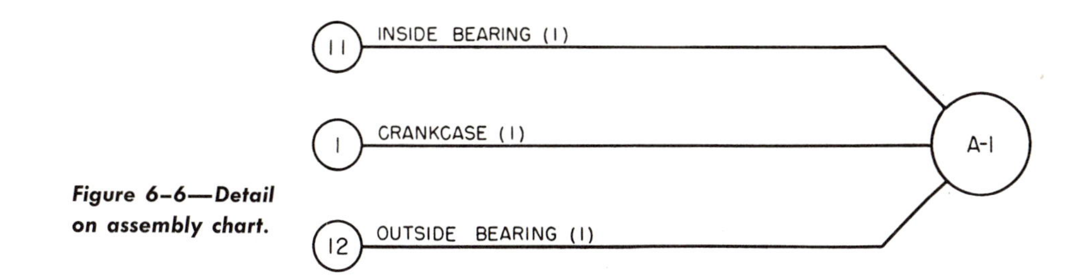

Figure 6–6—Detail on assembly chart.

been charted. All fabricated parts should be along the top, usually with the major component (chassis, base, etc.) to the right. All purchased parts should be within the body of the chart.

6. Subassemblies are handled in much the same manner as on the *Assembly Chart* (Figure 6-1), although they may *break out* differently with added information from the *Routings.*

   Do not try to relate the *Operation Process Chart* too closely to the *Assembly Chart,* which was based on the *Bill of Materials* or *Parts List* only, with no knowledge of the operations to be performed. When more detailed information regarding the product is available, as on the *Production Routing,* more specific relationships between parts can be shown.

   The accompanying sketches indicate what is meant in this connection. In a compressor crankcase, the bearings are pressed into position; this operation would appear on the *Assembly Chart* as in Figure 6-6. The proper way to depict the same operation on an *Operation Process Chart* is shown in Figure 6-7. This shows the interrelationship of parts more accurately.

   In other words, what was shown on the *Assembly Chart* as *Operation 1* in the assembly, actually becomes *Operation 40* in the fabrication of *Part 1.*

7. *Check chart against the Bill of Materials* and all *Routings* to insure that there are no omissions of parts or operations.

It will be seen from the completed *Operation Process Chart* that a definite flow pattern is beginning to shape up. In fact, with a little imagination, the layout will begin to form itself in the mind of the facilities designer. He can see which parts will present the biggest planning problems and which will be less important. If additional information is charted on each operation, the chart will then indicate where the most equipment will be concentrated. The chart also points out which parts are closely related to each other and should therefore be fabricated in adjacent areas, and shows where subassembly is desirable, as outlined in the previous discussion.

It is rather obvious that an *Operation Process Chart* would become unwieldy if made for a product containing a large number of components. For example, it has been calculated that a typical four-door automobile includes 13,512 parts—including bolts, nuts, washers, and cotter pins. The problem of complexity is overcome by showing operations and supplementary information for individual components on separate sheets. In this way, the writer has seen an *Operation Process Chart* for a diesel switch engine on a 36 × 72-inch sheet of paper.

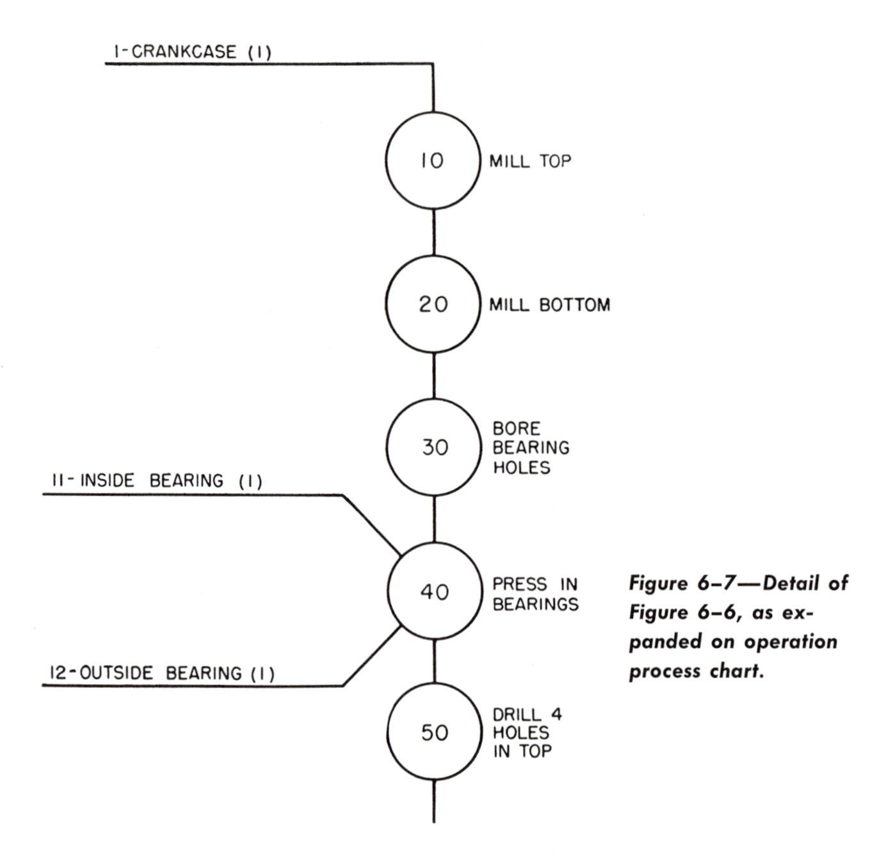

Figure 6–7—Detail of Figure 6–6, as expanded on operation process chart.

Multi-Product Process Chart

Closely related to the *Operation Process Chart,* is the *Multi-Product Process Chart.* (See Figure 6-8). It is particularly useful in showing production relationships (or lack thereof) between components of a product, or individual products, materials, parts, jobs, or activities. It is particularly helpful in job-shop-operations.

Construction of the Multi-Product Process Chart

1. Down the left side of a sheet of paper, list the departments, activities, processes, and machines through which the items or elements must pass. They may be listed from top to bottom: (a) in *geographical* sequence, as they occur, or might occur, in the facility; or (b) in as logical a sequence as possible, all factors considered (exact order is not necessary, since the construction of the chart is intended to show the proper arrangement or sequence).
2. Across the top, list the components, products, etc., being studied. For more than 20 to 25 items, divide them into groups of similar parts, and work with groups; work with major elements only; or use the *From-To Chart* (discussed later). The items should be arranged in a logical order, by similarity of operations required. Here too, constructing the chart will show up errors in the arrangement. (See also, Group Technology, on page 59.)

3. From the *Production Routings*, record the operations on each item, opposite the proper department, process or machine and under the proper item, by a circle containing the operation number from the *Routing*.
4. Connect the circles in sequence, even though back-tracking may be shown.
5. Study the resulting *Chart* for: (a) *back-tracking*—indicating the possibility of rearranging departments, etc.; (b) *similarity of "flow patterns*—indicating desirability of processing parts in the same area, at the same time, etc.; (c) *clues to arrangements* that will produce an efficient flow pattern.

The above three types of chart can be made much more valuable, if quantitative data are incorporated in their construction, showing number of pieces, weights, number of moves, and distances.

String Diagram

The *String Diagram* is a means of representing element flow on a layout of the area involved, using string, thread, yarn, etc., to show the paths of the elements as they move through the area (see Figure 6-9).

Construction of a String Diagram

1. Mount the layout on material into which pins, nails, or tacks can be inserted.
2. Insert pins through the layout and into the backing at each place an item stops along its path through the operations, activities, and processes; place pin in exact location such as in machine, on conveyors, in an aisle.
3. Starting with the entry of each item into the area being analyzed, connect the pins with string, by wrapping it around each pin, in correct sequence, stopping at the point where each item ends its travel or leaves the area.
4. Use different colors, or combinations of colors (by twisting strings), to represent different elements or material. Congested masses of string indicate areas of probable congestion.
5. If the layout is to scale, the string may be removed, measured, and converted to distance travelled.

Process Chart

The *Process Chart* is a tabular record of the steps in a process. It is one of the oldest and most common techniques for planning or analyzing material flow. For planning, it requires more knowledge of the proposed activity than the *Assembly Chart* or the *Operation Process Chart*, since it calls for identification of move or material handling steps. And, since it can be assumed that a move must be made between any two operations, its insertion on the *Process Chart* is insurance that the material handling steps will not be overlooked in subsequent planning. Figure 6-10 shows a typical *Process Chart*. Some uses of the *Process Chart are as follows:*

1. Provides a method of recording all steps in a process.
2. Forces detailed examination of the process.
3. Becomes the basis for analyzing the process, such as in: (a) Identifying all moves,

MULTI-PRODUCT PROCESS CHART FOR TOY TRAIN

MADE BY ALEX ALFARO ON OCTOBER 15, 1963

○ (bold) ROUGH LUMBER OPER.
○ FINISHING & ASSEM. OPER.

STOCK SIZE	$1\frac{1}{2}$" DIA.	$\frac{3}{4}$" DIA.	$\frac{5}{4}$" X 12' – 16'											$\frac{4}{4}$" X 12' – 16'			$\frac{10}{4}$" X 12'	NUMBER OF MACHINES
PART NUMBER	120	121	131	231	431	133	141	233	433	144	135	335	435	110	210	410	330	
CUT-OFF SAW			10	10	10	10	10	10	10	10	10	10	10	10	10	10	10	1
PLANER			30	30	30	30	30	30	30	30	30	30	30	20	20	20	20, 40	1
CIRCULAR SAW			20, 40	20, 40	20, 40	20, 40	20, 40	20, 40	20, 40	20, 40	20, 40, 50	20, 40, 50	20, 40, 50	30, 50	30, 50	30, 50	30	1
JOINTER	10		10	10	10	10	10	10	10	10				40	40	40	10	4
CIRCULAR SAW	20	10	20	20	20	20	20	20	20	20	10	10	10	10	10	10	20	5
DISC SANDER	30	20									20	20	20	30, 20	20	20	30	6
DRILL PRESS	40		30		30									50, 40, 60	40, 30	40, 30	50, 40	9

	ENGINE	GONDOLA	BOX CAR	CABOOSE	
BENCH TYPE 1	A110 (S), A120, A130	A210	A310	A410	7
DRYING RACKS	A135	A215	A315	A415	
DISC SANDER	A140	A220	A320	A420	4
PAINT BOOTH	A150, A160	A230, A240	A330, A340	A430, A440	2
OVEN	A155, A165	A235, A245	A335, A345	A435, A445	2
BENCH TYPE 2	A170, A180	A250	A350	A450	6

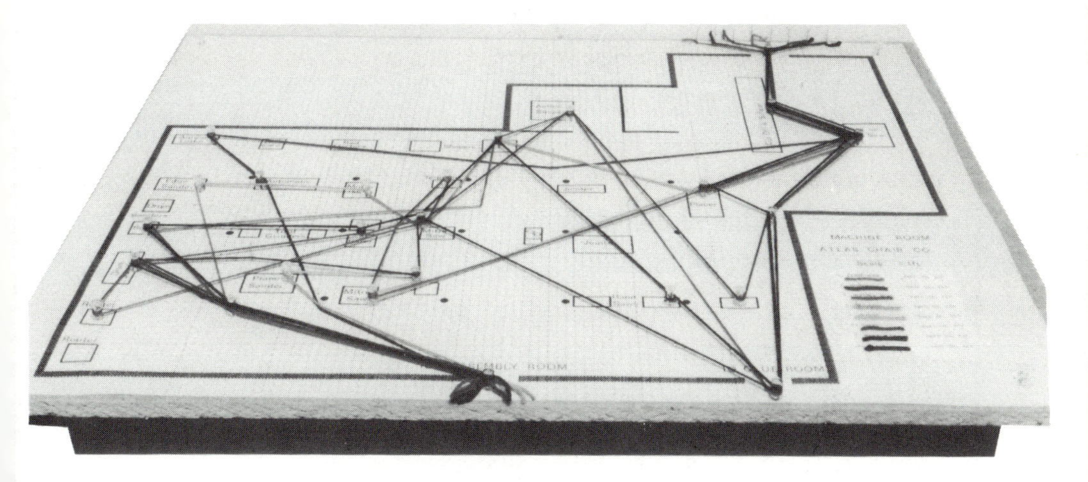

Figure 6-9—String diagram.

stores, delays; (b) Pointing out improvement opportunities; (c) Showing distances, equipment, manpower, etc.; (d) Raising questions about a process.
4. Familiarizes analyst intimately with the process.
5. Forms a basis for cost determination.
6. Forms a basis for comparison of alternative methods.

The basis for the *Chart* is in the *Process Symbols,* developed by F. B. Gilbreth in the 1920's. Figure 6-11 defines these symbols as they are used today.[2]

Construction of Process Charts

1. Most commonly, a printed form is used, with the process symbols printed in columns at the left.
2. Fill heading to properly identify the activity under observation (or synthesis).
3. Decide on item to be followed. Only one material, item, or person can be followed at one time on a *Process Chart.* If it is desired to follow several items (or persons), use the *Flow Process Chart* (explained later). *Do not skip from material to man to material.*
4. Decide on the type of information desired, and label the columns on the right for each type of data to be planned or analyzed, such as:
   (a) distance moved
   (b) number of men involved
   (c) type of container
   (d) time required
   (e) number of pieces handled
   (f) method of handling
   (g) frequency of move
   (h) time per move
   (i) department number
   (j) operation number
5. In the first column, on the first line, enter the *step number.*
6. Decide on the symbol that best represents the activity (or lack of it) at the very beginning of the process—most likely a Store.

[2]Maynard, *Industrial Engineering Handbook.*

PROCESS CHART

Page 1 of 2

PART NAME Gizmo

PROCESS DESCRIPTION Machine base and assemble, finish

DEPARTMENT Machine shop, Assembly, and Finishing

PLANT XYZ Products Co.

RECORDED BY I. M. Looking DATE

SUMMARY	NO.
○ OPERATIONS	
⇨ TRANSPORTATIONS	
□ INSPECTIONS	
D DELAYS	
▽ STORAGES	
TOTAL STEPS	
DISTANCE TRAVELED	

STEP	Operations / Transport / Inspect / Delay / Storage	DESCRIPTION OF PRESENT METHOD				
1	○⇨□D▽ (Storage 1)	in storage at receiving				
2	○⇨□D▽ (Transport 1)	to position at mach 2	walkie	6'		
3	○⇨□D▽ (Storage 2)	at mach. 2				
4	○⇨□D▽ (Transport 2)	into mach. 2	hand	4'		
5	○⇨□D▽ (Operation 1)	turn				
6	○⇨□D▽ (Transport 3)	to table	hand	4'		
7	○⇨□D▽ (Storage 3)	on table				
8	○⇨□D▽ (Transport 4)	to mach. 3	hand	4'		
9	○⇨□D▽ (Operation 2)	drill				
10	○⇨□D▽ (Transport 5)	to table	hand	4'		
11	○⇨□D▽ (Storage 4)	on table				
12	○⇨□D▽ (Transport 6)	into mach. 4	hand	3'		
13	○⇨□D▽ (Operation 3)	drill				
14	○⇨□D▽ (Transport 7)	to skid	hand	4'		
15	○⇨□D▽ (Storage 5)	on skid				
16	○⇨□D▽ (Transport 8)	to Assembly Dept.	walkie	10'		
17	○⇨□D▽ (Storage 6)	at end of assembly bench				
18	○⇨□D▽ (Transport 9)	onto bench to assy. position	hand	5'		
19	○⇨□D▽ (Operation 4)	assemble				
20	○⇨□D▽ (Transport 10)	to inspection position	hand	3'		
21	○⇨□D▽ (Inspection 1)	inspect				
22	○⇨□D▽ (Transport 11)	to skid at end of assy. bench	hand	8'		

Figure 6-10—Typical process chart.

Operation. An operation occurs when an object is intentionally changed in any of its physical or chemical characteristics; is assembled or disassembled from another object; or is arranged for another operation, transportation, inspection, or storage. An operation also occurs when information is given or received or when planning or calculating takes place. An operation symbol is also used to represent a person doing work.

Transportation. A transportation occurs when an object is moved from one place to another, except when such movements are a part of the operation or are caused by the operator at the work station during an operation or an inspection.

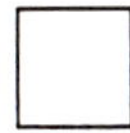

Inspection. An inspection occurs when a object is examined for identification or is verified for quality or quantity in any of its characteristics.

Delay. A delay occurs to an object when conditions, except those which intentionally change the physical or chemical characteristics of the object, do not permit or require immediate performance of the next planned action.

Storage. A storage occurs when an object is kept and protected against unauthorized removal.

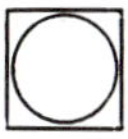

Combined Activity. When it is desired to show activities performed either concurrently or by the same operator at the same work station, the symbols for those activities are combined, as shown by the circle placed within the square to represent a combined operation and inspection.

Figure 6–11—Symbols used in the process chart.

7. Insert a small *number 1* inside the symbol. Each type of symbol is numbered consecutively to aid in tying in with other charts and data.
8. In the *description* column, enter just enough to indicate what is not told by other columns.
9. Fill in the remaining columns to the right, with pertinent data, as in *Step 4*, above.
10. Proceed through the entire process, or series of steps, until a logical or desired end point is reached. Remember, in selecting symbols and identifying steps, to follow either a person or an object—not both. When used as an analytical tool, the *Process Chart* is a record of an existing situation. When used as a planning tool, it is a record of what is intended to happen.
11. Fill in the summary box in the upper right-hand corner of the form.
12. Study the *Process Chart* for improvement possibilities, and make a *Chart* of the proposed improved method.
13. Use record of moves as basis for cost determination.

Flow Diagram

The *Flow Diagram* (see Figure 6-12) is a graphical record of the steps in a process, made on a layout of the area under consideration. It is frequently used as a supplement to the *Process Chart* (see Figure 6-10, which matches Figure 6-12).

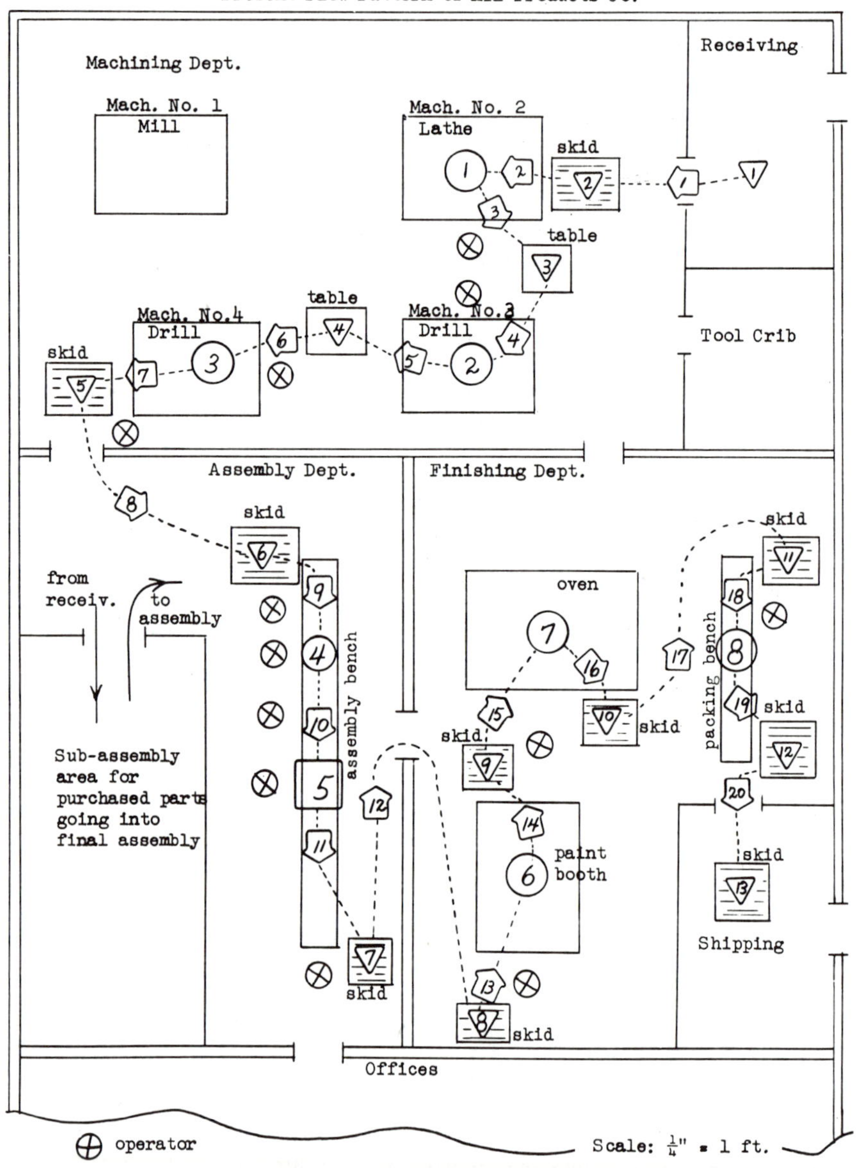

Figure 6–12—Typical flow diagram (to accompany Fig. 6–10).

Construction of the Flow Diagram

1. Obtain a layout of the area involved in the activity.
2. Record on the layout, as near the point of occurrence as possible, the process symbol that best describes each step of the activity, as with the *Process*

Chart—or if a *Process Chart* has been made, transfer the symbols from it to the *Flow Diagram*. Proceed, step-by-step, to the end of the process.

3. Number the symbols, in the same manner as with the *Process Chart*, to match the *Process Chart*.
4. Connect the symbols with a line, to show the path travelled by the object under observation.
5. Study the *Flow Diagram*, along with the *Process Chart*, for improvement possibilities.
6. Use the *Chart* for explaining the process to others.

Flow Process Chart

The *Flow Process Chart* (see Figure 6-13) is a combination of a *Operation Process Chart* (Figure 6-4) and a *Process Chart* (Figure 6-10) for each component of the product or assembly. It presents the most complete graphical representation of the overall process.

Construction of the Flow Process Chart

1. Obtain an *Operation Process Chart* for the process being studied.
2. Obtain a *Process Chart* for each component.
3. Re-draw the *Operation Process Chart*, inserting all process symbols from each *Process Chart* on the appropriate vertical line representing each component.
4. Enter any additional desired data alongside symbols, such as description, distance, quantity, time, cost, etc.
5. Study the resulting *Chart* for any possible improvement in the overall process, inter-relationships between operations, individual processes, etc.

From–To Chart

The *From–To Chart* (Figure 6-14) is one of the more recent techniques used in layout and material handling work. It is especially helpful where many items flow through an area, such as in a job shop, large general machine shop, office, or other facility. It is also useful anywhere there is a relationship among several activities and an optimum arrangement of the activities is desired. Some of its many uses and advantages are:

1. Analyzing of material movement.
2. Planning flow patterns.
3. Determining activity locations.
4. Comparing alternate flow patterns or layouts.
5. Measuring flow pattern efficiency.
6. Visualizing material movement.
7. Showing dependency of one activity on another.
8. Showing volume of movement between activities.
9. Showing interrelationship of product lines.
10. Pointing out possible production control problems.

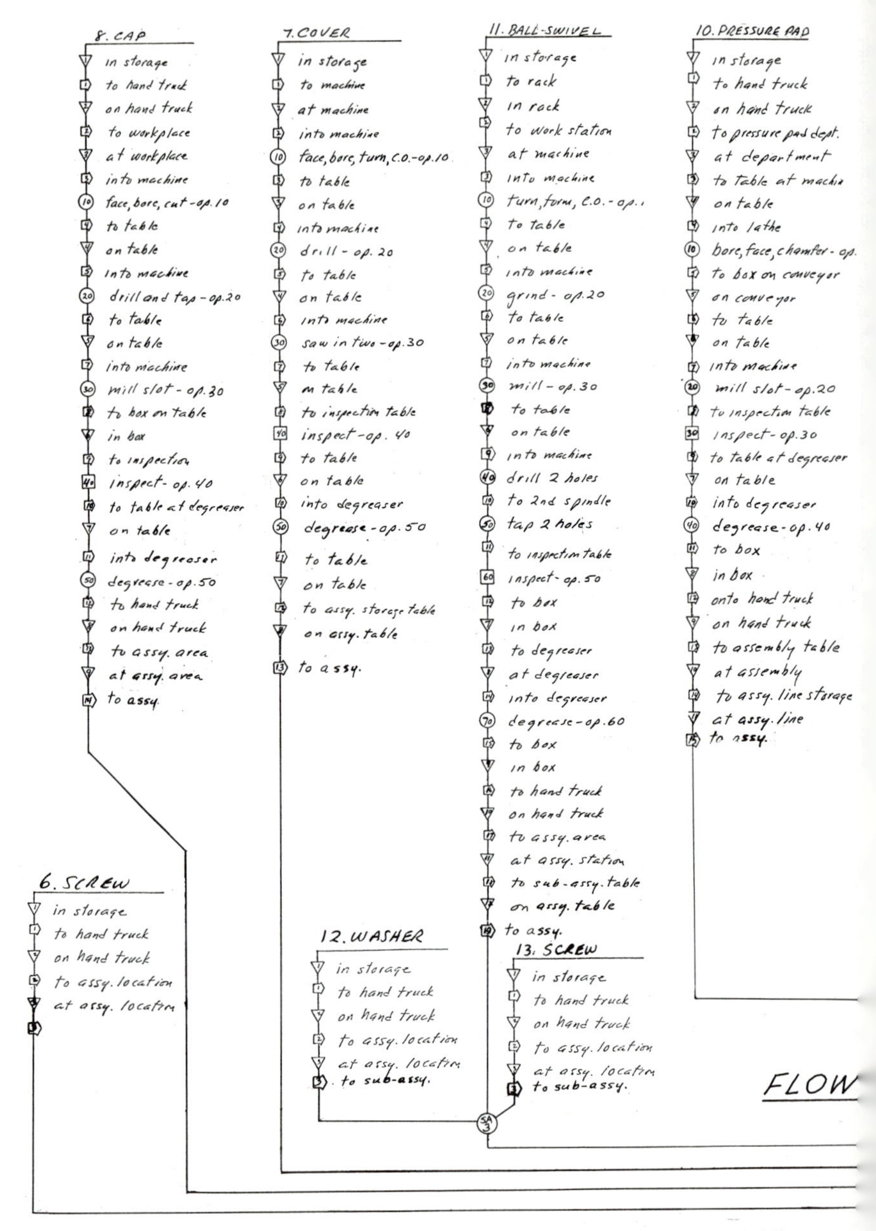

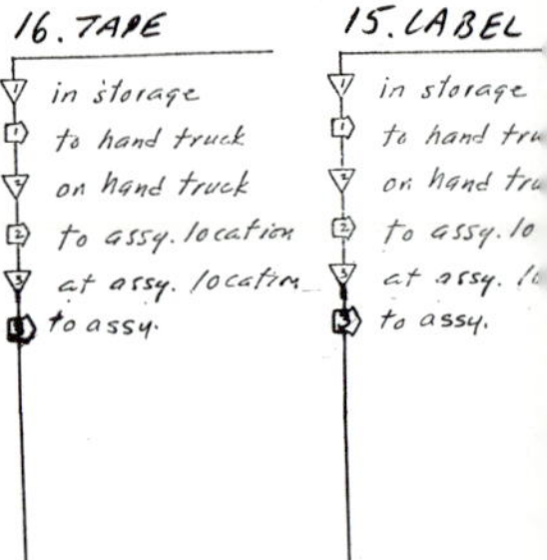

Figure 6-13—Flow Process Chart.

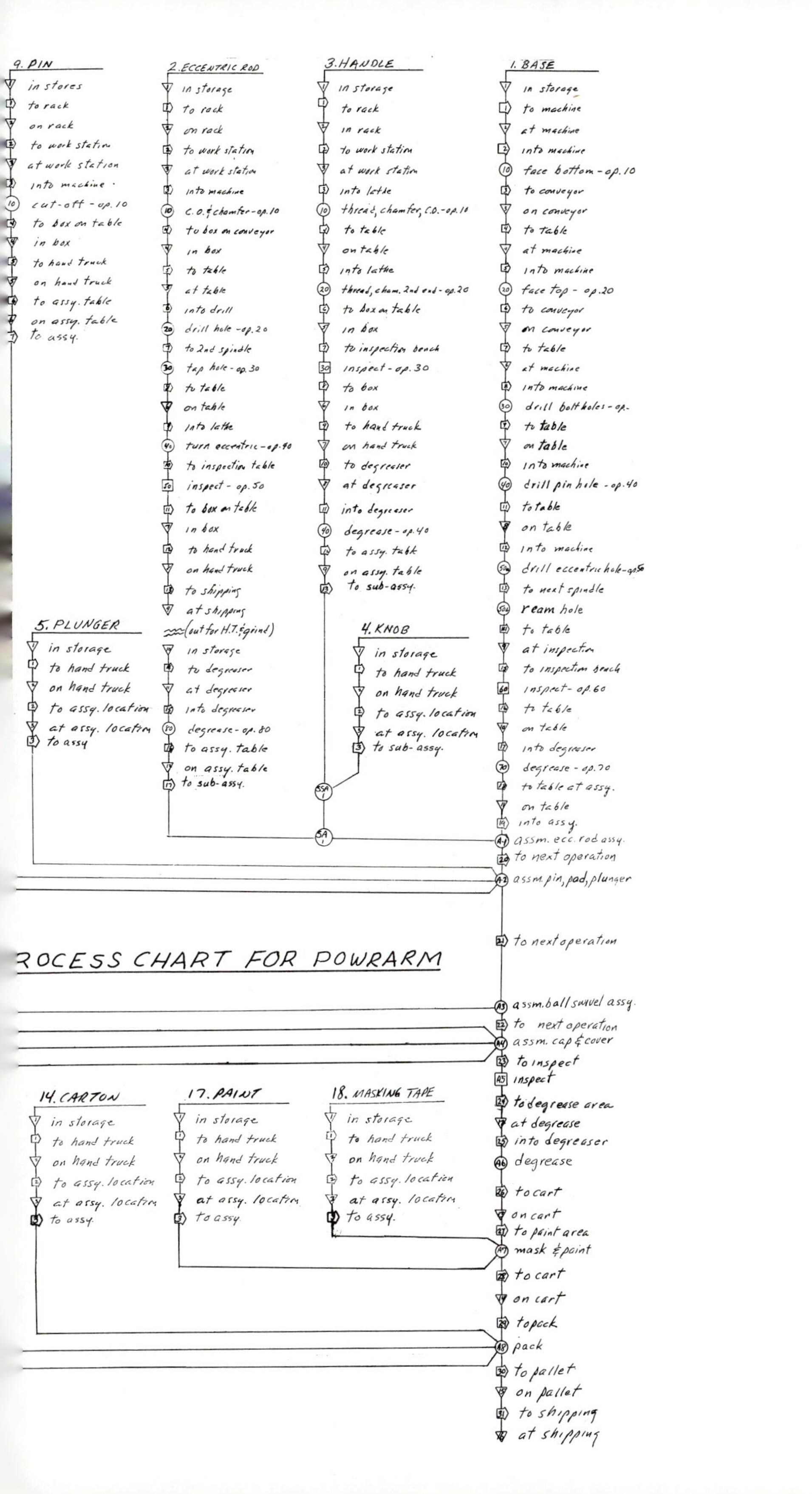

9. PIN
in stores
to rack
on rack
to work station
at work station
into machine
cut-off - op. 10
to box on table
in box
to hand truck
on hand truck
to assy. table
on assy. table
to assy.
2. ECCENTRIC ROD
in storage
to rack
on rack
to work station
at work station
into machine
C.O. & chamfer - op. 10
to box on conveyor
in box
to table
at table
into drill
drill hole - op. 20
to 2nd spindle
tap hole - op. 30
to table
on table
into lathe
turn eccentric - op. 40
to inspection table
inspect - op. 50
to box on table
in box
to hand truck
on hand truck
to shipping
at shipping
(out for H.T. & grind)
in storage
to degreaser
at degreaser
into degreaser
degrease - op. 80
to assy. table
on assy. table
to sub-assy.
3. HANDLE
in storage
to rack
in rack
to work station
at work station
into lathe
thread, chamfer, C.O. - op. 10
to table
on table
into lathe
thread, cham. 2nd end - op. 20
to box on table
in box
to inspection bench
inspect - op. 30
to box
in box
to hand truck
on hand truck
to degreaser
at degreaser
into degreaser
degrease - op. 40
to assy. table
on assy. table
to sub-assy.
1. BASE
in storage
to machine
at machine
into machine
face bottom - op. 10
to conveyor
on conveyor
to table
at machine
into machine
face top - op. 20
to conveyor
on conveyor
to table
at machine
into machine
drill bolt holes - op.
to table
on table
into machine
drill pin hole - op. 40
to table
on table
into machine
drill eccentric hole - op. 50
to next spindle
ream hole
to table
at inspection
to inspection bench
inspect - op. 60
to table
on table
into degreaser
degrease - op. 70
to table at assy.
on table
into assy.
assm. ecc. rod assy.
to next operation
assm. pin, pad, plunger
to next operation
assm. ball swivel assy.
to next operation
assm. cap & cover
to inspect
inspect
to degrease area
at degrease
into degreaser
degrease
to cart
on cart
to paint area
mask & paint
to cart
on cart
to pack
pack
to pallet
on pallet
to shipping
at shipping
5. PLUNGER
in storage
to hand truck
on hand truck
to assy. location
at assy. location
to assy
4. KNOB
in storage
to hand truck
on hand truck
to assy. location
at assy. location
to sub-assy.
SSA 1
SA 1
ROCESS CHART FOR POWRARM
14. CARTON
in storage
to hand truck
on hand truck
to assy. location
at assy. location
to assy.
17. PAINT
in storage
to hand truck
on hand truck
to assy. location
at assy. location
to assy.
18. MASKING TAPE
in storage
to hand truck
on hand truck
to assy. location
at assy. location
to assy.

PLANT *Acme Manufacturing Co.* TRIAL NO. *1* DATE *June 7*

FROM-TO CHART — TO (User) / FROM (Contributor)	1 *Rough Stores*	2 *Mill*	3 *Lathe*	4 *Drill*	5 *Bore*	6 *Grind*	7 *Press*	8 *Hone*	9 *Saw*	10 *Final Inspection*	TOTALS
1 *Rough Stores*		2	8			1	4		2		17
2 *Mill*			1	2			1			1	5
3 *Lathe*		2		4			1		1	3	11
4 *Drill*		1			1		2	1		5	10
5 *Bore*				1							1
6 *Grind*				1						1	2
7 *Press*				2						6	8
8 *Hone*										1	1
9 *Saw*			2			1					3
10 *Final Inspection*											
TOTALS		5	11	10	1	2	8	1	3	17	58 / 58

Figure 6–14—From-To chart, trial No. 1.

MILEAGE CHART

FROM \ TO	DETROIT	LANSING	GRAND RAPIDS	BATTLE CREEK	MUSKEGON
DETROIT					
LANSING	84				
GRAND RAPIDS	146	62			
BATTLE CREEK	113	48	62		
MUSKEGON	184	100	38	97	

FROM-TO CHART

FROM \ TO	STORES	SHEAR	BRAKE	PRESS	WELD
STORES		2	2	2	1
SHEAR			2	1	2
BRAKE				3	3
PRESS					1
WELD					

Figure 6–15—The similarity between a common mileage chart and the From-To chart.

11. Planning interrelationships between several products, parts, items, materials, etc.
12. Depicting quantitative relationships between activities and the related movement between them.
13. Shortening distances travelled during a process.

The *From–To Chart* is also sometimes referred to as a *Trip Frequency Chart,* a *Travel Chart,* or a *Cross Chart,* although sometimes the latter two are constructed and applied in a somewhat different manner.

The *From–To Chart* is actually an adaptation of the form of the mileage chart commonly found on a road map. Figure 6-15 shows the similarity between the two. In the *From–To Chart,* the numbers usually represent some measure of the material flow between the locations involved, such as number of unit loads, distances, weights, volume, or some other factor—or a combination of factors. (This will be seen in Chapter 13, where the *From–To Chart* becomes the basis for computerized layout procedure.)

The *From–To Chart* has great potential as an analytical tool, but has yet to be fully developed. Most current applications are relatively unsophisticated, tabulating only the number of moves and frequently the distances or volumes involved. The basic *Chart* is described below, followed by a brief discussion of possible variations.

Construction of the From–To Chart

1. Analyze basic data to determine activities, machine types, departments, buildings, etc. (see Figure 6-14).
2. Reduce basic data to usable form. Figure 6-16 shows the treatment of *Production Routings* for 17 parts through a general machine shop. Figures in columns represent operations on each machine type, and the sequence of operations, i.e., 5, 10, 20, 30, 40, etc.
3. Draw a matrix similar to the one in Figure 6-14, with as many rows and columns as there are activities under consideration.
4. Enter activity titles or names from *Step 1,* across the top, and down the left side, in the same order, as in Figure 6-14. The sequence may represent geographical arrangement in the plant, logical arrangement of process flow, or proposed sequence. If dealing with an existing situation, list areas in actual order of present flow of material, work, etc. As in the case of the *Multi-Product Process Chart,* the exact sequence is not important, as the use of the *Chart* will point out errors and suggest changes in sequence.
5. For each move of material, from one activity to another, enter a tally mark in the appropriate square of the matrix, such as from 5 to 10, from 10 to 15, etc.; that is, on *Part No. 1,* the first tally mark would go opposite Rough Stores, and under Mill (to show move from Rough Stores to Mill). The next would be opposite Mill and under Lathe, and so on. This must be done for each part, item, product, or material included in the analysis. Numbers in each square represent the total number of moves *From* and *To* the activity, or the data entered into the matrix can be in several other forms. Depending on objectives or desired results of the analysis, figures may represent: (a) number of moves between activities (as

	Rough Stores	Mill	Lathe	Drill	Bore	Grind	Press	Hone	Saw	Final Inspection
Part No. 1	5	10, 20 30	40	70, 80 90			50 60			100
2	5		10, 20	30, 40 50				60		70
3	5		10	20 30 40						50
4	5	70	20, 30 40, 50 60	80					10	90
5	5	50, 60		20, 40 80	30	10	70, 90			100
6	5		10, 20 30	40, 50 60						70
7	5		20			40			10, 30	50
11	5		10							20
12	5		10, 20							30
13	5	10		20, 30						40
14	5		10	20			30			40
15	5						10 20			30
17	5	20	10							30
18	5		10							20
19	5						10			20
20	5						10, 20			30
34	5						10, 20			30

Figure 6–16—Re-cap of operations on seventeen parts.

here); (b) quantity of material moved per time period; (c) weight of material moved per time period; (d) combination of quantity times weight per time period; (e) percentage of total through each activity to each subsequent activity; (f) move time; (g) move cost; etc.

6. Prove recording by totaling number of tally marks in each square and totaling each column and row. Each column should equal each comparable row (except first and last, which will be reversed). Grand totals should check. For the data in Figure 6-16, the *From–To Chart* would appear as in Figure 6-14.
7. Analyze the *From–To Chart.* Examination of Figure 6-14 will show that some of the entries are below the diagonal line. On the mileage table, this has no significance; but on the *From–To Chart,* these entries represent backtracking. The entries above the line represent moves directly from one department to another, along the line of normal travel. But if in leaving one department, a part travels back to a previous department, the entry appears below the diagonal line.

   It will also be noted that when the entries are in the squares just above the diagonal line, the parts have moved from one department to the next adjacent department. When they appear two spaces above the line, the part has skipped a department along the way. So, the objective is to have as many entries as close to the diagonal line as possible. This would indicate that most parts are moving from one department to the next in "line"—there would be a minimum of backtracking and of unnecessary travel.

By inspection then, it is seen that a better solution can be devised: by rearranging the columns and rows to put the larger numbers of tally marks closer to the diagonal, and smaller numbers below the line. Also those further away from the diagonal can be moved closer.

A more quantitative measure of the efficiency of the activity arrangement can be obtained by taking the torque of the system. This is done by totaling the values in the squares just above the diagonal line and multiplying by one, the values two squares above by two, etc. The same procedure is carried out for the values below the diagonal line. If desired, all of the backtracking moves might be multiplied by two to show backtracking as twice as bad as forward movement. *Torque* calculation on *From–To Chart,* Trial No. 1 (Figure 6-14) is:

Forward		*Reverse*	
1 × (2 + 1 + 4 + 1) =	8	1 × (2 + 1) =	3
2 × (8 + 2 + 1) =	22	2 × (1 + 1) =	4
3 × (2 + 6) =	24	3 × (2 + 1) =	9
4 × (1 + 1 + 1) =	12	6 × (2) =	12
5 × (1 + 1) =	10	Sub-Total	28
6 × (4 + 1 + 5) =	60		
7 × (3) =	21		
8 × (2 + 1) =	24		
Sub-Total	181		
		Grand Total =	209

PLANT *Acme Manufacturing Co.* TRIAL NO. *2* DATE *June 7*

FROM-TO CHART FROM (Contributor) \ TO (User)	1 *Rough Stores*	2 *Lathe*	3 *Drill*	4 *Mill*	5 *Press*	6 *Final Inspection*	7 *Saw*	8 *Bore*	9 *Hone*	10 *Grind*	TOTALS
1 *Rough Stores*		8		2	4		2			1	17
2 *Lathe*			4	2	1	3	1				11
3 *Drill*				1	2	5		1	1		10
4 *Mill*		1	2		1	1					5
5 *Press*			2			6					8
6 *Final Inspection*											
7 *Saw*		2								1	3
8 *Bore*			1								1
9 *Hone*						1					1
10 *Grind*			1			1					2
TOTALS		11	10	5	8	17	3	1	1	2	58 / 58

Figure 6–17—From-To chart, Trial No. 2.

Since Figure 6-14 shows several entries below the diagonal line (representing backtracking) and some larger entries several squares above the line, a rearrangement might be attempted in order to find a better arrangement, or sequence of areas. By inspection, it appears that *Press* and *Final Inspection* might be moved up, to get the larger values closer to the diagonal line. Recalculating the torque results in the improvement shown in Trial No. 2, Figure 6-17.

Forward		*Reverse*	
$1 \times (8 + 4 + 1 + 1 + 6)$	= 20	$1 \times (2)$	= 2
$2 \times (2 + 2 + 1)$	= 10	$2 \times (1 + 2)$	= 6
$3 \times (2 + 1 + 5 + 1)$	= 27	$3 \times (1)$	= 3
$4 \times (4 + 3)$	= 28	$4 \times (1)$	= 4
$5 \times (1 + 1)$	= 10	$5 \times (2 + 1)$	= 15
$6 \times (2 + 1)$	= 18	Sub-Total	30
$9 \times (1)$	= 9		
Sub-Total	122		
		Grand Total	152

This new arrangement shows considerable improvement, since the objective is the lowest practicable total value. These rearrangements might be tried over and over again, until an optimum solution is reached.

Although rather elementary, the torque method indicates that the second trial is more efficient, so far as the number of moves is concerned. It also indicates that there is a lesser degree of backtracking. As indicated previously, numbers in the cells can represent other values than just the number of moves. This would make the *From–To Chart* considerably more useful. For further details, the analyst should study recent research where the *From–To Chart* is used, as the basis for data arrangement in computerized plant layout algorithms (discussed in Chapter 13, Computerized Facilities Layout).

Procedure Chart

One of the more important aspects of facilities design is the communication of information, since it is frequently communication, or lack of it, that determines the efficiency of the activity. The *Procedure Chart* (Figure 6-18) is a technique used primarily to show the movement or flow of written or oral communications between activities, departments, and persons, and to show material flow as it is tied in with these communications. The *Procedure Chart* makes use of certain conventions and symbols in recording the flow, as in Figure 6-19.

Construction of the Procedure Chart

1. Prepare a form similar to the one in Figure 6-18.
2. The horizontal rows represent steps in the procedure. Ordinarily there should be only one dominant symbol in a row, except where action is being taken simultaneously at several places, or where the symbol, etc., is self-explanatory.
3. The vertical columns represent places of performance, activities, locations, persons, departments, etc. Symbols in a column indicate action taken, etc., by a person, department, etc.
4. The direction of flow is always in at the top of a symbol and out at the bottom, and generally from the upper left to the lower right of the *Chart.*
5. The number of lines leaving a symbol at the bottom must always equal the number of lines entering the symbol at the top, except for lines representing verbal communication.
6. Each line is identified by a code, which is explained in the legend at the bottom of the chart.
7. The routine *Filing* and *Form Destroyed* symbols are smaller than the dominant *Action Taken* symbol. As a general rule it is not necessary to amplify the smaller symbols with notes on the *Chart.* Codes alongside each will serve to identify forms filed or destroyed.
8. An alternate path may be shown by the same type of line as used for the regular route, with the alternate path identified as such.
9. Comments or notes are made on the left of the *Chart* when necessary to make the meaning of a step clear. These notes are as brief as possible and are located on the same horizontal row as the step being described. They are located at the left of the row in the space provided.

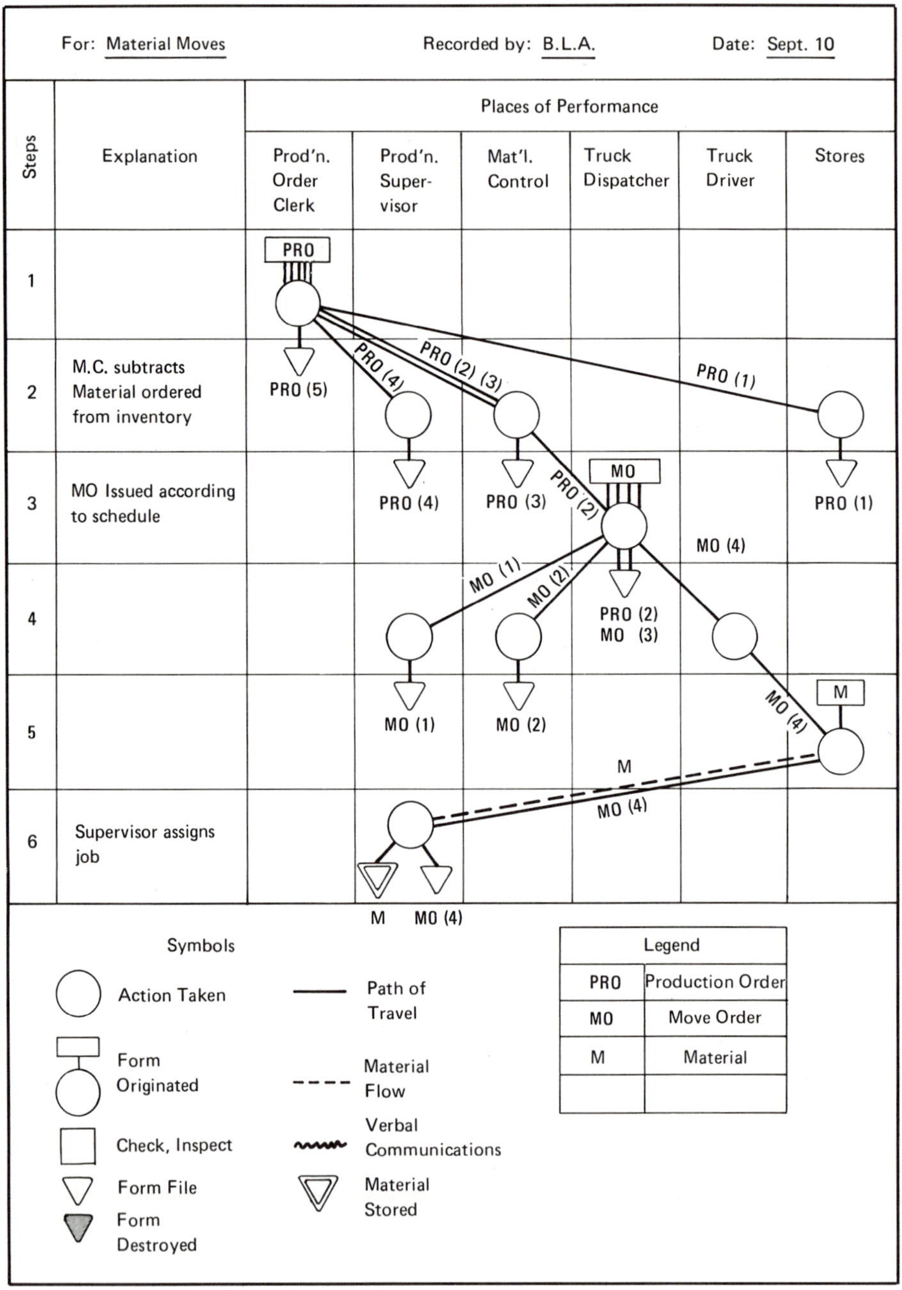

For: Material Moves Recorded by: B.L.A. Date: Sept. 10

Steps	Explanation	Places of Performance					
		Prod'n. Order Clerk	Prod'n. Supervisor	Mat'l. Control	Truck Dispatcher	Truck Driver	Stores
1		PRO					
2	M.C. subtracts Material ordered from inventory	PRO (5)	PRO (4)	PRO (2) (3)		PRO (1)	
3	MO Issued according to schedule		PRO (4)	PRO (3)	MO PRO (2)	MO (4)	PRO (1)
4			MO (1)	MO (2)	PRO (2) MO (3)		
5			MO (1)	MO (2)		MO (4)	M
6	Supervisor assigns job		M MO (4)		M MO (4)		

Symbols

- Action Taken
- Form Originated
- Check, Inspect
- Form File
- Form Destroyed
- Path of Travel
- Material Flow
- Verbal Communications
- Material Stored

Legend	
PRO	Production Order
MO	Move Order
M	Material

Figure 6–18—Procedure chart.

—— Solid line represents the movement of a written form of communication. A separate line is used for *each* piece of paper insofar as practical.

~~~~ Wiggly line is verbal communication.

- - - Dashed line represents the movement of a product, container, or equipment.

○ Circle (large) represents *action taken,* and should be larger than other symbols so as to be dominant.

▽ Triangle (large with small inside) represents *product stored.*

□ Square (large) represents *checking or inspecting.*

LS M R T A flag (with key initials in it) attached to a large "action" symbol represents a *form originated* at that time.

▽ Triangle (small) represents *paperwork filed.*

▼ Triangle (small and filled in) represents *paperwork destroyed.*

Figure 6–19—Symbols for procedure chart.

Some suggestions for checking the completeness and accuracy of the Chart are as follows:

1. Each form must be *originated.*
2. Places of performance should not be repeated.
3. Each copy of a form should eventually be either filed or destroyed.
4. There should be a line for each copy of each form.
5. The number of lines entering a *Place of Action* or *Temporary File* must equal the number leaving.
6. No two major *Action Taken* symbols should occur at the same step unless they are self-explanatory.
7. The *Chart* should read from top to bottom—no upward flow—and generally from upper left to lower right.
8. The number of items *Permanent Filed* and *Destroyed* should equal the number *Originated.*

The *Chart* should be a complete enough document in itself that it can be read by anyone familiar with it, or with only a brief explanation of this means of presentation, without referring to anything else. However, if the *Chart* is used in a report to be read by laymen, it may be desirable to include a *Discussion of Procedure*—a verbal description of the steps, referring to the steps on the *Chart.*

It is frequently surprising to find that making a *Procedure Chart* will show that the material flow or handling problem is a communications or paperwork prob-
~~~~

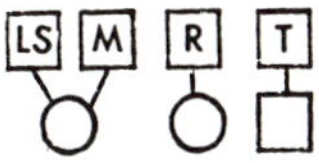

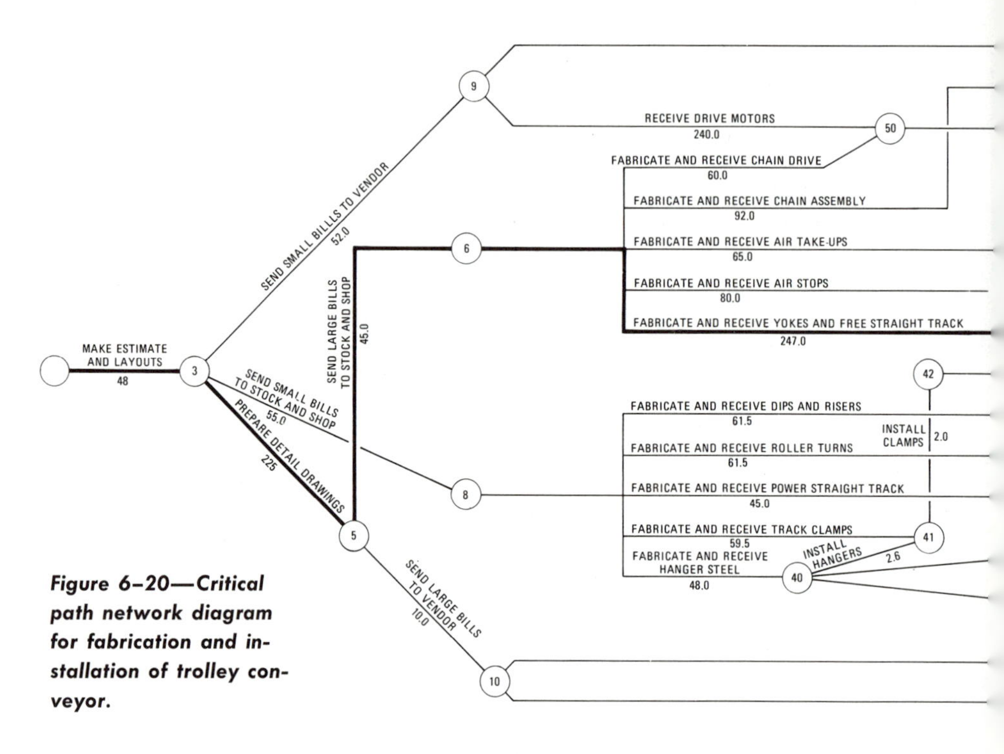

Figure 6-20—Critical path network diagram for fabrication and installation of trolley conveyor.

lem! While the *Procedure Chart* is not a commonly used technique, it is presented here in sufficient detail to permit its use—because it is felt that it is important in solving flow, handling, and related problems.

Critical Path Method (CPM)

One of the more recent analytical techniques that is basically a graphical tool, is the *Critical Path Method.* While it is not often used in analyzing material flow or handling problems, it is extremely useful in the planning of the installation of complex systems. *CPM* is actually a project management tool—an outgrowth of the *Gantt Chart.*

CPM uses a network diagram to graphically represent interrelationships between phases or elements of a project. Estimates of time to perform each event in the project are used to determine the *critical path*—the specific sequence of events governing the minimum total time required by the project. Figure 6-20 is an example of a *CPM* network.

A technique based in the *CPM* concept is *PERT* (*Program Evaluation and Review Technique*). Where *CPM* estimates are made at a single facility, dollar, or investment level, *PERT* estimates are over a range of levels, thereby resulting in a range of project times with a related range of project costs. The probabilities of

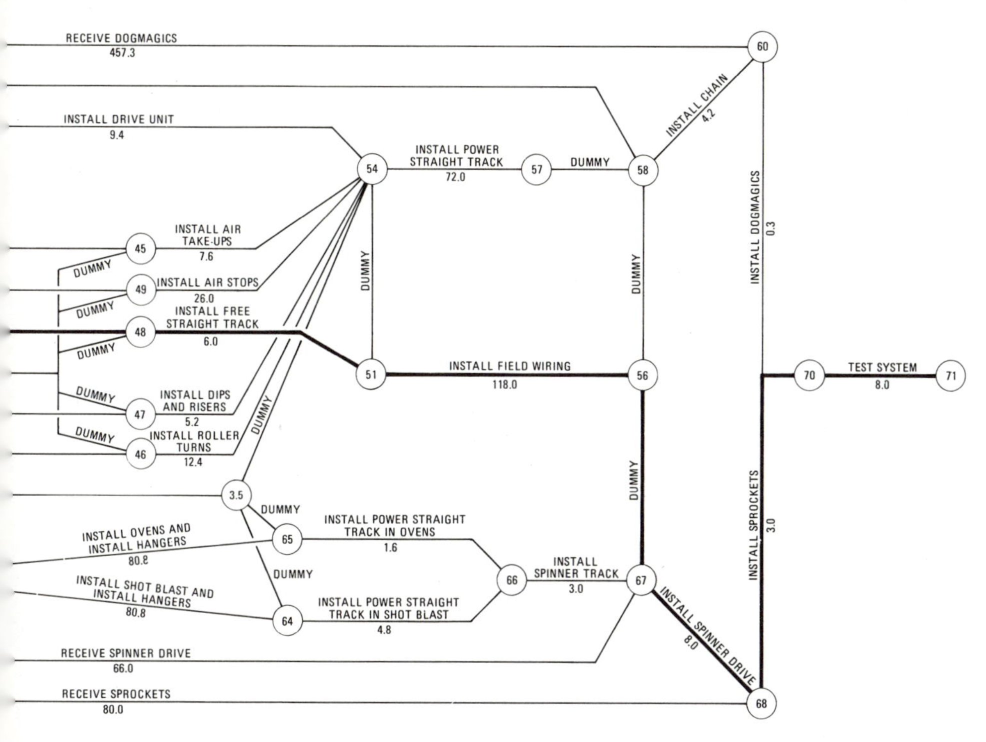

the various estimates, and the related uncertainties involved, must be taken into consideration. Statistical methods are used to develop optimum combinations, with calculations often programmed on a computer. (For additional details on *CPM* and *PERT* the reader should consult one of the several books covering the subject.)

Conclusion

This chapter has reviewed some of the more common graphical techniques useful in the analysis of element or material flow and handling problems. A single problem may make use of several of them, depending on the type of data on hand or the results desired from the analysis or planning efforts. Another area of analysis, making use of more sophisticated mathematics, is covered in Chapter 7, where several of the more commonly used techniques are presented briefly and may be categorized under the general heading of Operations Research.

Questions

1. Distinguish between *conventional* and *quantitative* techniques for analyzing flow.
2. What are some basic data sources for analyzing material flow?

Table 6-3. Operations on Powrarm

Operation No.	Operation Description	Machine Name	Pieces /hr	No. of Machines
		PART 1—BASE		
1	Face Bottom	14" LeBlond Eng. lathe	60	
2	Face top, turn OD, neck, drill and ream	Warner and Swasey #1A turret lathe[a]	23.8	
3	Drill three bolt holes	Cleereman drill press	84	
4	Drill pin hole	Delta drill press	238	
5	Drill and ream eccentric hole	2 Spindle Fosdick drill press	65.4	
6	Inspect	Bench	55.4	
7	Degrease	Detrex washer	143	
		PART II—ECCENTRIC ROD		
1	Cut off and chamfer	Oster #601	588	
2	Drill 5/16 hole	Delta drill press	250	
3	Tap 3/8-16 hole	Delta drill press	120	
4	Turn eccentric	LeBlond Eng. lathe	149	
5	Inspect	Bench	143	
6	Heat treat	Furnace (subcontract)	910	
7	Grind O.D.	Cinn. Centerless grinder (subcontract)	455	
8	Degrease	Detrex washer	455	
		PART III—HANDLE		
1	Thread, cut off, and chamfer	Warner and Swasey #1A	256	
2	Thread, & chamfer 2nd end	Warner and Swasey #1A	232	
3	Inspect	Bench	500	
4	Degrease	Detrex washer	600	
		PART VII—COVER		
1	Face, bore, turn, and cutoff	Warner and Swasey #1A[a]	60	
2	Drill 4-9/32 holes	Cleereman drill press	178	
3	Saw in two	Brown and Sharpe mill	125	
4	Inspect	Bench	250	
5	Degrease	Detrex washer	300	
		PART VIII—CAP		
1	Face, bore, and cut seat	Warner and Swasey #1A[a]	30.3	
2	Drill and tap	2 Spindle Fosdick DP	62.5	

(continued)

Table 6-3 (continued)

Operation No.	Operation Description	Machine Name	Pieces /hr	No. of Machines
3	Mill slot	Milwaukee vertical mill	83.5	
4	Inspect	Bench	100	
5	Degrease	Detrex washer	143	
		PART IX—PIN		
1	Cut off and chamfer	Oster #601	1000	
		PART X—PRESSURE PAD		
1	Bore, face, and chamfer	Warner and Swasey #1A[a]	40	
2	Mill slot	Brown and Sharpe mill	120	
3	Inspect	Bench	100	
4	Degrease	Detrex washer	600	
		PART XI—BALL SWIVEL		
1	Turn shank, form ball, and cut off	Warner and Swasey #1A	30.3	
2	Grind ball	Landis grinder type C	25	
3	Mill shank	Milw. Simplex mill	90	
4	Drill and tap two holes	2 Spindle Fosdick DP	62.5	
5	Inspect	Bench	83.3	
6	Degrease	Detrex washer	600	
		ASSEMBLY		
SSA-1	Knob to Handle		333	
SA-1	Handle Assembly to Eccentric rod		286	
A-1	Rod assembly to Base		200	
A-2	Plunger, Pin, and Pressure pad to Base		357	
SSA-3	Lock washers to Hex.-hd. screw		500	
SA-3	Hexhd screw assemblies to Ball swivel		350	
A-3	Ball swivel assembly to Base		90	
A-4	Cover and Cap to Base		100	
A-5	Inspect		178	
A-6	Degrease		143	
A-7	Mask and paint		80	
A-8	Pack		30	

[a]No stock rack necessary

3. Distinguish between: (a) *Assembly Chart,* (b) *Operation Process Chart,* (c) *Multi-Product Process Chart,* (d) *Process Chart,* and (e) *Flow Process Chart;* when is each particularly useful?
4. Describe the concepts of: (a) *From–To Chart,* (b) *Procedure Chart,* (c) *Critical Path Network;* what are the primary uses of each?
5. Draw, name, and describe the *Process Symbols.*
6. When and why would you use a *String Diagram?* How is it constructed?
7. What is the relationship of a *Flow Diagram* to a *Process Chart?*
8. How would you construct an *Assembly Chart* or an *Operation Process Chart* for a product composed of several hundred parts?
9. What is indicated on the *From–To Chart* when entries appear below the diagonal line? What is the procedure for correcting the situation?
10. What is meant by taking the torque of a *From–To Chart?* What does it accomplish?

Exercises

For Powrarm data applicable to some of the following exercises, see Table 6-3.

A. Make an *Assembly Chart* for a product familiar to you, with at least 10 to 12 parts.

B. Make an *Operation Process Chart* for a product for which you can obtain *Production Routings* for the parts.

C. Make a *Process Chart* of:
   1. A Coke bottle—from store shelf to the empty bottle, in your car, for return to store.
   2. You—getting up, dressed, etc. in the morning.
   3. A part—in a local plant.
   4. A gas station attendant—filling tank, checking under hood, etc., through payment by you.
   5. A hot-dog—at McDonald's, from delivery truck to serving counter.
   6. A shirt—in a retail store, from warehouse to your hand.
   7. Pick your own—must have at least 6 to 8 operations.
   8. Invent one for Table 4-4 or 4-5.

D. Make a *Flow Diagram* for the *Process Chart* you made for *Exercise C.*

E. Make a *Flow Process Chart* for an 8-to-10-part product in a local plant, or other data available to you.

F. Make a *From–To Chart* from the data you used in *Exercise E.*

G. Make a *Procedure Chart* for some procedure you are familiar with:
   1. Choose your own.
   2. Registration procedure for your classes.
   3. Procedure for dropping a course.
   4. Procedure for serving communion in your church.
   5. The receiving (paperwork) procedure in a local plant.
   6. Checking into (or out of) a hospital.

7

Quantitative Techniques for Analyzing Material Flow

In the previous chapter the more conventional techniques used in analyzing and designing material flow were presented. This chapter will review and briefly analyze some of the more quantitative approaches applicable to material flow analysis and planning. Each technique is evaluated as to its applicability, and each is presented with the hope that the reader can understand what it is and how it works. He should also be able to visualize its application to some phase of material flow or facility design. Finally, he should be able to compare it with the other approaches and evaluate it for possible use.

It is not the purpose of this chapter to present enough information for application, but to present the techniques and provide a background for the selection of the approaches applicable to analytical work in this field. The Bibliography indicates sources of additional information on both theory and application.

The approaches in this chapter apply primarily to activity location and material handling design. Some also apply to flow and system evaluation. The selection of approaches for presentation here is not to suggest a lack of relevance for other techniques and approaches, such as line balancing, machine loading, scheduling, sequencing, and plant location. Rather, it reflects a necessary limitation of the scope of this chapter.

Approaches Selected for Consideration

The approaches considered here are concerned primarily with the optimum location of equipment or with the movement of material. When discussing a particular method, the measure of effectiveness used is travel distance, unless otherwise indicated. Not all guarantee an optimal solution. Some of the procedures are based on classical mathematics, others have been developed from the

This chapter is adapted from Apple, *Material Handling System Design*, and draws on two theses submitted to the Georgia Institute of Technology: (1) J. R. Buchan, "An Evaluation of Selected Quantitative Methods Useful in Plant Layout," 1966; and (2) D. L. Totten, "Applications of Quantitative Techniques in Facilities Planning," 1967. The former was an attempt to identify, explain and evaluate selected quantitative methods felt to be useful in facilities planning; and the second surveyed the profession to determine the extent to which they had been used, and documented each by means of carefully selected and evaluated examples or case histories.

field of operations research. It is hoped that the analyst will be able to determine which can be presently applied to his layout problem, and which will require further study or development.

The approaches are roughly divided into four types: (a) linear–deterministic, (b) non-linear deterministic, (c) linear–probabilistic, (d) non-linear–probabilistic. The approaches to be discussed are:

1. Linear programming technique
2. Assignment problems
3. Transportation programming problems
4. Transshipment programming problems
5. The traveling salesman problem
6. Integer programming technique
7. Dynamic programming technique
8. Level curve technique
9. Queuing theory technique
10. Conveyor analysis
11. Simulation

This chapter draws upon a wide range of sources. Those directly quoted are referenced and listed in the Bibliography at the end of the book.

Linear Programming

Linear programming has become one of the most frequently applied of the quantitative techniques used in operations research. It is a mathematical method for determining the best allocation of limited resources or capacities to accomplish a desired objective—for example, minimizing cost or maximizing profit. It requires that the variables of a problem be fairly constant, and their values known. The relationship among the variables must represent a linear or straight-line relationship to one another, and is thus applicable mainly in static conditions.

Linear programming can be applied to optimizing non-automatic material handling, that is, almost any material handling activity other than that accomplished by conveyors. Several important prerequisites must be met before an attempt can be made to improve the handling activity with linear programming. They are:[1]

1. *An adequate inventory and material control system.* This implies a well-ordered storage and warehousing situation, with a designated place or area for everything.
2. *A measure of material handling between the various locations.* It is necessary to have a measure of handling, including loading, unloading, moving loaded, and moving empty, between the various departments, storage areas, warehouses, or internal shipping and receiving points in the facility, in order to develop the material handling schedules. Distance, cost, or time may be used as a measure, according to Metzger.
3. *An indication of average number of material movements per day.* The mathematical solution is based upon average material movements. The average moves per day can be simply a listing of origins, destinations, and number of moves.
4. *A daily indication of material requirements for the following day.* This is

[1]R. W. Metzger, *Elementary Mathematical Programming* (New York: John Wiley & Sons, Inc., 1958), pp. 212 *et seq.*

necessary preplanning information to allow optimum schedules to be prepared. In most operations the work for the following day is known, so most of the material requirements can be determined.

5. *The relationships involved in moving material must be linear.*

If these prerequisites can be met, the analysis can proceed, in the following five phases:

1. *Identify the various loading and unloading stations,* and determine the average number of material moves required per day.
2. *Prepare handling time data* between every possible origin and destination, and time requirements for loading and unloading at the various locations.
3. *Determine the minimum dead-heading* requirements via linear programming.
4. *Develop round trips and subsequent schedules* for the handling equipment.
5. *Prepare implementation and scheduling* to realize the continued benefits in optimum utilization of both handling equipment and personnel.

Linear programming is most commonly solved by the simplex method. The transportation, transshipment, and assignment methods are all special cases of the general linear programming procedure, and can also be solved by this procedure.

The general linear programming problem is to find an optimum solution to the linear function

$$Z = e_1x_1 + e_2x_2 + \cdots e_jx_j \cdots e_nx_n$$

subject to the linear constraints

$$x_j \geq 0 \qquad j = 1, 2, \ldots, n$$

which also satisfy m linear inequalities

$$a_{11}x_1 + a_{12}x_2 + \cdots + a_{ij}x_j + \cdots + a_{1n}x_n \leq b_1$$

$$a_{21}x_1 + a_{22}x_2 + \cdots + a_{2j}x_j + \cdots + a_{2n}x_n \leq b_2$$

$$\vdots$$

$$a_{i1}x_1 + a_{i2}x_2 + \cdots + a_{ij}x_j + \cdots + a_{in}x_n \leq b_i$$

$$\vdots$$

$$a_{mi}x_1 + a_{mx}x_2 + \cdots + a_{mj}x_j + \cdots + a_{mn}x_n \leq b_m$$

where a_{ij}, b_i, and e_j are given coefficients. If m is greater than n, the problem has an infinite number of solutions. Another way of expressing the general linear programming problem is to optimize the effectiveness function

$$E = \sum_{j=1}^{n} e_jx_j$$

subject to the constraints

$$\sum_{j=1}^{n} a_{ij}x_j \; \{\leq, =, \geq\} \; b_i \qquad (i = 1, 2, \ldots, m)$$

and

$$x_j \geq 0 \qquad (j = 1, 2, \ldots, n)$$

The simplex method is most commonly used to solve the linear programming problem and may also be used to solve the assignment, transportation programming, and transshipment problems. Although the mathematical computations in the simplex method are relatively simple, and essentially a mechanical repetition of a sequence of precisely defined steps, it may be long and tedious. The real task is to recognize that a problem can be solved by linear programming and to construct a model that will lead to a useful solution.

Evaluation. Linear programming has found wide use in distribution, location, allocation, replacement, product mix, and planning models. As in virtually every other mathematical model used in facilities design, linear programming lacks a valid unit of measurement. In most problems, time or cost is the value optimized but these values are usually arbitarily chosen—so it may be questionable whether time or cost was the best measurement to use.

The mathematical computations in linear programming are relatively simple and can easily be accomplished manually for small or medium problems. Large problems frequently require the use of a computer to handle the volume of computations.

Assignment Problem

The assignment problem is actually a special case of the general linear programming procedure, in which it is usually desired to assign n jobs to n facilities. The effectiveness of each facility for each job is given, and the objective is to optimize the measure of effectiveness in assigning each facility to one job only. Since one item can be assigned to one and only one job, the assignment problem can be expressed mathematically as that of optimizing the effectiveness function

$$E = \sum_{i=j}^{n} \sum_{j=1}^{n} e_{ij} x_{ij}$$

subject to the constraints

$$\sum_{i=1}^{n} x_{ij} = 1 \qquad (j = 1, 2, \ldots, n)$$

$$\sum_{j=i}^{n} x_{ij} = 1 \qquad (i = 1, 2, \ldots, n)$$

As in the linear programming technique, optimization requires the minimization or maximization of the measure of effectiveness chosen. The user has control of the assignments x_{ij}, but the effectiveness coefficients e_{ij}, are not directly under his control.

Although the assignment problem may also be solved by the simplex method, probably the most efficient routine is the one proposed by Kuhn, known as the *Hungarian method.* It requires the problem to be set up in the form of a square effectiveness matrix. However, problems of a matrix form that are not square, or have prohibited assignments, are often encountered and may be converted into a square matrix.

Example. The designer wants to add three new pieces of equipment, say A, B, and C, to an existing layout, and the load size is such that a lift truck can carry only one at a time. Truck movement is restricted by aisles to rectangular directions. Figure 7-1 shows the candidate locations for the possible new machines as X, Y, and Z. It also shows the existing machines and their locations as *1, 2, 3, 4,* and *5.*

Table 7-1 gives the expected traffic between the new and existing machines. Notice that some of the new machines do not have a handling relationship with all existing machines. Taking rectangular distances from Figure 7-1 yields the distance data of Table 7-2. Matrix multiplication of the traffic data and distance data develops the effectiveness matrix:

$$\begin{bmatrix} 25 & 8 & 4 & 0 & 30 \\ 0 & 7 & 10 & 12 & 8 \\ 8 & 5 & 60 & 0 & 16 \end{bmatrix} \begin{bmatrix} 2 & 1 & 6 \\ 2 & 3 & 4 \\ 4 & 3 & 2 \\ 6 & 7 & 2 \\ 9 & 6 & 3 \end{bmatrix} = \begin{bmatrix} 352 & 241 & 280 \\ 198 & 183 & 96 \\ 410 & 299 & 236 \end{bmatrix}$$

Traffic Matrix Distance Matrix Effectiveness Matrix

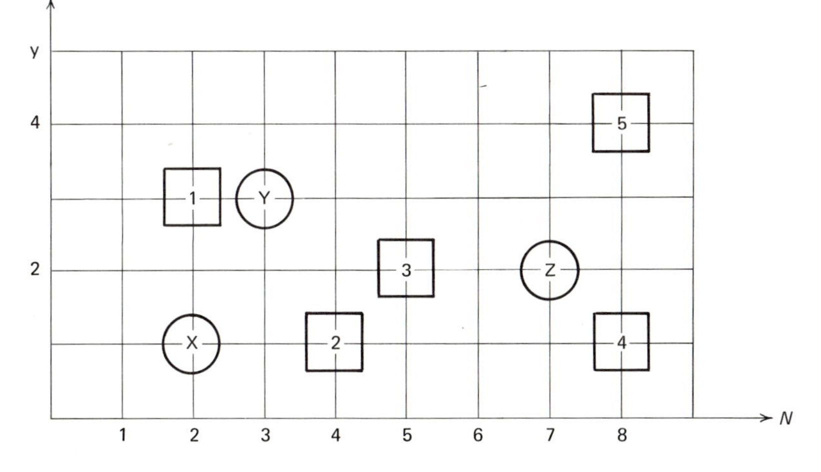

Figure 7–1—Squares show the locations of existing machines, the circles indicate the candidate areas.

Table 7-1. Traffic data (matrix)

New Machines	Existing Machines 1	2	3	4	5
A	25	8	4	0	30
B	0	7	10	12	8
C	8	5	60	0	16

Table 7-2. Distance Data matrix; developed from Figure 7-1

Existing Machines	Candidate Areas X	Y	Z
1	2	1	6
2	2	3	4
3	4	3	2
4	6	7	2
5	9	6	3

In terms of the layout problem, this matrix multiplication operation says that if *new machine* A is placed in *candidate area* X, the measure of effectiveness will be:

$$25\,(2) + 8\,(2) + 4\,(4) + 0\,(6) + 30\,(9)$$
$$= 352 \text{ handling unit-feet/time unit}$$

Figure 7-2 puts the effectiveness matrix into the context of the problem and shows the results of the application of the Hungarian method. A feasible and optimal layout solution is to:

1. Assign *A* to *candidate Y.*
2. Assign *B* to *candidate X.*
3. Assign *C* to *candidate Z.*

This procedure can easily be expanded to include any finite number of new machines to be located in the same number of candidate areas. The method is well suited to computer calculation.

Evaluation. The assignment problem has demonstrated value in certain types of handling and layout problems. The required mathematical data are relatively easy to apply. Situations involving unequal numbers of activity centers and candidate areas, as well as prohibited assignments, can be easily taken into consideration; but this technique does have limitations. It is not possible to consider cases where

New Machines	Candidate Areas *X*	*Y*	*Z*	*X*	*Y*	*Z*
A	352	241	280	9	[0]	39
B	198	183	96	[0]	87	~~0~~
C	410	299	236	72	63	[0]

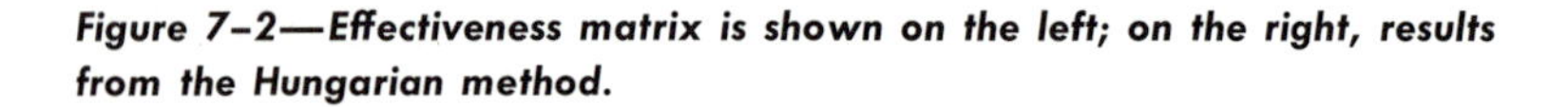

Figure 7-2—Effectiveness matrix is shown on the left; on the right, results from the Hungarian method.

activities are not independent. If two activities are to have direct material handling or flow contact, there is no way to incorporate this requirement into the model except by increasing generality.

Another shortcoming is the inability to consider the efficient use of floorspace or size constraints. A small facility may be assigned to a large area, resulting in a large amount of wasted floorspace, which can be partially controlled in the cost matrix.

Finally, the assignment problem is based on the assumption of deterministic traffic data. If the data are actually nearly deterministic, the procedure is satisfactory. However, if the traffic data are subject to relatively large random variation, the solution may not always be optimal, since the solution is sensitive to variation in traffic.

Transportation Problem

The transportation problem deals with the distribution of a single commodity from various sources of supply to various points of demand in such a manner that total transportation cost is minimized. The cost may be expressed in terms of distance, time, or dollars.

The technique for solving this type of problem allows movement only from sources to destinations. The problem is structured as follows, with the notation explained as the structure is developed. There are m origins or sources, with each source i possessing a_i items $(i = 1, 2, \ldots, m)$. There are n destinations, with destination j requiring b_j items $(j = 1, 2, \ldots, n)$. There are $(m)(n)$ costs, one associated with moving one item from each source to each destination. The objective of the transportation problem is to empty the sources and fill the destinations so that the total cost is minimized. The following conditions are required:

1. The total capacity of all sources must equal the total requirement of all destinations, or

$$\sum_{i=1}^{m} a_i = \sum_{j=1}^{n} b_j$$

2. The total number of items shipped to all destinations from any source must be equal to the capacity of that source, or

$$\sum_{j=1}^{n} X_{ij} = a_i \qquad (i = 1, 2, \ldots, m)$$

3. The demand of every destination must be fully satisfied by the total of the items shipped from all sources, or

$$\sum_{i=1}^{m} X_{ij} = b_j \qquad (j = 1, 2, \ldots, n)$$

4. The number of items shipped from any source must be non-negative, or

$$X_{ij} \geq 0 \quad \text{for all } i \text{ and } j$$

The cost equation is of the form

$$Z = \sum_{i=1}^{m} \sum_{j=1}^{n} C_{ij} X_{ij} = C_{11} X_{11} + C_{12} X_{12} + \cdots + C_{mn} X_{mn}$$

where C_{ij} is the cost of shipping one item from source i to destination j, and X_{ij} is the number of items shipped from i to j.

Example. Assume that an in-plant material handling system consists of identical forklift trucks that move identical containers from three storage areas to four work areas. The containers are handled one at a time by a truck and are not re-usable. The cost of handling is expressed in time units. A tabular statement of the problem is:

Source	*Containers Available*	*Destination*	*Containers Required*
1	10	1	9
2	5	2	4
3	8	3	3
	23 Containers	4	7
			23 Containers

Handling costs are as follows (in seconds)

$C_{11} = 62$	$C_{14} = 51$	$C_{23} = 66$	$C_{32} = 55$
$C_{12} = 18$	$C_{21} = 15$	$C_{24} = 54$	$C_{33} = 16$
$C_{13} = 39$	$C_{22} = 41$	$C_{31} = 37$	$C_{34} = 26$

The problem is put into matrix form for solution as shown in Figure 7-3. The numbers in the small cells are the handling costs, C_{ij}. All the stated conditions are satisfied and the problem is ready for solution. The solution procedure used below (Figure 7-4) is an adaptation of the *Vogel Approximation Method* as described by Sasieni, et al.[2]

Metzger approaches the in-plant material handling problem by making the same assumptions regarding the types of truck and container as in the previous illustration, except that the containers are not expendable and are subsequently removed from the delivery point for re-use, which becomes a second transportation problem. In this problem some departments always send out filled containers, whereas others always receive filled containers. The material handling activity must not only distribute the filled containers, but also take care of the empty containers. The problem is then stated as one of determining the most economical distribution of empty containers.

Evaluation. The transportation programming model was developed for calculating minimum cost schedules for moving goods from warehouses to destinations.

[2]M. Sasieni, A. Yaspan, and L. Friedman, *Operations Research—Methods and Problems* (New York: John Wiley & Sons, Inc., 1959), pp. 198 *et seq.*

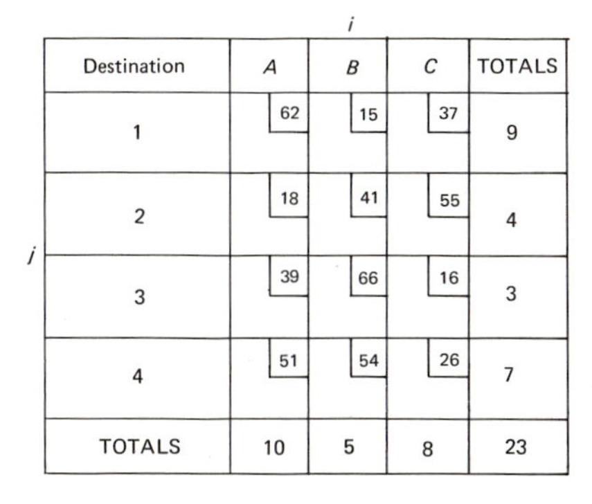

Destination (*j*) / *i*	A	B	C	TOTALS
1	62	15	37	9
2	18	41	55	4
3	39	66	16	3
4	51	54	26	7
TOTALS	10	5	8	23

Figure 7–3—Transportation problem in matrix form with costs included.

Destination (*j*) / *i*	A	B	C	TOTALS
1	62 3	15 5	37 1	9
2	18 4	41	55	4
3	39 3	66	16	3
4	51	54	26 7	7
TOTALS	10	5	8	23

Figure 7–4—Solution to the transportation problem.

Many enterprises have a similar problem on a smaller scale, such as distributing parts and supplies from storage areas to assembly areas. Manpower or personnel assignments, machine loading, and other problems have algebraic equations that are identical in form to those of the transportation problem. The situation as a distribution problem permits shipment only from source to destination, which may not always be realistic. The number of applications of transportation analytics to facility design problems has been limited, but the opportunities are there.

Metzger's approach is currently in wide use, as attested to by the many cases that have been published. The example used here illustrates a very important aspect of the transportation problem, namely, that mathematical programming is usually applicable to only a relatively small part of the total problem. One of the really serious shortcomings associated with the transportation problem, which is shared by virtually every other mathematical model in material handling, is the lack of a really valid unit of value measurement. *Time* was the value measure optimized in this example, but it is questionable if time is the best measure to use. The same question applies to the other readily available units of measure.

The demonstrated usefulness of the transportation problem solution justifies further research into its application to facilities design. The solution procedure is simple and can easily be done manually for most small-to-medium size problems. Large problems can be easily solved on a computer.

Transshipment Problem

The typical or standard transportation problem permits shipment only from source to destination. Transshipment problems include the possibility that any source or destination may be taken as an intermediate point in seeking an optimal solution. Shipments are not restricted to direct connections between origins and destinations but may follow any sequence. This is far more realistic in a distribution problem, since it is common for destinations to transship material to other destinations in order to meet abnormal demands in the other destinations. In such a case, a location may serve as both source and destination, and the distinction may not be clear.

An ordinary m-source and n-destination transportation problem, when converted into a transshipment problem—by removing the restrictions on the sources and destinations to permit them to receive as well as send—becomes one of $(m + n)$ sources and $(m + n)$ destinations.[3] Changing from an ordinary transportation problem to a transshipment problem makes a larger problem, but it can still be solved by the standard transportation method, with a few small changes.

To set up the transshipment problem, designate the sources and destinations as terminals (T). As before, the amount shipped from T_i to T_j is X_{ij}, with cost per unit c_{ij}. Obviously $X_{ii} = c_{ii} = 0$, since no terminal will ship to itself. Assume that at m sources, the total shipped out exceeds the total shipped in, and at the remaining n destinations, the total in-shipment exceeds the total out-shipment. Let the total inshipment at $(T_1, T_2, \ldots, T_m)$ be $(t_1, t_2, \ldots, t_m)$, respectively, and the total outshipment at $(T_{m+1}, T_{m+2}, \ldots, T_{m+n})$ be $t_{m+1}, t_{m+2}, \ldots, t_{m+n})$, respectively. The contraints for the transshipment problem are

$$\sum_{i=1}^{m} a_i = \sum_{j=m+1}^{m+n} b_j$$

where a_i is the excess of out-shipment over in-shipment at terminal i, and b_j is the excess of in-shipments over out-shipments at terminal j. The objective, or total cost, function is

$$Z = \sum_{i=1}^{m+n} \sum_{j=1}^{m+n} c_{ij} X_{ij}$$

The differences between the transshipment problem and an $(m + n)$ square transportation problem are:

1. In the transshipment constraints, there are no X_{ii} terms.
2. $b_j = 0$, for $j = 1, 2, \ldots, m$; and $a_i = 0$, for $i = m + 1, m + 2, \ldots, m + n$.

The t_i and t_j in these constraints may be considered as algebraic substitutes for X_{ii} and X_{jj}.[4] The transshipment problem has the form shown in Table 7-3. All t's are placed on the main diagonal with a negative sign so that the t's, as well as the X_{ij}'s, may be constrained to non-negative values. The procedure for solving this problem is identical with the one used for solving a transportation problem.

Example. The transportation problem previously used can be solved as a transshipment problem with the addition of the cost (time) data shown in Table 7-4. Figure 7-5 shows the tableau resulting from the addition of these data to the solution of the foregoing transportation problem. Computation reveals that the

[3] A. Chung, *Linear Programming*, (Columbus, Ohio: Charles E. Merrill Books, Inc., 1963), p. 273.
[4] Chung, p. 274.

Table 7-3. Format for the Transshipment Problem

Terminals		t_j				Capacity
		$1 \ldots m$		$m+1$	$m+n$	
t_i	1	$-t_i \ldots X_{im}$		$X_{i,m+1} \ldots X_{i,m+n}$		a_1
	.	. .		. .		.
	.	. .		. .		.
	.	. .		. .		.
	m	$X_{m,1} \ldots -t_m$		$X_{m,m+1} \ldots X_{m,m+n}$		a_n
	$m+1$	$X_{m+1,1}$	$X_{m+1,m}$	$-t_{m+1} \ldots X_{m+1,m+n}$		0
	.	.	.	.	.	.
	.	.	.	.	.	.
	.	.	.	.	.	.
	$m+n$	$X_{m+n,1}$	$X_{m+n,m}$	$X_{m+n,m+1}$	$-t_{m+n}$	0
Requirement		$0 \ldots 0$		$b_{m+1} \ldots b_{m+n}$		$\sum a_i = \sum b_j$

Table 7-4. Inter-Terminal Time Values for the Transshipment Problem

Destinations

From \ To	1	2	3	4
1	0	22	73	54
2	22	0	38	42
3	73	38	0	31
4	54	42	31	0

Sources

From \ To	*A*	*B*	*C*
A	0	33	59
B	33	0	66
C	59	66	0

TERMINALS	$T_1(A)$	$T_2(B)$	$T_3(C)$	$T_4(1)$	$T_5(2)$	$T_6(3)$	$T_7(4)$	CAPACITY
$T_1(A)$	0 / −0	33	59	62 / 3	18 / 4	39 / 3	51	10
$T_2(B)$	33	0 / −0	66	15 / 5	41	66	54	5
$T_3(C)$	59	66	0 / −0	37 / 1	55	16	26 / 7	8
$T_4(1)$	62	15	37	0 / −0	22	73	54	0
$T_5(2)$	18	41	55	22	0 / −0	38	42	0
$T_6(3)$	39	66	16	73	38	0 / −0	31	0
$T_7(4)$	51	54	26	54	42	31	0	
REQUIREMENT	0	0	0	9	4	3	7	23

Figure 7-5—Tableau of transshipment problem containing transportation problem solution.

old solution is not optimal for the transshipment problem. Figure 7-6 shows the optimal solution obtained by this technique.

A graphical representation of the optimum flow provided by the solutions to the transportation problem and the transshipment problem is shown in Figure 7-7. The total time (cost) required by the two solutions is not the same. The transportation problem required

$$
\begin{aligned}
Z &= \sum_{j=1}^{n} \sum_{i=1}^{m} X_{ij}\, C_{ij} \\
&= 3\,(62) + 5\,(15) + 1\,(37) + 4\,(18) + 3\,(39) + 7\,(26) \\
&= 669 \text{ time units}
\end{aligned}
$$

whereas the transshipment problem required

$$
\begin{aligned}
Z &= 8\,(18) + 2\,(39) + 5\,(15) + 1\,(16) + 7\,(26) + 4\,(22) \\
&= 583 \text{ time units}
\end{aligned}
$$

for a saving of 86 time units.

TERMINALS	$T_1(A)$	$T_2(B)$	$T_3(C)$	$T_4(1)$	$T_5(2)$	$T_6(3)$	$T_7(4)$	CAPACITY
$T_1(A)$	0 / −0	33	59	62	18 / 8	39 / 2	51	10
$T_2(B)$	33	0 / −0	66	15 / 5	41	66	54	5
$T_3(C)$	59	66	0 / −0	37	55	16 / 1	26 / 7	8
$T_4(1)$	62	15	37	0 / −0	22	73	54	0
$T_5(2)$	18	41	55	22 / 4	0 / −4	38	42	0
$T_6(3)$	39	66	16	73	38	0 / −0	31	0
$T_7(4)$	51	54	26	54	42	31	0 / −0	
REQUIREMENT	0	0	0	9	4	3	7	23

Figure 7-6—Tableau of optimum solution to transshipment problem.

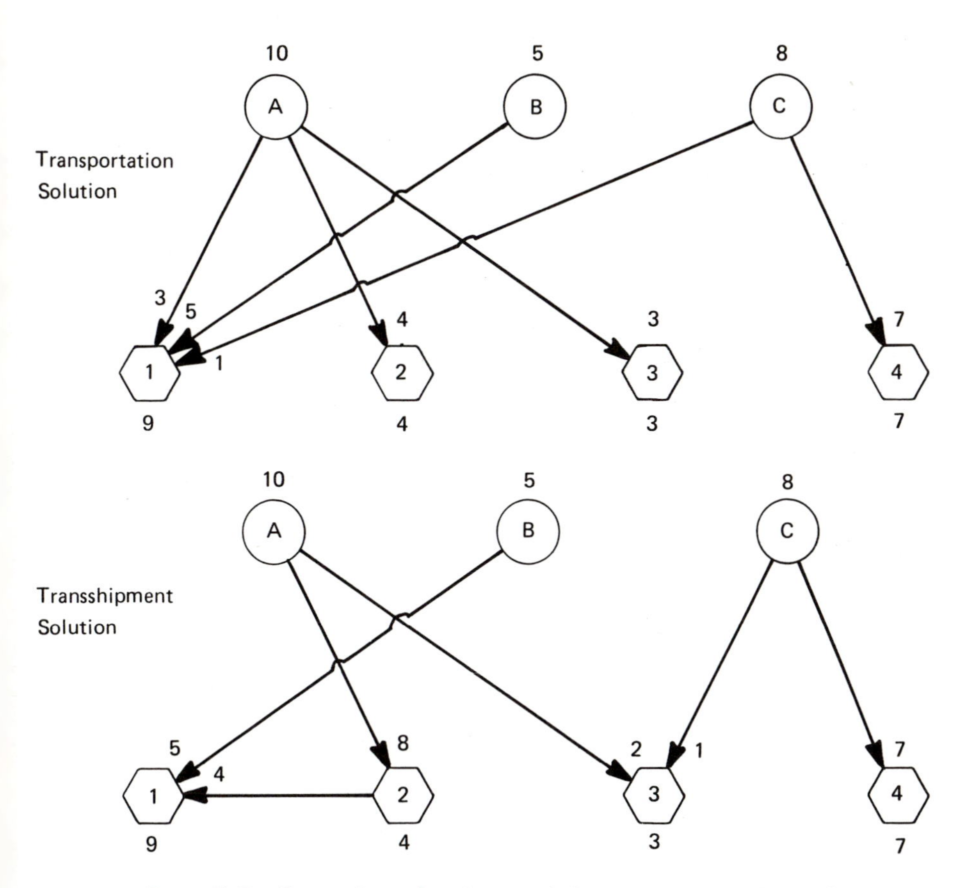

Figure 7-7—Comparison of optimum solutions to transportation and transshipment problems.

Evaluation. The greatest advantage the transshipment problem has over the transportation problem is its much more realistic statement of the distribution problem. It is very well suited for comparing alternative locations of activities, such as departments or machines, though a different tableau must be solved for each alternative. Similarly, it can be used to decide, with respect to some cost measure, which of a limited number of candidate areas to use as activity or machine locations.

The major disadvantage of the model is that it requires about twice as large a solution matrix as the transportation problem. This disadvantage is not serious, however, in view of the ease of finding a solution by computer.

In summary, the transshipment programming procedure appears to be an excellent tool for use in quantitative analysis of material flow and handling problems. Not much, however, has been published on real-world problems solved, or on applications of the technique.

The Traveling Salesman Problem

Solution of the traveling salesman problem involves finding an optimal route between a series of locations when each location is visited once, and then return is made to the point of origin, without backtracking. An optimal route is one for which either total distance traveled, total cost of travel, or total time of travel, is a minimum. The problem takes its name from the often-cited illustration of the traveling salesman who starts from home and visits $(n - 1)$ specified locations before returning home. It is a routing problem that involves sequencing the locations to minimize the travel between them.

The problem of locating facilities in some cases can be expressed as a variation of the traveling salesman problem. Other variations include the truck dispatching problem, and some machine sequencing problems.

Analysis. Although efficient methods and procedures are known for solution of problems of a similar nature, such as assignment and transportation problems, there is less known about efficient methods for solving the traveling salesman problem.

The problem may be stated as follows:

Determine $\quad X_{ij} \quad (i, j = 1, 2, \ldots, n)$

which minimize $\quad T = \sum_{ij} a_{ij} X_{ij}$

subject to $\quad X_{ij} = (S_{ij})^2 \quad$ (i.e., X_{ij} is either 0 or 1)

$$\sum X_{ij} = \sum_{j} X_{ij} = 1$$

$$X_{i_1 i_2} + X_{i_2 i_3} + \cdots + X_{l_{r=1} i_r} + X_{i_r i_1} \leq r - 1 \qquad X_{ii} = 0$$

where a_{ij} are specified real numbers representing cost or distance, X_{ij} is 0 or 1, and $(i_1, i_2, \ldots, i_r)$ is a permutation of the integers 1 through r. The problem is one of finding a permutation $P = (1, i^2, \ldots, i_n)$ of the integers from 1 through n with respect to which the value of the quantity

$$a_{1i_2} + a_{i_2 i_3} + \cdots + a_{i_n 1}$$

is minimized.

In general there are $(n - 1)!$ possible routes that can be taken when visiting $(n - 1)$ locations once each, and then returning to the point of origin. The number of possibilities for a 15-location tour would be 653,837,184,000 different routes. The real problem is to find an efficient method of selecting the optimum sequence, since even a modern high-speed computer would require an unreasonable amount of time to evaluate each permutation in a problem with as few as ten locations.

There are few mathematical results relative to this problem, and as yet no general method for obtaining an optimal solution has been found. There is, in fact, only one general theorem: *In the Euclidean plane, the minimal path does not intersect itself.*[5]

Even though no general method of solution exists, there are several procedures that will lead to near-optimal, and sometimes optimal solutions. One procedure, developed by Dantzig, et al.,[6] outlines a linear programming approach that sometimes enables one to find an optimum route and prove it so. Their approach is to start with a small number of linear constraints that are satisfied by all paths. The problem is then solved by the standard simplex technique. If the solution is not a *tour,* that is, does not visit every location once, without overlap or back-tracking, new constraints are added to eliminate the solution and obtain a new one. This continues until a near-optimal solution is reached, and then each remaining possibility is evaluated manually using a map.

Barachet[7] has developed an intuitive procedure that takes a feasible solution and obtains successively shorter routes by applying the following theorems:

1. *A route of minimum length does not cross itself.*
2. *If the protruding angles formed by three consecutive sections are obtuse angles, these three sections make up a route of minimum length.*
3. *If all the points make up a convex polygon, the route of minimum length corresponds to this convex polygon.*
4. *A circuit including $(N - k)$ consecutive sections that are common with a minimum circuit obtained for the case of K-consecutive sections is longer than the latter or equal to it.*

Barachet points out that the procedure is essentially intuitive and leaves doubt whether the minimum route will be obtained.

Croes[8] has an iterative procedure similar to the approaches used by Dantzig and by Barachet, though it is less intuitive than Barachet's method. It is cumbersome for computer computation.

Dacey[9] has developed a procedure for obtaining an initial near-optimal feasible solution that will occasionally also yield an optimal solution. If a solution is not optimal, better solutions can be obtained by one of the previous methods.

A rule of thumb for obtaining approximate solutions to the problem is to use

[5]R. L. Ackoff, *Progress in Operations Research,* vol. 1, (New York: John Wiley & Sons, Inc., 1961), p. 153.

[6]G. B., Dantzig, D. R. Fulkerson, and S. M. Johnson, "Solutions of a Large-Scale Traveling-Salesman Problem," *Journal of Operations Research Society of America* 2(1954):393–410.

[7]L. L. Barachet, "Graphic Solution of the Traveling-Salesman Problem," *Operations Research* 5(1957):841–45.

[8]G. A. Croes, "A Method for Solving Traveling-Salesman Problems," *Operations Research* 6(1958):791–812.

[9]M. F. Dacey, "Selection of an Initial Solution for the Traveling-Salesman Problem," *Operations Research* 8(1960):133–34.

the *nearest location* approach. The criterion is simply to select the nearest unvisited location as the next location to visit. Good results are often obtained by this method.

Example. This example is chosen to show the similarity and differences between the traveling salesman problem and a transportation or assignment problem. In the transportation problem the permutations need not be cyclic, but in the traveling salesman problem they must be. A solution to the assignment problem will also be a solution to the traveling salesman problem if, and only if, the assignment matrix is symmetric and the elements on the main diagonal are all zero.

Given the matrix of transportation costs in Figure 7-8,[10] it is desired to determine the minimum trip cost from some starting location through each other location once and back to the origin. The infinite costs on the main diagonal are to insure that no assignment will be made there. Subtracting the minimum from each row and column yields the solution shown in Figure 7-9.

The matrix is examined for some next-best solutions to the assignment problem, to try to find one that satisfies the additional restrictions. The smallest non-zero element is 1, so each element of value 1 has its row and column deleted and the resulting 4-by-4 matrix is searched for an optimal allocation solution and a feasible traveling salesman solution. The best solution is found to be the same as found for the Assignment Problem (Figure 7-9). The best sequence is

$$A_5 \rightarrow A_1 \rightarrow A_2 \rightarrow A_3 \rightarrow A_4 \rightarrow A_5$$

and the total cost is $1 + 2 + 3 + 4 + 5 = 15$.

Evaluation. The traveling salesman problem is useful in such situations as vehicle dispatching, determining the assignment of activities to candidate locations if the

[10] Sasieni, *et al.* pp. 266f.

From \ To	A_1	A_2	A_3	A_4	A_5
A_1	∞	2	5	7	1
A_2	6	∞	3	8	2
A_3	8	7	∞	4	7
A_4	12	4	6	∞	5
A_5	1	3	2	8	∞

Figure 7-8—Cost matrix or relationships between lacations.

From \ To	A_1	A_2	A_3	A_4	A_5
A_1	∞	1	3	6	0
A_2	4	∞	0	6	0
A_3	4	3	∞	0	3
A_4	8	0	1	∞	1
A_5	0	2	0	7	∞

Figure 7–9—Solution to assignment problem; and also for traveling salesman problem.

process is carried out in cyclic sequence, or for laying out a production line for minimized time or maximized output rate. Another valuable use is in determining the most economical sequence of n operations to be done in some sequence on a single facility. However, the lack of an uncomplicated general solution procedure that would guarantee an optimal solution has reduced the importance of the traveling salesman procedure in the area of facility design.

There are several procedures for solving this problem, but for practical size problems, it is not clear which, if any, achieves a good solution. One of the best is a case study by Hare and Hugli dealing with the production of kitchen units.[11]

Until a general solution may be obtained, the traveling salesman problem is likely to remain a secondary tool in facility planning, because of the extensive computation required, either manual or computerized. Often it is difficult, or impossible, to determine if a solution is optimal even after considerable effort. The technique does have a significant value in special cases, however, where the product passes through a series of operations without backtracking, and the problem is to determine the sequence that minimizes the distance, time, or cost of travel.

Integer Programming

The integer programming technique requires the variables to take on only integer values. The problem is called an *all-integer problem* if all the variables are restricted to integer values, and a *mixed-integer problem* when only certain specific variables must be integers. The integer problem may be linear, quadratic, or cubic in its constraint forms, but some or all variables are integers rather than continuous.

[11] Van Court Hare, Sr., and W. C. Hugli, "Applications of Operations Research to Production Scheduling and Inventory Control II," *Proceedings of the Conference on "What is Operations Research Accomplishing in Industry?"* (Cleveland: Case Institute of Technology, 1955), pp. 56–62.

Contrary to intuition, rounding the variables of an optimum simplex solution of a linear integer problem to the nearest integer does not provide an optimum integer solution,[12] for example, a routing problem solution that states that a particular truck should make 6.39 trips. Rounding the variables of an optimum simplex solution to the nearest integer will not necessarily result in an optimum integer solution. The problem of calculating feasible integral solutions remained unsolved until 1958, when Ralph Gomory[13] devised an algorithm for producing optimum integral solutions to all integer programming problems.

The add K machines model. The formulation of the layout program requires the definition of the integer variables W_{ij} and Z_{ij}, which are restricted to values of 0 or 1:

$$W_{ij} = \begin{cases} 1, & \text{if the } j\text{th new machine is located at } x = i \\ 0, & \text{otherwise} \end{cases}$$

$$Z_{ij} = \begin{cases} 1, & \text{if the } j\text{th new machine is located at } y = i \\ 0, & \text{otherwise} \end{cases}$$

The objective function for adding two new machines to an existing layout is

$$\text{Min } F = \sum_{j=1}^{K} \left\{ \sum_{i=1}^{N_x} C_{ij} W_{ij} + \sum_{i=1}^{N_y} d_{ij} Z_{ij} \right\}$$

where

N_x = Number of discrete units in the x-direction contained in the candidate area.

N_y = Number of discrete units in the y-direction contained in the candidate area.

C_{ij} = Total traffic-distance in the x-direction for the jth new machine located at $(x = i)$.

d_{ij} = Total traffic-distance in the y-direction for the jth new machine located at $(y = i)$.

This objective function is subject to restrictions that:

1. Insure that the variables take on only values of 0 or 1; and
2. Guarantee that no two machines are placed on top of one another.

The first restriction is stated as

$$\sum_{i=1}^{N_x} W_{ij} = 1 \qquad (j = 1, 2, \ldots, K)$$

[12] J. M. Danskin, "Linear Programming in the Face of Uncertainty: Example of a Failure," *Second Symposium on Linear Programming,* Paper No. 3.

[13] R. E. Gomory, "All-Integer Programming Algorithm" (IBM Research Center, Research Report RC-189, Jan. 1960).

$$W_{ij} = 1 \qquad (i = 1, 2, \ldots, N_x \quad \text{and} \quad j = 1, 2, \ldots, K)$$

$$\sum_{i=1}^{N_y} Z_{ij} = 1 \qquad (j - 1, 2, \ldots, K)$$

$$Z_{ij} \leq 1 \qquad (i = 1, 2, \ldots, N_y \quad \text{and} \quad j = 1, 2, \ldots, K)$$

The constraints insuring that no two machines are assigned the same location are

$$\sum_{j=1}^{K} W_{ij} \leq 1 \qquad (i - 1, 2, \ldots, N_x)$$

$$\sum_{j=1}^{K} Z_{ij} \leq 1 \qquad (i = 1, 2, \ldots, N_y)$$

There is an assumption that no new machine will have length longer than the discrete interval or width wider than the discrete interval. In other words, this procedure assumes that any new machine will require not more than one discrete interval.

Because of the current relative lack of applicability of this technique, no example will be given. Readers desiring more information should check the work of Gomory, Balinski, and Moodie.[14]

Evaluation. Integer programming is not widely used in facility design. Although it is a very promising tool, more efficient methods are needed to obtain solutions, since even the small problems generally require the use of a computer. It is useful, however, when many machines may be arranged in a limited number of candidate areas. Any type of movement may be used with this model, since it is not restricted to straight-line or rectangular movement. The model lends itself to computer calculation with existing algorithms, but more efficient computer programs are needed. Manual computation is quite tedious and cumbersome. Further work in developing more efficient computer algorithms should lead to an even more useful model. The model guarantees an optimum solution.

Dynamic Programming

Dynamic programming is a relatively new technique concerned with decision situations involving effectiveness functions of many variables that may be subject to constraints. An essential difference between dynamic programming and other approaches is that the dynamic programming approach changes one problem, in

[14] Gomory, "All-Integer Programming Algorithm," *Industrial Scheduling,* J. F. Muth and G. L. Thompson, eds. (Englewood Cliffs, N.J.: Prentice Hall, Inc., 1963), pp. 193–206; M. L. Balinski and R. E. Quandt, "On an Integer Program for a Delivery Program," *Operations Research* 12(March 1964):377–83; and C. L. Moodie and D. E. Mandeville, "Project Resource Balancing Techniques," *Journal of Industrial Engineering* 17(July 1966):377–83.

n variables, into n problems, each in one variable. As a result, problems exhibiting a large number of variables are reduced to a sequence of problems with only a single variable. The process is characterized by the determination of one variable, leaving a similar problem with one variable less. Dynamic programming is a computational technique developed to a considerable extent by Richard Bellman[15] and his associates at the Rand Corporation.

Consider a discrete process in one dimension, say x, with a finite number of stages, say n. Decisions are successively made, one at each stage, the process assuming a different state at each stage. Enumerate the stages in a fixed order $(1, 2, \ldots, n)$. The first decision, q_n, is taken when the process is in the nth stage. The state of the process is denoted by p_n. As a result of decision q_n, the process attains the state p_{n-1}, and $(n - 1)$ stages remain. When the process is in state p_i, there are i stages remaining. State p_i is a result of decision q_{i+1} when the process was in state p_{i+1}. State p_i may be expressed as a function of p_{i+1} and q_{i+1}:

$$p_i = T(p_{i+1}, q_{i+1})$$

The return function. The evaluation of a sequence of decisions requires a criterion as a basis of evaluation. A quantity called the return function, which may be return on investment, profit, savings in time, distance, money, or any given value measure, measures the return from a sequence of decisions. The return function is dependent upon a series of states and decisions, and may be expressed as

$$R(p_1, p_2, \ldots, p_n; \quad q_i, q_2, \ldots, q_n).$$

The problem is to maximize or minimize the return function. The total return from the sequence of decisions depends upon the returns associated with each individual decision, where the returns are denoted by

$$g_1(p_1, q_1), g_2(p_2, q_2), \ldots, g_n(p_n, q_n).$$

The following assumptions are made:

1. *The returns from all individual decisions can be interpreted in terms of a common parameter.*
2. *The total return of the sequence of decisions is a sum of the individual returns.*

Therefore,

$$R(p_1, p_2, \ldots, p_n; \; q_1, q_2, \ldots, q_n) = g_1(p_1, q_1) + q_2(p_2, q_2) + \cdots + g_n(p_n, q_n)$$

The functional equations. In place of a particular quantity of resources and a fixed number of activities, the return function considers the entire family of such problems, in which the number of resources may be any positive value, and n may be any integer value.

What seems to be a static process is artificially given a time-like property, by

[15] R. E. Bellman and S. E. Dreyfus, *Applied Dymanic Programming* (Princeton, N.J.: Princeton University Press, 1962), p. 15.

the requirement that the allocations be made one at a time. Viewed in this way, the allocation process appears to be dynamic.

The maximum of $R(p_1, p_2, \ldots, p_n; \; q_1, q_2, \ldots, q_n)$ depends upon p_n and n. Allowing p_n to range over $n = 1, 2, \ldots,$ a sequence of functions $\{f_n(p_n)\}$ is generated, defined by

$$f_n(p_n) = \max_{q_i}[g_1(p_1, q_1) + g_2(p_2, q_2) + \cdots + g_n(p_n, q_n)] \qquad (i = 1, 2, \ldots n, \quad n = 1, 2, \ldots)$$

Consider the decision q_n. Associated with the decision is the return $g_n(p_n, q_n)$. As a result of the decision, the process changes from p_n to p_{n-1} and now has only $(n - 1)$ stages. The return from the remaining $(n - 1)$ decisions must also be a maximum with an initial state p_{n-1}. This may be expressed as

$$f_n(p_n) = \max_{q_n}[g_n(p_n, q_n) + f_{n-1}\{T(p_n, q_n)\}]$$

which is a basic functional equation in a general form.

The optimality principle. The preceding development has depended on a very general technique called the *principle of optimality:*[16]

> An optimal policy has the property that, whatever the initial state and initial decision are, the remaining decisions must constitute an optimal policy with regard to the state resulting from the first decision.

The computational scheme. Allow p_n to take values at intervals, say Δ, over a certain useful range; p_n may take values $(\Delta, 2\Delta, \ldots, k\Delta)$. This is followed by computations of $f_n(p_n)$ $(n = 1, 2, 3, \ldots)$. For $p_n = k$, using the relation

$$f_1(k) = \max_{q_1}[g_1(k, q_1)]$$

$(f_1(\Delta), f_1(2\Delta), \ldots)$ are calculated with k taking values $(\Delta, 2\Delta, \ldots)$. The values of $(f_1(\Delta), f_1(2\Delta), \ldots)$ and the corresponding decisions $(q_1, q_2, \ldots)$ are recorded and retained.

Using the equation

$$f_n(p_n) = \max_{q_n}[g_n(p_n, q_2) + f_{n-1}(p_{n-1})]$$

the values of $(f_2(\Delta), f_2(2\Delta), \ldots)$ are computed, retaining the maximum value and the corresponding q_2 values. The procedure is repeated, with $(n = 3, 4, \ldots)$. The typical tabulation takes the form shown in Table 7-5.

Example. Assume a problem of the transportation type[17] with the addition of quadratic costs of shipping and a set-up cost for shipment from a source to a destination. *Set-up cost* is a cost that is independent of the quantity shipped, but is not incurred if nothing is sent. The cost and demand values are shown in

[16] Bellman.
[17] Bellman, p. 89f.

Table 7-5. Tabulation of Computational Algorithm

P_n	q_1	$f_1(p_n)$	q_2	$f_2(p_n)$	q_3	$f_3(q_n)$	$\cdots$	q_n	$\cdots$	$f_n(p_n)$
o										
Δ										
2Δ										
•										
•										
•										
kΔ										

Table 7-6. Each function $q_{ij}\,(X)$, where X is the number of items shipped, has the form

$$g_{jj}\,(X) = A_{ij}\,X + B_{ij}\,X + C_{ij}\,(X)$$

where

C_{ij} = Set-up cost ($C_{ij} = 0$ if $X = 0$; $= C_{ij}$ for $X > 0$)
A_{ij} = Cost of shipping one unit from source i to destination j.
B_{ij} = quadratic cost factor of shipping one unit from source i to destination j.

Thus, the cost of sending X items from *Source 1* to *Destination 3* is ($3X + 0.01X^2$), and from *Source 1* to *Destination 6* is ($5.0X - 0.10X^2$) + 10, if $X > 0$; or 0 if $X = 0$. For 100 units to be shipped from *Source 1*, and 80 units from *Source 2*, the optimal solution is shown in Table 7-7.

Evaluation. The dynamic programming technique offers considerable advantage over conventional mathematical procedures in solving multi-stage decision

Table 7-6. Cost Values Between Sources and Destinations

To	From Source 1			From Source 2			
Destination	Set-Up	x	x^2	Set-Up	x	x^2	Demand
1		1.0		2	3.1		10
2	1	2.0			4.1		25
3		3.0	0.01		2.1	4	45
4		1.5			1.1	0.1	15
5		2.5			2.6		5
6	10	5.0	−0.01		3.0		15
7		3.0		5	1.0	0.2	20
8		6.0			2.0		15
9	8	6.0	−0.05		2.0		10
10		6.0			5.0	0.01	20

Table 7-7. Solution to the Transportation Problem Using Dynamic Programming

To Destination	From Source 1	From Source 2	Cost	Cumulative Cost
1	10	0	$ 10.00	$ 10.00
2	25	0	51.00	61.00
3	5	40	99.25	160.25
4	15	0	22.50	182.75
5	5	0	12.50	195.25
6	0	15	45.00	240.25
7	20	0	60.00	300.25
8	0	15	30.00	330.25
9	0	10	20.00	350.25
10	20	0	120.00	470.25

problems. By its nature, the approach yields relative maxima or minima. In addition, the technique provides a family of optimization functions that can be used with a wide range of problems.

The functional equations technique provides a computational procedure that retains the advantages of the method of direct enumeration, and simultaneously eliminates the major bulk of the computations encountered in direct enumeration.

Most, if not all, problems encountered in facilities planning will be multi-dimensional; that is, more than one resource and constraint will be involved. The computations involved in multi-dimensional problems are tedious, and a computer is desirable for most problems of practical size. A number of computer programs are available.

The literature contains very few applications of dynamic programming to either material handling or facility layout activities. The technique appears to be particularly potent for work in this area, however, due to its ability to reduce the variables to a controllable number.

Level Curve Technique

The level curve technique developed by James M. Moore[18] is useful in solving the problem of locating new facilities in an existing layout. It measures effectiveness in terms of iso-cost lines, similar to contour lines, to minimize movement cost between (1) a new facility with those it serves and (2) the new facility and those served by it.

The level curve technique has two methods of solution: the straight-movement case, and the rectangular-movement case. Both methods assume (1) existing equipment will remain fixed, (2) cost is directly proportional to distance moved, and (3) equipment location is sufficiently described by defining a point on the floor

[18] J. M. Moore, "Mathematical Models for Optimizing Plant Layout." Unpublished Ph.D. Dissertation, Stanford University, 1965.

Table 7-8. Summary of Assumptions

Assumption	Model Applied to— Straight Line	Rectangular
1. It is considered undesirable to move any of the existing equipment.	x	x
2. The same type of material handling method, or an equal cost method, is used between all combinations considered.	x	x
3. Material handling costs are directly proportional to the distance moved.	x	x
4. Material flow between any two machines follows the shortest possible path.	x	
5. Material flow follows a rectangular pattern in a direction parallel to two given orthogonal axes.		x
6. It is sufficient to describe the location of all equipment by defining a point on the floor plan.	x	x

plan with rectangular coordinates. The assumptions in Table 7-8 provide the foundation for both.

Minimum straight-line movement. The existing facilities, numbered 1, 2, 3, . . . , n, are located at flow layout rectangular coordinates (X_1, Y_1), $(X_2, Y_2), \ldots, (X_n, Y_n)$. The objective is to determine the coordinates (X, Y) of a new facility, for which

$$M = \sum_{i=1}^{n} \sqrt{C_i (X - X_i)^2 + K_i (y - y_i)^2}$$

is a minimum, where M is the (inverse) measure of effectiveness of the new facility location.

As n increases, obtaining a graphical representation becomes more difficult. Figures 7-10, 11, and 12 show the graphical solutions for $n = 1$, 2, and 4. Figure 7-10 shows that the measure of effectiveness is the straight-line distance between two locations, shown as iso-cost lines consisting of a set of concentric circles centered at the location of the existing machine. Constants C_i and K_i are weighting factors that consider the relative costs of traffic frequencies and other items.

In Figures 7-10, 11, and 12, both C_i and K_i are equal to unity. Figure 7-11 shows the resulting curves to be ellipses, centered at the origin, with foci determined by the location of the two existing machines. The line ($-d$ *to* $+d$) may be thought of as a line of indifference, because preference for all points on the line is equal.

Figure 7-12 shows the location of the four existing machines at (0, 0), (1, 2),

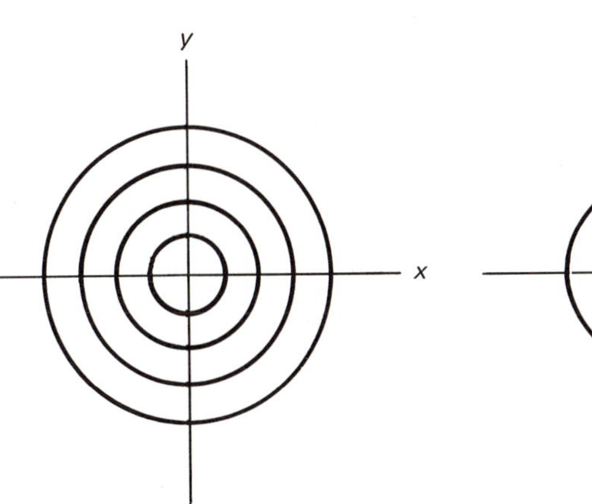

Figure 7-10—Level curves for one existing machine.

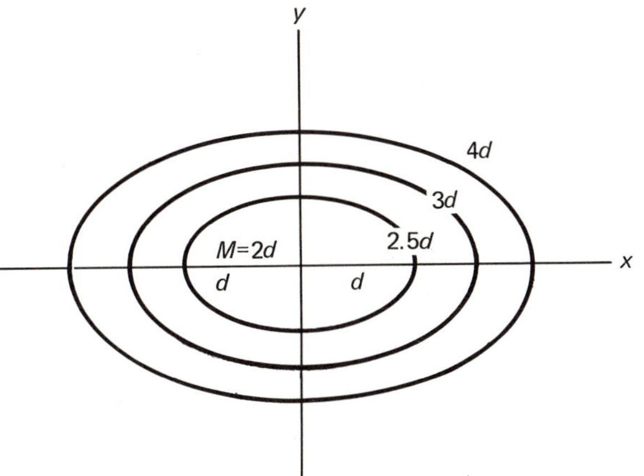

Figure 7-11—Level curves for two existing machines at points ±d.

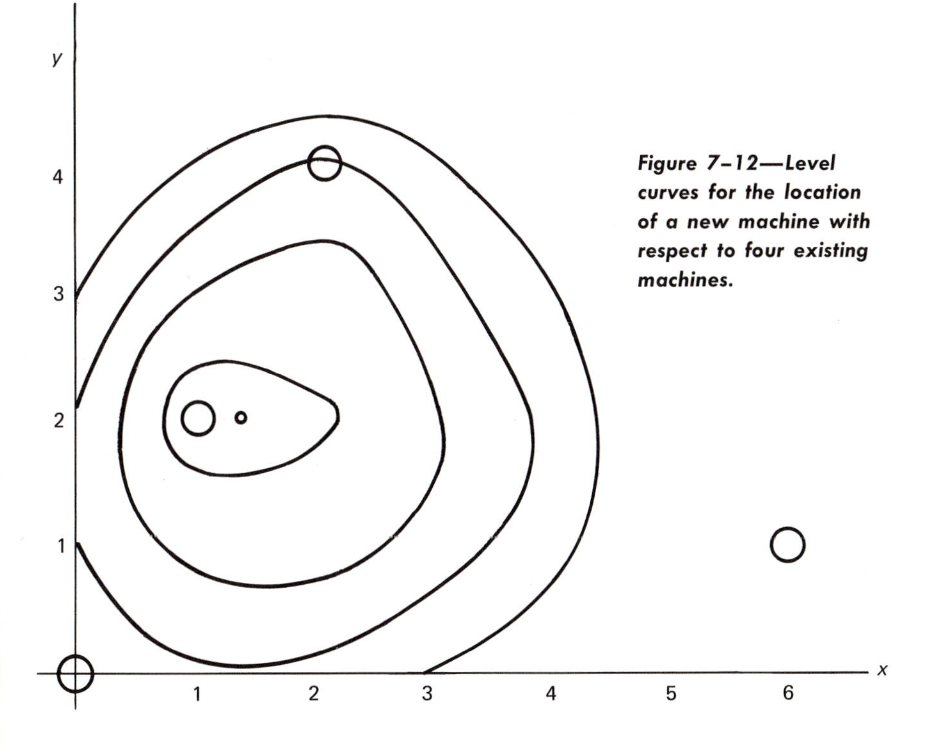

Figure 7-12—Level curves for the location of a new machine with respect to four existing machines.

(2, 4), and (6, 1). The computations were performed on a computer.[19] It is not necessary to perform the complete computation required to obtain the full set of iso-cost curves as shown in Figure 7-12 in order to solve a practical problem. Consider the expression for M shown above. The best theoretical location can be determined by taking the partial derivatives with respect to both X and Y and setting them equal to zero.

With these relationships satisfied, it can be concluded that the minimum measure of effectiveness values can be found. According to Moore, the recommended way is the *method of iteration* on a digital computer.

Minimum rectangular movement. The minimum-rectangular-movement case assumes:

1. *That all movement of materials is limited to the aisles or to directions parallel to the aisles;* and
2. *That the rectangular axes are so selected that they are parallel to the aisles;* that is, all aisles are set up on a rectangular basis.

The distance between any two points A and B of a layout, measured along directions parallel to a given pair of orthogonal axes, can follow an infinite number of paths. In all cases, however, the length of the path is the same.

The measure of effectiveness for the relationship between the new machine and existing machines is

$$M = \sum_{i=1}^{n} (|X - X_i| + |Y - Y_i|)$$

Rewriting this as

$$M = \sum_{i=1}^{n} |X - X_i| + \sum_{i=1}^{n} |Y - Y_i|$$

it can be shown that $\widehat{X}$, the median of the set of numbers $\{X_i, \ldots, X_n)$, yields a minimum value for $\Sigma_{i=1}^{n} | X - X_i |$; and likewise $(\widehat{Y})$, defined similarly, yields a minimum for $\Sigma_{i=1}^{n} | Y - Y_i |$. Therefore, $(\widehat{X}, \widehat{Y})$ will minimize[20]

$$M = \sum_{i=1}^{n} |X - X_i| + \sum_{i=1}^{n} |Y - Y_i|$$

Odd number of existing machines. Bindschedler[21] suggests a convenient graphical solution to the n-existing-machines problem. The graphical approach yields

[19] Moore, p. 117.

[20] H. Chevnoff and P. Moses, *Elementary Decision Theory* (New York: John Wiley & Sons, Inc., 1959), p. 319.

[21] A. E. Bindschedler and J. M. Moore, "Optimal Location of New Machines in Existing Plant Layouts," *Journal of Industrial Engineering* 12(Jan.–Feb. 1961):44.

different results, depending upon whether the number of existing machines is odd or even. The nature of the median is such that it is uniquely defined if n is odd, but not uniquely defined if n is even. The steps specified in Bindschedler's graphical approach are:

1. Delimit all discrete rectangular areas by tracing the lines $X = X_i$ and $Y = Y_i$ $(i = 1, 2, \ldots, n)$.
2. Determine the medians X and Y.
3. Determine a few representative slopes.
4. Then extend these values to other areas by symmetry.
5. Once the set of slopes has been determined, construct each level curve, starting at any arbitrary point.

Figure 7-13 shows the solution to the problem of three existing machines located at coordinate positions (8, 4), (15, 11), and (10, 14). Lines are drawn through the three points, parallel to the $\widehat{X}$ and $\widehat{Y}$ axes. The median X, is 11, and Y is 10. The location of the median point $(\widehat{X}, \widehat{Y})$ is an excellent starting point for plotting the

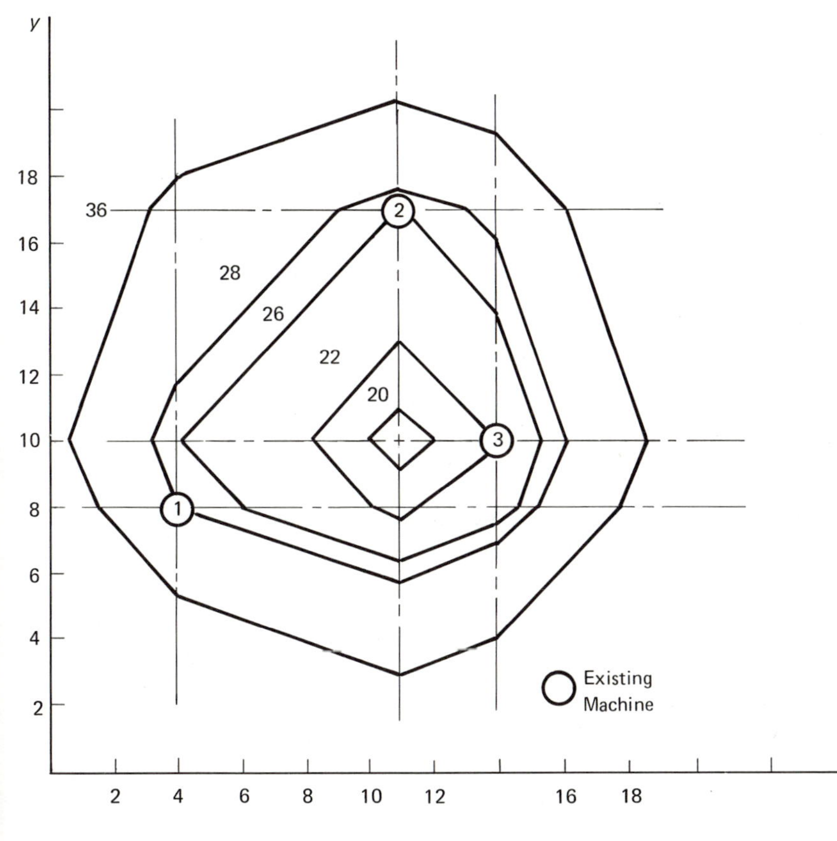

Figure 7–13—Three existing machines with unique point of optimum location.

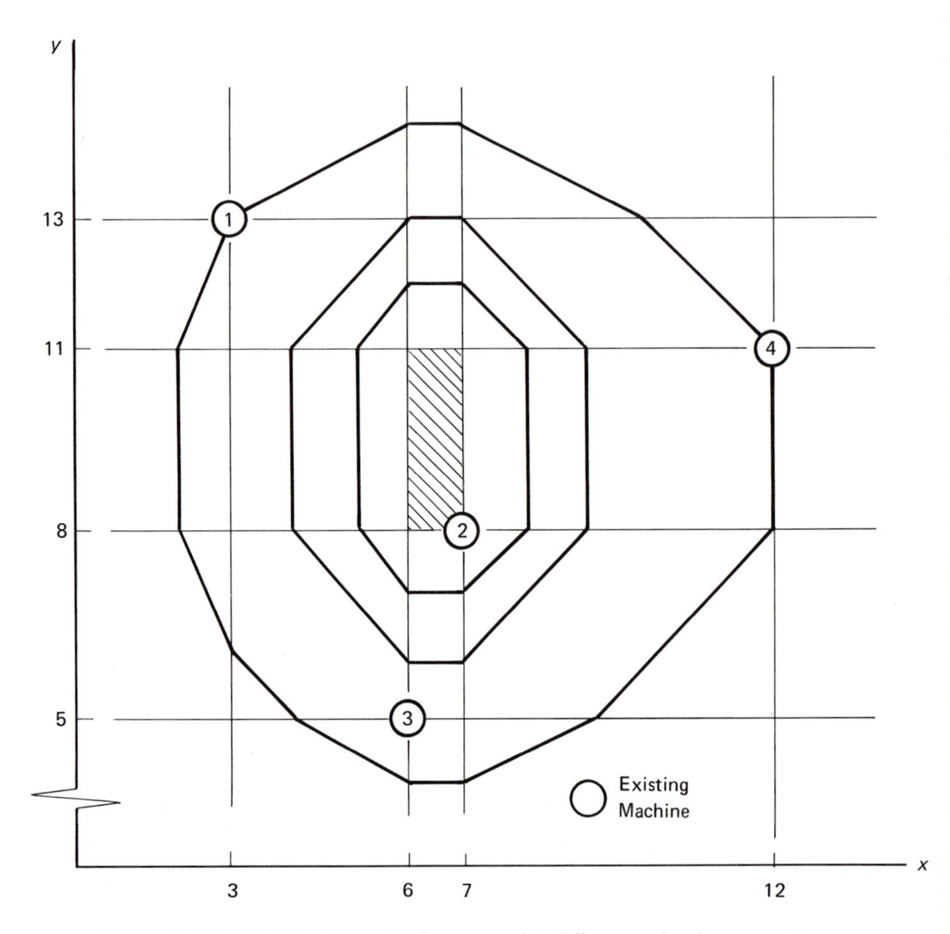

Figure 7-14—Hatched area is the area of indifference for four machines, an even number.

level curves, because all level curves will be closed lines enclosing the median point. This is a special property of problems involving an odd number of existing machines.

Even number of existing machines. When the number of existing machines is even, there appears to be no unique optimal location point for the new machine. In this situation an optimum indifference area develops in place of a unique optimum point (Figure 7-14). Whenever there are n existing machines, and n is an even number, the optimum location will include all points within a central rectangular indifference area.

Comparison of straight-line and rectangular movement. The rectangular-movement model provides a higher degree of flexibility for the location of new equipment than the straight-line model, because it often has large areas of indifference around the optimal location. This flexibility is useful if other location factors than the measure of effectiveness used here are considered.

In practice the optimum locations indicated by the straight-line model will be very close to the optimum locations determined by the rectangular-movement model. The optimum location is relatively insensitive to the path pattern chosen. Therefore, the rectangular-movement model is most useful, since it is considerably easier to use than the straight-line model.

Evaluation of the level curve model. The model has several advantages. One is that in solving many practical problems it is not necessary to obtain a complete solution. Only the solutions in the neighborhood of existing candidate areas need be considered, and this can usually be done graphically (this advantage is offset to some extent in that, if the minimum n value does not fall in one of the candidate areas, this approach does not solve the problem).

Another advantage is that the method is easy to apply, and adequate computer programs are available for large problems.[22] Also, it is relatively easy to add a limited number of weighting factors to give consideration to traffic frequencies and other items not included in the basic assumptions for the model.

The most serious disadvantage to the level curve concept is in the assumptions listed in Table 7-8; *assumption 1* is seldom satisfied in a practical situation.

The model is restricted in that it assumes an already existing layout of n machines, and it is limited to the addition of one new machine. If all assumptions are accurate, this method guarantees an optimal solution.

Queuing Theory

A number of different models of material flow systems include factors characterized by a distribution of random variables. The most general, and probably the most effective, analytical approach to such random flow systems is waiting line analysis, or queuing theory. Queuing theory refers to the mathematical and physical investigation of a class of problems characterized by several attributes: (1) there is an input of units entering the system, (2) the units moving through the system are discrete, (3) the units that have begun to require service are ordered in some fashion and receive service in that order, (4) a mechanism exists that governs the time at which a unit receiving service has its service terminated, and (5) at least one of the two mechanisms, arrival or service, is not completely determined, but can be considered a probabilistic system of some sort.

The reason the variables are random, rather than functionally dependent on time, is that arrivals are generally random events in time (that is, the exact instant they will occur is unpredictable) and service times also are random variables. Therefore, most applications of waiting line theory are concerned with averages,

[22]Bellman, p. 15.

such as the average length of the line at any instant, and the average idle time of the service facility in a given time period. Although times of arrivals and departures cannot be predicted with precision, they can be predicted in terms of means, variances, and probabilities.

The single-channel waiting line. A single-channel waiting line consists of a single line formed in front of a single-service mechanism. In a facility design system, this situation might be represented by units of product brought into live storage prior to an operation. In general, the mean time of an operation can be controlled, but the time for any particular cycle is subject to uncontrollable random fluctuation. Also, the mean rate at which units arrive can be controlled, but again, there will be random fluctuations beyond control.

The arrival process is analyzed in detail by Morris[23] and Feller.[24] The following is a tabulation of useful relationships taken from Sasieni,[25] where λ is the mean time between arrivals, and μ is the mean service time per unit.

$$E(m) = \frac{\lambda^2}{\mu(\mu - \lambda)} \quad \text{(average queue length)}$$

$$E(m \mid m > 0) = \frac{\mu}{\mu - \lambda} \quad \text{(average length of non-empty queues)}$$

$$E(n) = \frac{\lambda}{\mu - \lambda} \quad \text{(average number of units in system)}$$

$$E(w) = \frac{\lambda}{\mu(\mu - \lambda)} \quad \text{(average waiting time of an arrival)}$$

$$E(w \mid w > o) = \frac{1}{\mu - \lambda} \quad \text{(average waiting time of an arrival that waits)}$$

$$E(v) = \frac{1}{\mu - \lambda} \quad \text{(average time an arrival spends in the system)}$$

$$\sigma_n^2 = \frac{E(n)}{(\mu - \lambda)/\mu} = \frac{\lambda\mu}{(\mu - \lambda)^2} \quad \text{(the variance of } n\text{)}$$

All these relationships assume that (1) *arrivals occur in Poisson fashion,* (2) *service times occur in exponential fashion,* and (3) *the system is in a steady-state condition.*

A queuing model can be generalized to make λ and μ functions of the number of units in the system and also to describe a system in which the largest permissible number of units in the system is N. The procedure for generalizing is included in the reference by Morris. The generalized models are similar to, but more complex than, the model cited here. For example, the expected number of units in

[23] W. T. Morris, *Analysis of Materials Handling Management* (Homewood, Ill.: Richard D. Irwin, Inc., 1962), p. 63.

[24] W. Feller, *An Introduction to Probability Theory and Its Applications*, 2nd ed. (New York: John Wiley & Sons, Inc., 1957), pp. 400 *et seq.*

[25] Sasieni, *et al.*, p. 133.

a system that can contain at most N units is

$$E(n) = \left\{\frac{1 - (N+1)(\lambda/\mu)^N + N(\lambda/\mu)^{N+1}}{[\mu - \lambda]/\mu)[1 - (\lambda/\mu)^{N+1}]}\right\}\frac{\lambda}{\mu}$$

and the same assumptions hold as before. Unless N is relatively small, the generalized models will yield results little different from the simpler models given earlier.

Example. Whenever services are demanded in a non-regular manner, or arrivals occur in a non-regular manner, the system capacity can from time to time be overtaxed. An example is where the output of one or more facilities is the input of another facility. If the output of the first is approximated by a Poisson distribution, and the process is in steady state, these equations can be applied to determine the storage capacity required between them. Assume the first facility has an average output of 2.2 units per minute, and the output rate of the second is an average 2.5 units per minute. If there are no items lost in number two—that is, the input equals the output—then the servicing rate, μ, is 2.5 units per minute, and the arrival rate, λ, is 2.2 units per minute. Then

$$E(n) = \frac{\lambda}{\mu - \lambda} = \frac{2.2}{2.5 - 2.2} = 7.33 \text{ units}$$

the average number of units in the system. The variance is

$$\sigma_n^2 = \frac{\lambda\mu}{(\mu - \lambda)^2} = \frac{(2.2)(2.5)}{(0.3)} = 61.1$$

and the standard deviation is

$$\sqrt{\sigma_n^2} = 7.8$$

A storage area large enough to accommodate

$$E(n) + 2(\sigma_n) = 7.33 + 2(7.8) = 22.9 \text{ units}$$

will be large enough to handle storage of units between facilities 94 per cent of the time.[26] There will be, however, congestion in the storage area 6 per cent of the time.

Multi-channel waiting lines. The preceding model has the disadvantage of not being completely realistic, though it often serves well enough as an approximation. A more realistic model can be obtained by considering the case where a finite number of channels are available.

The assumptions used in obtaining the single-channel model hold, except that the number of service channels is greater than 1. If all service channels are busy, the arriving unit joins a waiting line and waits until a channel is freed. This means

[26]E. Richman, and S. Elmaghraby, "The Design of In-Process Storage Facilities," *Journal of Industrial Engineering* 8(Jan.–Feb. 1957):9.

that all incoming units have a common waiting line.[27] A waiting line will exist only if the number of units n awaiting, or in, service is greater than k, the number of service channels available.

As long as at least one channel is free, the situation is exactly the same as the single-channel model. For varying n, the following situations arise:

$$
\begin{array}{rl}
n = 0 & \lambda_n = \lambda\mu_n = 0 \\
1 \leq n < k & \lambda_n = \lambda\mu_n = n\mu \\
n \geq k & \lambda_n = \lambda\mu_n = k\mu
\end{array}
$$

In the special case of ($k = 1$), the equations on page 184 describe the system. For $1 \leq n < k$

$$P_n = (1/n!)\,(\lambda/\mu)\,P_0$$

and for $n \geq k$

$$P_n = \frac{1}{k!\,k^{n-k}}(\lambda/\mu)^n P_0$$

and

$$P_0 = 1\left\{\sum_{k=0}^{k-1}\frac{1}{n!}(\lambda/\mu)^n\right\} + \frac{1}{k!}(\lambda/\mu)^k\,\frac{k\mu}{k\mu - \lambda}$$

The above equations are valid only if $k < \lambda/\mu$; otherwise, stability is never achieved, and the queue builds up indefinitely. For the general case of k servicing channels, the probability of an arrival waiting is the probability that at a given instant there are at least k units in the system.

$$P\,(n \geq k) = \sum_{n=k}^{\infty} P_n = \frac{\mu\,(\lambda/\mu)^k}{(k-1)!\,(k\mu - \lambda)}P_0$$

The formulas below, expressed in terms of P_0 for brevity, are taken from Sasieni:[28]

$$E\,(m) = \frac{\lambda\mu\,(\lambda/\mu)^k}{(k-1)!\,(k\mu - \lambda)^2}P_0 \quad \text{(average queue length)}$$

$$E\,(n) = \frac{\lambda\mu\,(\lambda - \mu)^k}{(k-1)!\,(k\mu - \lambda)^2}P_0 + \frac{\lambda}{\mu} \quad \text{(average number of units in system)}$$

$$E\,(w) = \frac{\mu\,(\lambda/\mu)^k}{(k-1)!\,(k\mu - \lambda)^2}P_0 \quad \text{(average waiting time for arrival)}$$

$$E\,(v) = \frac{\mu\,(\lambda/\mu)^k}{(k-1)!\,(k\mu - \lambda)^2}P_0 + \frac{1}{\mu} \quad \text{(average time an arrival spends in the system)}$$

Table 7-9 gives a numerical illustration of the case $k = 3$ and $\lambda/\mu = 2$.

[27] Feller, pp. 400f.
[28] Sasieni, *et al.*, p. 138.

Table 7-9. Probability of n Units in System when $k = 3$ and $\lambda/\mu = 2$

n	0	1	2	3	4	5	6	7
Stations busy	0	1	2	3	3	3	3	3
Units waiting	0	0	0	0	1	2	3	4
P_n	0.111	0.222	0.222	0.148	0.099	0.066	0.044	0.029

Example. A production department has three identical machines in operation. Production units arrive in Poisson fashion at an average rate of 20 units per 8-hour day. The production rate of each machine has an exponential distribution, with mean service time of 40 minutes. Units are processed in the order of their appearance.

$$\lambda = 20/8 = 2.5 \text{ arrivals per hour}$$

$$\mu = 60/40 = 1.5 \text{ units serviced per hour}$$

$$P_0 = \frac{1}{1 + (5/3) + (1/2)(5/3)^2 + (1/6)(5/3)^3(9/4)} = 24/139$$

The expected number of idle machines

$$3P_0 + 2P_1 + P_2 = 3(24/139) + 2(24/139)(5/3) + (24/139)(1/2)(5/3)^2 = 4/3$$

The probability of one idle machine at a specified instant is $(1/3)(4/3) = 4/9$. The expected working time for one machine is $(1 - 4/9)(8) = (5/9)(8) = 40/9$ hours per day. The average time an arriving unit spends in the system is

$$E(v) = \frac{1.5(5/3)^3}{2[3(1.5) - 2.5]^2}(24/139) + \frac{1}{1.5} = 49.0 \text{ minutes}$$

Other values could be similarly determined and, with appropriate cost parameters, an optimum policy could be determined.

The models discussed here assume Poisson arrival and service distribution. However, there are other types of distributions associated with queuing lines. The Erlang family of distributions is probably the most common and is discussed in the cited literature.

Evaluation. The major problem involved in using queuing, or waiting line, theory is the assumption of Poisson input, or arrival rates, and output, or service rates. The question as to how much error is introduced by this assumption is difficult to assess, and where it occurs in the system is frequently hard to locate.

Waiting line analysis is a well-developed field, and what is presented here is only the barest introduction. The mathematical formulation of various models is widely published and understood. The application of queuing theory to facilities design problems is not extensively published, except in conveyor theory. One of the difficulties appears to be that most practical applications require considerable accurate cost information, which is difficult to obtain.

Useful application of queuing models requires, as do all mathematical models, a balance of practical experience, and experience in mathematical analysis, to produce effective recommendations for the improvement of facility systems. There are important insights into facility procedures to be gained from the use of analytical methods in addition to the improvement expected from application of the model.

In summary, in view of the vast quantity of literature published on queuing theory, the present most critical need is for the development of an understanding of the kind of cost information required for optimization techniques. Little of the literature uses cost figures in conjunction with queuing models; therefore it is difficult to determine if the true value of queuing models is proportional to the literature generated about it. Certainly it appears to be a fruitful area for investigation with specific objectives in mind.

Conveyor Analysis

Conveyors comprise a large portion of the linkages between production centers. Many are used for both transportation and in-process storage. Morris[29] classifies conveyors into four major categories:

1. Constant-speed irreversible belt conveyors that simply transport and store material.
2. Controlled movement systems that are reversible. These are operator-controlled, and move away from the work station for storage and into the work station for recovery.
3. Power and free systems, consisting of parts carriers that can be connected to and disconnected from the moving portion of the conveyor at will.
4. Closed loop, irreversible, continuous operating systems with parts carriers which cannot be removed.

The first three types may be called open-loop systems, and the fourth type a closed-loop system. The operating characteristics of the open-loop systems are quite easy to predict, on theoretical grounds at least, lending an aura of confidence to their use. Open-loop conveyors are generally quite flexible in use, whereas closed-loop systems are not. On the other hand, closed-loop systems are generally simpler and lower in cost per unit length than open-loop systems.[30]

Since most open-loop systems can be mathematically described quite easily, in

[29] Morris, p. 129.

[30] W. B. Helgeson, "Planning for the Use of Overhead Monorail Non-Reversing Loop Type Conveyor Systems for Storage and Delivery," *Journal of Industrial Engineering* 11(Nov.–Dec. 1960):488.

terms of arrival and service rates, this section will be devoted to closed-loop systems.

Design considerations. Kwo[31] lists three considerations basic to closed-loop conveyor operation. They are (1) the uniformity principle, (2) the capacity constraint, and (3) the speed rule. The *uniformity principle* states that it is necessary to try to load and unload the conveyor uniformly over the entire loop. The *capacity constraint* dictates that the accommodations furnished by the conveyor should be at least equal to the accommodations required. The *speed rule* defines a range of permissible speeds for the conveyor. The lowest speed is set by the higher of either the loading or unloading rate and the highest speed is restricted by the electromechanical limits of the conveyor or by the speed of the human material handler, whichever is lower.

Uniformity principle—Let

N_T = Number of elementary cycles considered.
T = Cycle times.
W = Revolution time.
N_W = 2, 3, . . . , integer.
T_L = Loading time in elementary cycle.
T_u = Unloading time in elementary cycle.

The conditions that must be met simultaneously are:

$$N_T T/W = N_W, \text{ a positive integer not equal to one, and } N_W \neq N_T$$
$$N_T T_L/W = \text{A positive integer, or}$$
$$N_T T_u/W = \text{A positive integer, with } T_L + T_u = T$$

The above indicates that after an interval of time $N_T T$, all things return to what they were at the start. This interval may be one elementary or N_T elementary cycles. Since neither accumulation by loading nor depletion by unloading can continue indefinitely, after some time lapse the accumulation must equal the depletion. The N_T cannot be equal to N_W, since if they were, any section of the conveyor would get the same loading–unloading effect from cycle to cycle, and uniformity would never be achieved. The same reasoning rules out $N_W = 1$. The second equation states that the total accumulation (or depletion) in the time interval $N_T T$ must be such that it can be shared evenly by the entire conveyor. This means that the time lapse over which there is net gain (or loss) of parts must be an integral multiple of the revolution time.

Capacity constraint—Let

m = Total number of carriers on the conveyor.
q = Capacity of a carrier.
v = Velocity of the conveyor.

[31] T. T. Kwo, "A Method for Designing Irreversible Overhead Loop Conveyors," *Journal of Industrial Engineering* 11(Nov.–Dec. 1960).

L = Length of the conveyor.
W = Revolution time of the conveyor.
S = Spacing between carriers.
K = Accommodations required per unit time.

$$mqv/L = mq/W = qv/S \geq K$$

The left-hand side of this equation is the number of racks passing by any point in a unit time. This is the potential capacity of the conveyor.

Speed Rule—Let

r_L = Loading rate.
r_u = Unloading rate.
t_L = Average time to load a part.
t_u = Average time to unload a part.
v_c = Maximum allowable speed of conveyor.

$$\max(r_L, r_u) \leq v/S \leq \min(1/t_L, 1/t_u, v_c/S)$$

The speed rule is generally invalidated for the upper and lower bounds in the above equation, since there is usually reserve carrier capacity available, and material handlers are allowed to move with the conveyor for a short distance when necessary. For these reasons the uniformity principle and capacity constraints are much more significant than the speed rule. In practice, the speed rule is used only as a check to insure feasibility of a suggested speed.

Kwo does not consider the effect of random variation on the individual station, but merely states that in practice the material handler is allowed to move with the conveyor when necessary to overcome delay. However, Reis, *et al.*[32] do consider the role of the individual station. They include graphs indicating delay time and bank sizes needed for various capacity stations.

Disney[33] discusses the conveyor system utilizing parallel loading and unloading stations with a sensor-controlled ordered entry discipline. There is apparently no current publication combining all these possibilities in a single system.

Example (from work done by Kwo[34]). Suppose the following is known:

1. The output of *loading area A* is 6 parts per minute.
2. The input to *unloading area B* is 2 parts per minute.
3. The distance from A to B is 1,200 ft.
4. Carrier spacing of 4 ft is to be used.
5. The reserve live storage capacity is to be 480 parts.

Assume that economic-lot-size considerations require A to produce in batches of 720 parts, requiring a 2-hr production run. The loading time will be 720 parts at

[32] I. L. Reis, L. L. Dunlap, and M. H. Schneider, "Conveyor Theory: The Individual Station," *Journal of Industrial Engineering* 14(Jul.–Aug. 1963).
[33] R. L. Disney, "Some Results of Multichannel Queueing Problems with Ordered Entry—An Application to Conveyor Theory," *Journal of Industrial Engineering* 14(Mar.–Apr. 1963).
[34] Kwo, *op. cit.*

6 parts/min = 120 min, and similarly the unloading time will be 360 minutes for an elementary cycle time of 120 + 360 − overlap of 120 = 360 min, or 6 hr. Then:

$$N_T = 1 \qquad T = 360 \qquad T_L = 120 \qquad T_u = 240$$

and

$$WN_W = 360 \qquad (N_W = 2, 3, \ldots, \text{integer})$$

From $N_T T_L/W$: $120/W$ = a positive integer, $240/W$ = integer; so W may be 1, 2, 3, 4, 5, 8, 10, 12, 15, 20, 24, 30, 40, 60, or 120. Assume that 60 ft/min is the maximum allowable conveyor speed. Since the distance between A and B is 1,200 feet, the conveyor must be at least 2,400 ft long. Therefore, $L/W = 2{,}400/60 = 40$. No value for W less than 40 need be considered.

Choose the revolution time, W, equal to 60 minutes, for simplicity. The operating capacity requirements are, for loading, 6 parts/min; unloading, 2 parts/min. The accumulation rate is $6 - 2 = 4$ parts/min for the first 120 min, and -2 parts/min for the remaining 240 min. The peak accumulation will be $120(4) = 480$ parts.

As a first try, let the conveyor length, L, be the minimum possible, 2,400 ft. The conveyor speed is then $v = L/W = 2{,}400/60 = 40$ ft/min, a reasonable figure. From the pre-determined conveyor spacing of 4 ft and chosen length of 2,400 ft, $m = L/S = 2{,}400/4 = 600$ carriers. The pre-determined maximum reserve capacity requirement, when converted to a 1-min interval basis, is $480/60 = 8$. The total capacity requirement, on a per-minute interval basis, is the sum of the maximum operating capacity requirement, the pre-determined reserve capacity requirement, and twice the smaller of the two rates. This is $K = 8 + 8 + 4 = 20$.

The total number of parts each carrier is to accommodate (from the expression for K) $q = KW/m = 20(60)/600 = 2$ parts/carrier, which appears to be a reasonable number.

The parameters of the conveyor are then: $W = 60$ min, $T = 360$ min, $L = 2{,}400$ ft, $m = 600$, $q = 2$, $S = 4$ ft, and $v = 40$ ft/min. These parameters should be applied to the various combinations of equipment available to arrive at a minimum cost facility that satisfies the parameters.

Evaluation. In general, the conveyor design problem is considerably more complex than the example given here. A regularly recurring recommendation throughout existing literature is, through simulation, to test the proposed conveyor system design over a wide range of system parameters. This recommendation apparently reflects the lack of adequate testing of the conveyor models developed. In addition, most conveyor models require consideration of a number of possibilities that result in poorly defined reactions within the system, such as interference and imbalance.

The existing mathematical models do not replace traditional design techniques but they do aid in the evaluation of cost of production, cost of delay, bank capacity, and conveyor accommodation requirements. With the models available, however, almost any type of closed-loop conveyor system can be planned with a reasonably high assurance that the system will do its intended job satisfactorily.

Simulation

For most practical problems, experimenting with full-scale potential solutions to layout problems or material handling systems is too expensive, and takes too much time. Simulation begins with a real system, and attempts to duplicate it on paper or with computers, by procedures that generate data representative of the effectiveness of various alternative solutions at relatively low cost and in a short period of time.

To simulate a system, a fairly good knowledge of the parts or components of the system and their characteristics is required, although knowledge of the parts of a system does not imply any knowledge of the system behavior. Simulation sums the behavior of the parts and helps to explain and predict the dynamic behavior of the system.

A major difference between one type of simulation and another is the presence or absence of chance or stochastic processes and their associated probability distributions in the simulation. Where probability distributions are explicitly included, the analysis is frequently called *Monte Carlo simulation,*[35] because the technique has to do with evaluating elements of chance, which have a major role in industry in the nature of rejects, breakdowns, and so on. Basically, Monte Carlo analysis is a simulated sampling technique, in which, when actual samples are unavailable, the sample data are synthetically generated.

The data are commonly generated by some random number generator, to replace the actual items to be sampled with their theoretical counterpart, as described by some probability distribution. The generator could be a random number table, a pair of fair dice, or any unbiased source of values. The data of interest are tabulated as a probability density function. If the data have a known probability distribution, the density can be computed mathematically and no real or observed information is needed. If the probability distribution is not known, it is necessary to take actual samples or to use historical information to approximate the probability density for specified values. For example, if records show that the number of breakdowns per hour in a group of machines is as follows:

Number of breakdowns/hr	0	1	2	3	4	5	6	7 or more
Frequency	893	227	189	101	41	11	2	0

The total number of observations is 1,464, and the relative frequency is:

$X =$	0	1	2	3	4	5	6	7 or more
Probability of X breakdowns/hr =	0.61	0.15	0.13	0.07	0.03	0.01	0.00	0.00

[35] E. H. Bowman and R. B. Fetter, *Analysis of Production Management* (Homewood, Ill.: Richard D. Irwin, Inc., 1961).

Then a set of random numbers used to generate sequences of breakdowns would have 61 per cent of its numbers indicating 0 breakdowns, 15 per cent indicating 1 breakdown, and so on.

Thus, a properly designed set of random numbers can be used to generate sequences of numbers that have the same statistical characteristics as the actual system being simulated. As long as the statistical model continues to represent the facts of the system, an infinite amount of simulated experience is available, usually at much lower cost than actual experience.

In summary form, the steps in the Monte Carlo simulation are:

1. Pick the measure of effectiveness.
2. Decide which variables significantly affect this measure.
3. Determine the proper probability density functions to represent the variables.
4. Choose the candidate solutions to the problem.
5. Generate a set of random numbers.
6. For each set of random numbers, enter the probability density function to determine the corresponding value of the data of interest.
7. Insert the values of the data of interest in the model of the measure of effectiveness, and compute.
8. From a run of figures computed in *Step 7*, select the best alternative.
9. Make some confidence statement concerning the choice in *Step 8*.

When a problem does not involve many variables, simulation can be done manually. In general, a computer is faster and more economical for problems of practical size.

Example. Schiller and Lavin[36] provide an example of simulation as a means of determining the warehouse dock space required by a department store chain that was planning to consolidate three warehouses into a single unit. The problem was to determine the dock space required to handle the truck traffic at the new warehouse, which fell into the framework of waiting-line analysis but was sufficiently complex to be more easily solved by simulation than by analysis.

A study of truck arrival patterns at the existing warehouses showed that arrivals were effectively described by a Poisson probability distribution, with no significant variation among days of the week or weeks of the month. Although there were significantly more arrivals in the mornings than in the afternoon, there was no difference in the number of arrivals in half-hour periods during the morning. The afternoon half-hour intervals were also homogeneous.

The study of service times at the dock showed that one class of truck was loaded only, and a second class unloaded only. There were two groups of unloadings, both negative exponentially distributed with means of 10 and 45 minutes. For trucks being loaded, service time was negative exponential after a mean holding time of 18 minutes.

The company wanted facilities that would be adequate for the pre-Christmas

[36]D. H. Schiller and M. M. Lavin, "The Determination of Requirements for Warehouse Dock Facilities," *Operations Research* 4(April 1956):231.

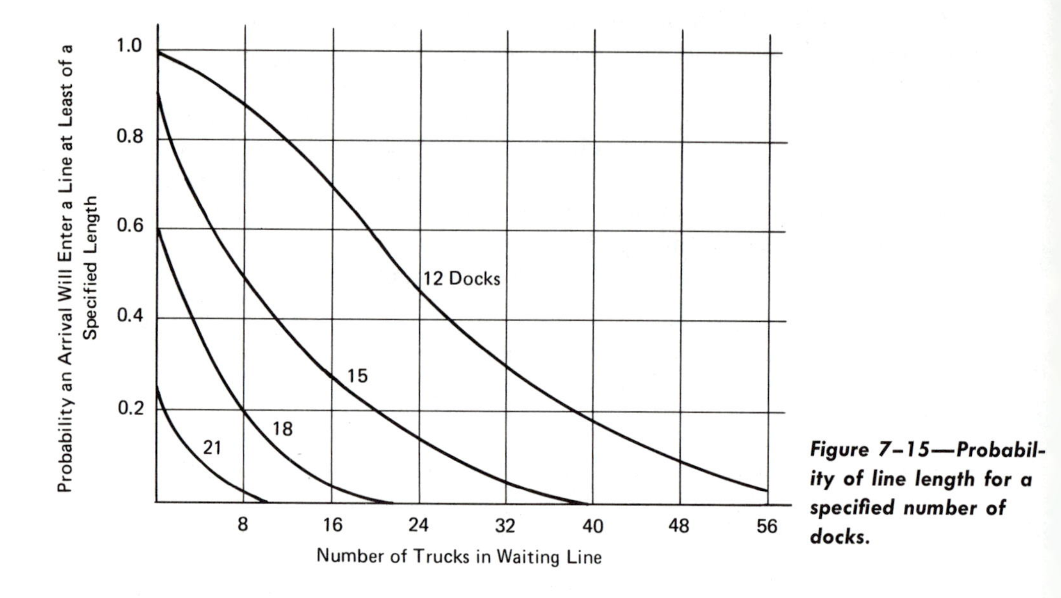

Figure 7-15—Probability of line length for a specified number of docks.

peak season. The arrival and service-time data were modified to reflect what was believed to prevail during this season. Warehouse facilities having 12, 15, 18, and 21 docks were then simulated and showed the rate at which the mean line length would build up during the morning, resulting in a heavy load on the system during the 11:00 A.M. to 12:00 A.M. period. The result of the simulation was the plotting of the probability distributions of line length and waiting time during this period, as shown in Figures 7-15 and 7-16. From these, management could make a decision as to the consequences of providing various numbers of docks at the new warehouse.

Evaluation. Simulation is most helpful where formal mathematical analysis is impossible or inconvenient, although it shares the disadvantage common to all models; that it focuses on certain characteristics of the system and largely ignores others.

Monte Carlo simulation has one outstanding advantage. In addition to being useful in problems for which no analytical solution exists, when the analyst finds one of the more elegant solutions to complex analytical problems beyond his mathematical training, the Monte Carlo methods of picking a solution can be applied.[37]

In general, it is useful to make a reasonable effort to treat a problem analytically before turning to simulation, since it is a sampling technique and is subject to the same difficulties encountered in any sampling plan. The factors to be tested, the level at which they will be tested, the number of samples to be taken, and the

[37]Bowman and Feller.

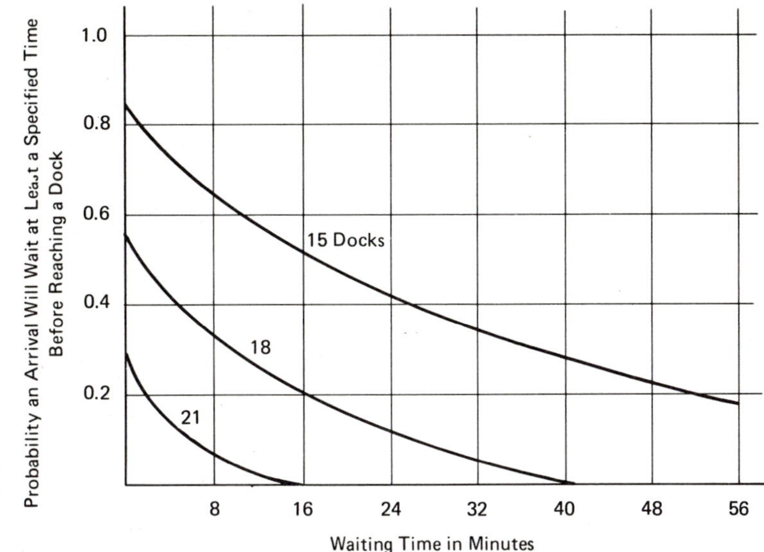

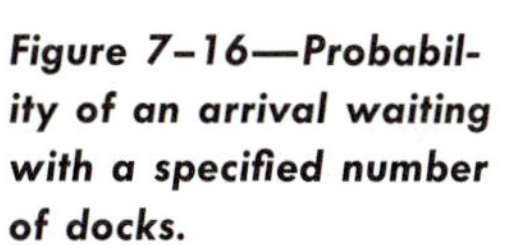
Figure 7–16—Probability of an arrival waiting with a specified number of docks.

analysis of variance must all be carefully planned in advance. Most simulation problems can be handled effectively on a computer, although this usually makes for inefficient use of computer time; small problems, however, can be done manually.

Summary

The quantitative techniques discussed in this chapter represent some of the most powerful tools currently available for use in facilities planning. However, the list of the techniques is by no means complete. Continued development of new concepts, together with quantitative analytical tools, will continue to expand the capabilities of the facilities designer.

At the present time, seven of the techniques covered in this chapter can be expected to insure optimal solutions when appropriate assumptions and constraints are satisfied. They are:

1. Linear programming technique
2. Assignment problems
3. Transportation programming problems
4. Transshipment programming problems
5. Integer programming technique
6. Dynamic programming technique
7. Level curve technique

The traveling salesman problem provides a solution that is near optimal and sometimes optimal. Queuing theory and simulation can not guarantee optimality due to their probabilistic nature. Conveyor analysis generally requires that its solution be tested by simulation or actual operation. The algorithms available for solving quadratic programming problems do not guarantee an optimal solution. In

Table 7-10. Characteristics of Solution Procedures for Each Technique

	Transportation Programming	Assignment Techniques	Transshipment Programming	Traveling Salesman Technique	Integer Programming	Dynamic Programming	Level Curve Models	Waiting Lines	Conveyor Analysis	Monte Carlo Simulation
Guaranteed optimal	x	x	x		x	x				
Computer solution required				x*	x	x				x*
Computer solution desirable	x	x	x	x		x	x	x	x	x
Number of added machines considered	m	m	m	m	m	m	l			m
Number of candidate locations Number of machines added		x	x		x	x				x
Consider prohibited assignments	x	x	x	x					x	

* Choice depends upon size of the problem.

all cases, a computer solution may be either required or desired for a problem of practical size.

As a possible guide to the selection of the most appropriate technique for analyzing a particular problem, Table 7-10 provides a tabular summation of the characteristics of each technique, and Table 7-11 lists the type of data each can use.

Even this group of techniques does not permit a solution to all facility design problems. Often one or more of them will be found useful as powerful tools to guide and assist in developing a solution or in improving an existing situation. The ingenuity of the analyst, however, is likely to continue to be the moving force in layout planning until these and similar techniques are more thoroughly tested in real-world problem situations, and analysts are trained to use them.

Recommendations for applications of techniques and problems. *Linear programming, assignment, transportation programming, and transhipment programming* are recommended when static conditions and straight line relationships exist. They may be used as allocation models for selecting among alternate ways of performing a task, or in choosing the best combinations of existing facilities or resources in order to optimize effectiveness. They are recommended when more

Table 7-11. Data Characteristics for Each Technique

		Transportation Programming	Assignment Techniques	Transshipment Programming	Traveling Salesman Technique	Integer Programming	Dynamic Programming	Level Curve Models	Waiting Lines	Conveyor Analysis	Monte Carlo Simulation
Handling between machines allowed				x					x	x	x
Routing fixed		x	x	x	x					x	
Movement direction:	straight line						x	x		x	x
	rectangular						x	x		x	x
	arbitrary	x	x	x	x	x	x			x	x
Candidate areas:	discrete	x	x	x	x	x	x	x		x	x
	continuous							x			
Measure:	deterministic	x	x	x	x	x	x	x		x	x
	stochastic						x		x	x	x

than one machine is to be added to a layout with discrete candidate areas. They are also useful when the number of candidate areas is greater than or equal to the number of new machines to be added. The computational procedures are relatively simple, and large problems can be readily solved on a computer. Each model guarantees an optimal solution, providing it is used correctly. Recommended applications include:

1. Optimum locations of facilities.
2. Optimum use of facilities or resources.
3. Improved lines or product mixes.

They may also be applied to transportation problems in which it is desired to:

1. Minimize transportation and distribution costs.
2. Establish an optimum route.

Integer programming is useful when there are more machines than candidate areas to place the machines. Any type of movement, straight-line, rectangular, or a combination of the two may be used with the model. It is not well suited to manual computation, but does guarantee an optimal solution when used correctly.

Dynamic programming is recommended for solving linear programming problems listed above when a multi-stage decision process is involved. Situations may be frequently encountered that require a series of decisions, with the outcome of each decision depending on the results of a previous decision in the series. Specifically, dynamic programming has been applied to procurement problems, equipment replacement policies, production and distribution problems, and inventory and allocation problems.

The traveling salesman problem is recommended for finding the optimal route from a given location, visiting a specified group of locations with no backtracking, and returning to the point of origin. The technique has also proved useful when scheduling a machine over a given set of repeated operations. Small problems can be solved with a combination of graphical and computational procedures, and larger ones can be solved on a computer. There is no known solution procedure, however, that always guarantees an optimal solution.

The level curve technique is recommended when a single facility is to be located in an existing complex. The location of a new machine in an existing layout is the usual application of this technique; several other analogous situations exist, however, such as the location of a warehouse or a tool crib.

Queuing theory analysis is recommended for situations in which some service is performed under random demand and waiting lines, or queues, form because of irregularity in demand or service. The technique may be used where the number of service facilities is subject to control but the timing of the customers is not. An example is in the determination of the number of dock facilities needed for loading and unloading trucks. Queuing theory may be used to solve scheduling problems where the timing of customer arrivals is subject to control but the number of service facilities is not. A sequencing problem in which the order of units to be processed so that the service time is minimized may be also solved by queuing theory. Queuing theory has also been used successfully to solve line balancing problems when the arrivals are subject to control and all units require the same multiple operations.

Waiting line models are widely documented and are recommended for the analysis of storage space requirements, material handling equipment requirements, the quantity of material required at a given point, and the scheduling and dispatching of material handling equipment within a production system, when the needs of the system occur in Poisson fashion and the needs are satisfied in an exponential manner. Most computations can be done manually, though large problems may require a computer. The probabilistic nature of waiting lines prohibits an absolute optimal solution, but a relative optimal solution can be achieved.

Conveyor models are developed primarily for closed-loop, irreversible conveyor systems. Most such models are useful as part of a simulation procedure, since the interactions of system components result in extremely complex mathematical relationships. Very few conveyor problems can be solved manually, and computer computation is almost mandatory. If the system relationships can be computed, an optimum or near optimum solution can be achieved.

Simulation is recommended whenever the number or complexity of system components creates a mathematically unmanageable problem. This technique is also a useful procedure for simpler problems where the mathematical ability of the analyst is limited. There is no guarantee of optimality except through an exhaustive repetitive procedure, which is seldom economical.

Conclusion

Although surveys have shown that there is comparatively little industrial application of the methods described here, their potential is extremely challenging. As further experimentation takes place, and results are documented, more analysts will be encouraged to use and experiment with these approaches. It is hoped that some readers of this text will do so and will report their successes to encourage further efforts along these lines.

For the student interested in the application of quantitative techniques to facility design problems, there are many interesting situations offering challenging opportunities for further investigation. Among them are:

1. Develop a procedure for designing material flow patterns.
2. Develop a model or technique for handling stochastic traffic–distance data.
3. Develop a procedure for evaluating the relative efficiency of alternative material handling plans or systems.
4. Develop a method of predicting and controlling material handling costs.
5. Develop useful relationships between distance traveled, handling equipment used, and total cost of material handling.

This chapter has explored the application of quantitative techniques to flow, handling, and facility design problems, and further details on these techniques can be found in the volumes cited in its footnotes.

Questions

1. In general, when and why would you use quantitative techniques?
2. What are their major advantages? —disadvantages?
3. How do they fit in with the conventional techniques, in Chapter 6?
4. Which quantitative techniques would you judge most useful in facility design? Why? How? Where?
5. Can you add some project or research ideas to those listed in the Conclusion above.

Exercises

A. Look up selected (or assigned) items in the Bibliography and footnotes and report in class.

B. Look up current articles or papers on selected techniques, and report in class.

8

Planning Activity Relationships

The several preceding chapters have dealt with the flow of material, or other elements, through the facility being designed. In addition to the element flow, around which the equipment and work centers are arranged, there is the problem of locating the many service or auxiliary activities. These should be located to serve the productive activity, but in varying degrees of proximity according to their relative importance to the activity.

The first task is to identify the service and auxiliary activities needed to support the major activity of the enterprise. Table 8-1 tabulates the types of service activities as related to several kinds of facility planning problem situations. As has been indicated or implied previously, the so-called *production* or *productive activity* will vary with the type of enterprise. In all cases, however, a number of supporting services are necessary—each related to the major function of the facility, but in varying degrees of importance.

Types of Activity

In the industrial facility, there are likely to be a much larger number of services than indicated in Table 8-1. A more detailed breakdown is shown in Table 8-2, where the activities are categorized as serving administration, production, personnel, and physical plant. As can be seen, with a large number of service activities the task of properly relating them to production, and to each other, can be rather complex. The first task is to identify them all, to insure that no activity of significance is overlooked or ignored.

Also, as pointed out, the locations of internal activities, as well as flow patterns, should consider the external relationships to the facility site and its characteristics.

Selection of Activity Centers

In choosing the activities or activity centers, the primary characteristics for consideration are:

1. Does a single, or specialized, or particular group of activities occur?
2. Does the activity require a significant amount of floor space—say 100 ft square,

Table 8-1. Activiity Centers for Various Types of Facilities

A. Home

1. Carport or garage
2. Kitchen
3. Dining room
4. Living room
5. Family room
6. Study
7. Deck/patio
8. Bedrooms—master, guest children
9. Utility room, storage, basement
10. Lavatory
11. Bathroom

B. Summer camp

1. Sleeping; tents, etc.
2. Toilet, washing
3. Dining, kitchen
4. Swimming
5. Boating
6. Campfire
7. Athletic field
8. Archery/rifle
9. Entrance
10. Parking
11. Garbage, etc.

C. Restaurant

1. Parking
2. Entrance
3. Coat room
4. Lounge
5. Rest rooms
6. Dining room
7. Private dining room
8. Kitchen
9. Receiving and delivery
10. Garbage, etc.

D. College campus

1. Library
2. Dormitories
3. Student center
4. Class rooms
5. Administration
6. Athletic facilities
7. Maintenance and utilities
8. Bookstore
9. Auditorium
10. Student parking
11. Parking—faculty & staff
12. Health center
13. Research

E. Hospital

1. Admission
2. Central supply
3. Laundry
4. Pharmacy
5. Kitchen
6. Food service—staff & visitors
7. Mechanical maintenance
8. Surgery
9. Laboratories
10. Administration and records
11. Out-patient
12. Patient rooms
13. Emergency
14. X-ray
15. Parking

F. High school

1. Entrance
2. Administrative office
3. Gymnasium
4. Teachers lounge
5. Cafeteria
6. Auditorium
7. Kitchen
8. Delivery and receiving
9. Garbage, etc.
10. Library
11. Supplies
12. Maintenance
13. Athletic field
14. Locker room
15. Faculty parking
16. Student parking

G. Manufacturing plant

1. Receiving
2. Material stores
3. Fabrication
4. Inspection
5. In-process stores
6. Assembly
7. Food service
8. Locker room
9. First aid
10. Offices
11. Warehousing—finished goods
12. Shipping
13. Parking

more or less? (Too small an activity, may get lost in the subsequent assignment of area to the activity.)

3. Does the activity have a lot of flow through it?

A study of the organization chart will help to identify activity centers, as will interviews with key personnel. Then, a study of the activity itself should be made, to become familiar with what goes on there. The result of the activity selection process should be a list, or lists, similar to Table 8-2.

Types of Relationships

Before dealing with specific activity interrelationships, it may be well to identify the types of relationship that exist among the several activities. In general, they are:

Table 8-2. Plant Service Activities

A. Administration

1. President
2. General manager
3. Sales and advertising
4. Accounting
5. Product engineering
6. Purchasing
7. Personnel
8. Product service
9. File room
10. Conference room
11. Vault
12. Reception room
13. Switchboard
14. Data processing

B. Production

1. Industrial engineering
2. Production control
3. Quality control
   a. Receiving inspection
   b. In-process (floor)
   c. Final inspection
4. Plant engineering
   a. General
   b. Maintenance
5. Receiving
6. Stock room (storage)
7. Warehousing
8. Shipping
9. Tool room
10. Tool crib
11. MH equipment storage
12. Supervision

C. Personnel

1. Health and medical facilities
2. Food service
   a. Kitchen
   b. Dining
   c. Vending machines
3. Lavatory
4. Smoking area
5. Lounge area
6. Recreation area
7. Parking
8. Time clock
   a. Bulletin boards
9. Fire escapes
10. Drinking fountains
11. Telephones—booths, etc.

D. Physical plant

1. Heating facilities
2. Ventilating equipment
3. Air conditioning equipment
4. Power generating equipment
5. Telephone equipment room
6. Maintenance shops
7. Air compressors
8. Scrap collection area
9. Vehicle storage
10. Fire protection
    a. Extinguishers
    b. Hoses
    c. Equipment
    d. Sprinkler valves
11. Stairways
12. Elevators
13. Plant protection

1. *Between two production activities* (this type of relationship has been dealt with in the discussion of production, material, or element flow).
2. *Between a production and a service or auxiliary activity.*
3. *Between two service activities.*

The latter two categories are the primary concern of the balance of this chapter.

Factors Affecting Relationships

As with so many other aspects of the planning process, there are a number of factors to be taken into consideration in planning activity relationships. Some of them are especially significant, such as:

A. *Special requirements* of specific activities, or departments, (as pointed out in Chapter 5):

B. *Building characteristics:*

1. Type
2. Size
3. Shape
4. No. of floors
5. Clear height
6. Column location
7. Column spacing
8. Door locations
9. Expansion directions

C. Building site:
1. Location
2. Size
3. Topography
4. Shape
5. Orientation of building
6. Weather (direction)

D. External facilities:
1. Transportation modes
2. Parking
3. Utilities
4. Auxiliary facilities

E. Expansion:
1. Future production flow and layout changes
2. Aisles—locations, width
3. Location of activities likely to expand; and sequence
4. Permanent equipment
5. Extra space; additional floors; etc.
6. Building shape
7. Column location and spacing

Degrees of Activity Interrelationship

In order to help decide which activities should be located where, a classification of degrees of closeness has been established, along with a code to identify each. These have been identified by Richard Muther[1] as:

A = Absolutely necessary—for the activities under consideration to be next to each other
E = Especially important—for them to be close
I = Important—that they be close together
O = Ordinary (closeness)—OK as they fall
U = Unimportant—for there to be any "geographical" relationships.

It should also be recognized that there may be a required degree of separation. That is the activities may best be separated somewhat, for such reasons as:

1. Dirt
2. Noise
3. Fumes, smoke
4. Odors
5. Vibration
6. Safety or health hazards
7. Interruptions
8. Distractions

The code for representing an undesired closeness is:

X = Undesirable—for the activities to be close together

These classifications and their codes are used in the *Activity Relationship Chart,* also developed by Richard Muther.

The Activity Relationship Chart

The *Activity Relationship Chart* (see Figure 8-1) is an ideal technique for planning the relationship among any group of interrelated activities. It is helpful in such

[1]Muther, *Systematic Layout Planning;* form used by courtesy of Richard Muther.

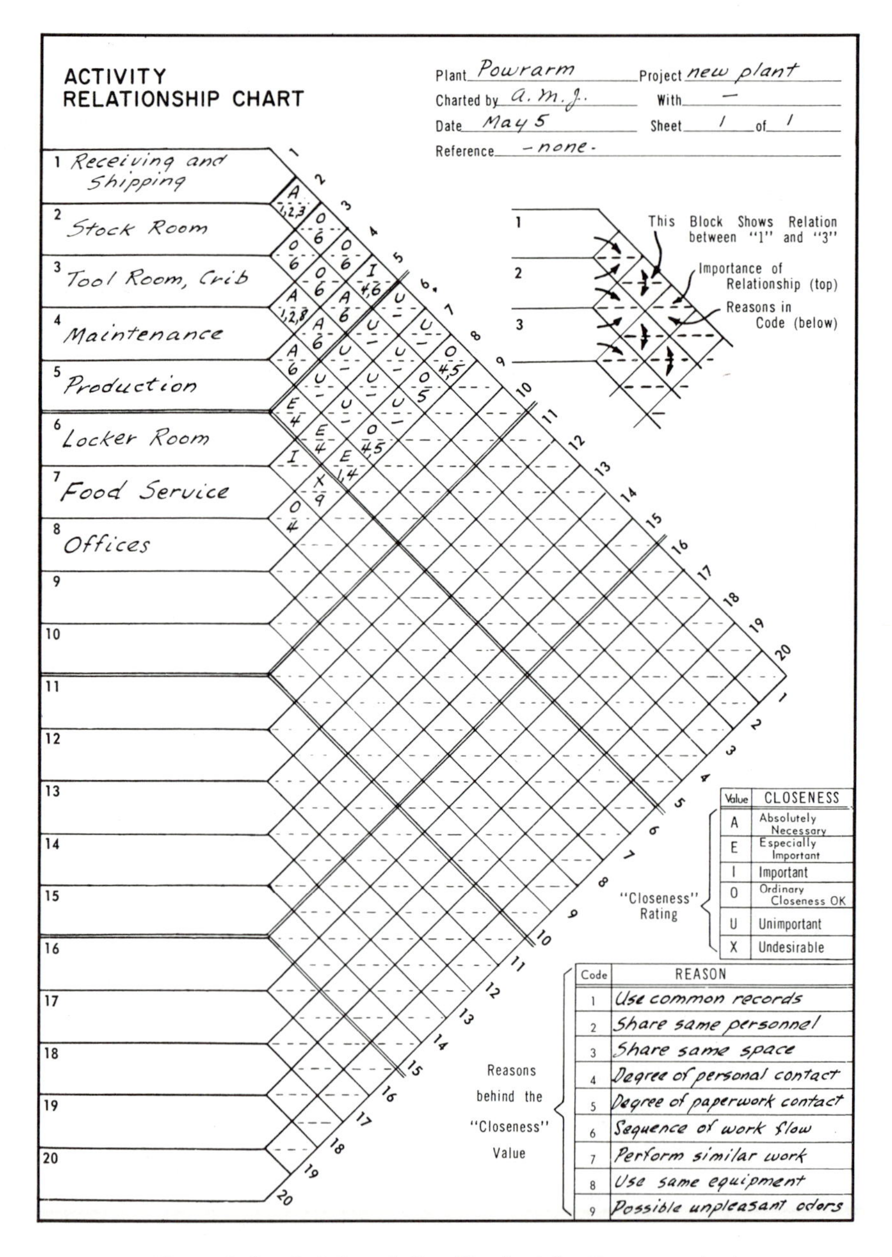

Figure 8–1—Activity relationship chart for Powrarm plant.

cases as:

1. Preliminary allocation of sequence for a *From–To Chart.*
2. Relative location of work centers or departments in an office.
3. Location of activities in a service business.
4. Location of work centers in a maintenance or repair operation.

5. Relative location of service areas within a production facility.
6. Showing which activities are related to each other, and why.
7. Providing a basis for subsequent area allocation.

The *Activity Relationship Chart* is similar to the *From–To Chart,* but only one set of locations is indicated. In fact it is again similar to some road map mileage tables; the distances are replaced by qualitative code letters, and numbers representing reasons for the letter codes. Figure 8-1 is a typical *Activity Relationship Chart.* The relationship code indicates which activities are related to each other and how important each closeness relationship is. Letters (*A, E, I, O, U,* and *X*) are entered in the top of the squares. Colors are sometimes used to represent the degrees of closeness. Code numbers are entered in the bottoms of the squares, representing the substantiating reasons for each closeness relationship. These codes are:

Closeness–Color Code:

A—Red—*Absolutely necessary*
E—Orange—*Especially important*
I—Green—*Important*
O—Blue—*Ordinary closeness*
U—Uncolored—*Unimportant*
X—Brown—*Undesirable*

Production Relationships
1. Sequence of work flow
2. Use same equipment
3. Use same records
4. Share same space
5. Noise, dirt, fumes, vibration, etc.
6. Facilitate material handling

Personnel Relationships
1. Share same personnel
2. Urgency of contact
3. Degree of personal contact
4. Normal path of travel
5. Ease of supervision
6. Perform similar work
7. Personal preference
8. Movement of personnel (traffic flow)
9. Interruption of personnel

Information Flow
1. Use common records
2. Degree of paperwork contact
3. Use same communications (etc.) equipment

On the *Activity Relationship Chart* blank form, the *Reason* column is left blank to permit entry of applicable substantiating reasons for each specific situation. The codes are as used in Figure 8-1.

Constructing the Activity Relationship Chart

The activity interrelationship planning process might proceed somewhat as follows:

1. *Identify all significant service or auxiliary activities* needed to support the major productive functions of the enterprise. Use the *Plant Service Activities List* in Table 8-2 as an aid in (1) drawing up one for the facility under consideration, or (2) editing it to fit the facility being planned.
2. *Separate into categories*—(a) production; (b) service (administration, production, personnel, physical plant).

3. *Collect data on flow* of material, information, personnel, etc.
4. *Decide which factors or sub-factors should determine relationships*—(a) material flow (production only), (b) equipment, (c) information flow, (d) personnel relationships, (e) physical relationships.
5. *Prepare a form* similar to that in Figure 8-1.
6. *Enter the activities* under analysis down the left hand side; the order is not important, although they may be placed in a logical sequence.
7. *Enter the desired (required) degree of closeness,* for every pair of activities, in the square at the intersection of the lines, as a letter (in the top) to represent the relative importance of the relationship—Care and judgement should be exercised in assigning letters to be sure there are not too many *A*'s, *E*'s etc., since this will cause difficulty later when the activities are arranged to satisfy the desired relationships, i.e., if everything must be close to everything else.
8. *A code number (in the bottom) to indicate the reason*—evaluations should be based on knowledge of the relationships among the activities under consideration, and the values of those relationships; it may be wise to discuss the evaluations with persons concerned, or use a form to collect pertinent data from those persons.
9. *Review the Activity Relationship Chart* with other people, to make sure there is some agreement as to the importance of relationships; it might be wise to obtain approvals from appropriate people.

If the analyst has worked out several charts with several people covering interrelated activities, someone may have to act as judge or arbitrator of any serious difference of opinion. Having made a chart the analyst has completed the task of recording the information. The next step is to make use of it in designing the physical interrelationships.

The Activity Relationship Diagram

While the *Activity Relationship Chart* is useful for planning and analyzing activity interrelationships, the resulting information is only useful if it is converted into a diagram. This is the objective of the *Activity Relationship Diagram,* which becomes a basis for planning the relationships between the material flow pattern and the location of service activities related to the production activity. The *Activity Relationship Diagram* is in reality a block diagram indicating approximate activity relationships, showing each activity as a single activity template. (There is no connotation of space at this stage of the planning process; that comes in the next phase, space allocation, in Chapter 9.)

The *Activity Relationship Diagram* is constructed, beginning with an analysis of *the Activity Relationship Chart* (Figure 8-1) and with the aid of the worksheet shown in Figure 8-2, as follows:

1. *List the activities* in the left hand column.
2. *Enter the activity number,* from the *Activity Relationship Chart,* in each column, to represent the degree of closeness with the activity on the line—for example, on the *Activity Relationship Chart,* Receiving and Shipping carries an *A*

WORKSHEET FOR ACTIVITY RELATIONSHIP DIAGRAM						
ACTIVITY	DEGREE OF CLOSENESS					
	A	E	I	O	U	X
1 Receiving and Shipping	2	-	5	3, 4, 8	6, 7,	-
2 Stock Room	1, 5	-	-	3, 4, 8	6, 7,	-
3 Tool Room and Tool Crib	4, 5	-	-	1, 2	6, 7, 8	-
4 Maintenance	3, 5	-	-	1, 2, 8	6, 7	-
5 Production	2, 3, 4	6, 7, 8	1	-	-	-
6 Locker Room		5	7		1, 2, 3, 4	8
7 Food Service	-	5	6	8	1, 2, 3, 4	
8 Offices		5		1, 2, 4, 7	3	6
9						
10						
11						
12						
13						
14						
15						
16						

Figure 8-2—Worksheet for activity relationship diagram for Powrarm factory.

relationship to *Activity 2* (Stock Room); an *I* relationship to *Activity 5* (Production); an *O* relationship to *Activities 6* and *7*. A check can be made by verifying that all activity numbers are recorded on each line (counting the number of the activity on the line; for example, on *line 1*, all activity numbers are included: *1, 2, 5, 3, 4, 8, 6, 7*).

3. *Continue the procedure*—for each line on the *Worksheet*, until all relationships have been recorded.
4. *Enter the identifying activity names* in the centers of the activity templates, using a form such as in Figure 8-3.
5. *Transfer numbers* from columns on the *Worksheet* to the corners of the activity templates, as shown in Figure 8-3; U's are not transferred, since they have been accounted for on the *Worksheet*, and are unimportant from now on.
6. *Cut out activity templates* from form.
7. *Arrange the templates into an Activity Relationship Diagram*, matching first the A's, next the *E*'s, etc., in the most appropriate arrangement; for example, *No. 1* (Receiving and Shipping) might be placed in the upper left-hand corner of the arrangement, as a start. Then, *No. 2* wants to be next to it. And then, *Nos. 1* and 5 want to be next to *No. 2* (*No. 1* already is), etc., etc. Figure 8-4 illustrates one possible arrangement satisfying most of the closeness requirements. As with

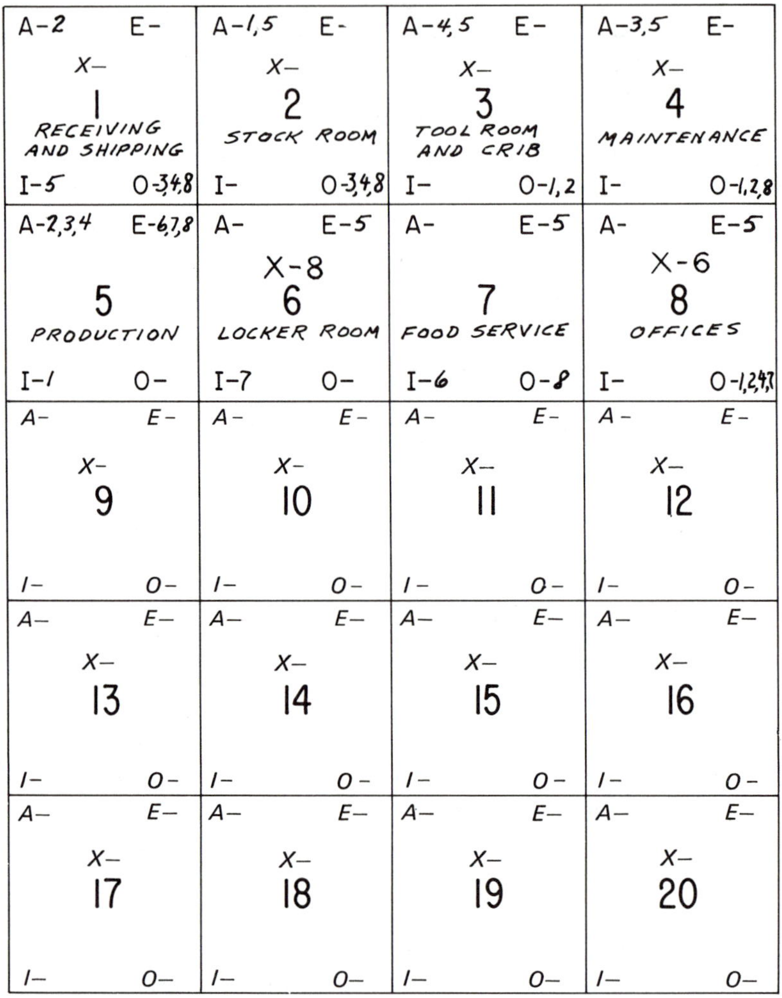

A-2 E- X– 1 RECEIVING AND SHIPPING I-5 O-3,4,8	A-1,5 E- X– 2 STOCK ROOM I- O-3,4,8	A-4,5 E- X– 3 TOOL ROOM AND CRIB I- O-1,2	A-3,5 E- X– 4 MAINTENANCE I- O-1,2,8
A-2,3,4 E-6,7,8 5 PRODUCTION I-1 O-	A- E-5 X-8 6 LOCKER ROOM I-7 O-	A- E-5 7 FOOD SERVICE I-6 O-8	A- E-5 X-6 8 OFFICES I- O-1,2,4,7
A– E– X– 9 I– O–	A– E– X– 10 I– O–	A– E– X– 11 I– O–	A– E– X– 12 I– O–
A– E– X– 13 I– O–	A– E– X– 14 I– O–	A– E– X– 15 I– O–	A– E– X– 16 I– O–
A– E– X– 17 I– O–	A– E– X– 18 I– O–	A– E– X– 19 I– O–	A– E– X– 20 I– O–

Figure 8–3—Activity relationship diagram activity templates for Powrarm factory.

many of the other techniques, there is probably no one best arrangement. Other trials should be made until all concerned are satisfied. Also, an adaptation of the *From–To Chart* could be constructed, and the relationships assigned numerical values (as shown under *From–To Charts*) to prove the best answer more quantitatively.

8. *Copy final arrangement* onto another cross-section sheet, as in Figure 8-4. This is the *Activity Relationship Diagram.*
9. *Draw a tentative flow pattern,* if desired, on the *Diagram.*

Actually, this relatively simple example does not utilize the technique as effectively as a more complex one. The simpler example is used to illustrate the

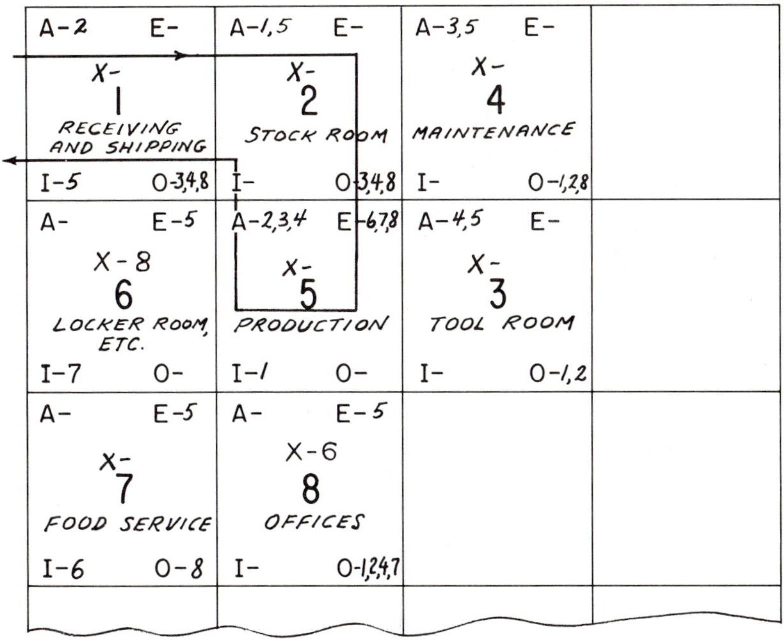

Figure 8-4—Activity relationship diagram for Powrarm plant.

procedure. Where there is a large number of activities and relationships, it may be desirable to divide them into groups of related activities and work first with the larger groups—as when the production function is made up of a large number of activities, or there is a larger number of service activities than in the accompanying illustration. Then, the larger functions may be more easily related to each other—and the process repeated with smaller activities within the larger ones.

An alternative technique, developed by Richard Muther,[2] uses a combination of lines, symbols, and colors, and results in a diagram as shown in Figure 8-5. Here, the symbols are the conventional process symbols, and the number of lines between symbols represent the importance of the closeness; that is, 4 lines = *A*, 3 = *E*, 2 = *I*, etc. Wiggly lines are used to show undesired relationships.

The Muther approach is not unlike that suggested by De Villeneuve,[3] and later by Hoffman[4] and Downs.[5] They have all developed similar flow diagrams,

[2]Muther, ch. 6.

[3]L. de Villeneuve, "The Quantitative Flow Chart." In *2nd Biennial Proceedings of the Packaging and Material Handling Institute* (University of Southern California, 1952).

[4]J. R. Hoffman, "An Evaluation of Quantitative Techniques in Plant Layout." Privately circulated.

[5]G. Downs, "Best Way To Layout a Job Shop," *Factory Management and Maintenance,* Nov. 1956.

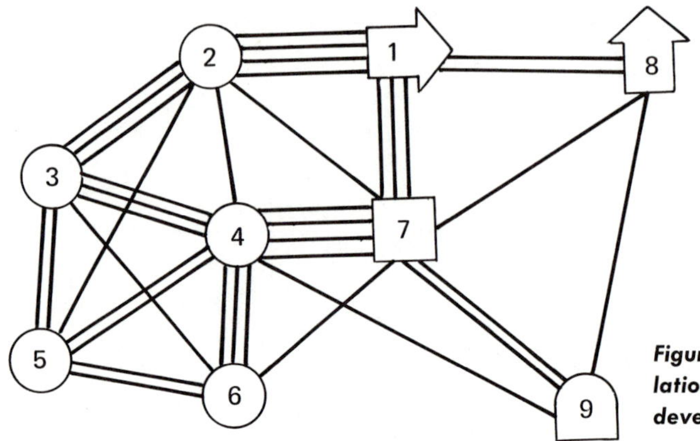

Figure 8–5—Activity relationship diagram as developed by Muther.

with activities connected by lines or bands of varying width or thickness. The width of the line indicates the volume of flow between activities and aids in properly interrelating them in the early stages of layout planning.

Conclusion

This chapter has covered both procedures and techniques for designing—or redesigning—interrelationships among a number of activities. It will be found equally useful in activity relationship planning for any of the types of enterprise referred to in previous chapters, ranging from schools to post offices to manufacturing plants. The next two chapters will be concerned with the details of selected service and auxiliary activities, followed by the determination of space requirements of the facility.

Questions

1. What are the four major categories of service and auxiliary activity? Name some in each category.
2. What are the three types of relationship among activities?
3. What are the categories of factors affecting relationships between activities? Name some factors in each.
4. Describe the *A, E, I, O, U,* and *X* degrees of closeness.
5. What are some reasons for the *X* category?
6. Describe the concept of the *Activity Relationship Chart* and discuss its use.
7. What are some of the reasons for desired closeness between activities?
8. Describe how the *Activity Relationship Diagram* is developed from the *Chart.*

Exercises

A. Make an *Activity Relationship Chart* for one of the facilities delineated in Figure 8-1, or choose your own.

B. Make an *Activity Relationship Diagram* for your solution to *Exercise A.*

9

Production and Physical Plant Services

The previous chapter dealt with the interrelationships among a number of activities serving the productive function in one way or another. Actually, it is difficult to totally define such interrelationships without some background in what they do. So this chapter will begin to discuss certain aspects of some of the activities, as a means of familiarizing the designer with their characteristics as they affect relationships with production and each other.

A fairly complete list of these services was shown in Table 8-2. There are nearly 60 such services or activities that may require space in the layout. Although the size of the enterprise, that is, the number of employees, will determine the relative importance of many of these functions, even the smallest enterprise must provide, in some way, for most of them. For example, a large industrial plant may have 40 to 50 or more persons in an Accounting, Product Engineering, Production Control, or Maintenance Department; whereas in a small plant, any one of the 60 functions might be handled by only a part of one employee's time. Nevertheless, each function must be accounted for in the development of total facility requirements. The service activities may be divided into four categories, serving:

1. *Production*—activities that primarily serve the productive function.
2. *Physical plant*—services that are primarily concerned with the needs of the physical facilities (building, equipment, utilities).
3. *Administration*—functions that serve the entire plant, consisting mostly of the general office areas and related activities.
4. *Personnel*—services required primarily for serving or handling the needs of people.

This chapter, because of space limitations, will discuss only:

Production services

1. Receiving
2. Storage
3. Warehousing
4. Shipping
5. Tool room/tool crib
6. Production (supervision) office
7. Handling equipment storage

Plant services

1. Parking area, drives, etc.
2. Scrap and waste disposal

(Chapter 10 will deal with administration service activities and personnel service activities.)

Before beginning the discussion of these activities, it is necessary to identify and distinguish some of them, as follows:

1. *Receiving*—getting the material, supplies, etc., into the facility.
2. *Storage*—safekeeping and issuing of material, etc., before and during the production operations.
3. *Warehousing*—safekeeping of finished goods.
4. *Shipping*—issuing and distributing the finished goods to customers, or other destination.

Since these functions are often closely interrelated, they are considered together in this chapter. They will be discussed separately, with one exception. Many aspects that must be considered in planning for one of them are either similar to or exactly the same as for others, such as docks, bay size, doors, and column spacing. These will usually be discussed only once—when they first appear as major considerations; but they must be considered important in all four of the activities. It might be well to study this entire chapter before attempting to plan any of the activities it covers, to insure consideration of the overlapping aspects of the activities involved.

Receiving

Receiving is concerned with the orderly receipt of all materials and supplies coming into the facility and their proper storage and disposition. It includes such responsibilities as:

1. Unload material from carriers.
2. Unpack shipping containers.
3. Identify and sort material.
4. Check receipts vs. packing slips.
5. Record receipt on receiving slip.
6. Note shortages, damage, defects, etc.
7. Maintain adequate records
8. Dispatch material to use area.

A function closely related to receiving, but usually not under its jurisdiction, is *receiving inspection,* which normally reports to the Quality Control or Inspection Department, and is responsible for the careful inspection of all incoming material and supplies. From a layout point of view, this means that the receiving inspection activity should most logically be adjacent to receiving. Receiving inspection will normally require space for:

1. Orderly temporary storage of material awaiting inspection.
2. Storage, handling, and work areas necessary for inspection operations.
3. Temporary storage of inspected items, awaiting delivery to points of use or to stores.

Developing receiving requirements. The planning of the receiving activity requires consideration of a large number of factors, as shown in Table 9-1. The actual or expected receipts should be analyzed in order to determine space needs, equipment requirements, work load, etc. (The determination of space requirements for selected activities will be covered in Chapter 11.)

Handling methods can be accurately determined only after having considered the physical characteristics of all items to be received. In most plants, the minimum equipment would consist of a two-wheeled hand truck and a forklift truck for unloading, handling, and storage; conveyors for movement of some items to or through receiving inspection; and space and storage equipment for holding, pending disposition.

As an aid in collecting or estimating data on the receiving activities, the form shown in Figure 9-1 may be useful. After enough data have been collected, it would be wise to re-cap, to extract such information as:

1. General characteristics of goods received.
2. Quantities of each type of goods, package, or container.
3. Average sizes of packages, containers, unit loads, etc., received.
4. Average weights of loads and shipments received.
5. Average number of shipments received per time period.
6. Average pieces per shipment received.
7. Number of receipts by each carrier.
8. Number of receipts required to complete an order.
9. Unloading method and time required.

Such information will be helpful in determining such things as:

1. Workloads
2. Unloading methods and facilities
3. Dock requirements
4. Receiving inspection needs
5. Temporary storage facilities
6. Handling methods and equipment
7. Space for receiving activity

Since averages do not always give an accurate picture, the analyst might want to make a distribution chart of the receipts for numbers of items by weights, etc.; or, a statistical analysis might be made as suggested under queuing theory, in Chapter 7. Such an analysis might indicate:

1. Unloading methods and equipment needs.
2. Handling methods and equipment needs.
3. Need for fewer but larger packages.
4. Need for suppliers to ship in unit loads.
5. Need for fewer but larger shipments.
6. Number of docks, doors, etc.

Another method of graphically portraying receiving data would be to plot such data as:

1. Arrivals
2. Waiting times
3. Spotting time
4. Unloading time
5. Volume unloaded from each carrier

against time, to show activity levels, and to give some idea of peaks to be expected and planned for—or possibly levelled out—by working with vendors or carriers.

Dock design procedure. The procedure for designing dock facilities might be as follows:[1]

[1] Adapted from *Shipper-Motor Carrier Dock Planning Manual* (Washington, D.C.: American Trucking Association).

Table 9-1. Factors for Consideration in Planning the Receiving and Shipping Activities

I. MATERIAL RECEIVED

A. Types

1. Raw
2. Purchased parts
3. Supplies
4. Scrap
5. Waste

B. Physical characteristics

1. Dimensions, size
2. Shape
3. Weight
4. Machine handleability
5. Palletizability
6. Crush resistance
7. Perishability
8. Unusual characteristics

C. Receipts

1. Number
2. Volume
   a. Normal
   b. Peaks
   c. Seasonality
3. Frequency
4. Schedule
5. Degree of containerization
6. Methods of receipt
   a. Truck
   b. Rail
   c. Other
7. Carrier characteristics
8. Vendor/user restrictions
9. Need for re-packing
10. Issue units
11. Condition of receipts
12. Pieces/receipt
13. Weight/receipt

II. SPACE

A. External

1. Yard
2. Approach
3. Apron
4. Dock
5. Auxiliary buildings
   a. Sheds
   b. Garages
6. Weather—direction of wind, rain, etc.

B. Internal

1. Ramp
2. Dock
3. Unloading
4. Inspection
5. Sorting
6. Temporary storage
7. Office
8. Equipment storage
9. Amount of space available, or needed

III. BUILDING CHARACTERISTICS

1. Size, shape
2. Dimensions
3. Clear height
4. Structural design
5. Construction type
6. Floor level
7. Floor load capacity
8. Overhead capacity
9. Floor condition
10. Elevators, ramps
11. Doors—No., size, location
12. Docks—No. openings, size, location, height, etc.
13. Lighting
14. Column spacing
15. Transportation facilities—existing; proposed
16. Unloading facilities
17. Location on site
18. Main aisle locations
19. Number of floors

IV. SPACE LAYOUT

1. Storage
2. Equipment space requirements
3. Activity interrelationships
4. Space available
5. Handling methods
6. Environment required
7. Aisle requirements—No., typ location, width
8. Flexibility
9. Expansion potential

V. EQUIPMENT

A. Handling and movement

1. Quantities
2. Frequencies
3. Distances
4. Handling units
5. Traffic
6. Running surfaces

B. Storage (see under I above)

C. General

1. Investment costs
2. Operating costs
3. Intangibles

Table 9-1 (continued)

VI. LOCATION		
1. Existing transportation facilities	8. Building shape	15. Need for more than one location for either
2. Desired transportation facilities	9. Main aisle locations	16. Relationship of receiving to other activities
3. Location of transportation facilities on the site	10. Door locations	17. Size of receiving area
4. Desired location of transportation facilities	11. Handling methods and equipment	18. Effective utilization of manpower and equipment
5. Orientation of building on site	12. Characteristics of receipts	19. Weather directions—wind, rain, sun, etc.
6. Traffic volume	13. Expansion potential	
7. Building size	14. Feasibility of separate receiving and shipping facilities	
VII. OPERATIONS		
1. Handling methods	3. Record keeping methods	4. Manpower requirements
2. Storage methods		

1. *Establish the annual workload:*
   a. Collect data on arrivals.
   b. Adjust test period volume to future peak load.
   c. Estimate vehicle arrivals (number).
   d. Estimate truck activity (schedule).
   e. Determine loading and unloading times.
2. *Calculate the number of doors.*
3. *Determine the required accumulation space:*
   a. Determine peak accumulation.
   b. Calculate space required.
4. *Design the dock:*
   a. Determine carrier characteristics.
   b. Determine vehicle parking requirements.
   c. Determine manuevering space.
   d. Plan traffic flow for (1) vehicle movement, (2) service roads, (3) roadway approaches and intersections, and (4) pedestrian lanes.
   e. Determine handling methods.
   f. Establish dock heights.
   g. Establish approach level (avoid slopes, ramps).
   h. Specify door height and width.
   i. Locate dock levelers.
   j. Establish area (inside building) for access to carriers.
   k. Plan weather protection.
   l. Plan dock accessories.

Receiving office. Depending on the extent of the receiving activity, the *receiving office* might consist of anything from a single stand-up desk to a full complement of office equipment. The details will have to be determined by the number of

RECEIVING DATA

COMPANY Acme Industries

Rec. Slip No.	Vendor	Carrier		Arrival Time	Unloading Time		
		Name	Type		Start	Finish	Elapsed
432	Bender	Ace	T/T	9:10	9:35	10:15	40
433	Green	Fox	Van	9:30	10:30	10:50	20

Figure 9–1—Receiving data collection form.

persons required for record-keeping and other paper work. (Office planning details are covered in the next chapter.)

Location of receiving activity. As mentioned previously, receiving is the beginning of the entire material flow pattern. It must usually be located conveniently to the external transportation facilities—highways, railroad tracks, or waterways. In the case of a new building then, the location of the transportation facilities may dictate the location of receiving. In some cases, the road or rails can be brought to the desired location. Considering the two alternatives of (1) locating receiving near transportation facilities, and (2) extending transportation facilities to the desired receiving location, Figure 9-2 illustrates several possible arrangements.

Where the manufacturing activity is large enough to warrant it, the local government unit and/or the railroad will frequently go out of its way to provide desired transportation facilities. When activity is small, it is more common to adjust the building plan to the existing transportation facilities.

Table 9-1 suggests a number of factors to be considered in locating the receiving activity (VI). Although this is an imposing if not frightening list to worry over in dealing with such a small problem as where to locate receiving, the importance of the task can be emphasized by asking oneself "Which factors are not important to me?"—and then gambling on ignoring them! How many could be honestly omitted, with no serious consequences? One way to consider such factors[2] is the approach used in Figure 9-3.

[2]This text contains a large number of lists of "factors to be considered." Considering them is not an easy task. The concept outlined in Figure 9-3 can be applied equally well to the other lists in the book.

COLLECTION FORM

COLLECTED BY A. M. James FROM Mar. 15 TO ______

Container			Pcs. per Con- tainer	Total wt.	Destination			—Remarks— Brief description of goods, condition of goods, etc.
					Rec. Insp.	Storage		
Type	Size	No.				Bin	Bulk	
pall.	48 x 40	6	48	8,000	1		2	2 Cartons damaged
ctn.	12 x 24 x 18	2	50	100	1	2		Need for back order

Sometimes it may be wise to plan for more than one location for the receiving activities. This can be true where there is a large volume of receipts, widely different locations where material is used or greatly different material characteristics—such as supplies vs. raw material.

It should be noted that shipping will frequently be located adjacent to receiving. There are two major reasons for this: existing location of transportation facilities, and efficient use of personnel in the smaller facility.

With such data and information as suggested, the analyst should be able to conceptualize the receiving function and get some feel for its activities and their location, size, and arrangement. (Space requirements will be covered in Chapter 11.)

Storage

The storage problem permeates the entire enterprise, from receiving, through production, to shipping. For planning, the overall storage concern can be broken down into the categories of:

1. *Receiving*—during the receiving process and before disposition.
2. *Stores*—safekeeping of raw material and purchased parts until needed by production.
3. *Supplies*—non-productive items used to support the productive function: (a) maintenance supplies and parts; (b) office supplies, etc.
4. *In-process*—material partially processed and awaiting subsequent operations.
5. *Finished parts*—awaiting assembly (may also be stored at an in-process area, or at assembly).
6. *Salvage*—material, parts, products, etc. for re-processing back into usable form.
7. *Scrap, waste, etc.*—accumulation, sorting, and disposal of items no longer useful.

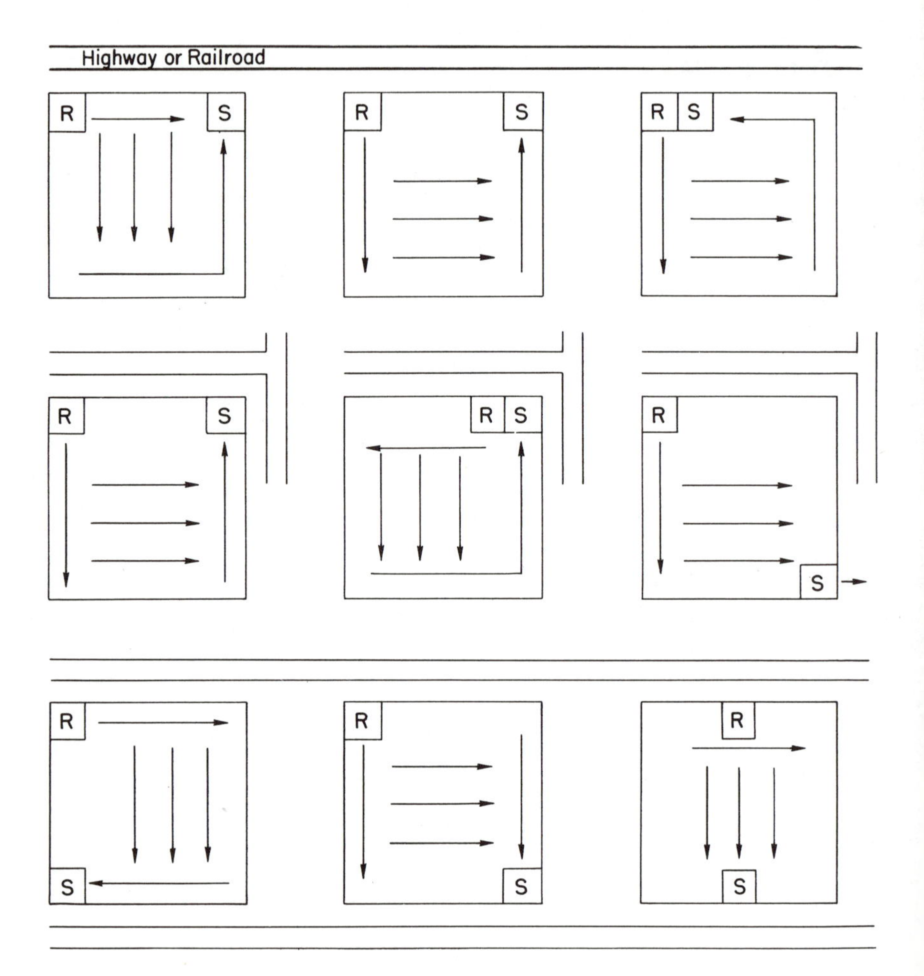

Figure 9–2—Possible arrangements of shipping and receiving areas, with transportation facilities on one side (above), on two adjacent sides (middle), and on two opposite sides (below) of a building.

8. *Miscellaneous*—unused equipment, tooling, containers, etc. for possible use at a later date.
9. *Finished goods*—(to be covered as *warehousing,* later in this chapter).

Receiving. As pointed out previously, space must be allowed for the storage of material at receiving (hopefully small, and temporary!) while awaiting (1) the receiving inspection processing, and (2) removal to the appropriate activity listed above.

Stores. This is the area assigned to the holding of any material to be used in production, until they are requisitioned out according to the production schedule.

SELECTED FACTORS FOR CONSIDERATION IN Locating the Receiving Activity				
SELECTED AND EVALUATED BY: T.A.F.	DATE: May 28			
Order of Impor-tance	Factors	Development of Implied Questions or Data for Consideration	Investigation To Be Made by—	Investi-gation Complete
1	Existing transpor-tation facilities	Now has only highway Will need rail in 1 to 2 yr	L.A.A.	OK 6/2
2	Location of transportation facilities	Roadway will have to be extended Rail siding can be provided	L.A.A.	OK 6/4
3	Orientation of building	Be sure Receiving faces East as both sun & storms are from West	M.M.F.	OK 6/15
4	Expansion potential	Don't buy property if we can't get adjacent piece to the North	M.M.F.	Decision due by 7/1
5	Feasibility of combining R&S	Can combine for now (1 to 3 yr) Design for separate facilities within 5 yr	K.E.H.	OK 6/18
6				
7				

Figure 9-3—Selected factors for consideration in locating the receiving activity.

Normally, for the majority of enterprises, it is indoors. The nature of some material permits outdoor storage, and a saving on storage cost, since no building cost is involved.

This function is often called the *stock room, raw material stores,* or, some other locally descriptive name, referring to the kinds of material stored. (The planning of storage space will be covered below.)

Supplies. As mentioned, supplies are of several kinds and are therefore frequently kept in different locations, near their place of use; production-related supplies would be near the production activity; maintenance supplies, probably in the maintenance shop area; and office supplies, probably in the office area. This would be especially true in larger enterprises, where the three activities might be a considerable distance apart. In the smaller business, all items might be kept in stores.

In-process stores. In the chapter on Process Design it was pointed out that in most manufacturing enterprises, a particular piece of material or part goes through several operations before it is completely processed. The procedure frequently requires that items that have been through one operation, be held before proceeding to the next, until a machine or operator is ready to perform the next operation or a sufficient quantity is on hand for an economical run through the next operation. In-process stores are usually kept in one of two locations:

1. *Small amounts* are usually kept between the workplaces or machines, or adjacent to a workplace.
2. *Larger amounts,* or those to be kept for a longer period of time, are frequently stored in a more remote allocated area until needed—to insure better overall control of the many items in process.

Finished parts. It is often impossible to schedule the production of parts so all will be ready at exactly the right time for assembly, and sometimes the economic production lot size may be larger than the desired assembly lot. In either case, items must be held until needed for the assembly of a specific number of products—which may be determined either by economic assembly lot sizes, or by a specific customer order. The latter may not be an economic lot size, but may have to be assembled to meet customer demand. So, finished parts may all be stored in a location close to the assembly area, or in a separate in-process storage area. The point-of-use storage concept locates all materials to be used at a specific location within that location, rather than in a distant storage location. (This will be covered in more detail in Chapter 12.)

Salvage. In most production processes there will be some material or parts that are mis-processed along the way. That is, they may not pass inspection as good parts, but are still subject to correction, or re-operation, so they can be saved and returned to the normal processing sequence. Also, the scrap from some parts may be large enough to be useful in making other smaller parts. For example, the piece blanked out for an automobile window can be used in making brackets and reinforcements. So an area must be provided for accumulation, evaluation, and re-issue to production. Unsalvageable items become scrap or waste.

Scrap and waste. Scrap normally implies material or parts mis-processed and not salvageable, whereas waste is the normal residue of production—chips, small pieces, ends—not usable for anything in the plant. Such material is usually collected, sorted and hopefully sold to someone else. Many businesses use the scrap, and waste from other businesses as their major, or only, raw material. Therefore, scrap and waste can be a profitable item, to collect and hold for subsequent sale rather than pay to have it hauled away—thereby converting an expense item to an income item. The reader should be able to recall many examples of this procedure, from old newspapers to old tires and scrap lumber.

Miscellaneous. It is common practice in many businesses to save items formerly used but no longer needed, such as machinery displaced for one reason or another, building materials from remodeling, replaced office equipment, containers not needed at present, and dies and tooling for old models of the product. Therefore, a space must be provided for holding them until the time they may be needed for some other purpose, or it is decided to dispose of them. They might then be processed through the salvage or scrap procedures.

It can be seen that all of the above will require space in the structure being designed. Therefore, the need for each must be determined, and the storage activity thought out, so it can be planned, and appropriate space allocated to it.

Storage Planning

Having identified the several potential kinds of storage problems[3] in an enterprise, it becomes necessary to consider the procedures for designing the needed space. At this point, all storage will be classified as just storage—since the data collection, analysis, and design processes are pretty much the same for all categories. The reader will have to translate the small differences required by a specific category. The general objectives of good storage methods are:

1. Maximum use of building cube.
2. Effective use of time, labor, and equipment.
3. Ready accessibility of all items.
4. Rapid, easy movement of materials.
5. Positive item identification.
6. Maximum protection of materials.
7. Neat and orderly appearance.

Consideration of the various items discussed below will aid materially in the achievement of these objectives.

Development of storage requirements. Before it is possible to determine specific storage requirements, it is necessary to consider a number of factors bearing on the space needed. It will be necessary to make many decisions in light of these factors, as the planning process develops. It should be emphasized again that much of the following pertains to all categories of storage, including finished goods warehousing.

Factors for consideration in the development of storage requirements are tabulated in Table 9-2. All of them will have an important bearing on the actual storage space requirements. After the layout designer has given consideration to

[3]For a much more detailed consideration of the storage and warehouse planning process, see Apple, *Material Handling Systems Design*, chs. 16 and 17.

Table 9-2. Factors for Consideration in Storage and Warehouse Planning and Design

I. COMMODITIES

A. Types

1. Raw
2. Purchased
3. Supplies
4. Scrap
5. Waste

B. Physical characteristics

1. Dimensions
2. Shapes
3. Weights
4. Stackable
5. Palletizable
6. Crush resistance
7. Perishable
8. Unusual characteristics

C. Receipts

1. Number
2. Volume
   a. Normal
   b. Peaks
   c. Seasonality
3. Frequency
4. Schedule
5. Degree of containerization
6. Sizes
7. Carrier characteristics
8. Vendor/user restrictions
9. Need for re-packing
10. Condition
11. Pieces/receipt
12. Weight/receipt
13. Types

D. Storage Units

1. Types
2. Number of units
3. Dimensions
4. Shapes
5. Weights
6. Machine handleable
7. Stackable
8. Palletizable
9. Crush resistance
10. Perishable
11. Quantities
12. Number of SKU's (stock-keeping units)
13. Inventory policies
14. Replenishment practices
15. Value

E. Activity

1. Popularity
2. Volume/time period (turnover)
3. Trends
4. Fluctuations
5. Frequency
6. Schedules

F. Orders

1. Number of customers
2. Orders/day
3. SKU/order
4. Quantity/SKU
5. Order mix
6. Need for repacking
7. Order reading time
8. Order filling time
9. Paperwork requirements

G. Issue units

1. Types
2. Volume
3. Popularity
4. Characteristics
5. Turnover

II. SPACE

A. General

1. Total amount
2. Future requirements
3. Expansion potential
4. Funds available
5. Handling methods
6. Storage methods
7. Activity interrelationships
8. Equipment (see IV below)

B. Storage

1. Commodities (see I)
2. Equipment factors (see IV)
3. Building factors (see III)
4. Pallet rack spacing
5. Pallet spacing on racks
6. Type
   a. Bulk
   b. Reserve
   c. Picking
   d. Security
   e. Environment
   f. Rack
   g. Bin
   h. Shelving
   i. Outdoor
   j. Etc.
7. Location in relation to other activities
8. Availability

C. Services

—**External**

1. Yard
2. Approach
3. Apron
4. Dock
5. Auxiliary buildings
   a. Sheds
   b. Garages
   c. Etc.
6. Weather—direction of wind, rain, etc.

—**Internal**

1. Ramp
2. Dock
3. Unloading
4. Inspection
5. Sorting
6. Temporary storage
7. Marshalling, accumulation
8. Packing
9. Loading
10. Shipping
11. Offices
12. Equipment storage and maintenance
13. Locker room
14. Washrooms
15. Food service
16. Equipment room
17. First aid
18. Others

Table 9-2 (continued)

III. BUILDING

1. Amount of space available/required
2. Dimensions; shape
3. Clear height
4. Structural design
5. Construction type
6. Floor load capacity
7. Overhead load capacity
8. Handling methods
9. Number of floors
10. Floor condition
11. Environment
12. Elevators, ramps
13. Doors—No., size, location
14. Docks—No., openings, size, location, height
15. Lighting
16. Column spacing
17. Aisle requirements—No., type, location, width
18. Loading and unloading facilities
19. Safety requirements
20. Location of roads, rails
21. Floor level
22. Location on site

IV. EQUIPMENT

A. Handling and movement

1. Quantity
2. Frequency
3. Origin and destination
4. Loading level
5. Unloading level
6. Distance
7. Unit handled
8. Area covered (scope)
9. Path
10. Cross traffic
11. Running surface
12. Clear height

B. Storage

1. Material characteristics (see I–A and B)
2. Quantity characteristics (see I–C and D)
3. Storage unit characteristics (see I–D)

C. Picking (see I–E, F, and G)

1. Issue unit characteristics
2. Order characteristics
3. Storage methods
4. Item destination
5. Order documentation

D. General

1. Functions to be mechanized
2. Potential equipment types
3. Investment costs
4. Operating costs
5. Intangibles
6. Pay-off policy
7. Savings

V. OPERATIONS

1. Storage methods (see II–B and IV–B)
2. Handling methods (see IV–A)
3. Stock location system
4. Order picking methods (see I–F and IV–C)
5. Communications system
6. Record keeping system
7. Information system
8. Training requirements

VI. RECEIVING AND SHIPPING

1. Receipts and shipments/day
2. Frequency distribution
3. Sizes of receipts and shipments
4. Schedules
5. R. & S. methods
6. Carrier characteristics
7. Vendor/customer restrictions

VII. COSTS

1. Investment required
2. Start-up costs
3. Operating costs
4. Return on investment
5. Space savings
6. Cost of capital
7. Depreciation policy
8. Indeterminate costs

VIII. OTHER FACTORS

1. Flexibility
2. Adaptability
3. Expansion
4. Long range plans
5. Maintenance
6. Obsolescence
7. Capacity
8. Possibility of dual system
9. Intangibles
10. Manpower requirements

the factors, he must convert the basic storage data into the actual spaces required for material storage. (Some of these calculations will be reviewed in Chapter 11, under Space Determination.)

Outdoor storage. Often, material can be stored outdoors. The primary reasons for this are:

1. Lower, or completely avoided, construction cost.
2. Protective coatings and covers can provide adequate covering.
3. Many kinds of material need no protection from weather.
4. Availability of improved vehicles for outdoor use.

Regardless of type of material stored outside, much the same planning must be done as for indoor storage. Different factors for consideration are the yard surface, access roadways, location, and amount of protection required. The requirements for various degrees of protection are shown in Table 9-3. An idea of the relative costs of various types of storage space can be gotten from Table 9-4.

An interesting variation on outdoor storage is the inflatable structure shown in Figure 9-4, which is supported by low air pressure from a blower, with the bottom held in place by sand in a tubular segment around the bottom edge.

Table 9-3. How Material Protection Requirements Affect Yard Storage

Material Protection Required	Elements Required in System							
	Land Area	Surfacing	Location or Identification	Protective Cover or Packaging	Fencing	Lighting	Watchman	Material Handling Equipment
None	R	N	R	N	N	N	N	R
Weather and Atmosphere	R	N	R	R	N	N	N	R
Theft	R	N	R	D	R	D	D	R
Weather, Atmosphere and Theft	R	N	R	R	R	R	D	R

R – Required N – Not required

D – Depends on value and ease with which it can be carried off

Source: *Modern Materials Handling.*

Table 9-4. Relative Storage Facility Costs

Storage Facility Type	Land	Building or Surface	10,000 Sq. Ft. of Storage
	Cost per Sq. Ft.	Cost per Sq. Ft.	Initial Cost,
	Dollars	Dollars	Dollars (Less Land)
Building – One story on slab with heat, light, offices and employee facilities	1.00	8.00–15.00	80,000–150,000
Building – Roof and lighting without sidewalls, heat, office space or employee facilities	1.00	4.00–8.00	40,000–80,000
Building – Rental with above facilities	–	2.00 yr.	–
Yard – Concrete surface	0.75	0.60	6,000
Yard – Black-top surface	0.75	0.15	1,500
Yard – Crushed stone surface	0.75	0.11	1,100
Yard – Undeveloped	0.75	–	–

Source: *Modern Materials Handling.*

Warehousing

Warehousing is concerned with the orderly storage and issuing of finished goods or products.[4] It includes such responsibilities as:

1. Receiving of finished goods from production.
2. Orderly, safe storage of goods.
3. Order picking for shipping.
4. Packing for shipment.
5. Maintaining proper records.

It is obvious at this point that many of the factors involved in planning for warehousing and shipping have been covered under receiving and storage.

Development of warehousing requirements. In general, the warehousing of finished goods or products will not involve as many different items as will the

[4]For further details, see Apple, chs, 16 and 17.

Figure 9–4—Inflatable warehouse sustained by low pressure air from blower. (Courtesy of C.I.D. Air Structures Co.)

storage of raw material or component parts, because many parts have been assembled into each finished product. Also, the finished products are more likely to be packaged into units lending themselves to easier handling and storage—such as cartons, boxes, and crates.

As with receiving and storage, the starting point for warehousing analysis is the physical characteristics of the items to be warehoused. Each item must be known, and certain facts about an item and its activity must be determined. In fact, most of what has been said about the storage activity applies equally well here. Before planning the warehousing activity, the reader should review the discussion of storage, and if more detail is desired, the reference in the footnote on page 225. (Space considerations are also covered in Chapter 11.)

Shipping

Shipping is concerned with the disposition of stock selected to fill orders, the packing of items for shipment, and loading them onto a carrier for delivery. As pointed out previously, the shipping function is frequently carried out in conjunction with the receiving and/or warehousing activity. Due to the close interrelationship between receiving, warehousing, and shipping, they must be considered together in the layout planning project, as evidenced by the coverage of many of the factors involved in planning the shipping activity, under either receiving or warehousing. In fact, it might be said that the shipping activity is the reverse of the receiving activity.

Planning shipping requirements. Since the space and work load determining factors and problems for the shipping activity are so nearly identical to those already covered, they will not be detailed again. Instead, the reader may want to refer back to receiving topics and adapt his thinking and planning to the shipping activity. Some of the factors to be reconsidered here are:

1. Physical characteristics of items handled.
2. Determination of work load: (a) number of shipments per unit of time, (b) pieces per shipment, (c) number and schedule of carriers arriving, (d) volume by truck and rail.
3. Design of docks and related equipment.
4. Office space for record keeping.
5. Handling methods and equipment.
6. Location of shipping area.

A brief review of the above topics in previous sections and references should enable the layout engineer to develop requirements for the shipping activity.

Of primary importance in planning are (1) the hold area in which items are placed awaiting packing and shipping, (2) the packing area itself, and (3) the additional truck or rail space—over and above that required for receiving. A recap of normal and peak shipping loads, using the techniques discussed earlier, should permit the development of space requirements adequate for planning. (Details of space planning will be found in Chapter 11.)

Interrelationship of receiving, storage, warehousing, and shipping. In review, the layout engineer should remember that the four activities discussed so far in this chapter are very closely interrelated. If a theoretical arrangement could be devised, it would place these activities in an arrangement somewhat similar to one of those shown in Figure 9-5. Actually, all factors must be taken into consideration—especially the location of external transportation facilities. Another very important factor is future expansion. These activities should not be so located in the proposed building that expansion possibilities are blocked by such permanent structures as roads, railroad tracks, truck wells, or other fixed items that would be costly to move.

Vendor and customer relationships in shipping and receiving. One last, but very important, factor remains for consideration—the close cooperation with both vendors and customers. Nothing can so disrupt the receiving and storage functions as to receive a shipment of castings in an old 55 gallon drum, when both handling and storage facilities are geared to the use of lift trucks, standard unit load containers, and pallet racks; or, to receive a large quantity of bearings or electric motors in individual boxes or cartons. It is only good business for purchasing and

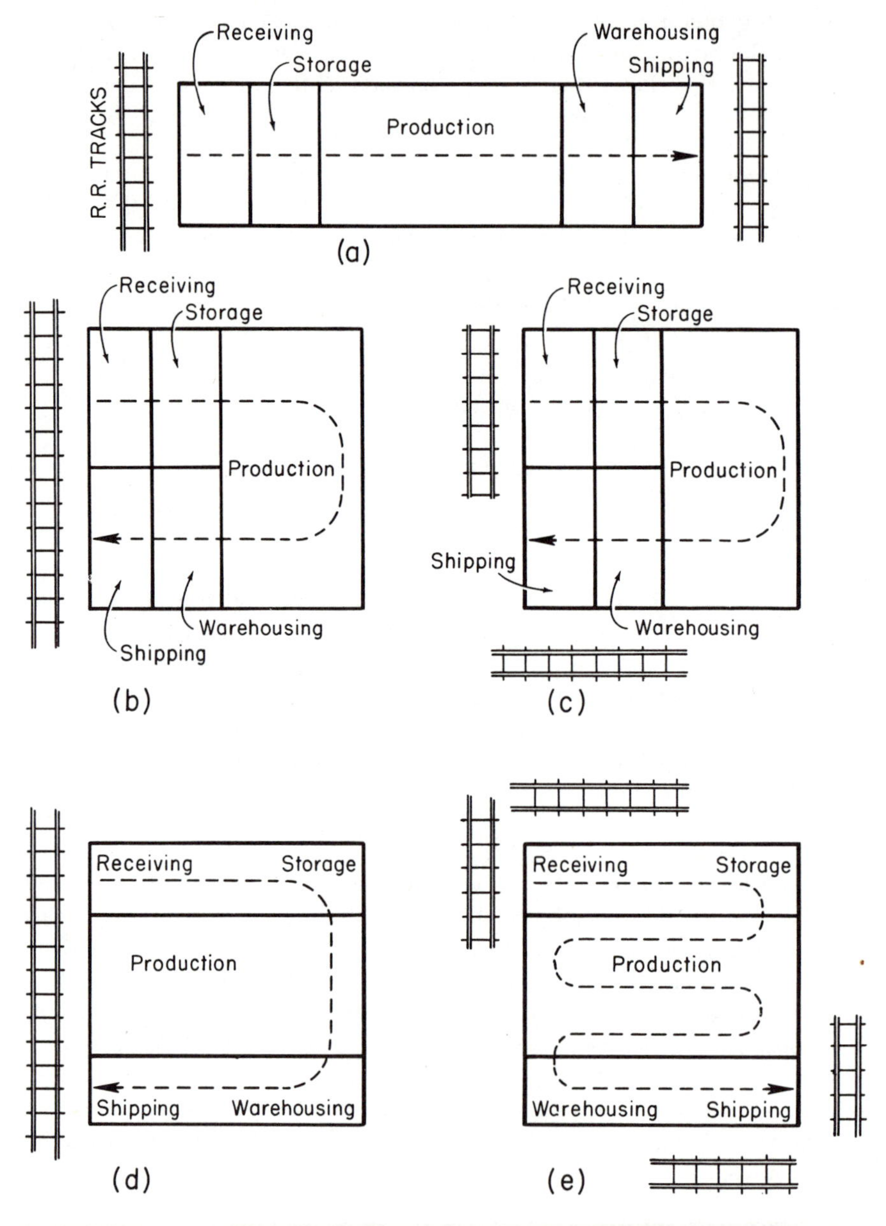

Figure 9–5—Theoretical space relationships.

sales departments to work closely with vendors and customers to establish packaging and packing methods and containers acceptable to both parties. Much time, effort, handling, and money can be saved by both parties if goods are received and shipped in containers suitable for use in subsequent handling operations.

Tool Room and Tool Crib

The tool room is charged with the responsibility of making and maintaining the tools, jigs, fixtures, dies, etc. used on, or with the production equipment. The tool crib is the repository for these things, plus such purchased items as drill bits, taps, chucks, and all manner of hand as well as small power tools. Their responsibility is with the storage of such items, their issue to the shop, and their control—in terms of safekeeping, checking in and out, and inspection to determine the need for sharpening, maintenance, and repair.

From a layout point of view, one of the first problems to be considered is that of centralization v. decentralization. The centralization of all such facilities into one area has the following advantages:[5]

1. Ease of control
2. Ease of supervision
3. Lower operating cost
4. Faster and better service

Centralization is common in smaller plants where one such area can conveniently serve all personnel. In larger plants, however, the use of one central location may not be practicable; the following factors indicate the desirability of decentralized areas;[6]

1. Multi-story building.
2. Multi-building plants.
3. Large plants where personnel would lose time in walking excessive distances.
4. Points of use of various tools are scattered about the manufacturing area.

Location of these areas is another important consideration. In general, a tool room is centrally located, and a tool crib is located where needed to issue tools to employees. Major considerations in the layout of tool rooms and tool cribs, to allow for easy, quick, and sure receipt, storage, and disbursement, are:[7]

1. *Do not store tools beyond reach,* generally not beyond 6 feet from the floor.
2. *Leave space for all sizes* of tools of a given type.
3. *Provide ample aisle space* between rows of bins for passage of personnel.
4. *Make shelf space adaptable* to accommodate the items to be stored and so that the shelving can be readily altered for other items.
5. *Maintain flexibility* in arrangement to permit expansion or contraction of space.
6. *Provide uniform storage equipment* to permit expansion of bins, racks, and boxes.
7. *Provide space and storage equipment* adequate in both size and load-bearing capacity for material to be stored.
8. *Provide for protection* against damage or deterioration.
9. *Provide means for identifying* and readily locating items.
10. *Provide for selection* of oldest stock first.

[5]Adapted from L. L. Bethel, F. S. Atwater, G. H. E. Smith, and H. A. Stackman, *Industrial Organization and Management,* 2nd ed. (New York: McGraw-Hill Book Co., 1952), pp. 280–81.
[6]Bethel, *et al.*
[7]Adapted from Bethel, *et al.*, p. 269.

Figure 9–6—Tool crib showing small tool storage racks and check-out window. (Courtesy of Lyon Metal Products Co.)

Storage equipment is usually of steel construction, capable of indefinite expansion, and made up of units of such a nature that they can be readily dissassembled and reassembled into a variety of sizes and arrangements. A typical example is shown in Figure 9-6. Where pallets are to be stored, space must be allowed for truck movement and the handling of pallets. Table 9-5 suggests a number of worthwhile guidelines for tool-crib planning.

Tool rooms usually contain a variety of machines to provide for the proper construction and maintenance of tools, jigs, fixtures, and the many special tools needed for the production operations. In smaller plants, the tool room or tool crib may be combined with a maintenance shop.

Production-Supervision Offices

In addition to the general or administrative offices in a production enterprise, there should usually be an area for production supervision to carry out planning and other office functions. This can range all the way from a stand-up foreman's desk, to a suite of offices, possibly at several locations throughout the production areas. Commonly, such an office is a small cubicle, say 10 to 12 ft square, located out in the plant. Fairly popular are the pre-fab, or portable office, and the

Table 9-5. First Steps in Planning a New Tool Crib

A. Location

1. Place the crib conveniently close to the machinery or work area.
2. Make a rough sketch of the proposed crib, showing dimensions as well as any windows, doors, columns, radiators, switches, sprinkler heads, and any other fixed obstruction that might interfere with setting up the crib.
3. Be sure that the crib can be supplied with lighting, electricity, or any other service it may require. Also, make sure that ceiling height is adequate. Plan for adequate lighting.

B. Layout

1. The layout should make best use of all of the alloted floor space using as few crib attendants as possible.
2. It should allow room for horizontal and vertical expansion.
3. Enough free space should be allowed for backup inventories.
4. All drawers, racks and bins should be accessible from the floor without stools or ladders.
5. Allow enough aisle width between cabinets for trucks, if necessary.
6. Eliminate dead ends to save steps and time.
7. Use floor areas with adequate capacity for anticipated loads.
8. Place tool issue openings or counters convenient to crib attendants and shop personnel.
9. Provide work areas for the crib attendants—one for record keeping and ordering, another for receiving goods.

C. Equipment

1. Where possible, use building block storage units. When racks, bins, and the other equipment are the same dimensions or multiples of an overall standard dimension, you can shift equipment without expanding the original space.
2. Bins, racks, drawers, and all outside surfaces should be easy to clean. All exposed surfaces should be the same color for appearance.
3. Equipment should protect tools and parts from dirt, grease, and pilferage. Cabinets holding valuable precision tools or parts should have locks.
4. Doors leading outside the crib should have locks to keep out unauthorized personnel.
5. Tool issue openings or areas should have counter tops of steel or other durable material, at waist height, to protect the counter top.
6. Record keeping pads or files should be conveniently located and easy to use.
7. To avoid damage to contents of drawers, bins, and racks, equip them with partitions, cradles, clamps, felt linings, or any other protection they may need.
8. Equipment should permit grouping of similar items.
9. Drawer partitions that may be shifted or removed add storage flexibility to the drawer space.

D. Storage and control

1. Items should be stored and identified for fastest, easiest retrieval.
2. The storage system should permit easy control of inventories to avoid the need for constant supervision by management.
3. Complete records should be kept to show how often each item in the crib is issued. Such records may indicate when an item may have to be replaced or serviced, how to change layout so the fastest moving items are nearest the issue point, or how many of a particular item ought to be available in the crib.
4. Store heavy items on the bottom, lighter items above, and the fastest moving items as close to the issue point as possible.
5. Keep complete records of lost, broken, or worn tools. Know where an item is at all times whether in the shop or in the repair department.

From *Mill & Factory*.

mezzanine or balcony arrangement (Figure 12-3). The latter provides two worthwhile advantages: (1) use of space above a production or other area, and (2) better surveillance of the area being supervised, as a means of identifying trouble spots.

Other Service Activities

Similar work areas, of appropriate size, may also be provided for such other production related activities as quality control, inspection, production control, and plant industrial engineering. Another function requiring space in the plant proper is maintenance, which will need a suitably equipped area for carrying out necessary activities, for both production equipment and the building, such as:

1. Preventive maintenance
2. Repair work
3. Alterations

In the smaller plant, these are combined; in the larger plant, separated—often placed in several different locations according to their functions.

Handling Equipment Storage

If the enterprise will require more than a very few pieces of mobile handling equipment, it is far better to provide specific places for their storage than to leave them in an aisle. Such items as lift trucks, tractors, trailers, extra pallets, and containers can be more easily controlled if returned to a central location. Also, the problem of their maintenance, along with battery charging and handling equipment, must be considered in determining space needs.

Parking Facilities

An increasing problem for today's plant is the provision of adequate parking space. This has become more acute in recent years with the trend toward suburban plant locations and suburban living, and the need to drive to work. It has been suggested that one parking space must be provided for a statistically computed n number of employees. This, of course, will vary with the plant location and the availability of public transportation. (Calculation of parking space requirements and layout of parking areas will be covered in Chapter 11.)

Conclusion

Production and plant service activities are extremely important to the operation of an efficient enterprise. Background information given here will permit the planner to more effectively determine space requirements and plan area allocation (covered in Chapters 11 and 12). The next chapter will deal with administrative and personnel service activities and their requirements.

Questions

1. Distinguish between receiving, receiving inspection, storage, warehousing, and shipping.
2. Why must these functions be considered before proceeding with the planning of the overall flow pattern?
3. What are some of the categories of factors to be considered in developing receiving requirements? Name some factors in each category.
4. What kinds of data do we hope to extract from a study of such factors?
5. Discuss the location of the receiving activity.
6. What are some important factors in planning docks? Briefly explain each.
7. What are the several types of docks? Explain.
8. Name the different categories of items for which storage must be planned in a typical enterprise.
9. Name some factors of concern in planning in-process storage.
10. Why, and how, are receiving and shipping functions sometimes interrelated?
11. What are some objectives of good storage methods?
12. What categories of factors must be considered in developing storage requirements? Name some factors in each category.
13. What are the several kinds of material that may be stored?
14. How does the container in which material is received affect storage and handling activities? Give examples—good and bad. How can the problem be minimized?
15. What are some uses and advantages of outdoor storage?
16. Explain how, and why, receiving, storage, warehousing, and shipping are interrelated and must be planned together.
17. What are some ways vendors and customers could cooperate to achieve mutual savings?
18. What are the advantages of centralized service areas? —decentralized service areas?
19. What are some of the major considerations in the layout of a tool crib?
20. Why has parking become more of a problem in recent years?

10

Administrative and Personnel Services

The previous chapter dealt with production and plant service activities. This chapter concerns itself with administrative and personnel services, sometimes called *offices* and *industrial relations*. And remembering that service activities frequently occupy over 50 per cent of a production oriented facility (and various other significant portions of most facilities), careful attention should be paid to detailing these important aspects of facility design. This is emphasized by the fact that their construction generally costs far more than factory space—in the neighborhood of 100 per cent to 200 per cent more—due to finishing requirements for floors, ceilings, wall treatments, and lighting.

Selected Service Activities

It would be impractical to go into detail on all the administrative and personnel services indicated in Table 8-2. So, this chapter, though including a few comments on less common personnel facilities, will deal primarily with some that are sufficiently general in nature and common in practice to warrant more extensive coverage:

1. Offices
2. Health and medical facilities
3. Food service
4. Locker room, lavatories, etc.

Offices

As can be seen in Table 8-2, office space is required for many people, and many purposes. In a smaller facility offices are commonly all combined into one area to facilitate communications among personnel. The area is frequently in the front of the building for the convenience of visitors. In a larger facility, the general or administrative office may be in the front of the building, and offices for services to production and personnel may be located within the production area. In many larger plants, general offices are located in a separate building, for closer coordination between various office functions, and to provide an appropriate environment for tasks not closely related to production.

In the layout of any office area, an analysis of the work to be done should be

Table 10-1. Guide to Office Area Layout

A. General

1. In laying out the departments, sections, working units, etc., remember that a straight line is the shortest distance between two points; and then, as nearly as is practical, have the flow of the work conform to this principle.
2. In planning the general layout, consider any electrical or structural need that must be conformed to in connection with mechanical equipment.
3. Remember that office space must be conserved, but not, of course, at the expense of appearance, production, or comfort.
4. Place related departments near each other.
5. Aisles should be at least 4 ft wide.
6. In assigning working space, provide for peak load rather than for bare minimum requirements.
7. Use the past annual increase in the volume of work handled as a basis for planning space requirements for future expansion, but remember that office work has a tendency to increase at a faster rate than factory space.
8. Group minor activities around major ones so that when more space is needed the major functions will be taken care of first.
9. The type of work to be done is the basis for departmentizing the office work.
10. Each employee, including desk, chair space, and share of the aisle, requires 75 to 100 sq ft of working space.
11. In the typical office type layout, in any given department, all employees should face in one direction with any natural light coming over the left shoulder or from the back.
12. Where employees must be placed back to back, it is well to leave at least 4 ft between chairs.

. Private offices

1. The tendency is distinctly away from private offices except where privacy is absolutely necessary.
2. Try not to construct private offices so that they cut off the natural light or (if no airconditioning) ventilation from those who work in the adjacent outer offices.
3. Those who do work that requires much concentration are entitled to privacy. Suggested office space requirements as follows:

   a. Senior executives, 300 to 400 sq ft
   b. Department heads, 150 to 300 sq ft
   c. Supervisors, 100 to 200 sq ft
   d. Staff personnel, 75 to 100 sq ft

4. By providing a general conference room where confidential meetings may be held, the need for private offices is often minimized.
5. For private offices, transparent or translucent panels can be used in the upper portion of 66-to-72-inch-high partial partitions.

C. The general office

1. One large office is a more efficient operating unit than the same number of square feet split up into smaller rooms, because (a) supervision and control are more easily maintained, (b) communication between individual employees is more direct, (c) better light and ventilation are possible. Centralization, however, has the disadvantages of the general noise, conversation, etc., that characterize large groups of people. There are also disturbances from aisle traffic.
2. Standard widths for main circulation aisles vary from 5 to 8 ft. Less important aisles vary from 3 to 5 ft.
3. Desks in general offices for clerical workers should not be less than 4 ft apart at chair spaces, with a minimum of 18 in. at the sides of the desks.
4. Passages between rows of desks or between desks and solid or window walls may be from 2 to 3 ft wide.
5. Rooms facing directly south or west are in general the least desirable.

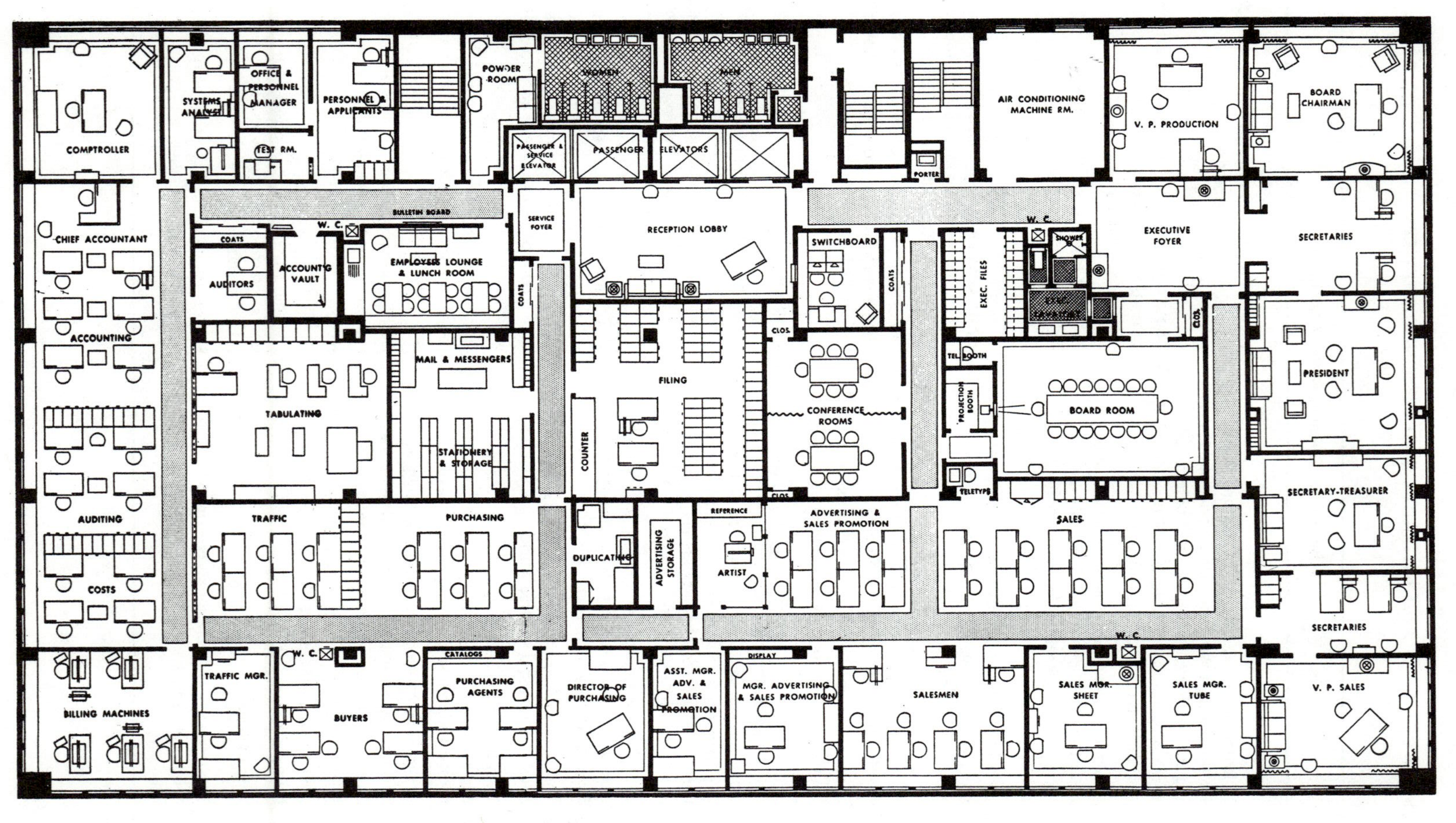

Figure 10-1—Typical office layout. (Courtesy of Rodgers Associates.)

made in much the same way as for the manufacturing processes, to find out what kinds of work are involved, and what special requirements of certain activities must be planned for. Such an analysis makes it possible to plan the special interrelationships of employees who must work in close cooperation with one another. At this point, an examination of the company organization chart is helpful to indicate or suggest logical groupings of related functions. Many of the techniques presented in Chapter 6 will be found helpful in the planning of activity relationships; especially useful are the *Activity Relationship Chart* and the *From–To Chart.* Some suggestions on office layout are listed in Table 10-1[1] as a general guide in planning the effective arrangement of office areas. A typical office layout is shown in Figure 10-1.

Office landscaping. One of the latest trends in office design and furnishing is referred to as *office landscaping* because of its irregular location of desks, tables, and chairs; free-standing screens in place of partitions; plants as a screening medium; and carpeting in place of floor tile. A major aspect of the trend is its attempt at arrangements that permit more efficient work flow and communication than achieved with the cubicle office arrangement. Proponents claim the layout should be designed more around the interaction of people than their place in the hierarchy. This often means breaking up traditional departmental lines, to put an accountant near an engineer—because they communicate regularly. The arrangement of screens and planters is planned to provide *channels of vision,* and carpeting and acoustical tile soften noise to allow more efficient work and decision making.

The idea of office landscaping originated in Germany, with the Quickborner Team of Hamburg, and has been adopted by many U.S. firms beginning in the late 1960's. Figure 10-2 shows a layout for a landscaped office.

One primary advantage of the concept is the flexibility it provides. Industry experts estimate that 50 per cent of office space requirements change within 5 years and that it costs about $5.00/ft to re-arrange an office layout that has standard fixed partitions; and that the landscaped office can be re-arranged for less than one-tenth of that cost.

The major emphasis in the design of a landscaped office is on interrelationships among people, and much attention is paid to defining such relationships. The *Activity Relationship Chart* and *Procedure Chart* in Chapter 6 are extremely helpful in documenting this type of information.

Noise. In the typical office layout, noisy equipment should be segregated as much as possible. With the current emphasis on human factors, however, office equipment is being designed to operate more quietly, and carpeting, drapes, and acoustic ceilings are also help toward quieter office areas.

[1] Adapted from *Office Standards and Planning Book* (Jamestown, N.Y.: Art Metal Construction Co., 1958), pp. 19–23.

Health and Medical Facilities

Some space must be provided in almost every facility for medical services of one kind or another. The simplest solutions for the smaller enterprise range from a first-aid kit to a room with a bed or cot, a chair or two, and first-aid equipment. These are minimum requirements in case of accident or sickness. Some large plants are equipped with complete hospitals, including operating rooms, X-ray facilities, and dental clinics. Most, however, will require facilities somewhere between. Examples of health facilities are shown in the layouts[2] in Figure 10-3, designed for a plant of about 1,500 employees, and Figure 10-4, for a plant employing less than 500 persons.

Food Service

Most plants provide an area where employees may eat meals in pleasant surroundings, away from their work. Some of the advantages of such an arrangement are:

1. Takes personnel away from workplaces, providing a break in routine.
2. Keeps food and related waste out of the plant proper.
3. Offers healthful, sanitary, and pleasant surroundings for meals.

[2] *Plan for an Industrial Medical Department* (Boston: American Mutual Liability Insurance Co.).

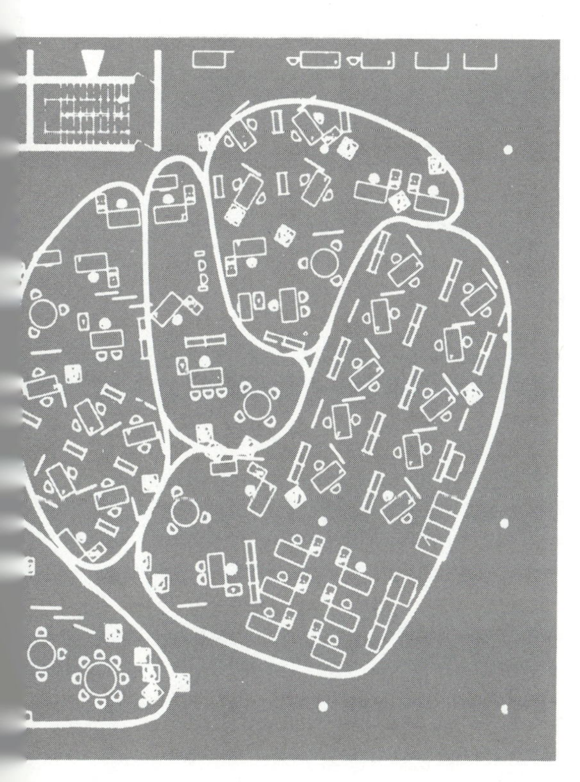

Figure 10–2—Above, typical office landscape installation, with screens of differing heights, planters, carpeting, and special treatments of ceiling and window walls. Below, typical office landscape floor plan, showing arrangement of work groups. (Courtesy of Quickborner Team, Inc.)

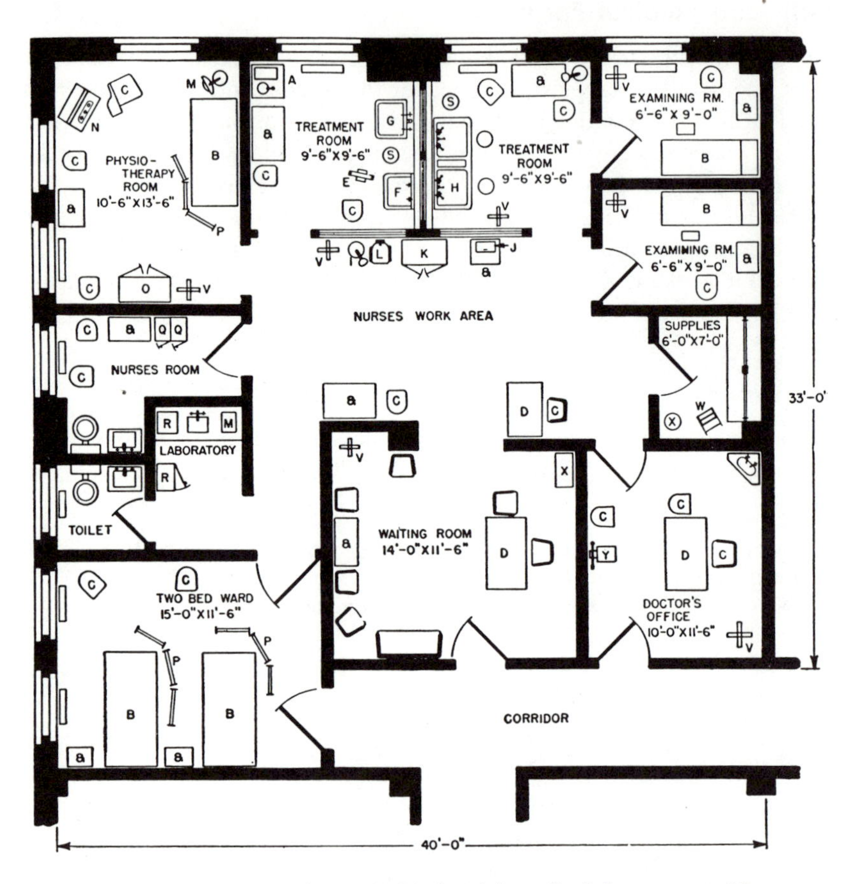

Figure 10–3—Layout of a typical industrial medical department. **(Courtesy of American Mutual Liability Insurance Co.)**

4. Makes possible the preparation of complete or partial meals to replace or supplement the cold lunches employees may carry.

Separate eating areas, however, are not a necessity. Various alternative eating facilities are described as follows.[3]

The snack bar. Organizations that do not have central lunchrooms may serve coffee, tea, milk, rolls, etc., at nominal cost, and some even offer such service free of charge. The installation of a snack bar requires but a small amount of space and equipment (counter, sink, coffee equipment, tableware, refrigerator, and storage cabinet). A small part of an employee's time is all that is required in its operation. The snack bar is usually open only during lunch periods and coffee breaks. The

[3]Adapted from *Lunchrooms for Employees* (New York: Policyholders Service Bureau, Metropolitan Life Insurance Co.).

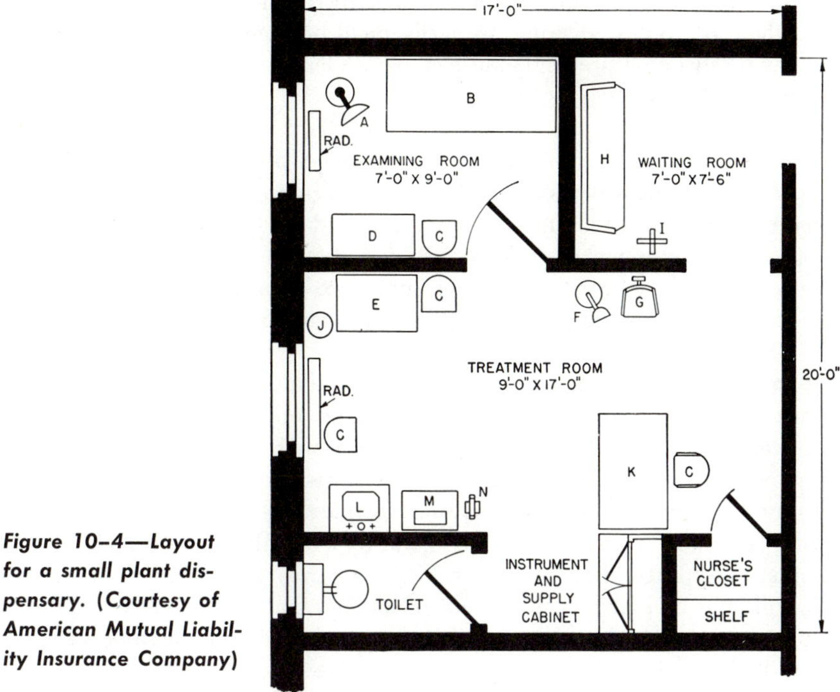

Figure 10-4—Layout for a small plant dispensary. (Courtesy of American Mutual Liability Insurance Company)

snack bar is often extended to include fountain service with cold drinks, ice cream, pastry, candy, and cigarettes. The snack bar may be a part of, or an adjunct to, a recreation room, where employees may go to smoke, relax, read, play games, or eat.

The rolling cafeteria. The traveling lunch truck or cart is used in plants and yards where conditions render a central kitchen or eating area impractical. In fact, the kitchen may be off the premises, with services being supplied by catering companies. Such a service is advisable for employees who should not leave the processes on which they are working for a sufficient length of time to go to a central eating facility. An advantage in the rolling cafeteria plan is that it can be expanded and contracted to meet changing needs more easily than a fixed form of food service. On the other hand, rolling cafeterias do not meet all needs. Close supervision is required so that time will not be wasted by employees meeting the trucks. Unless adequate provision is made for replacing the items carried on the trucks, employees may be tempted to go to the truck instead of waiting until it comes to them.

The mobile caterer. A fairly recent variation of the rolling cafeteria is the mobile caterer, or *chuck wagon*—a pick-up truck equipped with a body adapted to storing a wide variety of food items, hot and cold, ranging from drinks to sandwiches and salad plates. Caterers load such vehicles at a central location, and send them on routes covering primarily small business and "remote" locations, such as construction sites and industrial parks. Semi-schedules can be arranged to fit working hours.

Vending machines. A standard approach in many plants is in the use of vending machines for the entire food service. Modern vending machines handle literally everything *from soup to nuts.* Both hot and cold foods may be served, including soups, entrees, sandwiches, rolls, desserts, ice cream, hot and cold drinks, and fresh fruit. Vending machines relieve the enterprise of any responsibility for food service, except providing an eating area adjacent to the machines.

The cafeteria. Where many people are to be fed quickly, dining facilities served from a central kitchen are most satisfactory. Here full meals as well as lunches are obtainable. Such rooms are provided with enough tables and chairs that the employees may enjoy eating with their associates. In many plants, cafeterias are also used for recreation, and sometimes for meetings and other company functions. The main kitchen can be designed to serve all the eating facilities of a company—separate dining rooms, rolling cafeterias, and even facilities in outlying buildings or plants. The amount of kitchen equipment will depend on the number of people to be served and the type of service. Lunchrooms should be centrally located, if possible, to provide easy access for all. If the plant has several floors, however, the eating facilities may be located on the top floor to provide adequate ventilation and help prevent cooking odors from permeating the plant or offices. This, however, will require elevator service for incoming supplies, and for trash and garbage. The space requirements for a cafeteria will probably include:

1. Eating space
2. Kitchen
3. Dishwashing space
4. Storage areas
5. Service counter
6. Other facilities

Some suggestions on layout follow:[4]

1. *Dining room and service counters*—adequate to accommodate one sitting. Normally there are three sittings per meal period. Where the dining room is used for both factory and office people, it is customary to serve the factory people first and the office afterwards.
2. *Dining room area*—(a) 8 to 14 sq ft/person, exclusive of serving counters; (b) 17 sq ft/person, including serving counters and dishwashing units.
3. *Space in or adajcent to the dining room*—essential for people waiting to be served. A covered way between work locations and the lunchroom may be necessary during inclement weather.
4. *Table sizes*—30 × 30, 30 × 48, 30 × 72, 30 × 96 inches.

[4]Adapted from *Lunchrooms,* p. 9.

5. *Kitchen facilities*—sufficient to provide food for people working on the major shift, arranged for normal flow of production.
6. *Economical kitchens*—rectangular, length not more than twice the width; a long, narrow, or irregular shaped kitchen is costly to equip, difficult to operate, and requires more floor space.
7. *Kitchen manager's office and storage room*—located for the control of food, supplies, and personnel.
8. *Separate refrigerated space*—for meats, vegetables, fruit, dairy, products, and frozen foods. Walk-in type for large installations. The temperatures needed to keep the different types of food vary; also, certain kinds of foods flavor others unless isolated.
9. *Separate equipment*—for the preparation of salads or cold dishes.
10. *Refrigeration and disposal of* waste.
11. *Dishwashing space*—convenient to the dining area.
12. *Kitchen area*—(a) 22 to 35 per cent of total space available, exclusive of storerooms, office, washroom, and locker rooms; (b) by number of meals served during a lunch period; the sq-ft-per-meal factor can be: 100 to 200 meals, 5.00 sq ft; 200 to 400, 4.00 sq ft; 400 to 800, 3.50 sq ft; 800 to 1,300, 3.00 sq ft; 1,300 to 2,000, 2.50 sq ft; 2,000 to 3,000, 2.00 sq ft; 3,000 to 5,000, 1.85 sq ft; 5,000 to 8,000, 1.70 sq ft.

Illustrations of typical food service facilities are shown in Figures 10-5 to 10-8; Table 10-2 is from a survey by a leading periodical of the plant food service facilities used by plants of various size, and shows also more particular vending-machine placement data.

igure 10–5—Typical nack bar. (Courtesy of nternational Paper Co.)

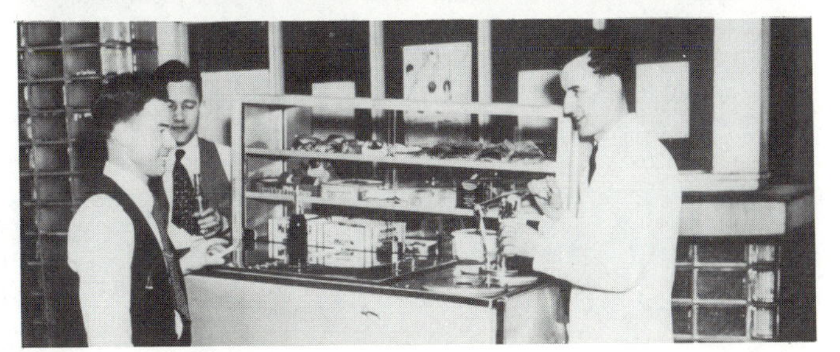

igure 10–6—Rolling afeteria for dispensing ood throughout the lant. (Courtesy of Inited Service Equip- ent Co.)

Figure 10-7—Vending machines serving as a cafeteria. (Courtesy of Vendo Company.)

Lavatories and Locker Rooms

In the small plant, these facilities are usually in the same locations because of the considerations of plumbing. In larger plants, however, facilities are necessary in several locations to be conveniently accessible as employees arrive or leave the plant. Frequently, locker rooms are close to employee entrances, with the time clocks placed along the path between the entrance and the locker room, but not so close that they cause a traffic problem at starting and quitting time.

Locker rooms should be so placed that they do not interfere with production. For this reason, and also to provide good ventilation, they are frequently located along an outside wall. In multi-story buildings, such facilities are located one above the other to reduce plumbing costs, although it has been said that you can *run* pipe, but you can only *walk* a man. In some one-story plants, locker rooms and toilets are placed in basement or on mezzanine levels, so that the locker

Figure 10-8—Typical plant cafeteria. (Courtesy of Albert Pick Co.)

Table 10-2. Food Service Facilities

Facility	Number of Employees: Under 250	250-499	500-999	Over 1,000	Average Percentage
Vending machines	87.5%	84.2%	81.0%	81.6%	83.5%
Cafeteria	26.8	42.9	66.9	83.4	55.0
Snack bar	14.3	16.4	14.1	22.2	16.8
Food carts	23.2	28.2	26.4	32.4	27.6
Other	7.1	4.5	3.1	0.3	4.5

Vending Machine Location	No. of Plants	Vending Machine Location	No. of Plants
Working areas	311	Special rooms with chairs and tables	56
Halls or corridors	282	Recreation rooms	47
Stations of two or more machines	145	Cafeterias	34

rooms and washrooms are located as conveniently as possible for access by employees. One authority has calculated that a toilet 150 feet out of the way, in a plant of 500 employees, could cost $20,000 per year in time lost in walking.

Many plants combine locker room, washroom, and toilet facilities. The major equipment includes such items as:

1. Lockers
2. Benches
3. Lavatories or washfountains
4. Toilet facilities
5. Showers
6. Drinking fountain

The arrangement of facilities involves problems little different from the plant layout itself. Some of the problems are:

1. Adequate space for the number of persons to be accommodated.
2. Spacing of equipment.
3. Provisions to handle peak loads of personnel at beginning and end of shift.
4. Effective utilization of space.

Sample layouts of typical washrooms, locker rooms, and toilet combinations are shown in Figures 10-9 and 10.

Miscellaneous Personnel Services

Many other kinds of employee service are offered by some companies, depending on their philosophies and finances. Not uncommon are recreation facilities, with space for Ping-pong tables, shuffle boards, excerise rooms, and even sauna baths. More unusual, but observed by the writer are bowling alleys, chapels, and even a beauty shop—provided by one plant, during World War II, to keep women from

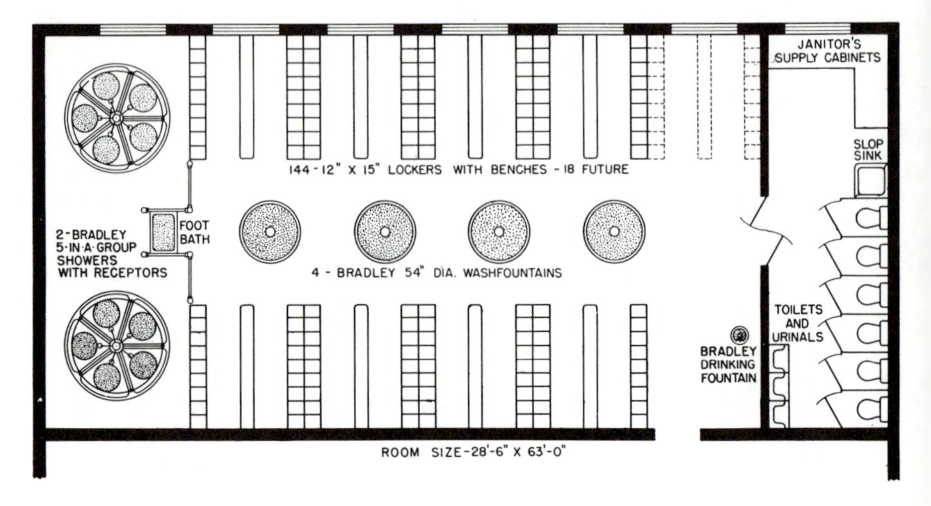

Figure 10–9—Washroom and locker room layout for 120 to 160 men. (Courtesy of Bradley Washfountain Co.)

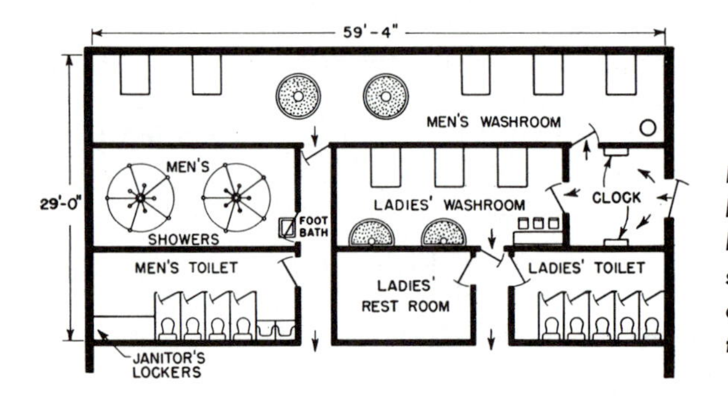

Figure 10–10—Typical layout for men's and ladies' washrooms for a smaller plant. (Courtesy of Bradley Washfountain Co.)

taking a whole day off for a permanent. Outdoor facilities may include ball diamonds, tennis courts, picnic areas, an occasional company owned park or resort, or even a service station.

Conclusion

Service activities are an important and integral part of the complete facility layout. They must be planned as carefully and accurately as the production areas. The layout planner should carefully study and plan the characteristics and requirements of the service areas considered in the last two chapters. (Space requirements for the auxilary and service activities will be covered in Chapter 11.)

Questions

1. What are some of the factors that would determine whether all office activities should be located in one central area, or placed in various locations about the plant?

2. What are some of the general considerations to be followed in planning office space?
3. What is meant by *office landscaping*? What are its characteristics? —its advantages?
4. What would be the minimum medical requirements in a plant too small to hire even a part-time doctor or nurse?
5. What are the advantages of providing a separate eating place for employees?
6. What are some of the methods of providing food for employees?
7. If you were building a new plant, what kind of eating facilities would you provide. Why?
8. If you were an employee, what kind of eating facilities would you prefer? Why.
9. What are some of the general locations in which a locker room might be placed? What would be the advantages of each?

Exercises

A. Obtain office planning literature from local office furniture dealers (or manufactureres) and discuss in class.

B. Have a local office designer speak to the class.

C. Look up some of the references on office landscaping in the Bibliography, and discuss pros and cons in class. (Instructor may choose to assign articles).

11
Space Determination

Previous chapters have dealt with problems involved in planning activity interrelationships and the characteristics of the various activities. It now becomes necessary to think more specifically, in terms of actual square feet of space available, required, or to be allocated to each activity or function. This chapter presents information and procedures for determining the amount of space needed for some of those activities requiring significant amounts of space. The activities covered here are:

1. Offices
2. Receiving
3. Storage
4. Production
5. Medical service
6. Food service
7. Lavatories, locker room
8. Tool room, tool crib
9. Maintenance
10. Warehousing
11. Shipping
12. Parking

Some will be covered in more detail than others, and the planner will no doubt find it necessary to obtain additional detail on many of them.

Factors for Consideration in Space Planning

Although the techniques presented in this chapter will aid in the development of space requirements, there are some factors (Table 11-1) the planner must be acquainted with, and must have sufficiently in mind to permit him to plan them into the layout.

Throughout the process of planning the area requirements, it would be wise to check the list of categories of factors in Table 2-1, as they pertain to each of the activities in the facility. Some have been covered in previous chapters; some will be in subsequent chapters; some are covered below, as they concern or affect the space determination problem. In any event, the planner should be impressed (or worried!) about the number of factors he should resolve into the space needs of the enterprise. Probably the best way to consider most of them, is to convert them into question format, and ask, at several points in the space planning and layout design process, "Have I considered——?" If so, fine. If not, pause and reflect on the specific factor and its relationship to the facilities design so far.

Determination of space requirements. The space required by a facility rather obviously bears a close relationship to equipment, material, personnel, and

Table 11-1. Factors for Consideration in Space Planning

A. General

1. Major space occupying activities
   a. Production (or equivalent)
   b. Serving production—administration, production, personnel, plant
2. Sales forecast
3. Number of products
4. Foreseeable technological advances in product or production processes
5. Possible changes in product or line
6. Long-range plans
7. Master plan
8. Expansion plans
9. Expansion vs. new site (or vice versa)
10. Flexibility desired (or required)
11. Number of shifts
12. Total number of employees
13. Male/female ratio
14. Economic trends

B. Production

1. Product size
2. Material size, nature
3. Production methods—line, job shop, etc.
4. Nature of processes
5. Number of operations
6. Work methods
7. Work standards
8. Production efficiency
9. Scrap percentages
10. Number of machines
11. Sizes of machines
12. Material flow pattern
13. Number of operators
14. Number of auxiliary and service personnel
15. Handling methods and equipment
16. Storage requirements
    a. Raw material
    b. Supplies
    c. In-process material
    d. Finished goods
    e. Inventory policy
    f. Storage methods and equipment

C. Building (new or existing)

1. Type
2. Construction
3. Number of floors
4. Clear height
5. Floor load capacity
6. Doors
7. Stairs; ramps; elevators; lifts
8. Aisles
   a. Types
   b. Location
   c. Width
9. Size
10. Shape
11. Condition
12. Availability
13. Column spacing
14. Possible use of mezzanines, balconies, basement, roof
15. Utilities
16. Plot size

D. Cost

1. Funds available
2. Interest rates
3. Economic trends

activities. This chapter deals with the determination of specific space needs of selected activities.

Offices

Offices and their characteristics were covered in Chapter 10, along with some suggestions as to the space needs. There are several ways of determining how much office space is required. One authority has said that the average employee requires about 200 sq ft—including his work space and his share of the amenities and related spaces (see Table 11-2). This suggests an overall office-plus-related-space requirement of say 200 sq ft, for 50 persons, or a total of 10,000 sq ft. This may be satisfactory for a rough estimate, but a more detailed analysis is to be advised before the facility can be detailed on the layout.

If the latter is desired, a form similar to Figure 11-1 can be used—or developed to suit—for the office portion. Also, help is available from office design firms and office equipment dealers. Related areas can be calculated individually, as in following paragraphs.

Table 11-2. Standardized Office Planning—Approximate Area per Employee (sq ft)

Average size of work place		106
Reception areas		2
Conference areas		2
Punched card & EDP		6
Internal services & storage		15
Total		137
Social facilities:		
Cloakrooms	2.5	
Washrooms	4.5	
Canteen & kitchen facilities	5.0	
Medical, welfare, library	2.0	
		14
Sanitary installations—		
Toilets, washrooms, showers		5
Communications areas—		
Stairs, lifts, transport facilities, corridors for general communications)		17
Technical areas—		
Air conditioning & heating plant, switch rooms, etc.; janitors' rooms, fixed installation, shafts, etc.		18
Construction areas—		
Pillars, outside walls, inside walls, etc.		6
Total		197

Receiving

The activities of this function were covered in Chapter 9. Planning the space requirements will again require some calculations—preferably with the aid of an appropriate form, as in Figure 11-2. To supplement the information given in Chapter 9, Table 11-3 deals with some of the considerations to be faced.

For general information regarding railroad track and building relationships, refer to Figure 11-4. It would be wise to check with the railroad serving the site for specific details.

An important factor to be remembered is that more than one receiving or shipping location may be advisable—especially if volumes are large or several varieties of goods are needed, or are shipped, at various locations in the facility.

Storage[1]

The space required for storage may be for any one of several rather different purposes, such as are listed on page 217. The method of determining space

[1]For further details on storage and warehousing, see Apple, *Material Handling Systems Design*, ch. 16.

needs, however, is just about the same for all. It consists primarily of listing the many items to be stored, and extending their physical characteristics into square or cubic feet required to store them. Again, a form will be found helpful in this task—and it is often a task to carry out such calculations. Often it is done by a computer—using information already stored for other purposes—with a relatively simple program. A *Storage Analysis Sheet* is shown in Figure 11-5. A few simple calculations will suffice to make an approximation of the storage space required at any one location.

Assume that production requirements call for the storage of a 2-week supply of a certain casting. Production is at the rate of 50 units per hour, or 4,000 units for two 40-hour weeks. If one casting is approximately $6 \times 6 \times 6$ inches, then eight castings will require 1 cu ft. A $3 \times 5 \times 2\frac{1}{2}$-ft container will hold:

$$(3 \times 5 \times 2\tfrac{1}{2}) \times 8/\text{cu ft} = 300 \text{ units}$$

then 4,000 castings can be contained in:

$$4{,}000/300 = 13\tfrac{1}{3}, \text{ or } 14 \text{ containers}$$

If, because of floor-load or ceiling-height limitation, containers can be stacked only 3 high, then:

$$14/3 = 4\tfrac{2}{3}, \text{ or } 5 \text{ stacks will be required}$$

As each stack requires 15 sq ft of floor space, then 75 sq ft of floor space will be required to store a 2-week supply of the castings. Similar calculations should be made for all parts of considerable size, whether they are raw material, parts in process, or finished parts.

By means of the *Storage Analysis Sheet* and such calculations as shown above, it can be determined how many square (or cubic) feet of storage space will be required for raw material (stores), in-process material, or finished parts (stock). All pertinent data should be entered for all inventory items of any significant physical size. That is, very small items, to be placed in hopper-type bins or drawers, can be estimated, but should still be listed to provide an accurate stock list and to be sure no space-occupying item has been omitted. A common method of selecting those items to be extended in developing storage space requirements is to apply the *A–B–C principle* to the sizes of the items in storage. Arrange inventory items (by computer or estimate) in descending order of cube required (size × quantity), from largest to smallest, and carry out the calculations for only the top 15 to 20 per cent—that is, the largest space occupiers. The totals at the bottom right of Figure 11-5 are simulated, or estimated totals from all sheets, to be used in determining total space requirements—not, however, before applying a factor to allow for other space occupiers within the storage area, such as; aisles, columns, and racks. A common figure is 40 percent of storage space. Using this figure on the total from Figure 11-5 will give: 140 per cent of $1{,}320 \text{ ft}^2 = 1{,}848 \text{ ft}^2$, to be entered on a *Total Space Requirement Work Sheet.* Figure 11-6 shows how a similar form was used for tabulating the data for the area shown in Figures 12-12 and 13, for a point-of-use-storage assembly area. The same form can be used for

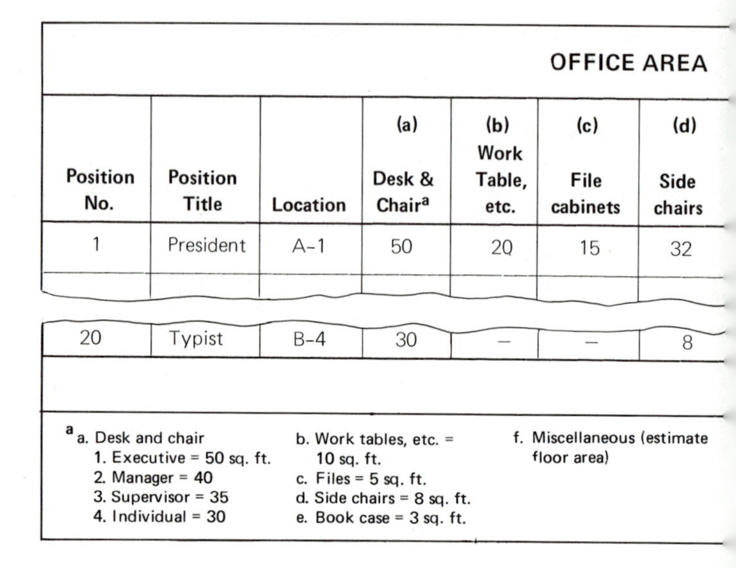

			OFFICE AREA			
Position No.	Position Title	Location	(a) Desk & Chair[a]	(b) Work Table, etc.	(c) File cabinets	(d) Side chairs
1	President	A–1	50	20	15	32
20	Typist	B–4	30	—	—	8

[a] a. Desk and chair
1. Executive = 50 sq. ft.
2. Manager = 40
3. Supervisor = 35
4. Individual = 30

b. Work tables, etc. = 10 sq. ft.
c. Files = 5 sq. ft.
d. Side chairs = 8 sq. ft.
e. Book case = 3 sq. ft.

f. Miscellaneous (estimate floor area)

determining space needs of the several types of storage areas listed at the upper left on Figure 11-5.

For in-process storage, that is, awaiting further processing, a number of factors should be considered that are somewhat different from ordinary storage situations. Among these factors are:

1. Line balancing
2. Production volume
3. Space requirements
4. Space available
5. Storage method and equipment
6. Clear height
7. Load size
8. Material characteristics
9. Distance from point-of-use
10. Handling method and equipment
11. Production rate
12. Product or process production
13. Process reliability
14. Process quality capability
15. Dunnage, packaging, etc. required
16. Environmental requirements
17. Storage time
18. Cycle time
19. Need for in-process storage
20. Flow direction
21. Storage cost
22. Storage volume required
23. Handling methods
24. Potential locations

(Further details on the several kinds of storage are covered in Chapter 9, and in *Material Handling System Design*, Chapter 15.)

Production

The determination of space required for productive activity is dependent to a great extent on the individual work places, and bears a proportional relationship to the sum of the areas of the individual work places. An allowance is usually added to provide for aisles, and such other non-productive areas that are too small

CALCULATION SHEET

(e) Book cases	(f) Other	Sub-Total	No. of Positions	Totals: Per Position No.	Totals: Now at 100% allowance[b]	Totals: 5 years from now (19XX)
6	–	123	1	123	246	246
–	–	38	4	152	304	608
			GRAND TOTAL		3,000	4,800

[b] 100% allowance for access, leg room, aisle space, walls, columns, etc.

Figure 11-1—Office area calculation sheet.

to warrant individual calculation. In the manufacturing enterprise, as well as many others, a *work place* is the location at which an operation (or operations) take place, as developed below.

Work place design. In Chapter 4, a *manufacturing operation* was defined as *a unit process, or group of related unit processes—and related activities—that modifies the material or part, and is usually done at one location.* Usually they are identified on the *Production Routing* as in Figure 11-7.

The manufacturing operation was technically described in more detail on the *Manufacturing Operation Planning Sheet,* and defined in scope to include:

1. Material—raw, finished, scrap
2. Tools
3. Machines and auxiliary equipment
4. Operator
5. Auxiliary services (utilities, etc.)

The work place design activity delineates the operating details of the method to be used in carrying out the operation. It deals with the integration of the operator into the operation, with the overall objective of determining how personnel should be integrated into the production process to most effectively contribute to the enterprise objective. The importance of this activity is emphasized by the steadily increasing cost of labor, not only in actual wages, but also in job training costs ($500 to $1,000 per employee) and fringe benefits ranging from a small percentage to (in some cases) 100 per cent of the basic hourly wage—usually averaging 20 to 30 per cent of hourly wages. It can easily be recognized that a minute lost to an inefficient work method equals $0.01/min. for every $0.60 of hourly wage plus any fringe benefit allowance. So a person whose *basic* wage is $3.00/hr, with a fringe benefit of 20 per cent—for a total of $3.60—is wasting $0.06 for every minute lost to an inefficient method.

RECEIVING AND SHIPPING AREA PLANNING SHEET

Company Ajax Mfg Co. Analyzed by AMJ. Date Apr. 14

RECEIVING		SHIPPING
Units	Estimated Weekly Activity	Units
81	Number of Items	4
60	Number of Shipments	5
10,000 lbs	Total Weight	9,000 lbs
1,000 ft³	Total Physical Volume	2000 ft³
20	Total Man Hours	½

RECEIVING	Areas in Square Feet		SHIPPING
Unloading platform (In) (Max. area required to maneuver largest item received)	400	600	Accumulation (In) (Awaiting packing)
Unpacking and sorting	200	100	Packing and labelling
Storage at receiving	400	200	Storage at shipping
Receiving inspection	100	300	Marshalling (Out) (awaiting pick-up)
Marshalling (Out) (awaiting delivery to production or storage)	100	200	Loading platform
Truck well (14' x 40' per truck & trailer)	560	560	Truck well
Siding (R.R.)	—	—	Siding (R.R.)
Ramp	—	—	Ramp
Aisles, etc. (add 50% of Storage)	200	100	Aisles, etc. (add 50% of storage)
Handling equipment storage	80	(same)	Handling equipment storage
Office	100	(same)	Office
TOTALS	2140	2060	TOTALS

Figure 11-2—Receiving and shipping area planning sheet.

Table 11-3. Considerations in Planning Receiving Activity

1. **Number of truck spots.** This problem calls for a careful analysis of activity in and out, as described in Chapter 9. Basically, the number of spots must be sufficient to handle peak periods (receiving and shipping) without causing undue delay. For each truck spot needed, allow 10 to 14 ft of lineal dock frontage.
2. **Length of dock.** The total length will be about 12 to 14 ft, times the number of spots needed—possibly a little less, if double doors are used.
3. **Depth.** The dock depth should be about 12 to 15 ft behind any dock boards installed in the floor, to allow maneuvering space.
4. **Column spacing.** Two 12-ft spots can be placed in a 25-ft bay; two 14-ft spots in a 30-ft bay; narrower spots result in delays when spotting trucks and trailers.
5. **Dock height.** Optimum height for highway van and trailer-type trucks is 46 in.; pick-up and city-delivery types, 24 to 30 in. If traffic warrants, consider 2, or even 3 different heights. In most cases, adjustable dock boards should be installed for safe and efficient operations.
6. **Enclosure.** Obviously, enclosed docks are more convenient for all concerned, and are preferred, if possible. If not, a canopy of 6 to 8 ft out from a flush dock (door-way), or 6 to 8 ft beyond the edge of the dock, if of the platform-type, is advisable.
7. **Loading area.** The area directly in front of the dock should be a minimum of 65 ft forward of the dock face.
8. **Maneuvering area.** The distance necessary for maneuvering a tractor-trailer into the loading area and up to the dock is an additional 40 ft (if traffic flow is counterclockwise, and at least 100 ft for a clockwise traffic pattern.
9. **Service roads.** Roads servicing the dock should be 12 ft for one-way, and 22 ft for two-way—plus 6 ft if pedestrians use the same roadways for walks.
10. **Right-angle intersections and curves.** Minimum turning radius of 35 feet; 50 feet desirable.
11. **Space needed.** This topic was covered in Chapter 9.
12. **Ceiling height.** A clear height of 15 ft should be allowed under lead edge of a canopy or roof. Inside, the building could even be more, to get best value from the floor and roof, and make optimum use of (1) the cube and (2) modern storage, handling, and stock picking equipment.
13. **Doors.** Overhead type, 9-by-9-ft minimum
14. **Lighting.** Minimum of 20 foot-candles
15. **Ramps.** Equipment type and maximum recommended grades.

    a. Power operated hand trucks—3%
    b. Powered platform trucks—7%
    c. Low-lift pallet trucks—10%
    d. Electric fork trucks—10%
    e. Gasoline fork trucks—15%

16. **Dock boards,** etc. (for bridging gap between truck and dock)

    a. Portable
    b. Built-in to dock—mechanical or hydraulic
    c. Truck leveling devices—built into truck well or pit
    d. Wedges—on pit floor, for backing truck up on, to match dock height

17. **Dock type** (see Fig. 11-3). One last, or maybe first, item is the type of dock. Choices are:

    a. Flush (i.e., door-in-wall)
    b. Platform (extended from building)
    c. Drive-through (building)
    d. Drive-in

    Probably the flush-dock is most common, since it is cheaper. But a covered dock is to be preferred. If considering a sawtooth, or 45°-angle dock, remember the difficulties caused and the lost floor area in all the unusuable triangular pieces of space—at both front and back of vehicle.

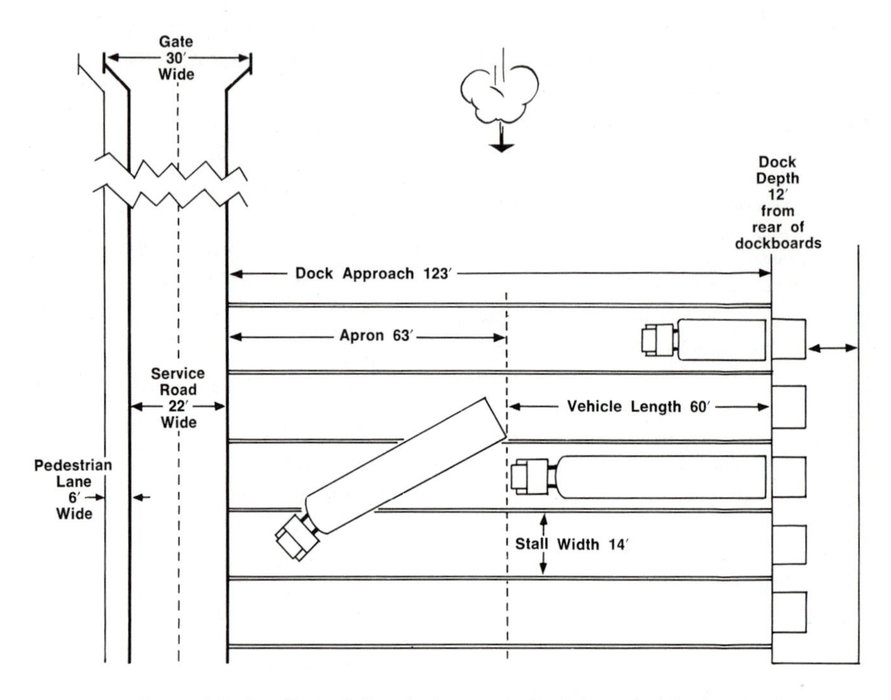

Figure 11–3—Typical details for truck dock layout. (Courtesy of Transportation and Distribution Management Magazine.)

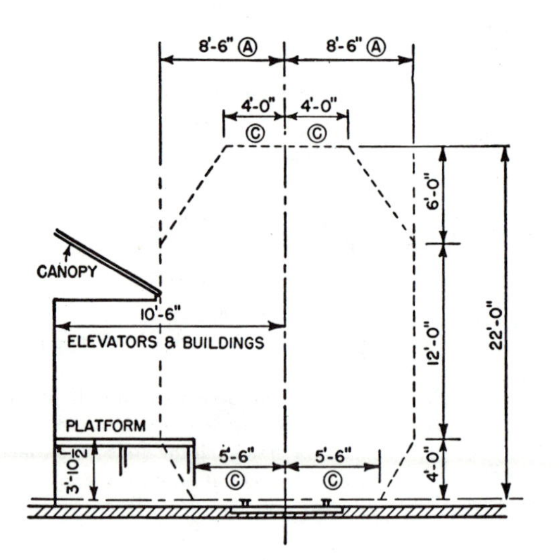

Figure 11–4—Necessary clearances for a railroad siding. (By permission of Material Handling Engineering.)

The work place design activity involves the (1) analysis, conceptualization and design of the most economical method of performing a task, (2) standardizing that method, and (3) subsequently assisting in training the operator in the prescribed method.

What is a work area? Up to this point in the facilities design process, the major concern has been with the broad, general phases of flow patterns, larger area relationships, and the plant-wide movement of material. Attention must now be given to individual operations or work areas, so they can be carefully positioned within the framework of the overall flow pattern.

In general, a work area is the space occupied by a machine or work bench, necessary auxiliary equipment, and the operator; or it may contain a group of smaller or a group of similar machines, and may require more than one operator. Or it may be merely a piece of floor space where an operator works alongside a conveyor, as in an assembly operation. Figure 11-8 shows a typical work area.

Work methods design and material flow. Since each work place represents a location along the overall flow pattern, it should be designed to fit into the overall material flow, as shown in Figure 11-9. So the flow through a work place should be planned as in Figure 11-9*b*, not *d*.

Factors for consideration in work methods design. In view of the facilities designer's overall function, and his involvement in integrating work methods into the layout, he should carefully question and consider the factors influencing or involved in the work method design process as shown in Table 11-4. It can easily be seen that work place design is no simple task if it is desired that the operations be performed in the one best way.

Guides to work area planning. If the work place is to be designed for best efficiency, the designer should consider the principles of motion economy and work place design, which are concerned with good practice, and are based on experience. These principles are discussed in considerable detail in any of the many books on motion economy. Table 11-5 contains a list based on those principles and adapted to the work place environment.

Because they reflect the experience of many people over a period of years, such principles should be carefully observed in the work design process—and incorporated into the methods as appropriate.

Planning an efficient work place. It should be remembered that each work place is a miniature factory, with its own receiving, production, and shipping areas. One of the big problems in planning work places is that of properly designing each one for optimum efficiency, and then fitting each into the overall flow pattern. To aid in this, the following general procedure will serve as a guide:

1. *Determine the direction of general flow* of material or activity through the plant or department, from the overall flow pattern.

☒ Incoming Stock
☐ In-Process Materials
☐ Finished Parts
☐ Finished Product

STORAGE ANALYSIS SHEET

Analyzed By: A. M. J.
Date: Mar. 17

COMPANY Powrarm PRODUCT Model 1 ANNUAL PROD'N. 100,000

ITEM OR PART							QUANTITY			RECEIPT		HANDLING UNIT							SPACE REQUIREMENTS							STOR. LOC.
			Size in inches																Bulk or Pallet							
No.	Description	No. per assy.	L	W	Ht	Wt	Max. Inv.	Mo. Reqt.	Norm. Rec't.	Freq.	Carr. Type	Type	L	W	Ht	Wt	Items/ Hdlg. Unit	Hdlg. Units For Max. Invent.	no. hdlg. units high	no. base units	sq.ft. per base unit	sq.ft. for max. no.hdlg. units	cu.ft. for max. no.hdlg. units	Shelf	bin	
1	Base	1	6"	6"	6"	10#	4000	8000	2000	1/wk	truck	pal-let box	5'	3'	2½'	3000 #	300	14	3	5	15	75	·562	-	-	Rec.
																			(TOTALS)			1,320	10,430	—	—	—

Figure 11-5—Storage analysis sheet.

STORAGE ANALYSIS SHEET FOR 610 – 620 **QUARTERLY SALES** 610–355 620–200 **PRODUCTION LOT SIZE** 610–120 620–65

CODE NO.		ITEM DESCRIPTION	QUANTITY PURCH.	QUANTITY QT.	QUANTITY LOT	ITEM SIZE L	ITEM SIZE W	ITEM SIZE H	PACKAGE TYPE	PACKAGE SIZE	PACKAGE NO. PCS.	PACKAGE PER PALL	PALLET LOAD SIZE L	PALLET LOAD SIZE W	PALLET LOAD SIZE H	NO. PALLETS PURCH.	NO. PALLETS QUART.	NO. PALLETS LOT	NO. PALL. HIGH	BULK PURCH. SQ. FT.	BULK PURCH. CU. FT.	BULK QUART. SQ. FT.	BULK QUART. CU. FT.	BULK LOT SQ. FT.	BULK LOT CU. FT.	BIN – SHELF PURCH. SIZE	BIN – SHELF PURCH. # SP.	BIN – SHELF QUART. SIZE	BIN – SHELF QUART. # SP.	BIN – SHELF LOT SIZE	BIN – SHELF LOT # SP.	WORKPLACE (STACKBINS) PURCH. BIN #	WORKPLACE PURCH. # BIN	WORKPLACE QUART. BIN #	WORKPLACE QUART. # BIN	WORKPLACE LOT BIN #	WORKPLACE LOT # BINS	LOCATION BULK	LOCATION BIN	LOCATION WORKPL.
213804	1	Grommet	5000	555	185	5/8	5/8	1/2	Carton																															
627802	1	Sheet CR/SL	300	555	185	33-1/2	27	3/16				1000	36	36	62	1	1	1		Mez. & Whse.																				
627804	1	Sheet DC/SL	300	555	185	25-3/8	22-1/4	3/16				1000	36	36	62	1	1	1		Mez. & Whse.																				
628603	1	Tank-10 Gal-6100	~~600~~ 300	355	120	16-7/8	16-7/8	20	Carton	17x17x21	1	12	39	39	55	50/25	30	10	Flow Rack																					
630204	1	Adapter-Tk.-63	900	555	185	3-3/4	2-1/2	1-3/4	Carton	17x17x18	250															#1x20	2	#1x20	1							#5	1			
630206	1	Adapter-Bag	650	555	185	16-3/8	16-3/8	1-1/2	Carton	17x17x18	20	130	36	36	36	5	5	1	5	9	135	9	135	9	27															
630208	1	Adapter-Blower-6-3	350	555	185	3-3/4	2-1/2	1-3/4	Carton	17x17x18	250															#1x2	2	#1x20	1							#5	1			
630602	1	Filter-Bag	800	555	185	13	11	1/2	Carton	22x14x21	150															#1x24	3	#1x24	2	#1x24	1									
630804	2	Bracket-Hole 63		1110	370	2-5/8	7/8	1	Carton	17x17x18	1000																	#2x20	1							#4	1			
630806	1	Bracket 60		555	185	2-1/2	2-1/4	2	"	17x17x18	500																	#2x20	1							#4	1			
630808	1	Bracket-CSTR-63	750	555	185	6	3-3/4	4	Carton			300	36	36	36																									
631103	4	Bushing	3000	2220	740	9/16	9/16	7/16	Carton	12x12x12	500															#3x14	1	#3x14	1							#2	1			
631106	1	Bushing Valve Cup	1000	555	185	1-1/2	1-1/2	1/4	Carton	12x12x12	1000															#3x14	1	#3x14	1							#1	1			
631702	1	Carton		555	185	59-1/2	50-3/4	5/8				150	59-1/2	50-3/4	75		4	1		Mez. & Whse.																				
632002	4	Clamp - 63		2220	740	2	5/8	5/8	Carton	17x17x16	2500																	#2x18	1							#3	1			
632004	2	Clamp - 63	1500	1110	370	3-1/2	2	1-1/4	Carton	17x17x18	400															#1x20	2	#1x20	2							#5	1			
632006	1	Clamp-Motor	600	555	185	15-1/2	15-1/2	3	Carton	16x16x24	22	440	41	41	87	1	1	1	1	12	85	12	85	12	85	4 Layed 5H.														
632008	1	Clamp Valve Tube	1000	555	185	3-1/4	3-1/4	1/16	Carton	10x10x6	1000															#3x8	1	#3x8	1							#1	1			
632102	4	Retainer Clip	2500	2220	740	1	1/4	1/16	Carton																	#4x10	1									#1	1			
632302	1	Cover-Lid	700	555	185	16-1/8	16-1/8	2-3/4	"			60	36	36	36	10	10	3	5	18	270	18	270	18	81															
632304	1	Cover-Motor	700	555	185	8-1/4	8-1/4	4				200	36	36	36	3	3	1	3	9	81	9	81	9	27	#2x20	1	#2x20	1							#4	1			
632402	1	Valve Cup	600	555	185	2	2	1-3/8	Carton	17x17x18	800																													
632404	1	Cage	1000	555	185	10-1/2	10-1/2	10-1/2	Carton	28x24x36	25	200	56	48	72	5	1	1	2	75	900	18-1/2	112	18-1/2	112															
632410	1	Can Valve	700	555	185	6	6	6-1/8	Carton			180	36	36	36	4	3	1	4	9	108	9	8	9	27															
632712	1	Tank-Deflector		555	185	4-1/2	3-3/8	2	Carton	17x17x14	400																	#2x16	1							#2	1			
633402	1	Gasket		555	185	2-3/4	2-3/4	1/16	"	9x9x8	500															#4x10	1	#4x10	1							#2	1			
633404	2	Gasket-Motor	1000	110	370	8	8	3/4	"	36x36x36	700	700	36	36	36	2	2	1	2	9	54	9	54	9	27											#5	2			

Figure 11–6—Storage analysis sheet. (Courtesy of Clarke Floor Machine Co.)

Operation No.	Operation Description
10	Cut to length
20	Straddle mill both ends
	(etc.)

Figure 11–7—Production routing.

2. *Determine the desired direction of flow,* based on the above, through the workplace: left-to-right, right-to-left, front-to-back, or back-to-front.
3. *Determine items to be contained in the work place,* such as machine, bench, stock containers, and conveyors.
4. *Make a rough sketch of the major pieces of equipment* in the work place in their approximate desired positions, and indicate direction of material flow—from *Step* 2 above—by an arrow.
5. *Indicate sources of material* used in the work place, and the directions in which it must go.
6. *Indicate the destination of the material* from the work place, and the direction in which it must go.

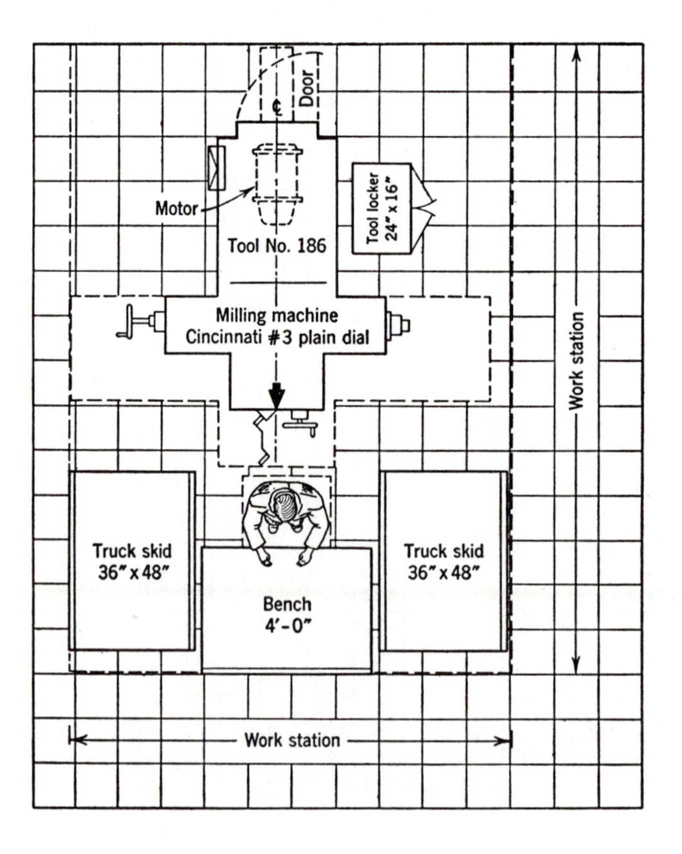

Figure 11–8—A typical work area.

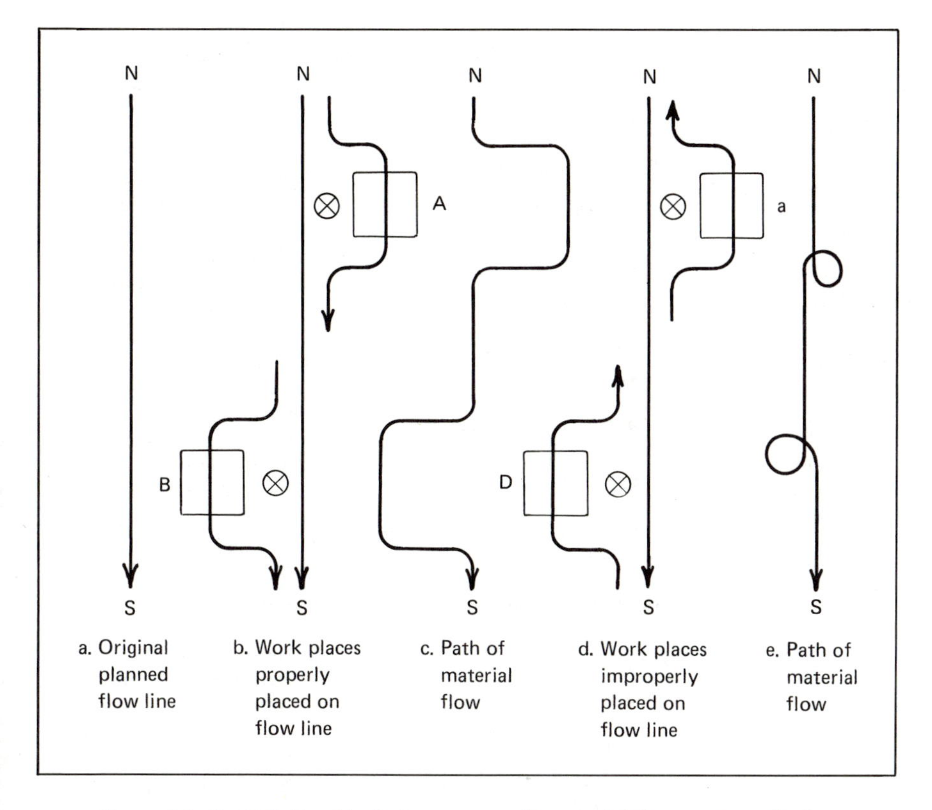

Figure 11–9—Relationship between overall material flow and work place orientation.

7. *Indicate method of waste or scrap disposal,* if applicable, and direction in which it must go.
8. *Sketch in any material handling equipment* serving the work place or area.
9. *Check the sketch* against the principles of motion economy and work place planning, and the factors in Table 11-4.
10. *Indicate distance between items* in the work place on the sketch.
11. *Record work place plan to scale and in detail* on an *Operation Analysis and Planning Chart* (Figure 11-10), or similar form.
12. *Indicate method of operation,* if appropriate, at this point in the design process.

Adhering to above procedure should help insure the proper integration of each work place into the overall material flow pattern. Further design effort is required to integrate the several work places into a segment of the overall flow pattern.

Material handling at the work place.[2] Material handling at the work place may be defined as handling that would normally be done after the material has been

[2]For further details, see Apple, ch. 14.

Table 11-4. Factors for Consideration in Work Place Design

A. Purpose of the operation

B. Product design (via value analysis)

C. Material-related

1. Space required
2. Quantity, total
3. Size
4. Characteristics, incoming
5. Characteristics, finished
6. Scrap; waste
7. Production rate
8. Amount in work place
9. Inspection requirements
10. Precision required

D. Equipment-related

1. Size
2. Utility requirements
3. Service requirements
4. Over-travel
5. Tools required
6. Auxiliary equipment
7. No. of machines
8. Nature of process
9. Height of work surface, controls
10. Noise level
11. Type of equipment
12. Pollution potential
13. Equipment utilization
14. Maintenance requirements
15. Level of mechanization
16. Safety hazards
17. Flexibility
18. Life of equipment
19. Alternate equipment
20. Access for repair and maintenance
21. Access to safety stops
22. Stability of process

E. Quality requirements

1. Precision required
2. Inspection requirements

F. Material handling methods

1. Raw materials
2. Finished parts
3. Scrap removal

G. Space-related

1. Physical characteristics
2. Aisle space requirements
3. Area required vs. available
4. Ceiling height (clearance)
5. Expansion requirements
6. Space utilization
7. Nearby safety hazards
8. Accessibility of work place
9. Floor load capacity

H. Operator-related

1. Space required
2. Sit or stand
3. Comfort
4. No. operators
5. Movement during cycle
6. Man or woman
7. Operator efficiency
8. Safety hazards
9. Supervision required
10. Training required

I. Working conditions; environment

1. Noise level
2. Lighting required
3. Heating and ventilation
4. Dust, etc.
5. Building characteristics
6. Vibration
7. Window location

J. Method-related

1. Direction of flow
2. Location of items
3. Related work places
4. Delivery of materials to—
5. Movement of materials in—
6. Removal of materials from—
7. Principles of motion economy
8. Hazards, safety requirements
9. Overall flow
10. Mechanization possibilities
11. Line possibilities
12. Process or product layout
13. Overall MH system
14. Relative distances in work place
15. Operation sequence
16. Standard time
17. Expansion possibilities
18. Paperwork involved
19. Tool location
20. Material location
21. Location of work place in material flow
22. Height of work place
23. Relationships to preceding and following operations
24. Production rate
25. Batch or continuous production
26. Economic lot size
27. Flexibility
28. Cycle time
29. No. of duplicate work places
30. Auxiliary services required
31. Column locations
32. Items required in work area
33. Effect of breakdown

K. Miscellaneous

1. Funds available
2. Total quantity to be produced
3. Production control methods
4. 2nd shift?
5. Costs involved
6. Legal requirements
7. Union attitude

Table 11-5. Guides to Work Area Planning

1. Plan for tools, gages, materials and machine controls to be located close to and in front of operator.
2. Plan a definite place for tools, gages, and materials.
3. Plan to use gravity when possible to feed and remove materials.
4. Plan to pre-position materials and tools within the work place.
5. Plan for delivery of materials directly to point of use.
6. Arrange material so operator is not required to prepare or re-position it.
7. Plan for prompt and efficient removal of materials from work place.
8. Provide adequate means for planned scrap removal.
9. Plan location of materials within work place to permit obtaining them with the most efficient sequence of motions.

0. Plan proper height relationships between material supply, point-of-use, and disposal.

1. Plan each work place in proper relationship to preceding and following operations.

2. Plan storage for a practicable minimum of incoming materials at work place as well as for finished work awaiting removal.

3. Leave sufficient space at work place for efficient material delivery, storage, and removal.

4. Select appropriate handling equipment.

5. Be sure work place handling equipment is properly integrated into overall handling system.

16. Avoid placing material directly on the floor, without a pallet or other support underneath.
17. Plan to use the same container throughout the system; avoid frequent changes.
18. Provide necessary clearances in and around each work place for proper performance of the operation.
19. Place product on packing base (pallet, skid, etc.) as early in the process as is practical.
20. Combine operations in order to eliminate intermediate handling.
21. Make judicious use of manual handling.
22. Plan to minimize walking.
23. Use containers, racks, etc., to hold items so as to prevent damage to work already completed.
24. Allow for the over-travel of machine parts such as milling machine tables, etc.
25. Allow for the projection of work, such as the bars of stock fed through a turret lathe.
26. Include floor conveyors, chutes, stock tables, etc., in the plan.
27. Allow enough space to permit placing parts in machines and removing them with ease.
28. Do not place machines or auxiliary equipment so that access to parts requiring adjustments, lubrication, frequent repair, etc., is difficult or impossible.
29. Plan carefully for quick access to safety stops.
30. Arrange automatic or semiautomatic machines so that one operator can attend more than one machine.
31. Consider the location of columns, aisles, and elevators.

delivered and set down for use at the work place and before it is picked up to be moved to the next operation. The following characteristics of material handling at the work place suggest some of the unique aspects that separate it from traditional material handling:

1. The distances involved are relatively short.
2. The number of moves is usually high, that is, it is a function of the number of cycles performed.

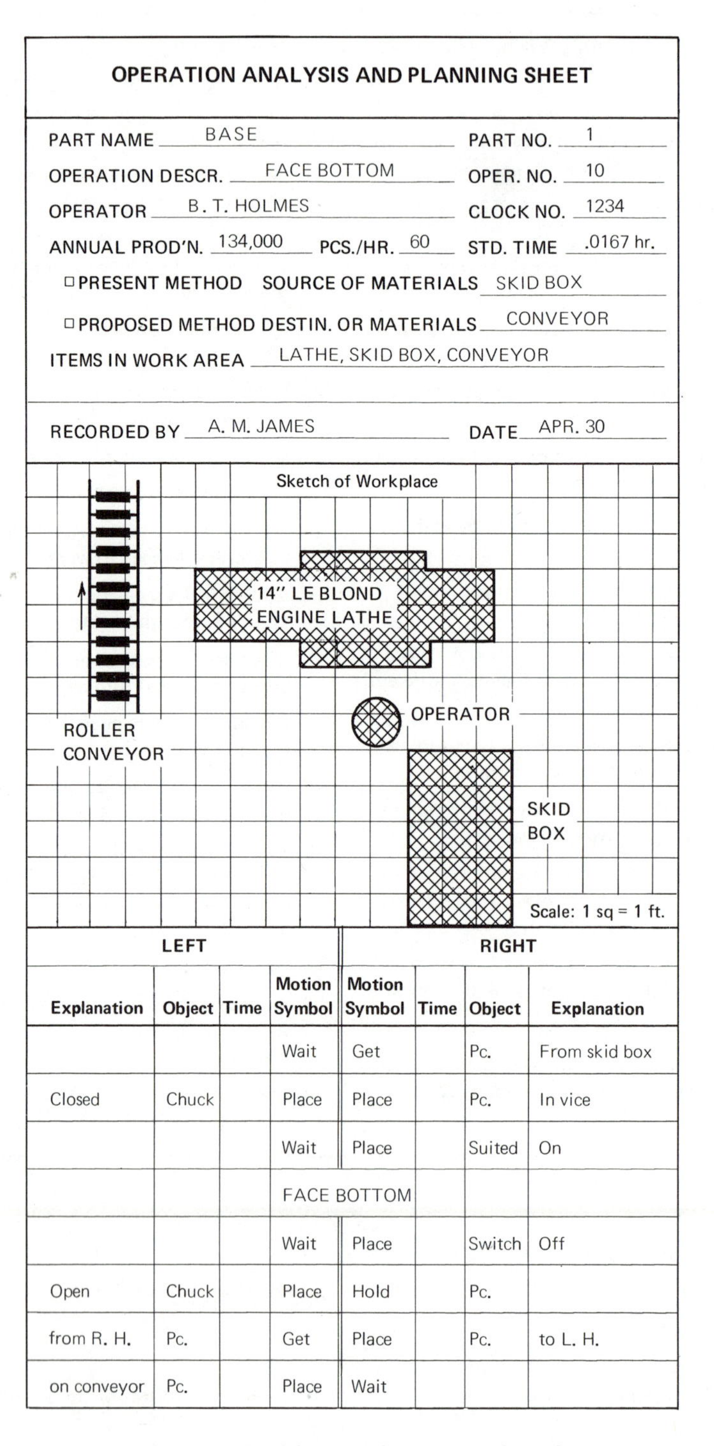

OPERATION ANALYSIS AND PLANNING SHEET

PART NAME BASE PART NO. 1
OPERATION DESCR. FACE BOTTOM OPER. NO. 10
OPERATOR B. T. HOLMES CLOCK NO. 1234
ANNUAL PROD'N. 134,000 PCS./HR. 60 STD. TIME .0167 hr.
□ PRESENT METHOD SOURCE OF MATERIALS SKID BOX
□ PROPOSED METHOD DESTIN. OR MATERIALS CONVEYOR
ITEMS IN WORK AREA LATHE, SKID BOX, CONVEYOR

RECORDED BY A. M. JAMES DATE APR. 30

LEFT				RIGHT			
Explanation	Object	Time	Motion Symbol	Motion Symbol	Time	Object	Explanation
			Wait	Get		Pc.	From skid box
Closed	Chuck		Place	Place		Pc.	In vice
			Wait	Place		Suited	On
			FACE BOTTOM				
			Wait	Place		Switch	Off
Open	Chuck		Place	Hold		Pc.	
from R. H.	Pc.		Get	Place		Pc.	to L. H.
on conveyor	Pc.		Place	Wait			

Figure 11-10—Final layout of a proposed work area.

3. The greater portion of the cycle time required at an individual work place is taken up by material handling.
4. Much of this material handling is commonly classified as direct labor!
5. Much of the handling activity at the work place is subject to improvement by traditional material handling techniques and devices.

Although the distances involved in handling material at the work place are generally relatively short, the number of these moves is the factor that emphasizes their importance. For example, at the start of his shift a fork truck or conveyor may deliver a container of 1,000 items to an operator in a matter of a few minutes. But, the handling and processing of these 1,000 pieces may occupy the operator for an hour, a day, or a week. Since it is probable that each individual item must be handled by the operator—into, through, and out of the work place—the total time involved in handling the material at the work place may run into many hours. Therefore, the analyst should pay close attention to the material handling aspects of work place design.

Work standards. After the designer has established the work place arrangement for an operation, it is necessary to determine the time required to perform the operation. This is usually done by time study personnel, with the results posted to the *Production Routing,* in terms of hours/piece, or pieces/hour (see Table 4-6). It is this value that is used as the basis for determining the number of work places, machines, and operators, as was shown in Chapter 4.

Production space requirements. Having established the number of work places, machines, and operators for each operation or activity, the total production space must now be determined. One way of doing this is by calculating the estimated space required for each piece of plant equipment or related group of equipment. This can be accomplished with the help of the *Production Space Requirement Sheet,* as shown in Figure 11-11. It is filled out as follows:

1. *Enter data from Production Routings* (or equivalent) in *Columns 1, 2, 3,* and *4* to identify the work areas involved.
2. *Enter space requirement* estimates in *Columns 5, 6, 7,* and *8* (all entries in square feet):
   Column 5 = Maximum length × Maximum width of machine.
   Column 6 = Maximum length × Maximum width of auxiliary equipment such as tables and benches.
   Column 7 = Maximum length of machine × 3 ft for operator working area.
   Column 8 = Actual size of stock containers, etc.
3. *Add Columns 5, 6, 7,* and *8;* enter total in *Column 9* for subtotal per machine.
4. *Multiply subtotal* by 150 per cent. This allows access space for material handling, maintenance and personnel movement, columns, work space share of necessary aisle area, etc. Enter this figure in *Column 10.*
5. *Enter number of machines* of each type in *Column 11* (from calculations shown previously). Multiply this number by the figure in *Column 10,* and enter the result in *Column 12.*
6. *Follow the same procedure* for all operations and determine the total for each area, department, activity, etc. Enter area totals in *Column 13.*

PRODUCTION SPACE REQUIREMENT SHEET

(1) NO.	(2) Activity, Dep't., Area or Item	(3) Oper. No.	(4) Machine or Equipment	SPACE REQUIREMENTS (5) Machine, etc. L × W = A	+ (6) Auxil. Equip. L × W = A	+ (7) Operator Space 3' × L = A	+ (8) Material Space L × W = A	= (9) SUB-Total	(10) Sub-total × 150% Allowance	× (11) No. of Mach.	= (12) Total Sq.Ft. per Operation	(13) Total per Area
I	Base	10	LeBlonde Eng. Lathe	3x6=18	2x8=16	3x6=18	3x5=15	67	100	1	100	
		20	W & S Turret Lathe	4x7=28	2x20=40	3x7=21	(incl.)	89	134	3	402	
		30	Drill Press	3x4=12	2x2=4	3x3=9	(incl.)	25	38	1	38	
		40	Drill Press	3x4=12	2x2=4	3x3=9	(incl.)	25	38	1	38	
		50	Drill Press- 2 spindle	3x5=15	2x2=4	3x3=9	(incl.)	28	42	1	42	
		60	Inspection Bench	2x6=12	2x2=4	3x6=18	(incl.)	34	51	1	51	
		70	Degreaser	3x10=30	-	3x10=30	(incl.)	60	90	1	90	761
III	Handle	10	W&S Turr. Lathe (with	4x15=60	-	3x15=45	3x12=36	141	212	1	212	
			bar stock attachment									
		20	W&S Turret Lathe	4x7=28	-	3x7=21	-	49	75	1	75	
		30	Inspection Bench	2x6=12	-	3x6=18	-	30	45	1	45	
		40	Degreaser	(uses same equipment as Part No. I)						-	-	332

Figure 11-11—Production space requirement sheet for Powrarm factory.

LAYOUT PLANNING CHART

Part No. 1 | Part Name Powrarm Base | PCS./ASSY. 1 | PCS./HR. REQ. | SHEET 1 OF 2
Assy. No. | Assy. Name | ASSY./JOB | PROD. HRS./DAY | PREPARED BY F. J. Wealkes
Material Aluminum alloy | Size 6" x 6" x 5-3/8" | PCS./DAY | MODEL M-2 | DATE 7/11

ST. NO.	O T I D S	DESCRIPTION	OPER. NO.	DEPT. NO.	T. S. REG. NO.	TIME PER PC.	PCS. PER HOUR	TOTAL LOAD HRS.	OPER. PER MACH.	TOTAL MAN-POWER	MACHINE OR EQUIPMENT	NO. MACH. REQ'D.	DIST. MOVED	HOW MOVED	TYPE OF CONT'R	REMARKS
1	○ ⇨ □ D ▽	gondola													Gon	
2	○ ⇨ □ D ▽	to Operation 1											6'	Hand		
3	○ ⇨ □ D ▽	face bottom	1	2		.016	60		1	1	14" LeBlond engine lathe	1				
4	○ ⇨ □ D ▽	to table or roller conveyor											4'	Hand		
5	○ ⇨ □ D ▽	on table									Table & roller conveyor					
6	○ ⇨ □ D ▽	to Operation 2											3'	Hand		
7	○ ⇨ □ D ▽	face top, turn O.D., neck, drill and ream 5/8" hole	2	2		.042	23.8		1	3	3-Warner&Swasey turret lathes	3				
8	○ ⇨ □ D ▽	to roller conveyor											3'	Hand		
9	○ ⇨ □ D ▽	change of carrier									Roller conveyor					
10	○ ⇨ □ D ▽	along roller conveyor											10'	Mech.		
11	○ ⇨ □ D ▽	on table									Table					
12	○ ⇨ □ D ▽	to Operation 3											4'	Hand		
13	○ ⇨ □ D ▽	drill 3 bolt holes	3	2		.012	83.4		1	1	21" Cleeveman drill press	1				
14	○ ⇨ □ D ▽	to table											4'	Hand		
15	○ ⇨ □ D ▽	on table									Table					
16	○ ⇨ □ D ▽	to Operation 4											3'	Hand		
17	○ ⇨ □ D ▽	drill pin hole	4	2		.0042	238.0		1	1	Delta drill press	1				
18	○ ⇨ □ D ▽	to table											2'	Hand		
19	○ ⇨ □ D ▽	on table									Table					
20	○ ⇨ □ D ▽	to operation 5											2'	Hand		
21	○ ⇨ □ D ▽	drill & ream 3/4" eccentric hole	5	2		.0153	65.4		2	2	#4 Fosdick 2-spindle drill press	1				
22	○ ⇨ □ D ▽	to table											3'	Hand		

Figure 11–12—Layout planning chart for Powrarm base.

This procedure will give the estimated total number of square feet required for each department, activity, or production area. Rather obviously, no one is going to design a department of 761 sq ft—but the figure developed is far better than the most common method—*guestimating*. Also, space for future expansion should be added to the 761 sq ft. And, some consideration should be given to whether the area recommended (department or total) will be cut down later, in the approval or cost estimating processes. For example—the required 761 sq ft could easily be cut by 10 per cent (often rather arbitrarily) to a rather tight and therefore inefficient 685 sq ft—and this, without consideration for future expansion.

A simpler alternative method for preliminary calculation is to use a predetermined number of square feet to represent a typical machine or piece of production equipment. This, of course, would vary with the type of plant and production process. A quick calculation from a present layout (Total square feet allocated for production ÷ Number of machines) would give an adequate figure for planning. For a typical machine shop operation, it has been found practicable to use 100 to 150 sq ft/machine.

It should be recognized that in re-designing an existing facility, the present condition may be crowded. Therefore, another allowance should be made to loosen it up, before making the type of calculation suggested above. For example, assume that the present layout contains 100 work areas in 10,000 sq ft, or 100 sq ft/work area. If a 20-per cent increase is desired to overcome the crowded conditions in the present layout, use 120 sq ft/work area, and then add any space required for new activities as well as for expansion.

Coordinating the work areas. At this point in the planning process, a sizable amount of detailed work has been done that may appear to be unrelated. Actually it is all very closely related. The need is for some means of tying it all together, such as the *Layout Planning Chart* (Figure 11-12), which is somewhat of a combination of a *Process Chart* and a *Production Routing* with other pertinent data added to aid in the planning process. Many companies use a form similar to the *Layout Planning Chart*, although it may vary somewhat in detail. It is frequently found that certain processes or kinds of work are better recorded if the form is designed specifically for use in a given plant or department. Columns can be changed, omitted, or added to suit the particular needs of the type of problem under consideration. Items other than those commonly shown, which might be desirable on the *Chart*, are:

1. Machine or equipment number
2. Tool names and numbers
3. Floor space requirements
4. Utilities needed, etc.

It will be seen that the left-hand side of the *Layout Planning Chart* is the common *Process Chart*. It is used here as a check on previous planning to be sure that each and every step in the process has been given proper consideration, and that a method of performing each step has been determined. A *Process Chart* can not be properly made without recording plans for each step; so, at this point such a check is made. If plans have not been made during the operation planning phase,

Table 11-6. Procedure for Layout Planning Chart

1. Fill in the heading from the information on the *Production Routing* or available from other sources.
2. Referring to operation planning sketches previously made, decide which process symbol represents the first step in the operation. It will probably by *Storage* because the object must be picked up from where it has been stored, awaiting the process.
3. In the *first column*, insert *step number*, and within the appropriate symbol, a symbol sequence number.
4. In the *description column*, enter just enough information to indicate what is not told by the other columns of the chart. In this case, the information would probably be merely an indication of where the part had been stored, such as "in gondola."
5. Fill in any other column pertaining to the step just recorded. In this case, an important one is the *Machine or Equipment* column, where "gondola" (gon.) and its specifications could be entered.

   Other columns are explained as follows:

   Operation number. Pertains only to *operation* or *inspect* steps and should show same number as on the *Routing* or *Operation Process Chart.*

   Department number. Identification of department, plant, or area performing the operation.

   Time study register number. Serial number of time study from which time per piece is taken.

   Time per piece. Standard time in minutes or hours per piece.

   Pieces per hour. Number of pieces expected per hour at 100 per cent performance, according to standard time in previous column.

   Total load hours. Number of hours required to produce the *pieces per day* indicated in the heading of the chart.

   Operators per machine. Number of operators required for each machine.

   Total manpower. Number of operators to man the machines required to produce the daily requirement.

   Machine or equipment. Brief description, name, or number of machine or equipment necessary.

   Number of machines required. Number of machines or pieces of equipment needed to handle the daily production.

   Distance moved. Estimated distance the average part will move from one step to the next.

   Type of container. Designation of container to be used in handling the object.

   Remarks. Any other pertinent data.
6. Repeat *steps 2, 3, 4, 5* until every step of each operation in the planned process has been recorded.
7. Repeat *steps 1 through 6* for each part to be processed.

then they must be made here. This is actually a way of forcing the planner to consider the steps or activities in the process that occur between the operations listed on the *Production Routing.* Here, the interrelationship between operations must be considered, and non-productive as well as productive activity must be planned. To make a *Layout Planning Chart,* see the suggested steps listed in Table 11-6.

It will be seen from the completed *Layout Planning Chart* (Figure 11-12) that many of the loose ends of the planning process have been tied together. At this point a check has been made, by means of the *Chart,* on the completeness and accuracy of the planning. This is where any errors or omissions in the planning should be corrected.

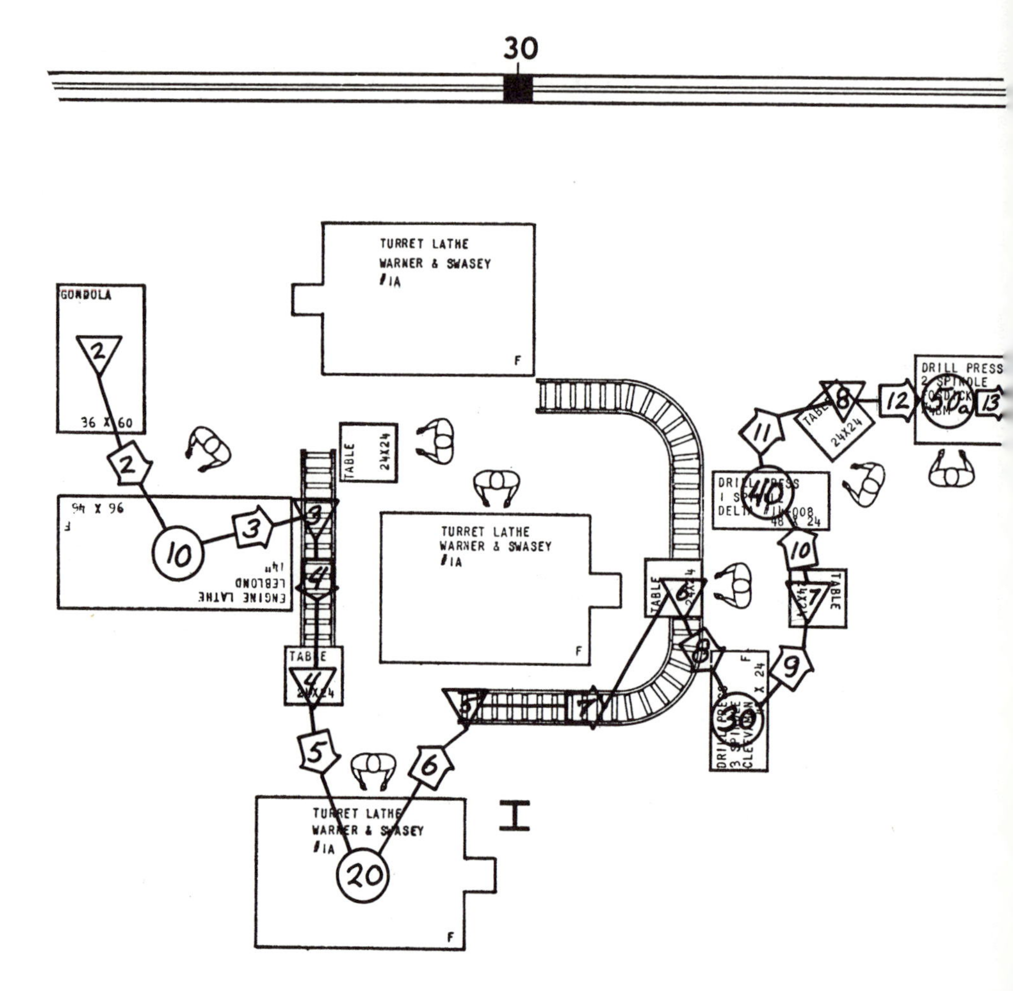

The Flow Diagram. Closely related to the *Process Chart* (and therefore the *Layout Planning Chart*) is the *Flow Diagram*, presented in Chapter 6 as an analytical technique. In the planning process, it becomes the next step in the coordination of the individual work areas, and must be synthesized. Actually, the *Flow Diagram* (Figure 11-13) is merely a consolidation or integration of the individual *Operation Analysis and Planning Charts* made previously.

The completed *Flow Diagram* may be adjusted to suit the more specific nature of the actual space allocated. In performing this step, either templates or scale models of the equipment and facilities may be used—or the diagram may be just a rough sketch.

Line production. One topic related to the coordination of a sequence of several work places is line production—a method of making full use of equipment and manpower. Line production, simply defined, is the physical arrangement of work

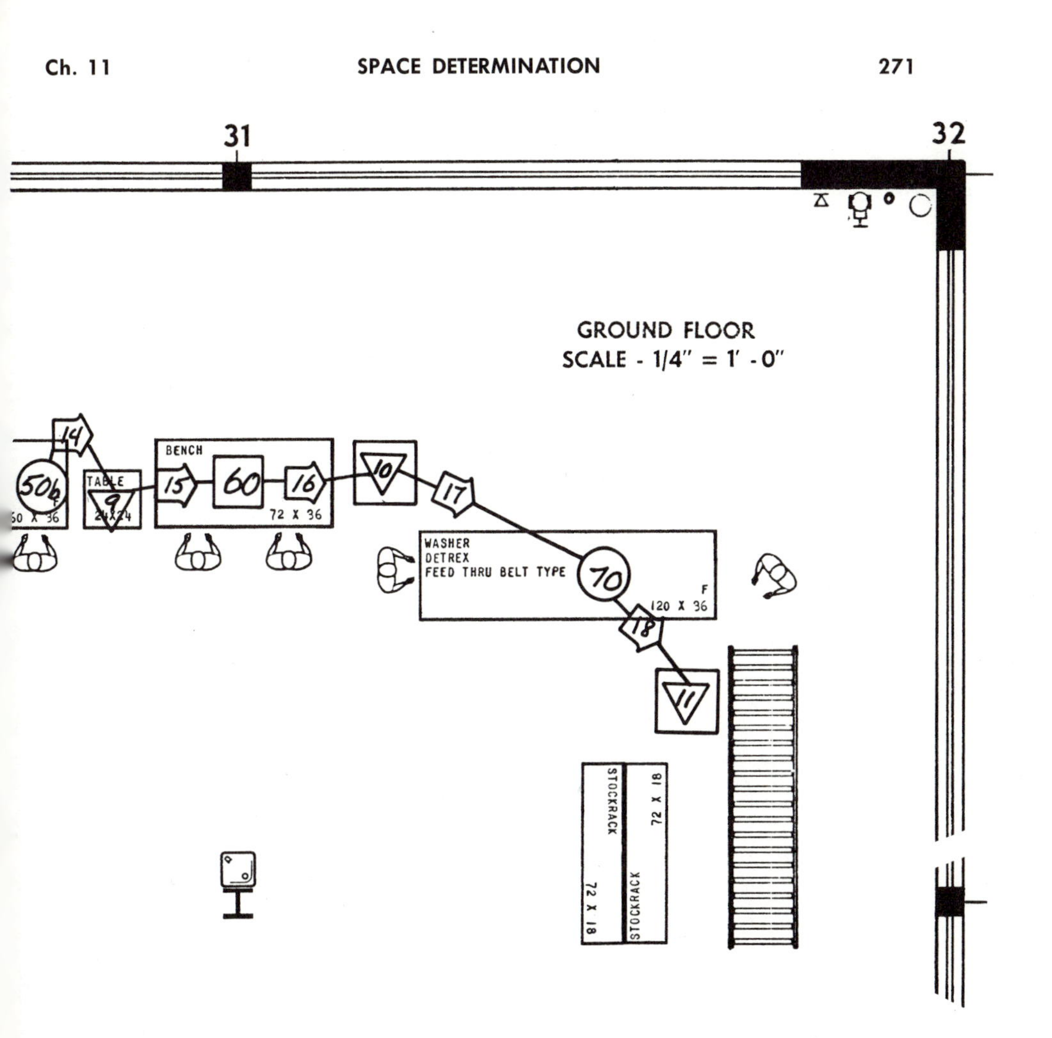

Figure 11–13—Flow diagram for Powrarm base.

places or operations in the sequence called for on the *Production Routing*—or any similar sequential list of operations or activities to be performed. Work areas are so planned and arranged that the material moves continuously, at a uniform rate. Operations are performed simultaneously at all work stations; each area turns out the same number of pieces per unit of time, and the material flows in a reasonably direct path.

Line production of this sort is usually attained by establishing a production pace in terms of units per hour. This pace may be governed by the speed of a conveyor, by the output of a certain machine or work area, by the stimulus of incentive pay, or by the slowest man. In practice, perfect pacing is most closely realized with the transfer-type machine.

This concept, however, is probably best known by the type of assembly line commonly associated with the production or assembly of automobiles, refrigerators, radios, or similar mass-produced items. The technique is applied equally well to the production of such unit parts as machined castings, sheet-metal stampings, wood parts, or even sandwiches or food trays for airlines.

In order for line production to be at all successful, each work station must feed material to the next station at a relatively fixed rate of units per hour. That is, if 60 units per hour are wanted at the end of the line, then each work station must turn out 60 units per hour, regardless of the actual time required to perform its own operation. This means that if a certain work station turns out only 30 pieces per hour, two stations will be required. If it turns out 15 pieces per hour, four will be required; and if a machine or operator turns out 120 pieces per hour, that workplace will be busy only one-half the time. The balancing of operations in the line is one of the biggest problems in line production. It should be the concern of the designer to see that the major factors—men, machines, and material—are furnished in balanced quantities to each work station. The work should be so planned that each operator's share of the overall job is a multiple of the cycle time set for the line. That is, if the cycle time is set at 1 minute, then an operator's share should be 1 minute or a multiple thereof. This concept will be illustrated later in this chapter.

Although line production may appear to be the best way to produce any item, there are many times when line production would not be the answer to the production problem. The following are some of the prerequisites necessary to the successful application of the line concept of production:

1. There must be sufficient volume to justify the product-type machine or the work place arrangement.
2. The work must be of such a nature that the operations to be performed on the line can be broken down into units sufficiently small:
   —To permit each to be learned in a relatively short time by a relatively unskilled operator, thus facilitating the shifting of personnel necessary when production volume increases or decreases.
   —To permit the elements of work to be recombined into time units of about equal length for each operator in the line.
3. Jigs and fixtures must be used to make sure that each operation is performed in exactly the same manner on each part or assembly.
4. The line, after it is set up, must not be so inflexible as to prevent minor alterations that might result from design, model, or methods changes on the part or product. In spite of the application of the production-line technique to the *n*th degree to automobile assembly lines, one manufacturer could turn out 1,000,000 variations of his product on one line. At some assembly plants, several different cars are turned out on one line! (e.g., Chevrolet, Pontiac, and Oldsmobile).
5. Material must be continuously supplied to the line at the required places in order that a material shortage will not cause the line to be shut down. Production control and material control must be properly worked out and must function well.
6. There must be enough operations to be done on the part or product to warrant the line.

7. The production of each kind of unit must extend over a sufficient period of time. A line cannot be set up if a month's supply can be turned out in a few hours.
8. The line must run a sufficient portion of the working time to be economical.
9. The job must last long enough to justify setting up the line.
10. The product design must be fairly well frozen or standardized so that changes will not disrupt the line too often.
11. Parts must be interchangeable.

It would seem that if all these prerequisites are necessary, line production would be next to impossible. However, a glimpse into almost any plant will show the line production principle in use to some degree. It is in plants making items in large quantities that the line production principle really finds its most efficient applications.

A production line may or may not be conveyorized. If it is conveyorized there will be certain advantages, such as:

1. Uniform rate of production
2. Simpler cost keeping
3. Pacing of operators
4. Less handling
5. Less time for work-in-process
6. More effective use of labor
7. Easier production control
8. Easier supervision
9. Less congested work areas

There are also some disadvantages. One of the most commonly discussed is that of monotony. That is, continuous repetition of the same job. Job enlargement, or job enrichment, is considered to be an antidote to the complaint. However, it is not certain that a good many people don't prefer this monotony to the headaches of planning, organizing, and supervising necessary in a group. Experiments thus far have not proved conclusive in either direction.[3]

Balancing line production. Stated briefly, the problem is one of so planning the work to be done that each work station will carry an equal work load and turn out the same number of pieces in a unit of time. Because situations within individual lines cause each problem to be different, various ways have been developed to attain balance. These are:

1. Banks of materials.
2. Moving or shifting operators.
3. Grouping or subdividing work elements.
4. Improving the operation.
5. Improving operator performance.
6. Having the operator work on subassemblies during available time.

Probably the easiest, but not at all the best, method of balancing a line is to build up banks of materials ahead of the slower work areas. These areas must then work overtime or take on additional personnel. This is not an equitable solution and usually results in a large float of material in the line, with accompanying bottlenecks, wasted space, and idle time.

A second method is to have an operator move along the line so that he covers

[3]The interested reader is invited to study the growing volume of literature on *job enlargement, job enrichment,* and *work structuring.*

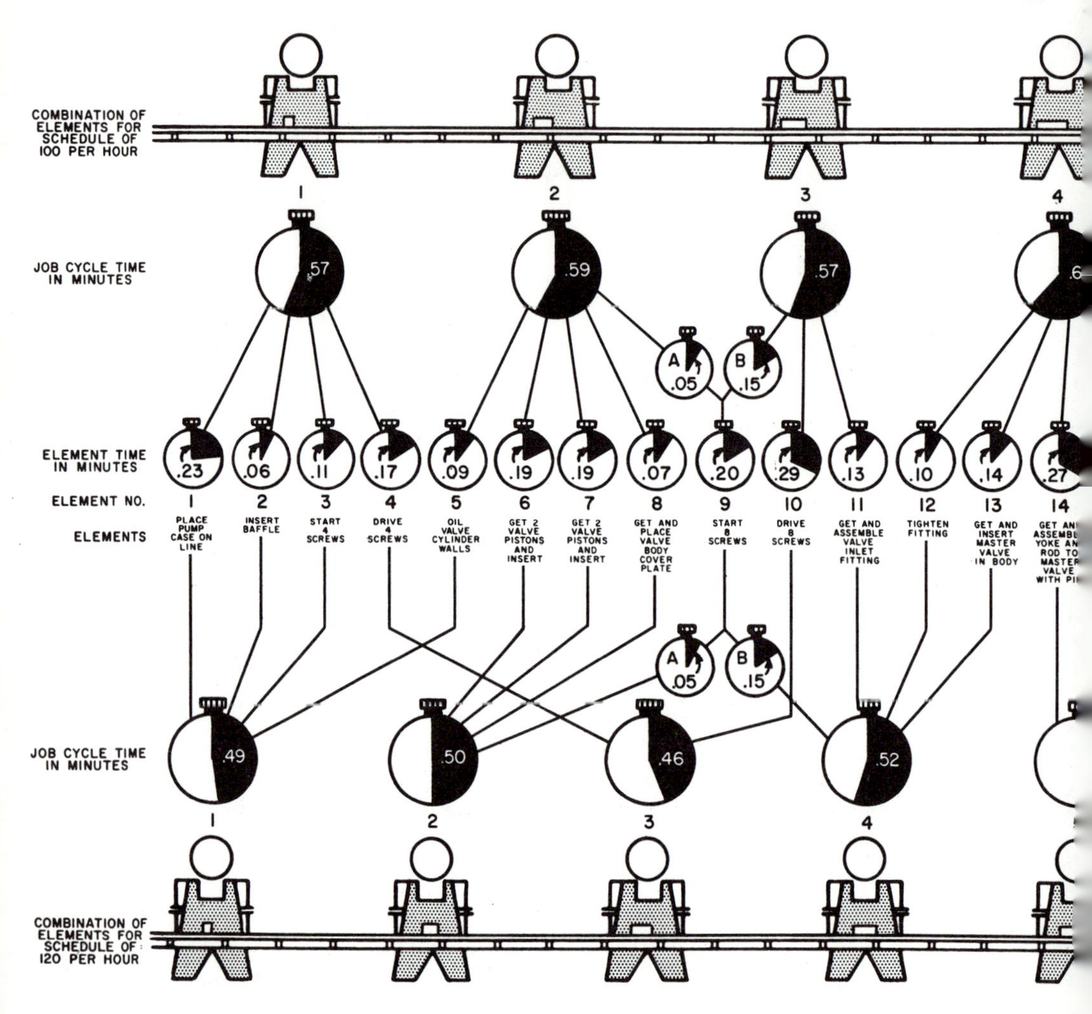

more than one operation when his work assignment is shorter than the others; or if it is longer than the others, he may be instructed to work on only every other unit. A second operator takes the alternate unit.

A third method is probably the most common. It involves subdividing the work to be done into elements, and then recombining the elements into tasks of equal length. This method is widely used and is reasonably successful. However, there are places where it is not applicable; for instance, when the work is of such a nature that it cannot be subdivided, or if it requires a particular skill, or if it would be too costly to work out the methods and equipment for accomplishing the work in subdivisions of the whole. Figure 11-14 and Table 11-7 demonstrate the combining of work elements into uniform tasks of equal length.

A fourth method, and one that should always be tried, is improving the operations. This method is especially useful if the line is already set up and several operations appear to be slower than the others. Studying the method will fre-

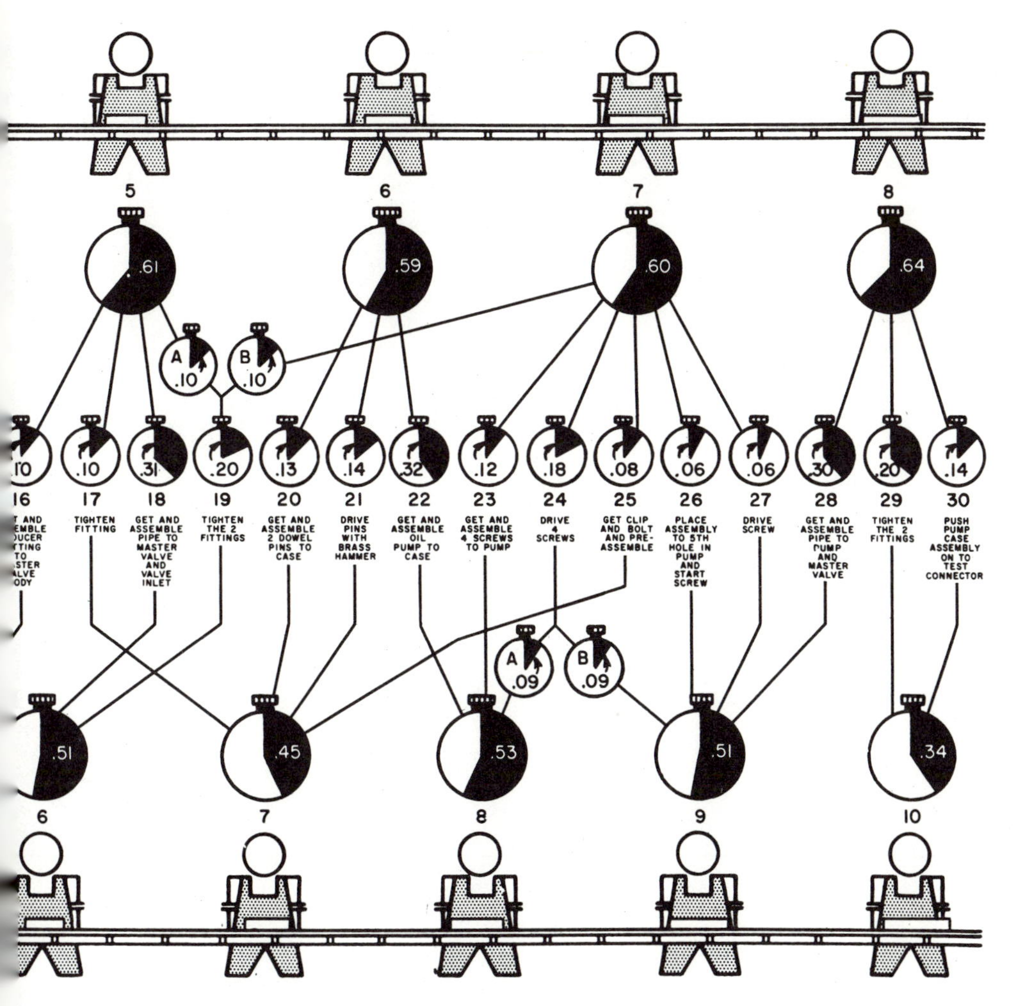

Figure 11–14—Illustration of equitable distribution of work on conveyorized assembly line for each of two production schedules. (Courtesy of General Motors Institute.)

quently result in discovering a better way to do the work, thereby reducing the time required so it will fit in with the others.

The fifth method is very closely related, and concerns itself with improving operator performance. This can often be accomplished by additional training for the person doing the work or, sometimes, by substituting an operator better fitted for the task at hand.

The sixth method is to have operators with idle time making up subassemblies when not occupied on the line. This is more successful on slowly moving lines, with long cycle times, than it is on rapidly moving lines with short cycle times.

Table 11-7. Theory of Assembly Line Balancing

Operations from Routing	Operation Elements from Time Study		Task Assignments at 100% Production (elements combined to make tasks of approximately equal duration)	Task Assignments at 50% Production
		.14 min.		
		.26	.... .48 min. = Task No. 1	
1	.80	.08		
		.11		= .98 = Task No. 1
		.21	.... .50 min. = Task No. 2	
		.06		
2	.46	.12		
		.28	.... .46 min. = Task No. 3	
		.18		
		.25		= .97 = Task No. 2
		.10	.... .51 min. = Task No. 4	
3	.91	.16		
		.09		
		.13	.... .47 min. = Task No. 5	
		.09		. . . etc.
4	.25	.16		

It can be readily imagined that in many cases, several of these methods are used together to attain balance on a particular line. It should also be pointed out that some types of lines are relatively easy to set up, and balance becomes more serious when the number of variations of the product increase, and when the size of the object increases.

It should be noted that a considerable body of quantitative knowledge has been built up in the area of line balancing. The interested student should investigate such sources as the *A.I.I.E. Transactions* (American Institute of Industrial Engineers), as well as O.R.S.A. (Operations Research Society of America) and T.I.M.S. (The Institute of Management Sciences) publications, which present mathematical treatments, modelling, simulation, and computer programs useful in analyzing the line balancing problem.

Warehousing

The space required for the warehousing of finished products can usually be determined by the procedures used in arriving at storage requirements. These procedures were covered earlier in the chapter, and should be followed—adapting as necessary—in calculating finished goods warehousing space needs.

Shipping

The determination of space needs for the shipping activity is closely related to that for the receiving activity. Therefore, space requirements should be determined by referring to the previous information on receiving, and adapting it as necessary to accommodate the slight differences between the two, as far as space determination procedures are concerned.

Parking

The problem of parking space is an increasingly expensive one to resolve. As businesses move out to suburban areas, usually more and more parking spaces must be provided. And industrial parks (see Chapter 19) commonly regulate parking area design to conform with both ecological and appearance standards, which are growing tougher each year and usually require more space than the typical parking lot. The information included here deals with conventional parking practices. The designer will have to check carefully with state, county, city, and subdivision or industrial park requirements, which may be far more stringent in terms of green space and plantings. These suggestions will prove helpful in the layout of a parking area:[4]

1. The width of the parking area is determined by the angle of the parking stall, as in Table 11-8 (see Figure 11-15 for details of stall size).
2. As the angle of the stall increases, the number of feet of aisle space increases.
3. The wider the stall, the sooner the driver can start turning, thus reducing the aisle width.
4. A greater number of cars can be parked on 90 degrees than on 60 degrees, using the same stall width.

Figures 11-15, 16, and 17 show details of parking area designs. It will be seen that the 90-degree parking arrangement requires only a little over 300 sq ft/car (19,220 sq ft ÷ 64 cars) compared with over 360 sq ft/car for the 60-degree arrangement. For rough planning purposes, the 300 or 360-sq ft figures can be used for estimating parking lot requirements. Space for drives and roadways to service them must be added.

Other Auxiliary and Service Activities

Most of the medical, food, lavatory, locker room, tool room, tool crib, and maintenance area requirements can be estimated by the same general procedure as others previously covered. The *Auxiliary and Service Activities Detail Planning*

[4]Based on *How To Lay Out a Parking Lot* (Chicago: Western Industries, Inc.).

A	B	C	D	E	F	G
0°	8'0"	8.0	12.0	23.0	28.0	–
	8'6"	8.5	12.0	23.0	29.0	–
	9'0"	9.0	12.0	23.0	30.0	–
	9'6"	9.5	12.0	23.0	32.0	–
	10'0"	10.0	12.0	23.0	32.0	–
20°	8'0"	14.0	11.0	23.4	39.0	31.5
	8'6"	14.5	11.0	24.9	40.0	32.0
	9'0"	15.0	11.0	26.3	41.0	32.5
	9'6"	15.5	11.0	27.8	42.0	33.1
	10'0"	15.9	11.0	29.2	42.8	33.4
30°	8'0"	16.5	11.0	16.0	44.0	37.1
	8'6"	16.9	11.0	17.0	44.8	37.4
	9'0"	17.3	11.0	18.0	45.6	37.8
	9'0"	17.8	11.0	19.0	46.6	38.4
	10'0"	18.2	11.0	20.0	47.4	38.7
45°	8'0"	19.1	14.0	11.3	52.2	46.5
	8'6"	19.4	13.5	12.0	52.3	46.5
	9'0"	19.8	13.0	12.7	52.5	46.5
	9'6"	20.1	13.0	13.4	53.3	46.5
	10'0"	20.5	13.0	14.1	54.0	46.9

A	B	C	D	E	F	G
60°	8'0"	20.4	19.0	9.2	59.8	55.8
	8'6"	20.7	18.5	9.8	59.9	55.6
	9'0"	21.0	18.0	10.4	60.0	55.5
	9'6"	21.2	18.0	11.0	60.4	55.6
	10'0"	21.5	18.0	11.5	61.0	56.0
70°	8'0"	20.6	20.0	8.5	61.2	58.5
	8'6"	20.8	19.5	9.0	61.1	58.2
	9'0"	21.0	19.0	9.6	61.0	57.9
	9'6"	21.2	18.5	10.1	60.9	57.7
	10'0"	21.2	18.0	10.6	60.4	57.0
80°	8'0"	20.1	25.0	8.1	65.2	63.8
	8'6"	20.2	24.0	8.6	64.4	62.9
	9'0"	20.3	24.0	9.1	64.3	62.7
	9'6"	20.4	24.0	9.6	64.4	62.7
	10'0"	20.5	24.0	10.2	65.0	63.3
90°	8'0"	19.0	26.0	8.0	64.0	–
	8'6"	19.0	25.0	8.5	63.0	–
	9'0"	19.0	24.0	9.0	62.0	–
	9'6"	19.0	24.0	9.5	62.0	–
	10'0"	19.0	24.0	10.0	62.0	–

A. PARKING ANGLE
B. STALL WIDTH
C. 19' STALL TO CURB
D. AISLE WIDTH
E. CURB LENGTH PER CAR
F. CENTER TO CENTER WIDTH
G. OF DOUBLE ROW WITH AISLE BETWEEN
F. CURB TO CURB
G. STALL CENTER

Figure 11–15—Parking lot dimensions table. (Courtesy of Western Industries, Inc.)

Table 11-8. Parking-Lot Width

No. of Rows of Cars	Lot Width-Feet		
	90° Parking	60° Parking	45° Parking
1	43	39	33
2	62	60	50
3	105	99	79
4	124	120	99

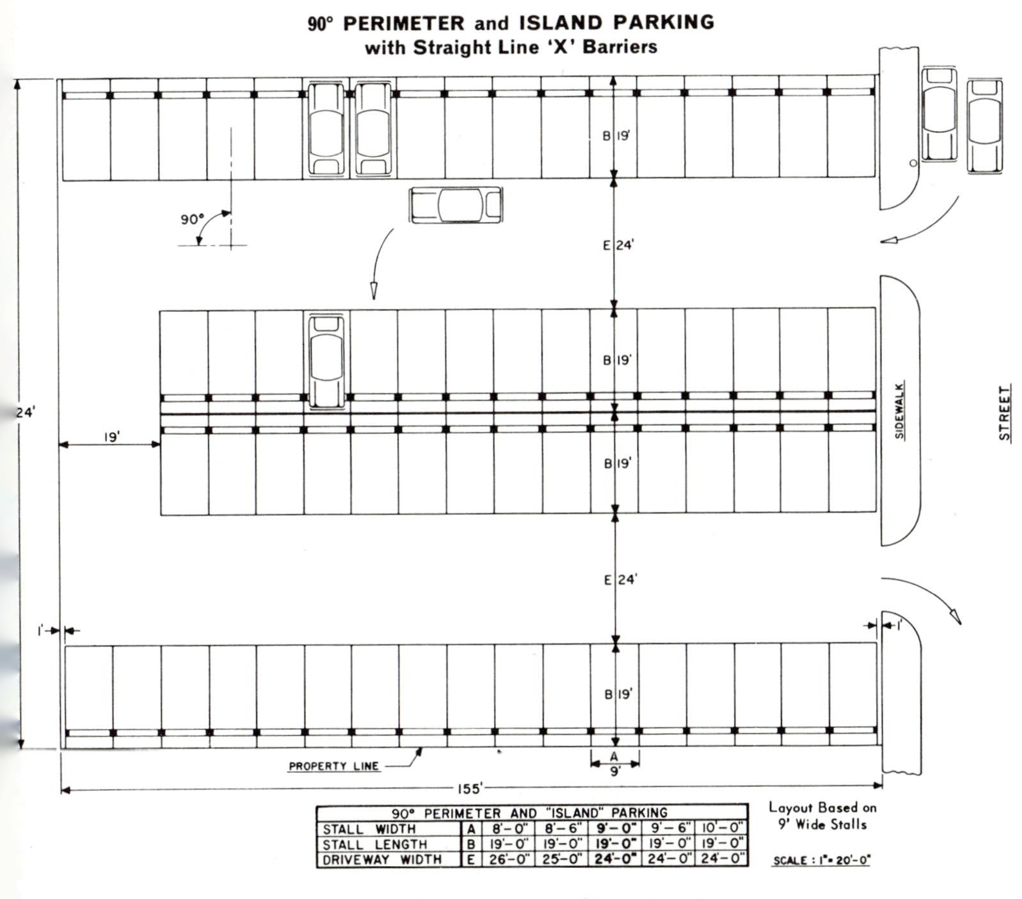

Figure 11-16—90° perimeter and island parking. (Courtesy of Western Industries, Inc.)

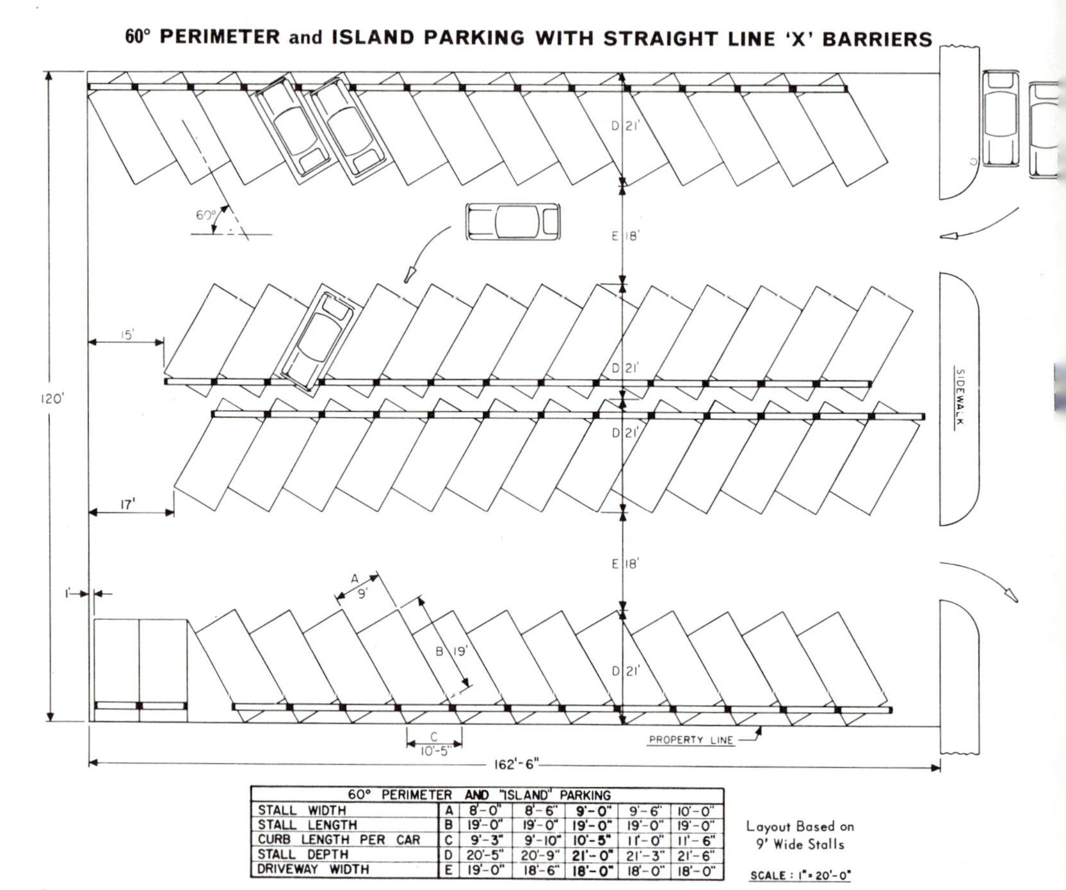

60° PERIMETER AND "ISLAND" PARKING						
STALL WIDTH	A	8'-0"	8'-6"	**9'-0"**	9'-6"	10'-0"
STALL LENGTH	B	19'-0"	19'-0"	**19'-0"**	19'-0"	19'-0"
CURB LENGTH PER CAR	C	9'-3"	9'-10"	**10'-5"**	11'-0"	11'-6"
STALL DEPTH	D	20'-5"	20'-9"	**21'-0"**	21'-3"	21'-6"
DRIVEWAY WIDTH	E	19'-0"	18'-6"	**18'-0"**	18'-0"	18'-0"

Figure 11–17—60° perimeter and island parking. (Courtesy of Western Industries, Inc.)

Sheet (Figure 11-18) can be used to accumulate the details and reach a total for each activity. Under the major headings, such as *Medical,* list the items to be included in the area, such as:

1. Cot
2. Screen
3. Lavatory
4. Examining table
5. Desk
6. Chair
7. Supply cabinet
8. Clothes tree

Extend each item across the sheet to a subtotal or total.

Establishing Total Space Requirements

The preceding has suggested methods of estimating the space needs of selected activities required in the facility. Those not covered must be estimated in somewhat the same fashion. The *Auxiliary and Service Activities Detail Planning Sheet* (Figure 11-18) can be used for any of the 60 or so activities suggested in the *Plant Service Area Planning Sheet* (Figure 11-19).

Next comes the task of determining the total space required for the facility. The figures from both the *Auxiliary and Plant Service Activities Detail Planning Sheet,* and the several individual space calculation forms can be re-capped on the *Total Space Requirements Work Sheet* (Figure 11-20). (The figures will be used in developing the *Area Templates,* in Chapter 12.)

Conclusion

This chapter has dealt with the problems involved in determining the amount of space required for the several activities of the enterprise. It has suggested procedures and dimensional information on some of them. The general approach to determining space needs for any enterprise are much the same, and the thought process developed here will serve—with proper translation—to aid in determining the space requirements for any of the many non-industrial enterprises referred to throughout this book. The next chapter will be devoted to the allocation of the spaces for individual activities to their appropriate locations in the facility being developed.

Questions

1. What are some of the factors to be considered in space planning under each of these categories? (a) General. (b) Production. (c) Building. (d) Cost.
2. In general, what is involved in determining the space needs for a specific activity?
3. What space-occupying items must be planned for in an individual work place?
4. Draw a sketch to illustrate the relationship between a work place and the overall flow pattern.
5. Name some of the method-related factors of concern in work place design.
6. Give some of the guides to work area planning.
7. Outline the general procedure in planning an efficient work place.
8. Discuss the concept of material handling at the work place.
9. Of what concern are scrap and level of efficiency in determining the proper number of pieces of equipment and operators? Discuss.
10. What technique is especially useful in coordinating individual work areas? How, and why, does it help?
11. What is meant by *line production?*
12. What prerequisites are usually necessary to justify it?
13. Give some of the methods of balancing the work in production lines.

AUXILIARY AND SERVICE ACTIVITIES

PLANT: Acme Mfg. DEPT.: Tool Rm & Tool Crib PLANNER: M.M.F.

Activity and/or Item(s)	Sq. ft. per Unit	No. Units	Estim'd. Sq. ft. Req'd.	Allowance[a] (25%)	Sub-totals
Equipment					
Engine Lathe	32	2	64	80	
Milling Mach.	40	1	40	50	
Drill Press	20	3	60	75	
C.O. Saw	45	1	45	56	
Band Saw	12	1	12	15	
Bench	18	4	72	90	366
Mock-up Area	—	—	250		250
Storage Area					
Racks	40	1	40	50	
Sheet, Plate	50	1	50	63	
Shelves	9	10	90	113	
Rot-a-bin	10	4	40	50	
Misc.	—	—	—	—	276
Counter Area	16	2	32	40	40
Office Supvr.	100	1	100	125	
Clerk	75	2	150	188	313

[a]For flexibility, aisles, elbow room, etc.

Figure 11-18—Auxiliary and service activities detail planning sheet.

DETAIL PLANNING SHEET

DATE: Nov. 10

Total Now 1975	Future Needs			Sub-totals	Build for 1980	Remarks
	Equipment Area	Personnel	Prob. Expansion by 1980			
	3		112			
	2		100			
	4		100			
	1		56			
	1		15			
	5		113	496		
	—		250	250		
	2		100			
	1 1/2		75			
	15		169			
	8		100			
			—	444		
	3		60	60		
		1	125			
1,245		4	375	500	1,750	

PLANT SERVICE AREA

Plant (or Area) Powrarm

General	Est. sq. ft.	Production Services	Est. sq. ft.
1. President 2. General Manager 3. Sales and Advertising	400	16. Industrial Engineering a. plant layout b. materials handling c. methods d. standards e. packaging f. process engineering g. tool design	
4. Accounting a. general b. cost c. payroll d. credit 5. Product Engineering a. research b. development c. design d. drafting e. testing and experimental 6. Purchasing 7. Personnel a. general b. employment c. training d. credit union e. safety	400	17. Production Control a. planning b. scheduling & dispatching c. traffic d. follow-up 18. Quality Control a. receiving b. in-process (floor) c. final 19. Plant Engineering a. general office b. maintenance shops	
8. Product Service		20. Receiving	
9. File Room		21. Stock Room	1400
10. Conference Room		22. Warehousing	800
11. Vault		23. Shipping	
12. Reception Room		24. Materials Handling Equipment Storage	500
13. Switchboard		25. Tool Room	
14.		26. Tool Crib	
15.		27. Product Supervision	
		28.	
		29.	
		30.	
Total General Office Area	800 sq. ft.	Total Production Service Area	2700 sq. ft.

Figure 11–19—Plant service area planning sheet.

PLANNING SHEET

Estimated by A. M. James Date Nov. 15

Personnel Services	Est. sq. ft.	Physical Plant Services	Est. sq. ft.
31. Health and Medical Facilities		46. Heating facilities	
32. Food Service a. kitchen b. dining 50 x 15 c. vending machines	750	47. Ventilating Equipment 48. Air conditioning Equipment 49. Power Generating Equipment	
33. Lavatory a. showers 5 x 20 = 100 b. locker room 50 x 15 = 750 urinals 10 x 10 = 100 w. basin 5 x 20 = 100 stools	1050	50. Telephone Equipment Room 51. Maintenance Shops	500
34. Smoking area		52. Air Compressors #51	
35. Lounge Area		53. Scrap Collection Area	
36. Recreation Area		54. Vehicle Storage	
37. Parking (outside)		55. Fire Protection a. extinguishers	
38. Time Clock a. bulletin boards		c. equipment d. sprinkler valves	
39. Fire Escapes		56. Stairways	
40. Drinking Fountains		57. Elevators	
41. Telephones		58. Plant Protection	
42.		59.	
43.		60.	
44.			
45.			
Total Personnel Service Area	1800 sq. ft.	Total Physical Plant Service Area	500 sq. ft.

TOTAL SPACE REQUIREMENT WORK SHEET

FOR Powrarm PLANT

BY A. M. James DATE Nov. 15

Activity or Area	Estimated Square Feet		Module Size 20 X 20 = 400	
	Individual Areas	Sub-Totals	No. Mod.	Size of Area Templates
A. General Services	800			
		800	2	20 x 40
B. Production Services				
20, 22, 23 Rec., Shipping, Whsg.	800		2	20 x 40
21 Stock	1400		$3\frac{1}{2}$	20 x 70
24, 25 Tool Room, Tool Crib	500	2700	$1\frac{1}{4}$	20 x 25
C. Personnel Services				
32 Food Service	750		$1\frac{7}{8}$	20 x 50
33 Locker Room	1050		$2\frac{1}{2}$	20 x 50
		1800		
D. Physical Plant Services				
51 Maintenance	500		$1\frac{1}{4}$	20 x 25
		500		
E. Production	6500	6500	$16\frac{1}{4}$	60 x 108
F. Other				
GRAND TOTALS	12,300	12,300	$30\frac{5}{8}$	

Figure 11-20—Total space requirement worksheet for Powrarm factory.

Exercises

A. Develop a list of space-occupying activities in a home.

B. Develop a form for calculating the space needs of a home, which might be used by a family or an architect in planning a new house.

C. Perform the above for a facility of your choice (or instructor's assignment).

D. What are some ways of implementing in-process storage? Explain each.

E. Sketch a workplace incorporating some of the guides from *Question 6*. Indicate where each is included in the sketch.

F. Calculate the number of machines and operators required for the following sequence of operations:

Production Routing

Operation No.	*Description*	*Pcs/hr*	*Expected Scrap*
1	Cut to length	60	2%
2	Drill two 1" holes	25	4%
3	Chamfer two holes	45	3%
4	De-burr	30	1%

Required output is 120 pc/hr. Operation efficiency is expected to be 80 per cent. How many pieces must be stared at *Operation 1* to obtain 120 good pieces out of *Operation 4?*

12
Area Allocation

Activity interrelationships and space requirements are integrated in the area allocation process. And the *Area Allocation Diagram* is the basis for the detailed layout and the building design. In fact, it is not until this stage of the facility planning process that the architect or builder need be seriously involved, since they both must have a reasonably fixed layout configuration from which to begin their work.

The overall objective of the area allocation process is to design an efficient arrangement of the space units required by each activity into an integrated whole. The resulting arrangement, inasmuch as possible, should respect the activity interrelationships previously determined, and reasonably maintain the area requirements of each activity. Some advantage and uses of the area allocation process are:

1. Systematic allocation of activity area.
2. Facilitates the layout process.
3. Permits a more accurate layout.
4. Helps to avoid overlooking an activity.
5. Provides a total area estimate.
6. Provides an easily understood preliminary idea of plant activity arrangement.
7. Minimizes waste space.
8. Suggests alternative arrangements.
9. Forces detailed consideration of individual activities.
10. Translates area estimates to a preliminary arrangement in visual form.
11. Insures adequate space.
12. Shows relative sizes of activities.
13. Basis for further planning.
14. Aids in presentations.

It can be seen that area allocation is a key phase in the facility design process, and should be done very carefully, checked by appropriate personnel, and approved by management before further detailed layout work is performed.

Factors for Consideration in Area Allocation

As with many of the previous steps in the facilities design process, there are a number of important factors to be considered in carrying out the area allocation procedure. Of those listed in Table 12-1, some have been discussed in previous chapters (particularly in Chapter 5, some are discussed below, and others will be treated in following chapters).

Table 12-1. Factors for Consideration Area Allocation

1. Relationships between internal and external material flow
2. Expansion plans, directions, growth (by activities)
3. Flexibility to meet changing needs
4. Building characteristics—type, construction, size, shape, restrictions
5. Special requirements of certain departments—environmental requirements, undesirable characteristics, etc.
6. Allowance for later trimming (during approval process)
7. Possible use of mezzanine, balcony, basement, roof, etc.
8. Probability of product or process changes
9. Aisle locations and sizes
10. Relative importance of individual activity areas
11. Personal preferences of people involved
12. Column spacing and location
13. Material flow pattern
14. External facilities and their relationships to internal activities
15. Building orientation on site
16. Activity interrelationships
17. Area requirements of each activity
18. Value of space
19. Site—size, topography, orientation
20. Space available
21. Transportation facilities
22. External MH facilities
23. Funds available
24. Building code
25. Zoning ordinance
26. Location of service and auxiliary activities
27. Need for shared facilities
28. Building restrictions
29. Storage requirements

Expansion

One of the most perplexing problems facing both management and the facilities designer is the question of expansion. In a well-run, progressive, successful enterprise, expansion should be inevitable. However, the vagaries of the future, and the possibility of either over- or under-building can have serious economic consequences. Under-building can result in tight quarters, inefficient production, and the need for disruptive additional construction, too soon. But, if finances can be arranged, overbuilding may be the least serious, as the excess space may be leased until needed.

Therefore, pre-planning for potential expansion is an important part of the facility design process, from site selection to building construction. Of interest here, however, are the factors and problems involved in preparing for the potential expansions. The need for expansion may come from any one of a number of reasons related to a need for additional volume, products, parts, processes, or services. For example:

1. It may not be possible to meet sales demand, due to inadequate plant capacity.
2. New parts may be added to products.
3. New processes may be required.

4. Additional operations and services may be needed.
5. Activities previously sub-contracted may be pulled back.

These and many other considerations may give rise to the need for additional space, which can be more easily provided if it has been planned for.

Factors for Consideration in Planning for Expansion.

Again, there is the need for being aware of the many various factors concerned with the problems of expansion. Table 12-2 can be used as a check sheet during

Table 12-2. Factors for Consideration in Planning for Expansion

A. General

1. Time schedule
2. Extent of expansion required (desired)
3. Disruption of present activities
4. Master plan
5. Sales forecasts
6. Possible changes in products and lines
7. Advantages of expansion over a new site
8. Need for facilities all at one location
9. Advantages of move now vs. add some now and move later
10. Disadvantages of new facilities at remote location
11. Disruption caused by actual move
12. Possibility of sub-contracting some operations
13. Logic of separating activities on multiple sites
14. Effect of present expansion on future expansion
15. Foreseeable technological advances in production methods
16. Possibility of changing from process to product layout
17. Possibility of changes in product or product line

B. Cost

1. Capital available
2. Interest rates
3. Cost of addition vs. new building—relative payoff, etc.
4. Economic trends
5. Economic feasibility of adding to present building
6. Possibility of recovering cost of expansion if move is made later
7. Cost of construction now vs. later
8. Cost of moving into new space

C. In relation to site

1. Availability at present site vs. new location
2. Cost now vs. future
3. Distance from other company operations
4. Roads, etc.
5. Legal restrictions
6. Topographical features
7. Transportation facilities
8. Availability of utilities
9. Expansion potential of site
10. Parking
11. Placement of building on site
12. Zoning
13. Aesthetic aspects
14. Relationship of receiving and shipping in plant, and to site
15. Size of site
16. Orientation of building on site
17. Shape—plant, site
18. Adjacent structures, etc.

D. In relation to the building

1. Type of structure
2. Type of construction
3. Size of building
4. Shape of structure
5. Movability of walls
6. Building code, etc.
7. Additional floors vs. additional area on ground
8. Construction cost
9. Directions of expansion feasibility
10. Location of building on site
11. Expansion of service areas to match production
12. Possibility of leasing; or build and lease back
13. Time required to build
14. Expansion potential of expanded facility
15. Number of floors
16. Clear height
17. Column spacing
18. Door (etc.) locations
19. Foreseeable technological trends in building construction

Table 12-3. Future Space Requirements

Activities	Available at present	Needed at present	Future needs. . . 1-2 yr	4-5 yr	8-10 yr
Receiving	2,000	2,600	3,000	3,800	5,000
Production	15,000	17,000	20,000	25,000	35,000
Assembly					
Warehousing					
Offices					
Shipping					
Totals	25,000	30,000	35,000	50,000	80,000

SIMULATED DATA

the expansion planning phase of the facilities design process. Some of the factors are discussed below.

An early consideration for expansion was covered in the discussion of flow patterns, in Chapter 5, where it was suggested that the flow pattern be designed for easy expansion, so that new directions of flow will have been planned in advance.

A second consideration is the amount of future space required, which can be at least approximated, by estimating in terms of specific activities, somewhat as in Table 12-3.

Certainly such an exercise will better prepare the enterprise for future space needs than just waiting to see what happens and then hastily throwing up an addition, which can prove very inadequate in many ways—such as in size, configuration, and location.

Where to add the space. Along with what and how much, comes the question of where the added space should be located. This has been partially answered by the material flow pattern—but more specific plans should be made. That is, as the layout for current needs is being designed, the areas likely to be needed in the future should be shown in their respective planned locations and sequence. This will assure their proper relationships to present activities when the new areas are added. As someone has said, "It's wise to know where you're growing."

Factory Magazine has presented six ways to expand a plant—as shown in Figure 12-1, with the pros and cons of each. The basic ways are:

1. Mirror image
2. Straight-line flow
3. T-flow
4. U-flow
5. C-flow
6. Change-no-walls concept

Modifications, adaptations and combinations of these will offer even more possibilities.

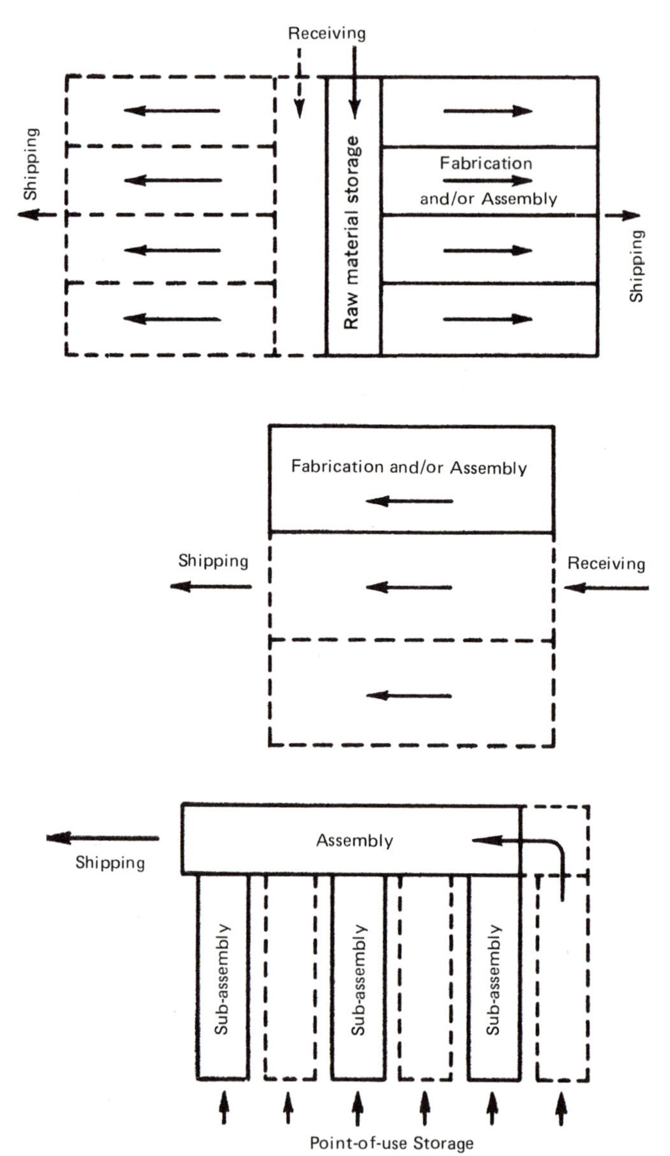

Mirror image expansion

PRO: Doubles production space by dupli cating original layout. Simple to execute Sets up one receiving area and centra corridor storage for entire plant. Produc tion flow is free of bottlenecks.
CON: Limited to one-time expansion Two separated shipping areas.

Straight line flow expansion

PRO: Unlimited expansion by adding c additional bays. Simplest of all to ex cute. Low "add-on" building costs. Mir mum column interference. Straight-lir flow. Well suited where overhead cran are required.
CON: Hard to selectively expand son departments. Land must be level.

"T" flow expansion

PRO: Selectively adds departments wit out disrupting flow. Utilizes adjace building columns. Provides point-of-u receiving and storage, thus minimizi material handling.
CON: May force extension of main a sembly, and eventual relocation of ce tain departments.

Figure 12–1—Six ways to expand a plant. **(Courtesy of Factory Magazine.)**

"U" flow expansion

PRO: Expands in concentric layers about U-core to match growing needs. Excel-ent concept when removable or trans-plantable walls are used. Combines receiving and shipping at one centralized ocation.
CON: Eventually becomes unwieldly, after successive expansions.

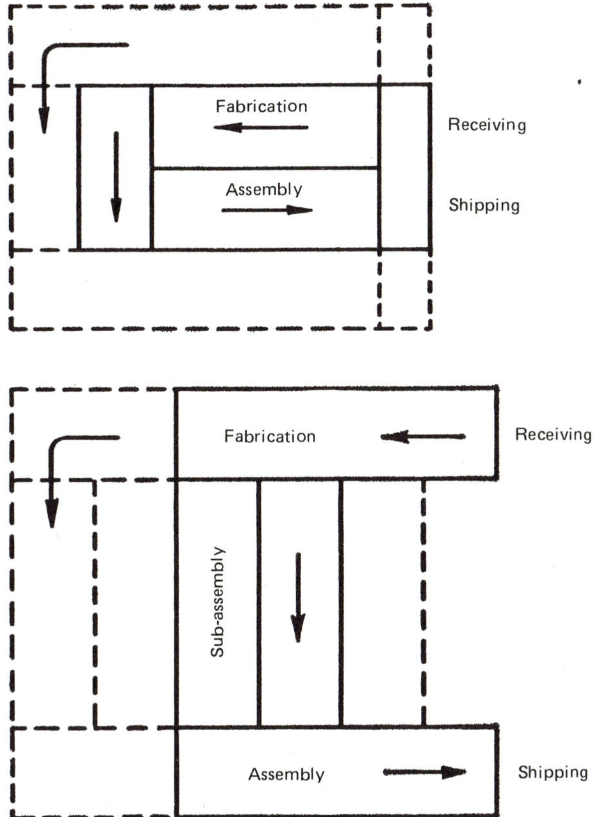

"C" flow expansion

PRO: Allows selective addition to several departments without disruption of flow or building lines. Accommodates in-plant freight car sidings. Ideal for overhead cranes, conveyorization, lift trucks, etc.
CON: Original design or earlier changes may preclude its use.

"Change no walls" expansion concept

RO: Often overlooked when expansion eems indicated are techniques like hese, which free space and may bring onus savings. Usually cheaper than onstruction alternatives.
ON: Management may resist these temporary" solutions, preferring the ase and prestige of adding space.

Conveyorization may eliminate in-process storage points and speed up the manufacturing cycle. Multiple-loop towline system with self-dispatching carts overcome problems of criss-cross work flow.
A B C production scheduling. "A" items (fast movers or the 10%-20% of products that commonly account for 80% or more of sales) are made to stock by automated machines. Other parts or products are made only to order in separate areas.
Numerically controlled machine tools may reduce the need for departments, such as jigs and fixturing, fabrication, and sub-assembly. Product is made almost completely on one N/C machine.

Table 12–4. Considerations in Planning for Expansion

A. Designing a new facility

1. Plan future layout in at least enough detail to show major features.
2. Pre-plan for expansion in at least two directions.
3. Plan aisle arrangement for ease of extension.
4. Locate activities most likely to expand in best position for expansion.
5. Design flow pattern for logical extension.
6. Consider the possibility of new processes (as well as future developments).
7. Locate service departments for convenient use in expanded (future) layout.
8. Arrange equipment in such a way as to permit inserting additional equipment easily.
9. Use wider aisles—for the present.
10. Use wider machine spacing—for the present.
11. Allow extra space in interior departments.
12. Locate permanent equipment in fixed locations—i.e., not to be moved later—because of special foundations, utilities, installation problems, etc. (such as washrooms, offices, heavy machines, and ovens).
13. Allow up to 25% more production space than presently needed.
14. Plan for up to 50% more office space than presently needed.
15. Leave outside walls free of permanent installations.
16. Plan building shape for ease of expansion.
17. Use a minimum of partitions.
18. Plan utility location, arrangement, and capacity for ease of extension (water, electricity, plumbing, air, heating, ventilation, air conditioning, sewer, drain, etc.).
19. Place columns and carefully space them for ease of expansion.
20. Plan larger building than needed and lease or rent extra space.
21. Locate such activities as receiving, shipping, parking, walks, roads, and utilities for minimum re-arrangement or re-location in expansion.
22. Make use of roof for ovens, air conditioning, etc.
23. Locate building on property to facilitate additions.
24. Use movable walls, partitions, etc.
25. Use modular construction.
26. Design foundation to allow for extra floors.
27. Install intermediate footings to facilitate addition of extra columns to support possible mezzanine or balcony.
28. Plan adequate height (20 ft minimum) for addition of mezzanine or balcony.
29. Purchase from 3 to 10 times present land needs.
30. Plan for additional parking.
31. Locate large exterior doors to serve as passages (to aisles) to new additions.
32. Use quick-connect utility fixtures.
33. Plan for outer wall of first construction phase to become fire wall after expansion.
34. Support roof with columns, not walls.
35. Use sandwich or panel walls and partitions to facilitate their relocation.
36. Locate receiving and shipping (especially exterior facilities) for convenience after planned expansion.
37. Plan for longer trailers (in future).
38. Attempt to foresee technological advances in your industry.
39. Attempt to predict changes in product, product line, or product mix.
40. Purchase site in advance of need (i.e., future availability and cost vs. present).

Table 12-4 (continued)

B. "Expanding" in existing space

1. Re-locate aisles.
2. Narrow the aisles.
3. Reduce inventory—new materials: receive later, oftener; finished goods: ship earlier.
4. Store materials outside—shed, outdoors, other locations.
5. Stack materials higher (make use of building height).
6. Move machines closer together.
7. Re-locate equipment.
8. Re-orient equipment.
9. Sub-contract selected activities, functions, processes (or rent outside space for them).
10. Use overtime for production.
11. Add another shift.
12. Remove partitions to increase free space and flexibility.
13. Add a mezzanine or balcony.
14. Drop products where sales do not justify production.
15. Eliminate processes not fully utilized (and sub-contract).
16. Make use of roof for ovens, air conditioning, etc.
17. Examine possibility of product (vs. process) layout, if volume is sufficient.
18. Increase productivity.
19. Use narrow-aisle handling equipment.
20. Use collapsible containers to save space required for empties.
21. Use palletless handling techniques to eliminate *lost cube* in pallets.
22. Use pallet size that makes best use of space between columns.
23. Use overhead space for storage, conveyors, other equipment.
24. Dispose of obsolete materials, equipment.
25. Double-deck, in-plant service areas—or use low ceiling areas for them.
26. Eliminate honey-combining in storage areas.
27. Eliminate assigned storage spaces.
28. Reduce space between pallets, containers, racks.
29. Use space over aisles, especially in storage area.
30. Plan for a minimum of material to be stored at work place.

Suggestions for facilitating expansion. In Table 12-4 are listed a number of ideas and suggestions, which if heeded, will facilitate expanding the building. Most relate to the design of the present facility, with expansion in mind.

Expanding within the walls. Table 12-4 suggests some ways to expand within an existing building, where for some reason actual physical expansion of the structure is not feasible.

Mezzanines and balconies. A relatively inexpensive method of providing additional square feet of space is by the use of mezzanines or balconies. Such areas can be pre-planned, to make use of clear space over facilities not requiring as much headroom as production. For example, above production or other floor-level

Figure 12-2—Mezzanine washroom facilities. (Courtesy of The Austin Co.)

activities, double-decking can be used to locate such activities as:

1. Offices
2. Toilets
3. Carton storage
4. Slow moving stock or stores
5. Subassembly operations
6. Any low floor load activity

Examples are shown in Figures 12-2, 3, and 4.

Flexibility

Closely related to the problem of expansion, is the need for flexibility in the planning of efficient facilities, to provide for possible future developments. In fact,

Figure 12-3—Mezzanine office area over a washroom. (Courtesy of The Austin Co.)

Figure 12-4—Mezzanine store room. (Courtesy of DeLaval Separator Co.)

one of the nation's largest concerns, in extolling its progressive nature, has said: 80 per cent of the products they are making today were unknown 10 years ago, which presents the formidable challenge to the facility designer—design a plant today, to provide for our needs from 5 to 50 years in the future.

Flexibility is required for many of the same reasons mentioned above under Expansion. And many of the demands for future flexibility can best be met if they are anticipated in the original planning.

Just as true today as when written—and likely to hold true for the future, is the comment by consultant Robert P. Neuschel,[1] "Today's unsettled conditions and rapid technological progress make plant and facilities planning a market research as well as a manufacturing problem." He felt the balance of specialization and flexibility to be one of management's "most perplexing problems:" new plant vs. modification of exiting layout; specialized vs. multi-purpose plant vs. "strike a happy medium." With the ever accelerating technological advances, product, equipment, and demand are in such a state of flux as to infinitely complicate the work of the planner. Specialized plants and facilities leave an enterprise vulnerable to great expense in making successive changes to keep up the competitive pace, where far-sighted, though perhaps at the time forbiddingly costly, planning and construction for flexibility could have minimized much of the later time and expense of operational adaptations. And there are no sweeping formulas to guide the planner; from the oil refining industry where opportunities

[1]Believe it or not, most of the thought here and in the following list of several related considerations derives from a *1951* article, by R. P. Neuschel, "How Flexible Shall We Make Our Plant and Facilities?" *Advanced Management,* Jan. 1951:16.

for flexibility are extremely limited, to high-volume line-production operations, to the production of short-lived products, each enterprise must find its own way through the middle ground between flexibility and specialization. As a general guide-line, however, he suggests the following considerations:

1. *Basic Cost-to-Selling-Price Margin in the Industry*—A small margin calls for specialization for minimum manufacturing cost; a greater gross margin allows more leeway in designing for flexibility.
2. *Projection of Present Operations*—changes in a product or a short expected demand, with possible related plant obsolescence, calls for more flexibility.
3. *Compare Building and Installation Flexibility*—If operations must be highly specialized, it still may be relatively inexpensive in the long run to construct a more standard multi-purpose building, for both possible unforeseen major changes in layout and possible future plant disposal.
4. *Re-evaluate Planning Premises*—Both personal and industry-wide attitudes and concepts can interfere with planning for balance between specialization and flexibility; the best process flow of the moment may be outdated sooner than thought possible;[2] complete obsolescence is made up primarily of individually insignificant day-to-day changes—unforeseen and unplanned-for.

Group technology. Group technology (discussed on page 59) is an approach to providing flexibility, and should be referred to in this regard, during the planning process.

Methods of obtaining flexibility. In designing the facility, there are a number of actions that can be taken to insure the varying degrees of flexibility required by future demands. A number of these are listed in Table 12-5[3] and some are discussed below.

A common way of facilitating the rearrangement of productive equipment, when the building is constructed, is to install utility systems to which service connections are easily tied. Good examples are the electrical ducts and the cutting-compound pipe lines which may be installed overhead, down the centers of bays. Such arrangements permit machines to be plugged-out, moved to a new location, and plugged-in almost at will. Figure 12-5 shows the use of such an overhead system to simplify the disconnecting and reconnecting of a machine to a power source. Still another method is to install cellular steel beams as an integral part of the floor. Through the cells or channels utility lines are run. The cells may be tapped into with very little trouble at almost any place in the floor. Figure 12-6 shows a cut-away view of this cellular type of flooring.

An interesting story is told concerning the flexibility of such systems. An employee in a certain plant left his machine during his lunch hour. Upon returning after his meal, he discovered that the machine and all traces of it had vanished. In its place was a new machine, and another employee already at work.

[2]Chapter 5 may not have completely prepared for this rather critical concern; it should be remembered that material flow is but one of many factors in layout design.

[3]Some duplication will be noted here, from the lists on expansion in Chapter 12, as many items are applicable in both areas.

Table 12-5. How to Plan for Flexibility

1. Use as large a bay size as is practicable.
2. Use rectangular bays to permit alternate equipment arrangements.
3. Provide clear heights of 15 to 20 ft for production; 25 to 35 ft for storage.
4. Provide for mezzanines or balconies.
5. Install utilities (especially electricity and sprinklers) on a grid basis.
6. Provide for uniform lighting over entire plant area.
7. Locate light fixtures between bar joists or beams (i.e., not below).
8. Locate sprinklers between or through bar joists (or beams).
9. Locate unit heaters over aisles so as not to lose stacking height in open spaces.
10. Use quick-connect utility connections or "capped-off" utility lines.
11. Use a minimum of partitions.
12. Use movable partitions.
13. Provide wide doorways.
14. Plan for extra space—25% for plant; 50% for offices.
15. Provide roof support with columns—not walls (so wall-partitions can be removed if desired).
16. Install a ramp between ground level and floor level.
17. Provide for future mezzanines or balconies by installing extra footings between columns.
18. Design adequate floor-load capacity for future needs.
19. Provide adequate storage space between work areas.
20. Use standard production equipment.
21. Avoid use of specialized equipment.
22. Use flexible, adaptable material handling equipment.
23. Plan adequate utilities and service facilities.
24. Use modular construction.
25. Avoid roof openings, floor pits, special foundations, etc.
26. Use high and wide doors.
27. Provide equipment with built-in lifting lugs or skid supports for easy re-location by crane or lift truck.
28. Use machine mounts instead of lagging equipment to the floor.

Figure 12–5—Electric power connections placed overhead for flexibility in machine relocation. (Courtesy of Trumbull Electric and Manufacturing Co.)

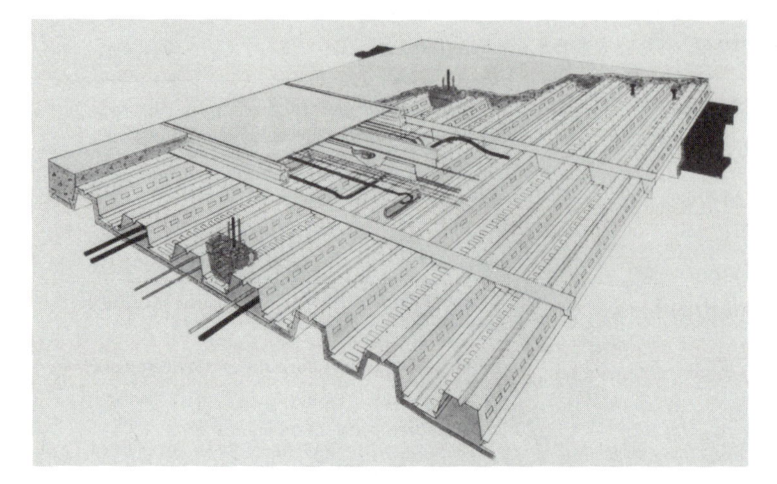

Figure 12-6—Q-flooring construction for utility distribution. (Courtesy of H. H. Robertson Co.)

The first machine had been moved to a new location several hundred feet away, and was there ready for him to continue his work.

Mobile equipment is used by a company that makes sheet metal parts of various kinds in limited quantities on one line of large presses. Smaller presses for intermediate operations are mounted on skids and moved in and out, as called for, by a lift truck. This concept is shown in Figure 12-7.

Another sheet-metal plant has installed a special mounting pit along which any of the many presses required may be mounted in the order required. A pit for large and small presses is shown in Figure 12-8.

Use of All Levels

It should be remembered that a plant contains six levels of activity (discussed in Chapter 5). One of these levels is the roof. It is commonly used to support drying

Figure 12-7—Small presses mounted on skids for spotting between larger presses by fork lift trucks. (Courtesy of American Seating Co.)

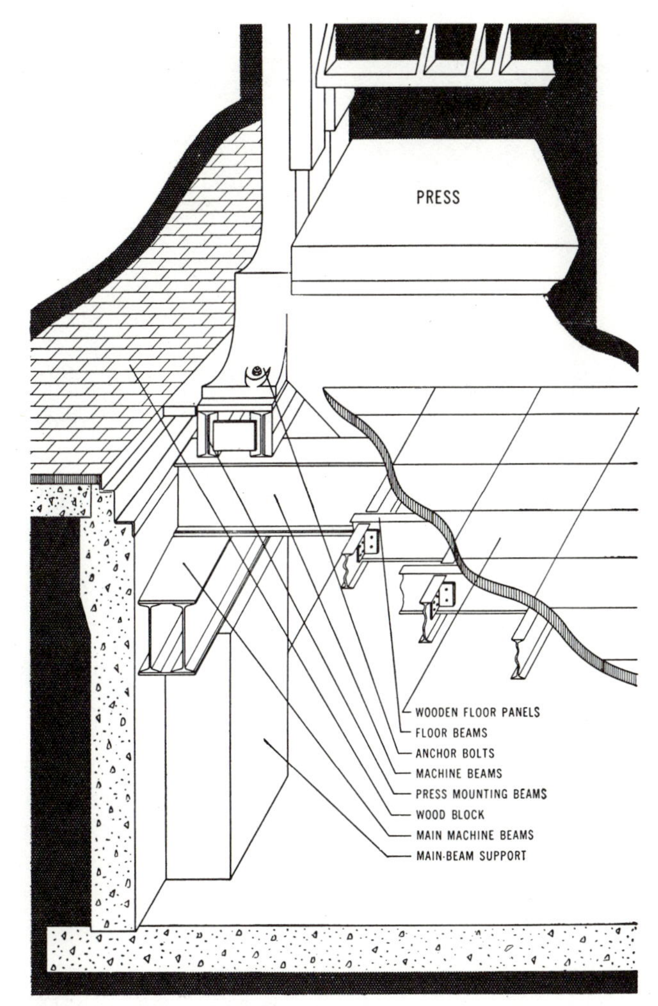

Figure 12–8—Cut-a-way section of a large press pit. (Courtesy of Factory Magazine.)

ovens, water tanks, and electrical and air conditioning equipment. Also, do not overlook the possibility of other levels, for example, within the trusses, in a basement, or in a utility tunnel. Some plants have personnel corridors, employee and service facilities below the manufacturing floor, as in Figure 12-9. A central, cross-wise, or perimeter tunnel may also be adequate for utility lines. This is very common in modern sprawling one-floor, school buildings.

Inter-Building Handling

Even the fact that plant buildings may be widely separated need not interfere too much with material flow between them. Although surface transportation is most

Figure 12-9—Underground tunnel between buildings for utilities and personnel passage. (Courtesy of The Austin Co.)

common, many buildings are connected, either underground or overhead. Figures 12-10 and 11 show such possibilities.

Point-of-Use Storage

An important factor in area allocation is the application of point-of-use storage—vs. centralized storage–of material, parts, and supplies. Basically, the point-of-use storage concept calls for the storage at a work place or as near by as possible, of items used in a work place. This eliminates intermediate storage locations, minimizes material handling operations, and in theory reduces the number of moves to two: one from carrier into storage; and the second from storage into the product. Often, even receiving inspection can be eliminated if the vendor is reliable in terms of both quality and quantity.

Figure 12-10—Underground passageway for employees, from plant entrance to various production areas, via stairs and service areas on underground level. (Courtesy of The Austin Co.)

Figure 12–11—Overhead handling between buildings. (Courtesy of American Monorail Co.)

Figures 12-12, 13, and 14 illustrate one application of this concept. Figure 12-12 shows material from receiving being placed in a subassembly bay (one of several, along a final assembly line). The entire inventory of all items used in an assembly are stored in this bay. The material above normal-reach height is all reserve storage. Figure 12-13 shows the portable work bench in place and in use. Notice the air and electrical connections hanging down for connection to the portable work bench. Figure 12-14 shows the portable cart (upon which finished subassemblies are placed in Figure 12-13) alongside the final assembly line.

Figure 12–12—Restocking point-of-use material supply. (Courtesy of Clarke Floor Machine Co.)

Figure 12–13—Taking advantage of point-of use storage, in the area shown in Figure 12–12, with mobile work station in place and air and electrical connections refitted. (Courtesy of Clarke Floor Machine Co.)

Some advantages of point-of-use storage are:[4]

1. *Production planning easier* to accomplish.
2. *Tighter controls* can be maintained, and more information is available for the customer on short notice.
3. *Tighter security* against pilferage and damage is possible.
4. *Better rotation* of perishable items is permitted.
5. *Recall is much easier* for defective or obsolete parts.

[4] R. J. Craig and W. C. Turner, "Point-of-Use Storage," *Industrial Engineering,* Oct. 1973:25.

Figure 12–14—Typical assembly line work station using portable tray-trucks to furnish sub-assemblies made in bays at rear and as shown in Figures 12–12 and 12–13. (Courtesy of Clarke Floor Machine Co.)

Some advantages of decentralized storage[5] are that it:

1. *Minimizes production interruptions* due to delayed delivery of raw material from central storage.
2. *Provides easy visual inventory* of stock.
3. *Reduces paper work and clerical costs* in maintaining essential records.
4. *Utilizes idle floor space* in the production area.
5. *Minimizes delays* due to delivery to the wrong point.
6. *Usually reduces the number of inventory clerks and material handling personnel* needed.
7. *Adds to the authority and flexibility of the foreman* as well as increasing his responsibility; at the same time, storage supervision may require less of his time if major problems have been encountered in the past in delivery of material to his department.
8. *May reduce material handling costs.*

Resolving Factors in Area Allocation

One of the most unique current concepts in retaining flexibility, providing for expansion, applying the point-of-use storage concept, and at the same time making use of the building cube, is the installation of work places within a high-rise storage system at General Electric's Appliance Park—East, in Columbia, Md. Two basic elements were involved: (1) cross-traffic delivery systems, and (2) random-access storage and delivery units. This concept,[6] shown in Figure 12-15, arranged cross-traffic delivery systems at right angles to major flow, at the boundaries between major functional areas; storage-and-delivery units were related directly to facilities, and provided (1) access to material and parts, (2) point-of-use storage, and (3) part containers for cross-traffic delivery.[7] The layout was somewhat like the layout of boulevards and cross-streets, with facilities on the lots defined; the flow structure is easily expandable in any direction without interference with existing operations. Thomas M. Harrison, who guided the project, states the principles of handling and layout upon which the flow-through system was based:[8]

1. *Design plants as groupings of manageable units,* or blocks of responsibility.
2. *Provide precise physical and administrative control* over all material in the manufacturing system.
3. *Provide straight-through flow,* with provision for buffers where required.
4. *Provide random access* to all material stored at point-of-use.

[5]For a similar concept used in a lift truck assembly plant of Allis–Chalmers, see "Point-of-Use Components Storage Transforms Plant Layout," *Factory,* Nov. 1971:43.

[6]For further information on this unique installation, see (1) "Preview of G.E.'s New Appliance Park," *Modern Materials Handling,* Sept. 1971:68; (2) T. M. Harrison, "A Break-Through in Job-Shop Layout," *Modern Materials Handling,* July 1971:30; (3) Harrison's "Basic Principles for Flexible, Efficient Handling," *Automation,* July 1972:32; and (4) G. F. Schwind, "Appliance Park East: Total Design for Material Handling," *Material Handling Engineering,* Sept. 1971:85.

[7]Harrison, "A Break-Through. . . ."

[8]Harrison, "Appliance Park East. . . ."

Figure 12–15—Basic city-block layout concept; major flow is in one direction, through storage-and-delivery units and associated process units in all manufacturing areas; cross-traffic aisles permit two-way flow to link an appropriate series of units for a variety of product lines. (Courtesy of Modern Materials Handling Magazine.)

5. *Design* the major processing units in this order: (a) final assembly; (b) subassembly, (c) contributing parts areas (finishing, then fabrication)[9]
6. *Provide for incremental expansion* without distruptive rearrangement of facilities or interruption of major flows.
7. *Standardize* handling and storage equipment.
8. *Ensure that systems and equipment are compatible* with eventual on-line computer applications.

Figures 12-16 and 12-17 further describe the material flow and interrelationships among the types of material, the personnel, and the storage facilities.

Aisles

One of the most important factors in the allocation of space is a very careful consideration of aisle location and width. One authority has stated that a typical

[9]It will be noted that this principle follows the idea of the assembly chart, and subsequent detailing in this book.

layout may show 1.2 sq ft of aisle for every 1.0 sq ft of manufacturing and storage space, and that production equipment commonly accounts for only about 25 per cent of the entire area of the facility. Aisles are primarily used for:

1. Material handling
2. Personnel movement
3. Finished product handling
4. Scrap and waste removal
5. Equipment re-location and replacement
6. Fire fighting equipment access

Aisles may be classified as:

1. Main (transportation)
2. Cross
3. Departmental
4. Personnel
5. Service, maintenance, etc.
6. Miscellaneous; access to:
   —Elevator
   —Air compressor
   —Electrical panel
   —Sprinkler valves

An excellent picture of the importance of aisles in layout by Reibel[10] points up the following considerations:

1. *Flow economy*—In many plants aisles form the routes for all movement of personnel and material.
2. *Space economy*—Aisles often take so much space in a layout that their careful design can have a direct effect on profitability.
3. *Sequence of Design*—Main aisles, for flow between departments or out of a plant, should be designed first; then service facilities can be placed and secondary aisles laid out.
4. *Large plant space economy*—Whereas a small building 20 ft wide may have one aisle 5 or 6 ft wide, taking 25 to 30 per cent of the available floor space, a building 600 ft wide may have three main aisles (down center and sides, 300 ft apart) each 10 ft wide, taking only 5 per cent of the space, with secondary aisles raising this to 10 or 12 per cent given to aisles in a large plant.
5. *Backbone aisle*—There should usually be a main aisle through the center of a building, straight if at all possible, starting and ending at outside entrances, or connecting an outside entrance and a main cross aisle (though aisles on upper floors may dead-end).
6. *Interior aisles*—These are often offset, dead-end, or quite narrow.
7. *Work space as aisles*—Often work area around machines is sufficient for movement of workers and material.
8. *Aisle width to Repair*—Expected frequency of overhaul or breakdown should help determine whether aisles should be wide enough to take the biggest machines.
9. *Hazardous conditions*—These call for aisles to be clear at all times for quick passage of fire-fighting and first-aid equipment.
10. *Clear aisles*—General design should try to foresee and forestall the kinds of compromises that may be made with aisle width when unforeseen demands for space arise.
11. *Aisle width*—In large plants, main aisles may be 12 to 20 ft wide; a 10-ft passageway will usually accommodate two loaded lift trucks passing, plus

[10] S. Reibel, "Aisles Can Affect Efficiency of Handling and Storage Methods," *Industry and Power*, Oct. 1940:78.

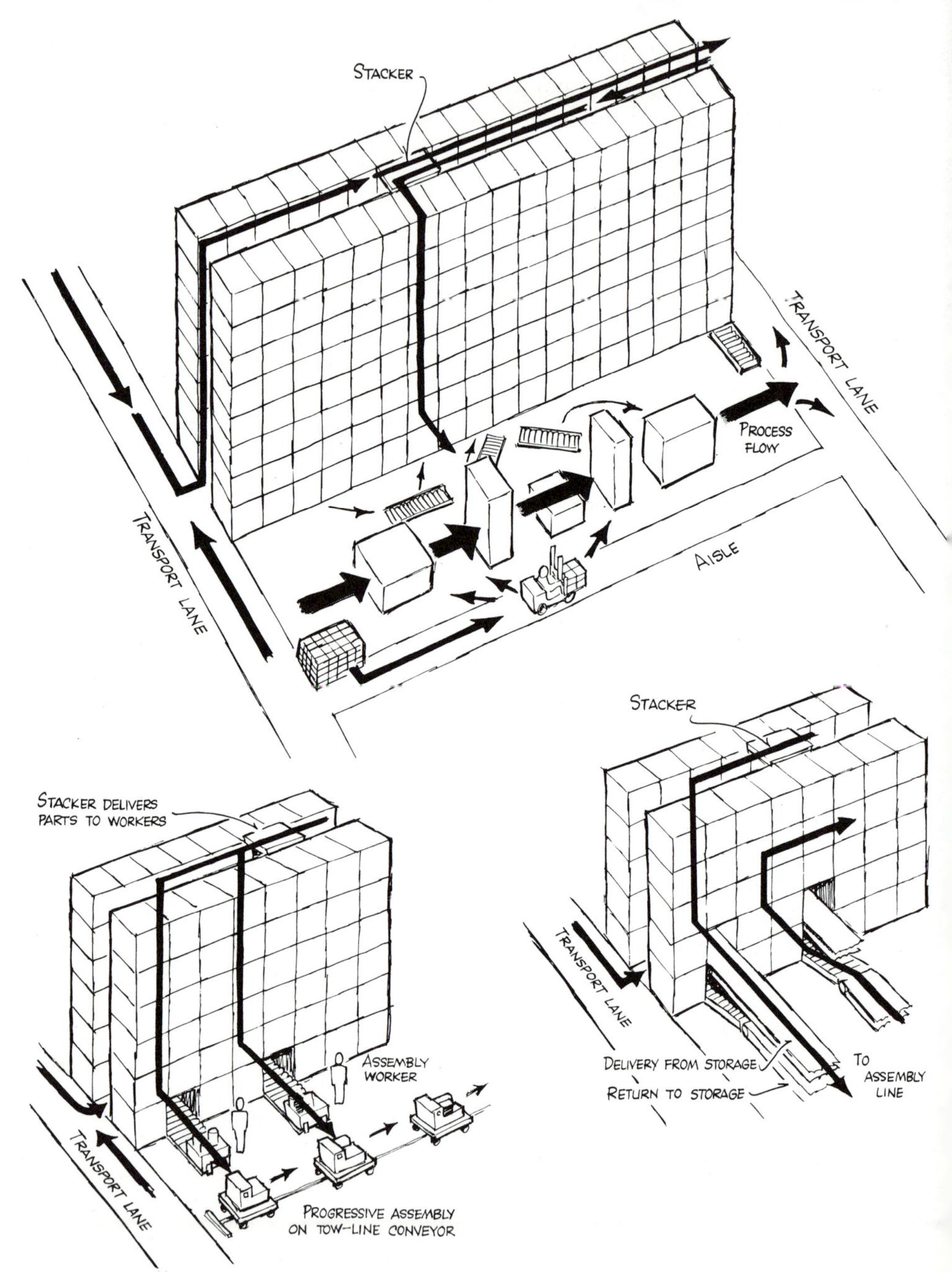

Figure 12–16—Random access storage facilities are located to store necessary items at point of use alongside process equipment. Here stacker system provides high rack storage of incoming parts and work in process for production line made up of individual machines connected by short sections of conveyors where possible. (Courtesy of Automation Magazine.)

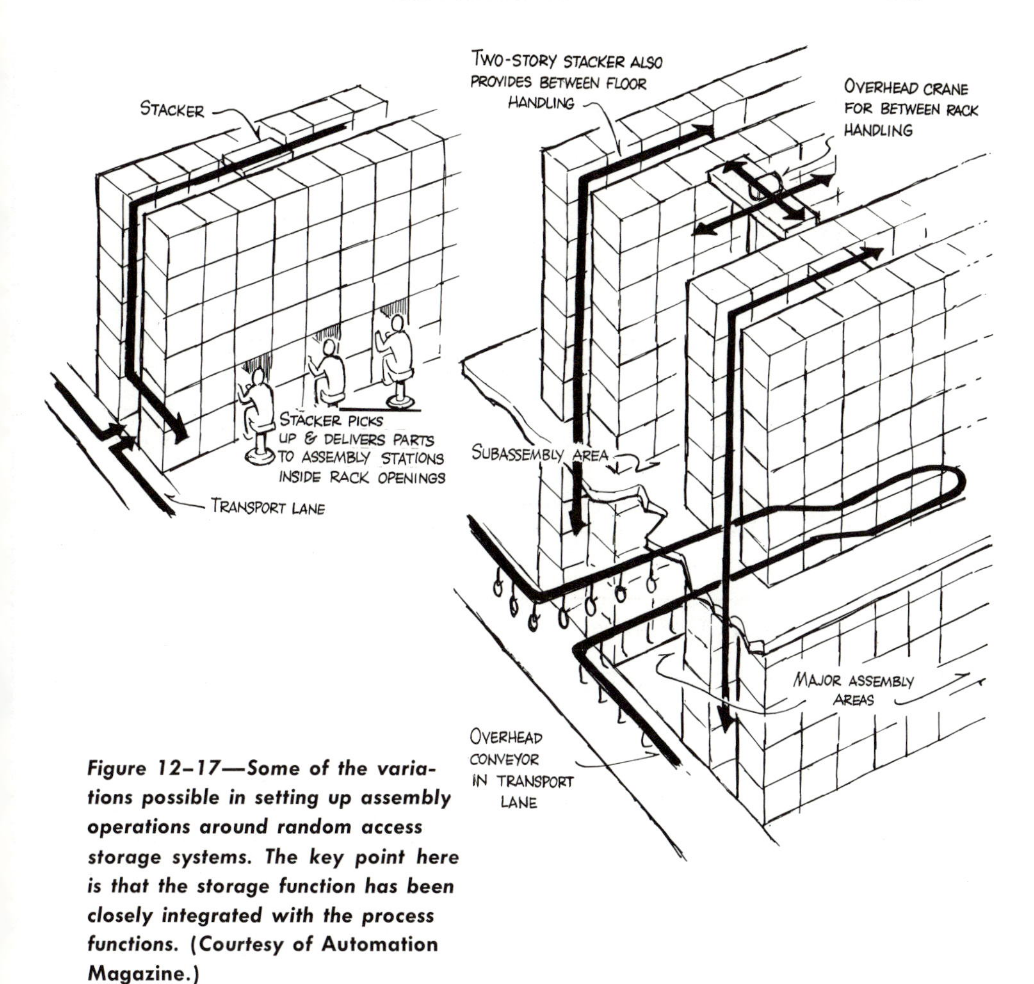

Figure 12–17—Some of the variations possible in setting up assembly operations around random access storage systems. The key point here is that the storage function has been closely integrated with the process functions. (Courtesy of Automation Magazine.)

clearance for a person walking; 8-ft aisles may be crowded, especially if loads are more than 40 inches wide; personnel and interior aisles may be as narrow as $2\frac{1}{2}$ or 3 ft, but movement in them will be constricted. Lift-truck turning radius, and unit-load size, are factors to be considered; by-pass widenings may be used where low-frequency traffic does not warrant two-way aisles.

12. *Interfloor traffic*—Elevators are a special case of aisles where the periodic flow calls for off-setting from a main aisle for holding material on its way to or from an elevator without obstructing main-aisle traffic.

Figure 12-18 indicates some of the relationships listed above; and Table 12-6 lists factors to be considered in determining aisle width, location, and spacing. It can be seen that the proper width and placement of aisles calls for much careful study.

If tractor–trailer trains will be used in a facility, aisle width becomes a major factor in their effective utilization, since trains of trailers require an adequate

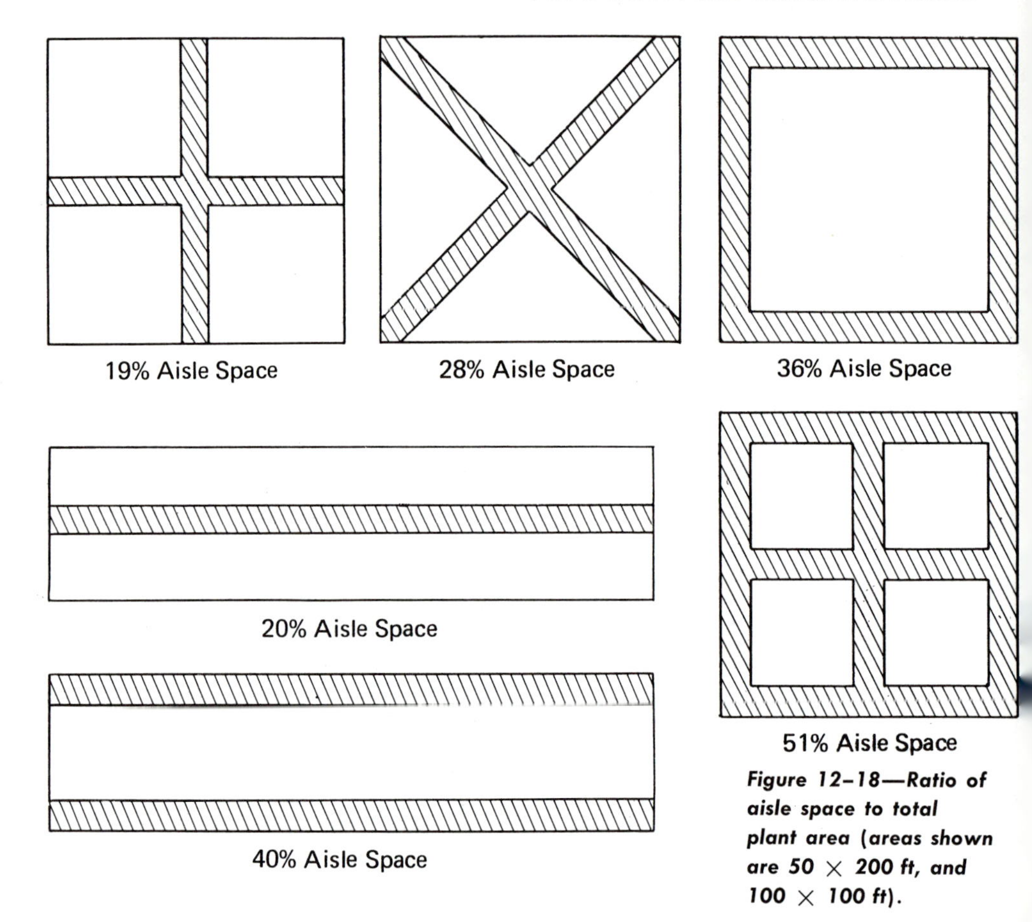

Figure 12–18—Ratio of aisle space to total plant area (areas shown are 50 × 200 ft, and 100 × 100 ft).

turning radius. This is true whether the tractor is operator-driven or electronically controlled. It is best to check with the tractor or trailer manufacturer for specific recommendations in relation to trailer size, train length, and aisle width.

Column Spacing

In allocating area and locating aisles, column spacing, or bay size is another important factor. Most buildings, to be economically constructed, require columns or posts to hold up the roof or provide necessary supports for overhead equipment. In larger new structures, bay sizes are increasing over those built prior to 1940. One authority states that, before World War II, the average plant had 150 interior columns for 100,000 sq ft of floor space. This average has been reduced to about 40 columns per 100,000 sq ft by using bay sizes up to 100 ft in one direction. Some wide spans, of course, are necessary because of the nature of the product, but require more expensive structural work. In some large aircraft plants, 300-ft spans are not unusual. In general, wide column spacing is desirable

Table 12-6. Factors for Consideration in Designing Aisles

A. Width

1. Sizes of products stored
2. Handling equipment
3. Storage practices
4. Cost of floor space
5. Traffic type
6. Traffic volume
7. Traffic directions
8. Safety
9. Ease of accessibility desired
10. Pallet size

B. Location

1. Distance and accessibility to doors
2. Lot sizes stored
3. Fire wall locations
4. Column spacing
5. Location of service areas
6. Floor load capacity
7. Elevator and ramp locations
8. Rack or pallet location
9. Access to elevators, service areas, etc.
10. Road and rail locations

C. Width and location

1. Equipment locations related to aisles
2. Building size
3. Future space requirements
4. Future aisle requirements
5. Fire regulations
6. Access by fire equipment
7. Use of aisle
8. Accessibility to equipment
9. Flexibility (for re-layout)
10. Building shape

D. Spacing

1. Equipment size
2. Item size—in warehouses
3. Lot size—in warehouses
4. Storage unit size
5. Building size
6. Building shape
7. Building type
8. Structural design
9. Type of construction
10. Aisle location
11. Aisle width
12. Number of aisles
13. Types of aisles required
14. Pallet size
15. Rack type; size
16. Floor load capacities
17. Load on roof or trusses
18. Type of columns
19. Column loading
20. Number of floors
21. Location of auxiliary functions
22. Flexibility desired
23. Expansion plans
24. Fire wall location
25. Construction cost
26. Money available
27. Layout
28. Building code

for maximum efficiency in handling material, arranging equipment, and providing for future flexibility. In today's plants, common spacing of columns varies from 30 to 50 ft, with some up to 100 ft or over; most common spacing would be about 30 by 50 ft, 40 by 60 ft, etc. Actual distances are best determined by consulting contractors and checking available sizes of structural members in the locality.

As most items in the facility are rectangular, rectangular bays are to be preferred over square bays. Therefore, if an item will not fit one way, it may be turned 90° for a possible better fit. One authority[11] feels that 30-by-40 and 50-by-20-ft column spacing is most often short-sighted, and that 100-by-40 and even 120-by-60-ft column spacing, though entailing somewhat higher initial cost, can repay this many times over in the duration of useful occupancy and the cost of reduced future changes. Possible future disposal is another recommendation for wider column spacing. Each of the column-spacing factors in Table 12-6 should be questioned before settling on a bay size; in no case should the size be arbitrarily set by the random suggestion of an architect or builder. Good architects and builders should inquire and assist in determining the size desired by the client; but

[11] A. T. Waidelich, "Trends in Modern Industrial Buildings," *Mill and Factory*, May 1958:3–4.

if there is any doubt, the question should be thoroughly investigated, and the width carefully calculated. In a warehouse, for instance, pallets and racks should fit between columns, so as not to leave a partial pallet or rack width empty for the length of the building.

The Area Allocation Procedure

In general, the area allocation procedure consists of constructing a space template for each activity to represent its gross space needs. The templates are then arranged in proper relationship to each other—usually into a rectangular shape—complying with requirements and restrictions implied by the *Activity Relationship Diagram* and the many related factors above. This is the last preliminary planning step prior to the detailed planning of material handling methods, individual work stations, and the final plant layout. The bases for the allocation process are:

1. *Production flow*—(a) materials, (b) equipment.
2. *Activity Relationship Chart*—(a) information flow, (b) personnel flow, (c) physical relationships.
3. *Space requirements.*
4. *Activity Relationship Diagram.*

In carrying out the area allocation process, the following are some of the criteria that should be kept in mind (see also Table 5-2):

1. Planned interrelationships between activities.
2. Economical utilization of space.
3. Ease of expansion.
4. Potential for vertical expansion.
5. Possibility of adding mezzanines, balconies, etc.
6. Potential flexibility.
7. Reasonable basis for future planning.
8. Good tie-in with external transportation facilities.
9. Reasonable configuration for building structure.
10. Logical aisle arrangement.
11. Proper orientation and relationships to site features.
12. Activities with specific requirements properly located.
13. Ease of supervision.
14. Easy to maintain control of production.
15. Complies with health and safety requirements.
16. Reasonable column spacing.
17. Complies with building code and zoning ordinance.
18. Adequate area for each function.

The actual area allocation process might proceed somewhat as follows:

1. *Fill out the Total Space Requirement Work Sheet* (Figure 11-19) as follows:
   a. Enter activities in left column.
   b. Enter space requirements for each.

c. Establish a module size for each activity, based on bay size and column spacing—or a convenient or logical subdivision of the bay size (such as 20-by-20, 30-by-30, or 50-by-50-ft) and enter at top of the *Module Size Column.* A square module is suggested at this stage in the planning; it can be changed to a rectangular shape later.
d. Reduce each activity area to: (1) a number of modules, (2) an area template size.

In the Powrarm example, and assuming a 20-by-20-ft module, these template sizes would be as follows (activity numbers are from activity templates on the *Activity Relationship Chart* in Figure 8-1):

1. Receiving and shipping, 20-by-40 ft
2. Stock room, 20-by-70 ft
3. Tool room and crib, 20-by-25 ft
4. Maintenance, 20-by-25 ft
5. Production, 60-by-108 ft
6. Lavatory, etc., 20-by-50 ft
7. Food Service, 20-by-50 ft
8. General office, 20-by-40 ft

Incidentally, if too great a module size is selected, there will not be enough area templates to conveniently arrange into a reasonable configuration; and if they are too small, there will be too many to manage easily.

2. *Mark off area templates* (to scale) on blank template sheets, or cross-section paper, as shown in Figure 12-19.

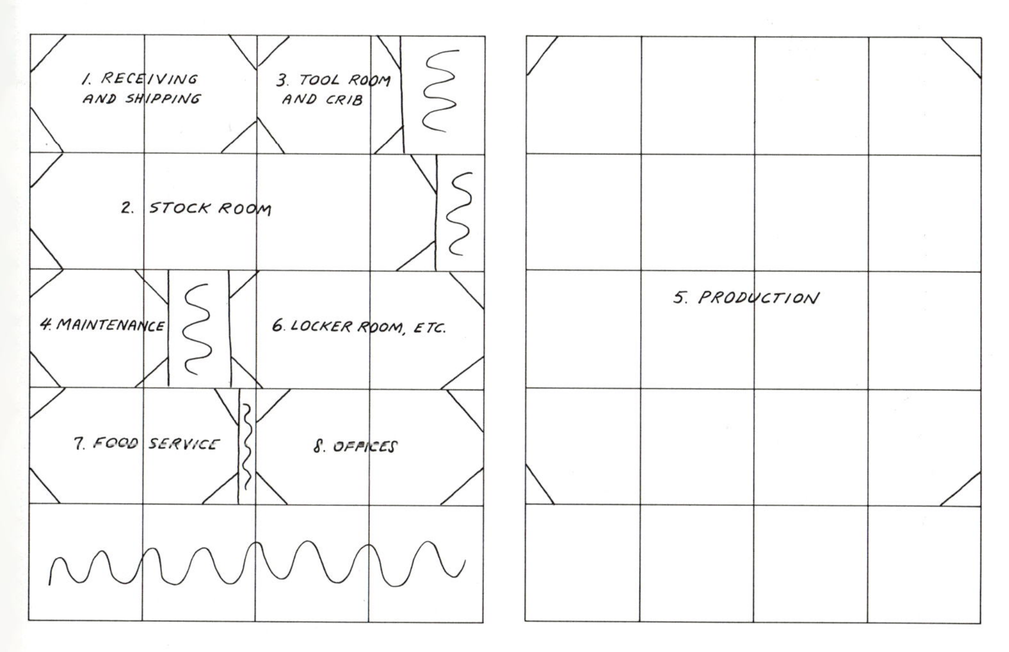

Figure 12-19—Area templates for Powrarm factory.

3. *Cut out area templates.*

4. *Arrange templates* so that they will match the *Activity Relationship Diagram* (Figure 8-4) as closely as possible.

This procedure will probably require some compromises and changes in area shapes and sizes, and will probably not be able to meet all the *Activity Relationship Chart* priorities. However, it is usually only possible to meet the A and some of the *E* requirements with any degree of satisfaction. The *I* and *O* relationships must usually give way to A and *E*—or other judgment-oriented reasons. A possible arrangement of the preliminary Powrarm *Area Allocation Diagram* is shown in Figure 12-20.

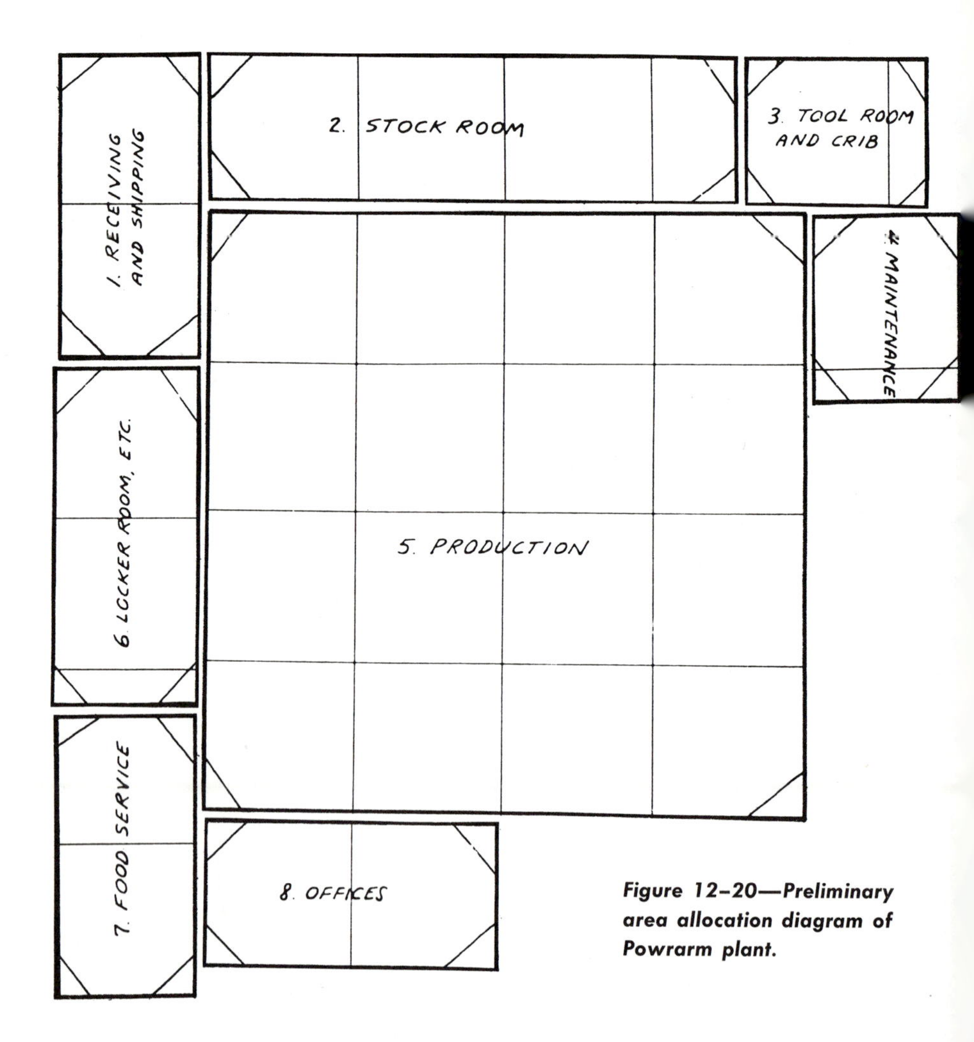

Figure 12-20—Preliminary area allocation diagram of Powrarm plant.

5. *Adjust the preliminary Area Allocation Diagram* to satisfy judgment and other criteria (as listed on page 312).

The preliminary diagram will probably contain several voids, and will also result in some odd-shaped activities. Also, it may be desired to adjust the diagram to convert it from the module size used in planning, to the bay size desired in the actual layout.

6. *Draw finished Area Allocation Diagram to scale*—adjust and round off (up) as required.

The voids between the area templates must now be evaluated and eliminated by shifting the templates, or splitting them between adjacent areas. This will often result in giving an activity more area than was originally planned; but this is usually more desirable than giving them less. Such adjustments can usually be made without much difficulty, and detailed changes can be made in the final layout process.

7. *Draw in the material flow pattern.*

This should be based on the *Assembly Chart, Operation Process Chart,* or other representation of the flow pattern, as developed in Chapter 5. The final *Area Allocation Diagram* might appear somewhat as in Figure 12-21.

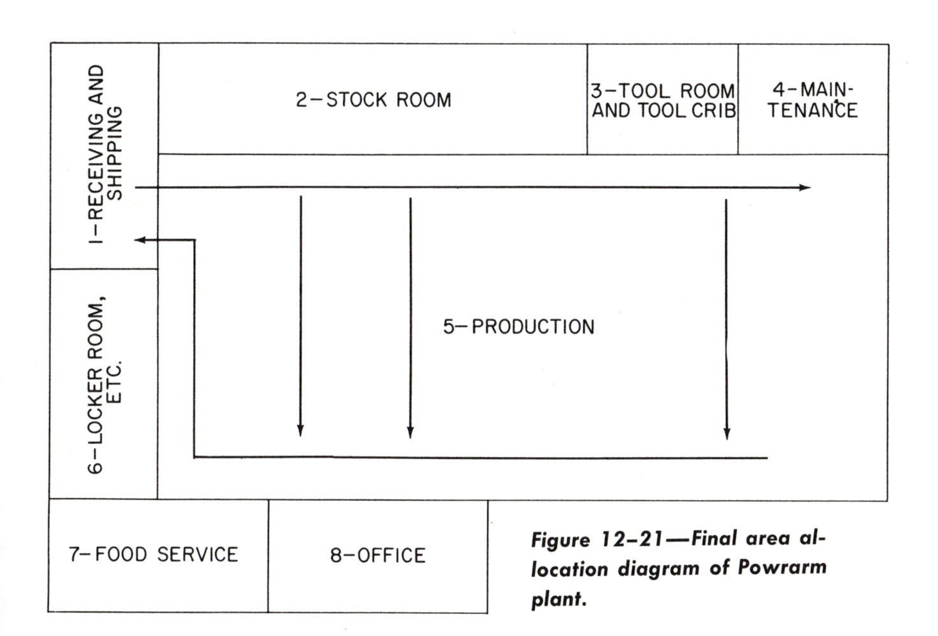

Figure 12–21—Final area allocation diagram of Powrarm plant.

The area allocation procedure will be found ideal as a method for arriving at an allocation of activity areas, and as a basis for the detail planning activities to be covered in subsequent chapters.

A similar technique for use in space allocation is the unit area template, developed by Richard Muther[12] to extend his *Activity Relationship Diagram* (Figure 8-5). The unit area template concept is similar to the block template except that a specific block is broken up into a number of smaller units (unit areas). That is, one 20-by-20-ft block would become a number of smaller rectangles, or squares, such as the following indicates:

1 block template, 20-by-20 ft—400 sq ft
4 unit area templates, 10-by-10 ft—400 sq ft
16 unit area templates, 5-by-5 ft—400 sq ft

The outstanding advantage of the unit area template is that it allows a greater degree of freedom and flexibility in adjusting the various areas to the shape of the space available or desired. As used by Muther, the unit area templates are made of $1/32$-inch plastic, or even wood or tiles; he also adds a color code and letters or numbers to identify the areas involved.

Computerized area allocation. Much has been said and written about computerized plant layout. In reality, this is primarily a method of area allocation—the activity interrelationships having already been evaluated and ranked. Techniques for computerized allocation will be discussed in the next chapter.

The Plot Plan

One of the most important relationships in the entire facilities design process is the one between the internal flows and the external flows. This relationship is depicted on the *Plot Plan*—a drawing or sketch upon which are shown the physical features of a piece of property and the items (buildings, roads, etc.) to be placed on the property. The objectives of the *Plot Plan* are:

1. To make the best possible use of the land.
2. To place the buildings, etc. in their most effective positions on the land.
3. To pre-plan for future expansion needs.
4. To relate internal to external material flow.
5. To insure proper orientation of the building in relationship to the elements (wind, rain, light, etc.) and thereby suggest interior arrangement of functions that would be affected by such orientation.
6. To insure equitable allocation of the available space to the necessary functions and activities.
7. To aid in planning a pleasing appearance and arrangement of the proposed structures on the available land.
8. To provide a beginning step for the architect or builder in his work on the facility.

[12] Muther, *Systematic Layout Planning.*

Factors for consideration in designing the Plot Plan. As suggested earlier, in relation to the factors in area allocation (Table 12-1), a number of them are directly related to the piece of property on which the facility is (to be) built. Table 12-7 lists some of the more important factors.

Constructing the Plot Plan. The *Plot Plan* is commonly made on tracing paper or other medium suitable for making prints for the several people who will need one, such as architect, builder, major contractor, sub-contractors, and layout designer. In making the *Plot Plan*, include such features as:

1. Building location on the site, in conformance with zoning ordinances.
2. Other exterior facilities—such as sheds, tanks, sub-stations.
3. Roads and right-of-ways—access, interior.
4. Other transportation facilities—rail, waterways.
5. Driveways and truck aprons.
6. Sidewalks.

Table 12-7. Factors for Consideration in Designing the Plot Plan

A. Regarding piece of property

1. Size, shape (plant and site)
2. Location of property
3. Topographical features
4. Elevation
5. Orientation of site
6. Transportation facilities
   a. Available
   b. Required; now, later
7. Utility location
8. Availability (now vs. future)
9. Cost (now vs. future)
10. Geological factors
11. Possible pollution
12. Legal restrictions
13. Aesthetic aspects
14. Adjacent structures; features

B. Relationship to *Area Allocation Diagram*—maintain them as much as possible

C. Relationship between internal and external materials flow pattern and handling

D. Expansion (see also Table 12-2)

1. Select location so as to avoid natural barriers to expansion—RR, rivers, hills, highways, buildings, poor soil, topography, etc.
2. Plan directions and priorities, by activities
3. Purchase enough land—should be 3 to 10 times present *total* needs, to properly allow for probable growth, for up to 10 years
4. Locate building on site in correct position for expansion, by stages
5. Plan for additional parking

E. Zoning ordinance

F. Building code, etc.

G. Regarding the building itself

1. Type
2. Size
3. Shape
4. Construction
5. Restrictions
6. Orientation on site
7. Number of floors
8. Expansion
9. Location on site
10. Floor loading

H. Requirements of certain departments

I. Parking requirements

7. Natural physical features—such as elevations, ditches.
8. Landscaping, green space, recreation areas.
9. Fences, gates.
10. Parking lots: (a) types—office, visitors, factory employees; (b) size—300 sq ft/car, 1 car/$1\frac{1}{2}$ employees.
11. Receiving and shipping locations—relationship to such items as roads, parking, tracks.
12. Utility sources—water, sewage, gas, electricity, other.
13. Expansion (see also Table 12-3): (a) areas: plant—1st, 2nd, 3rd additions, and parking—1 parking space per/300 sq ft of additional plant or office; (b) size; (c) shape; (d) location; (e) logical sequence; (f) logical directions.

In designing the *Plot Plan:*

1. Establish size and features of plot (or obtain drawing).
2. Sketch overall plot shape, size, etc.
3. Sketch in roads, drives, etc. (right of way: 60 ft; pavement: 12 ft/lane).
4. Sketch in other features, as listed above.
5. Prepare finished *Plot Plan.*
6. Review and obtain approvals.

In developing the *Plot Plan,* there are three *alternatives* facing the designer:

1. *Ideal*—select a potential site for the proposed new plant.
2. *Practical*—use present site (or location) and re-design facilities to implement optimum flow in existing structures.
3. *Long-Range*—design optimum layout for existing (or new) site and plan to gradually implement it over a period of 5 to 10 years.

An example of the *Plot Plan* is shown in Figure 12-22 (for the Powrarm plant).

Long-Range Planning

In regard to *Alternative 3* above, it may be found that such pre-planning will save much more time and money than it will cost to do it piecemeal, over a period of years. For example, one plant purchased a small building (60-by-160 ft) on a large site (14 acres) and pre-planned the construction program and move from an older downtown plant, in the stages shown in Figure 12-23. The plan was developed in 1960, for completion in 1970. It was not fully implemented until 1972, but the program was carried out in stages, and accomplished with relative ease. Each subsequent addition was pre-planned—first in the generalities shown, and later in conventional layout detail.

Conclusion

This chapter has dealt with many varied problems, factors, and procedures related to the allocation of space for the individual activities of the enterprise, within the confines of the plant site. The procedures have resulted in the development of a

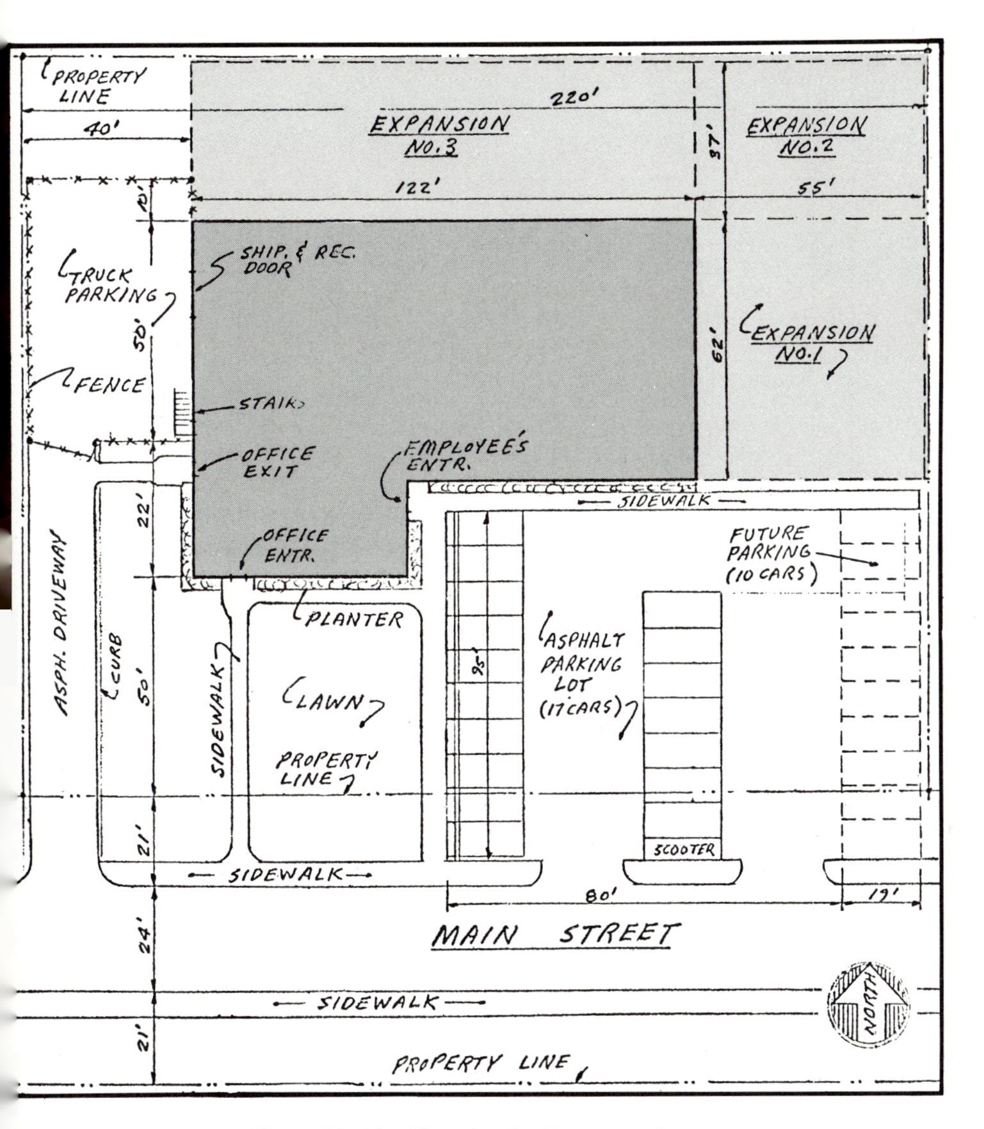

Figure 12–22—Plot plan for Powrarm plant.

fairly definite allocation of activity areas—definite as to both relationship and size. Although the procedures may have appeared somewhat lengthy and detailed, they have been orderly and as accurate as practicable with the amount and type of data commonly available.

Now that activity, area, and space relationships have been worked out, it is necessary to plan the details within each area—both service and production. Subsequent chapters will consider information and procedures necessary for this detailed planning.

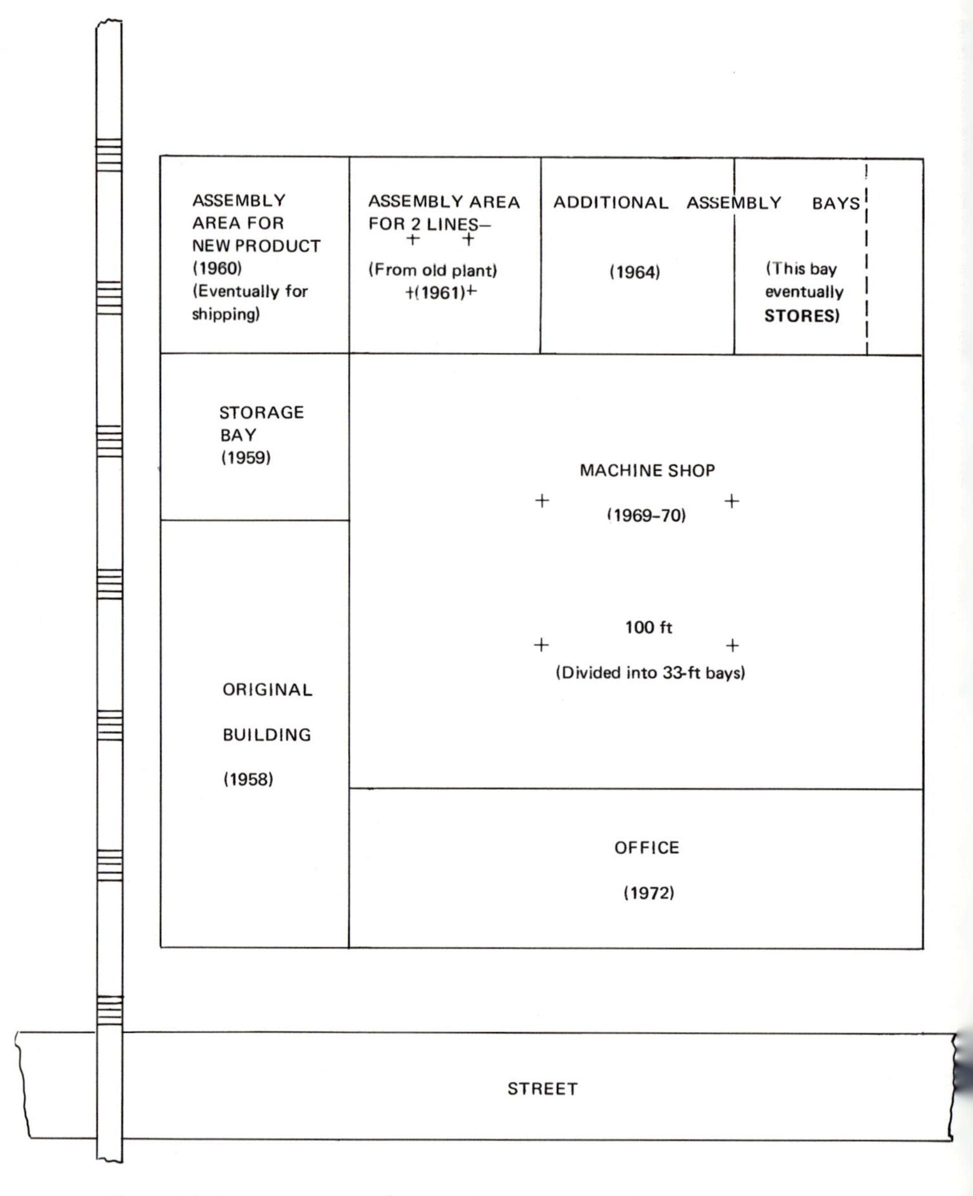

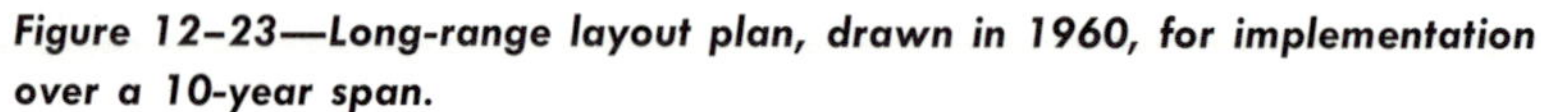

Figure 12–23—Long-range layout plan, drawn in 1960, for implementation over a 10-year span.

It will first be necessary, however, to take a couple of side trips: (1) into computerized plant layout, and (2) into the field of material handling, to become acquainted with the equipment and techniques necessary to implement the flow pattern. Chapter 13 will deal with computerized plant layout, and Chapters 14,

15, and 16, with material handling. Chapter 17 will return to a consideration of the details of constructing the plant layout.

Questions

1. What is meant by *area allocation?*
2. Why is possible future expansion a factor in area allocation?
3. What are some of the major factors to consider in the area allocation process?
4. What are the advantages of the area allocation process in layout design?
5. What are some of the ways in which future expansion can be allowed for in the original planning process?
6. Differentiate between expansion and flexibility. What are some similarities?
7. In what ways could one provide for flexibility in the planning of a factory or commercial building?
8. Sketch some of the general ways in which one might plan to expand a facility.
9. List some ways the designer can plan the facility to facilitate expansion. Discuss as necessary.
10. What actions can be taken to expand within-the-walls of a facility? List and explain several.
11. How can the *levels of activity* concept (see page 106) contribute to the area allocation process?
12. What uses can be made of the "space" on a building roof to free interior space for other purposes?
13. Why might mezzanines or balconies be considered in planning a facility? Give several examples you have seen—not in an industrial enterprise. Discuss reasons.
14. How does the group technology concept (see page 59) facilitate flexibility in a layout design?
15. Explain the point-of-use storage concept. What are some advantages of point-of-use storage?
16. Briefly explain the concept used by General Electric, in their Columbia, Md., plant, to facilitate the movement of material in and out of production and assembly areas.
17. What are the several uses that are made of aisles?
18. Name some factors to be considered in planning: (a) aisle width; (b) aisle location.
19. Discuss the significance of column spacing (bay size) in a facility.
20. List some factors for consideration in determining column spacing.
21. What is meant by an *Area Allocation Diagram?* —an area template?
22. Briefly describe the area allocation procedure.
23. Distinguish between an *Activity Relationship Diagram* and an *Area Allocation Diagram.* What is the specific use, or purpose, of each?
24. How does the activity template differ from the area template?
25. What is a *Plot Plan?* What are some of its objectives? What does it include?
26. Name some of the factors for consideration in developing the *Plot Plan.*
27. Discuss interrelationships between the area allocation process and the *Plot Plan.*

28. Discuss the three alternatives facing the designer of the *Plot Plan.*
29. Discuss the implications of long range planning as it pertains to the *Plot Plan.*

Exercises

A. On the basis of *Exercise B* for Chapter 11, estimate the areas for each activity (or for some other facility).

B. Sketch representations of the concepts in *Questions* 11, 14, 15, and 16.

C. Look up the articles in *Footnote 6* (p. 305) and present findings in class.

D. For the data in *Exercise A* develop an *Area Allocation Diagram* (use or make an *Activity Relationship Chart* and *Diagram;* see Chapter 8).

E. Develop a plot plan for your problem in *Exercise D.*

13

Computerized Facilities Layout

The preceding twelve chapters have covered a great amount of detailed preliminary work necessary for designing the *Area Allocation Diagram.* In a sense, computerized layout permits the computer to develop the *Diagram,* based on an heuristic program that hopefully considers enough factors for the resulting printout to be acceptable. Often, however, the connotation of *computerized layout* has been much inflated. No computer program is capable of considering all the facts, factors, and interrelationships covered in these chapters, and resolving them into an optimum layout. In fact, more than one person has said, "If I had all the data I needed for a layout computer program, I wouldn't need a computer!" And herein lies the problem. Much of the real data required by most programs is not readily available, because it is not a product of the typical accounting system or it is too difficult to obtain. Consequently much of the data is approximated. In all honesty however, it should be said, that if layouts are being made, by any method, with less data than the computer requires, the result will be extremely questionable.

A computer program (mathematical model, simulation, etc.) can nevertheless be a tremendously powerful tool in the facility designer's hands, if it is properly used, and the printout is not blindly accepted as *the* layout just because it was done on a computer. The computerized algorithm is an extremely powerful device for making comparisons of alternative arrangements of activity areas, in terms of selected criteria and available data. And it should be noted here that in most cases, the computer printout is only an *Area Allocation Diagram*—it is not a layout.

Historical Background

The term *quantitative techniques* has come to mean techniques that rely on, or are oriented toward, mathematical, statistical, and modelling approaches to problem solving. Although the computer has been used in applying these techniques only since the early 1960's, other mathematically related techniques were used in plant layout work some 20 years earlier. The advent of the computer, and its free use by researchers in the colleges and universities, turned on a literal flood of computerized routines.

Prior to that, a number of practitioners and researchers made attempts to apply mathematics and statistics to layout problems. Some of the earlier efforts, in chronological sequence from 1947 to 1957, were:

1. *1947*—A study of a layout problem based on the reduction of handling as the major objective and used it as the basis for the proposed more efficient layout. An *Interdepartmental Development Chart* showed handling by lines—with greater amounts of handling shown by more lines.[1]
2. *1949*—A proposed method for finding the proper unit of measurement for packaged commodities developed the *H* (the weight of a commodity divided by its specific gravity) as an aid to identifying and defining material handling problems.[2]
3. *1950*—An extensive survey of the literature encountered no quantitative measures of evaluating plant layouts. From good plant layout practice, as found in the literature, eleven measurable indices were then developed and proposed as criteria of effective physical plant utilization, and a critique was made on their applicability to continuous, intermittent, and jobbing layouts.[3]
4. *1951*—An extension and validation of the proposals in *Item 3* above, concluding them to be valid criteria of physical plant utilization.[4]
5. *1951*—One of the first mentions of the *Travel*, or *From-To Chart* technique applied it to the analysis of activity interrelationships.[5]
6. *1952*—It was suggested that an optimum layout can be achieved by tabulating annual volume of items, and ease of handling, and calculating the product as a weighted volume for each interdepartmental move.[6]
7. *1954*—A mathematical program (complex) was suggested for evaluating the effect of type of layout on expected investment or loss.[7]
8. *1955*—The sequence analysis technique was presented as a means of developing schematic diagrams depicting optimal relative positions of work centers.[8]
9. *1956*—A technique was presented for making a flow study of a job shop by tabulating (a) flow pattern data on all movements, (b) flow pattern data by material handling code, and (c) flow pattern by load centers, with results entered on *load center summary tickets* and these used in relating departments to each other.[9]
10. *1957*—Criteria for plant layout proposed by the contemporary plant layout authorities were reviewed and critically evaluated. Traditional operations research techniques that might be applicable were considered, and a technique was described for minimization of the criterion of interwork-area material

[1] J. Freeman, "Optimum Transportation Cost as a Factor in Plant Layout." Thesis, Purdue University, 1947. (Most theses can be borrowed via inter-library loan from the school of origin.)

[2] W. B. McCelland, "A Unit of Measurement," *Modern Materials Handling*, Dec. 1949:15.

[3] S. P. Gantz, "A Proposal of Criteria for the Evaluation of Industrial Physical-Plant Utilization." Thesis, Purdue University, 1950.

[4] R. B. Pettit, "A Further Evaluation and Extension of Criteria of Physical Plant Utilization." Thesis, Purdue University, 1951.

[5] M. L. Levy, "Let the Travel Chart Simplify Your Material Movement Problems," *Mill and Factory*, May 1951.

[6] S. A. Weart, "Mechanical Methods for Determining Plant Layout," *Advanced Management*, April 1952.

[7] M. F. Shakun, "A Quantitative Methodology for Decision-Making in Plant Layout." Thesis, Columbia University, 1954.

[8] E. S. Buffa, "Sequence Analysis for Functional Layout," *Journal of Industrial Engineering*, March–April 1955:12.

[9] G. Downs, "Best Way to Layout a Job Shop," *Factory*, Nov. 1956:10.

movement by proper equipment location within the intermittent manufacturing situation is described; a summary problem was included.[10]

In a way the last item above marks the end of an era, as it was a first attempt to model the location process in the current context of modelling. Although the approach was later questioned, the resulting controversy probably instigated the subsequent efforts to develop computerized plant layout models. The general area of quantitative approaches to plant layout was much investigated and discussed over the next several years,[11,12,13] and a 1967 survey indicated that quantitative techniques in the following list were being applied, with *waiting lines* the most commonly used:[13]

1. Waiting Lines
2. Monte Carlo Simulation
3. Transportation Programming
4. Conveyor Analysis
5. Materials Analysis
6. Assignment Technique
7. Dynamic Programming
8. Transhipment Programming
9. Integer Programming
10. Quadratic Programming
11. Traveling Salesman Technique
12. Level Curve Model

For almost every technique, the consultant group was by far the greater user, in most cases by a margin of two-to-one. The major development during the above period was the first layout program capable of printing out a rough version of an *Area Allocation Diagram;* the program, called CRAFT (Computerized Relative Allocation of Facilities Technique), is discussed later in the chapter.

Why Quantitative Techniques

It should be evident from the first 12 chapters that the facility design process is not as simple a matter as it may seem. Although it is often passed over lightly, given little technical support, and frequently carried out by people almost totally unaware of its complexity, it is fraught with dangers of both omission and commission. For instance, consider these characteristics of the facility design problem:

1. Its myriad interrelationships
2. Its overall complexity
3. The great number of factors
4. The wide range of factors
5. The intangibleness of many factors
6. Flexibility requirements
7. The human aspects of the layout problem
8. The importance of experienced judgement
9. Personal preferences
10. The economic consequences of a poor layout

[10] R. J. Wimmert, "A Quantitative Approach to Equipment Location in Intermittent Manufacturing." Thesis, Purdue University, 1957.

[11] R. C. Wilson, "Facilities Planning—the State of the Art." Privately published survey, 1963; updated 1965.

[12] J. R. Buchan, "An Evaluation of Selected Quantitative Methods Useful in Plant Layout." Thesis, Georgia Institute of Technology, 1966.

[13] D. L. Totten, "Application of Selected Quantitative Techniques in Facilities Design." Thesis, Georgia Institute of Technology, 1967.

It is because of such characteristics that the layout problem has been treated largely by non-quantitative, heuristic methods. The techniques presented in Chapter 6 are typical of the degree of sophistication of the analytical tools commonly applied to layout problems. However, on behalf of these techniques, it must be said that they are extremely useful and sufficiently successful for analyzing a large majority of layout and handling problems. In fact, they are far beyond the means used in most facility planning done even today.

In carrying out the early research in modelling the layout process, the experimenters[14] found the problem so overwhelming in its requirements that it was impossible to structure it in mathematical terminology. To facilitate their effort they made the necessary assumptions to cut it down to size. More often than not, such assumptions were false, unrealistic, or deceptively simplified. Or, it might be said that the complicating, real world aspects were wished away—with the inevitable result that the answer obtained was actually an *unanswer to a non-problem*. Some of the misleading assumptions might have been:

1. All departments (activity centers) are square (or symmetrical).
2. Material flows between the centers of departments.
3. Material handling cost is directly proportional to distance.
4. All data on material flow are known, and are deterministic in nature.
5. Waste or scrap material need not be considered.
6. All travel is in two dimensions.

Obviously, all of these are likely to be false, and their use as bases for a model will result not only in a sub-optimum solution, but in a highly misleading representation of the actual situation. Many of the existing algorithms make at least some of these, or similar, assumptions, before they proceed to solve the problem.

The above is not meant to discredit attempts to design layouts by computer, but to warn the user of the relative validity of results and to encourage him to obtain as much valid data as possible. If a serious attempt is made to obtain good data on most of the relevant factors, the result of a computerized layout algorithm can be extremely useful in:

1. Exploring a great many potential relationships not otherwise accessible.
2. Permitting the designer to learn from the data collection process.
3. Providing an insight into the problem by watching the print-out process.
4. Better defining the problem, which is necessary for computerization.

Once sufficient data have been accumulated and a model developed and run, the results of the program should provide an excellent basis for developing a better layout than might have been achieved otherwise.

Criteria for a Computerized Layout Program

In view of the difficulties in developing layout models, and in improving their usefulness, it seems wise to consider desirable criteria for such a model. The

[14] Adapted from Apple and Deisenroth, "A Computerized Plant Layout and Evaluation Technique (PLANET)."

following list of desired criteria came out of a Seminar discussion and evaluation of several existing models:

1. Reliability.
2. Use of real data.
3. Ability to weight inputs.
4. Elimination of subjective evaluation of solutions.
5. Better configurations of activity centers.
6. Allowance for fixed activity locations.
7. Honoring of building restrictions.
8. Usability for multi-story layouts.
9. Consideration of cost incurred in generation of alternate layouts.
10. Provision of more realistic cost evaluation.
11. A minimum of restrictions—to retain flexibility.
12. Ability to extract desirable features from a specific layout for insertion into another.
13. A more realistic graphical print-out.
14. Elimination of manual adjustment of graphical print-out.
15. Ability to handle undesirable (negative) interrelationships.
16. Applicability to detail layout—i.e. machine layouts, etc.

While it is unlikely that any program will meet all the above criteria, they do serve as a guide to the researcher. Some of the common models do attempt to meet some of the criteria.

In spite of the above admonitions, the introduction of the quantitative techniques and the advent of the computer have encouraged experimenters to search for better methods of analysis of the increasingly complex facility design problem. This has led one writer to conclude:[15]

> . . . the combination of man and computer can lead to accomplishments that neither is capable of alone.
>
> Beyond the technological capabilities of computers and heuristic programs is the question of economics. Sunlight is free, and we know how to convert it into electricity; but it has not replaced coal as a source of energy for generating power. Likewise, computer technology has outpaced its economic feasibility in some applications. Humans are still less expensive decision makers than computers are for many problems that both are capable of solving. We should add, however, that computers are decreasing in cost (per computation) while humans are increasing in cost. The break-even point is changing continually.

Computerized Layout Programs[16]

In this section, four of the more commonly known and widely used programs are described and discussed briefly in chronological order of their appearance in the literature. The basic concepts of each are as follows:

[15] J. D. Wiest, "Heuristic Programs for Decision Making," *Harvard Business Review*, Sept.–Oct. 1966:138.

[16] Adapted from Muther and McPherson, "Four Approaches to Computerized Layout Planning," p. 39.

CRAFT—interchanges activity locations in the initial layout to find improved solutions based on material flow. Successive interchanges lead to a sub-optimum, least cost layout.

CORELAP—locates the most-related activity, and then progressively adds other activities, based on rated closeness desired, and in required size, until all activities have been placed.

ALDEP—selects at random and locates the first activity. Subsequent activities, in required size, are selected and placed: (a) according to closeness desired, or (b) at random, if no significant relationships are found. Alternative layouts are generated and scored.

PLANET—utilizing interdepartment flow data, computes the "penalty" cost associated with separating departments. Three heuristic algorithms are available for generating alternative configurations to be manually evaluated and adjusted.

Input Requirements.[17] All four programs require the fundamental inputs of relationships and space.

CRAFT uses material flow data as the base for developing closeness relationships, in terms of some unit of measurement (pounds per day, units per year, skid-loads per week) between pairs of activities to form a matrix for the program.

Other input data permit the entry of the cost of moving, per unit moved, and per unit distance. When such cost input is not available, or is inadequate, it can be neutralized by entering 1.0 for all costs in the matrix.

Space requirements are the third input. They take the form of an existing layout. For new layouts, a rough layout must be developed. In either case, activity identification numbers, in a quantity approximately scaled to their space requirements, are entered in an overall area of definite configuration. The location of an activity can be fixed in the overall area.

PLANET requires two basic types of input data: department information, and material flow information. Each is identified and the area requirements stated.

The basic approach to the analysis of the movement of material within the facility is to examine all material as it moves from department to department. Such information as the frequency of move, method of movement, cost of movement, and sequence of movement are of utmost importance in establishing the material flow cost.

Such characteristics of material as size, shape, weight, and durability must be considered in selecting handling method and estimating cost; before the cost estimate can be made, the method must be selected. And finally, there must be provided the sequence of movements associated with each part.

CORELAP and ALDEP—the other two programs—use the vowel-letter closeness relationships as input data, based on material flow and other factors.

[17]Adapted from Muther and McPherson.

CORELAP requires the amount of space for each activity to be known and also a maximum building length-to-width ratio. ALDEP requires the size of each activity and a representation of building dimensions, including assignments of specific building features, such as aisles and stairwells, and preassigned activity locations.

How the programs work. CRAFT computes the product of the flow, handling cost, and distance between centers of activities. Then it considers exchanges between locations and examines two-way and three-way exchanges. An exchange involving the greatest cost reduction is made, and a new total cost determined. The process is repeated until no significant cost reduction can be found. The program is path-oriented, so not all possible exchanges are examined. Therefore, what must be called a sub-optimum layout is reached. To consider all possible exchanges in arriving at an optimum is not feasible at present.

CORELAP calculates which of the activities in the layout is the busiest or most widely related. Sums of each activity's closeness relationships with other activities are compared, and the activity with the highest total (TCR) is located first in the layout matrix. Next, an activity that must be close to it is selected and placed as nearly adjacent as possible. This activity is denoted as *A* (closeness *absolutely necessary*). A search of remaining relationships for more *A*-related activities is then made. These are placed, again, as nearly adjacent as possible. When no *A*'s are found, the same procedure is followed for *E*'s (closeness *especially important,*) *I*'s (closeness *important,*) and *O*'s (*ordinary* closeness acceptable), until all have been placed. CORELAP also puts a value on the *U* (closeness *unimportant*) and *X* (closeness *not desirable*) relationships.

ALDEP uses a preference table of relationship values to calculate the scores of a series of randomly generated layouts. If, for example, *activities 11* and *19* are adjacent, the value of the relationship between them would be added to that layout's score. A modified random selection technique is used to generate alternate layouts. The first activity is selected and located at random. Next, the relationship data are searched to find an activity with a high relationship to the first. It is placed adjacent to the first. If none is found, a second activity is selected at random and placed next to the first. This procedure is continued until all activities are placed. This process is repeated to generate another layout.

PLANET utilizes information about the material flow patterns, with the algorithm establishing a layout by asking:

1. *Which department should be selected for placement next?*
2. *Where should this department be placed?*

It then fixes each department in the layout in such a way as to keep the material handling cost as low as possible. Three alternative methods evaluate the relationships between departments not yet selected for placement and those that have been selected. Strong interrelationships between department pairs or within a department group will imply early selection. A search routine finds a location for each department as it is selected—the location that will have the smallest *placement penalty* with respect to the then existing partial layout. Based on the input data, the relationships and penalties are a function of material handling volumes, methods, and costs.

How the programs present the layout. CRAFT prints a layout in basic rectangular form. Each activity appears on the printout, as a certain number of square feet or square meters. CRAFT's output indicates activities by letter. While the resulting overall configuration is rectangular, individual activity shapes tend to be irregular and must be adjusted into practical shapes. The total cost is calculated for each, and the differences between it and the initial total cost is shown as a saving.

CORELAP prints a layout of facilities in an irregular format. Neither the individual activities nor the total layout is in any practical rectangular shape, so further adjustment is imperative for a workable layout. Each activity number on the printout represents a certain portion of that activity's total space requirement.

ALDEP prints a layout contained within a given rectangular area boundary, although individual activities tend to be irregularly shaped. Activities are placed or located by means of vertical scan, so activity shapes tend to be rather elongated. As with the other programs, each activity number represents a certain portion of the activity's total space.

ALDEP offers the advantage of being able to print the layout on a maximum of three different floor levels. This is beneficial where multi-story problems are encountered, but no between-floor relationships are considered, and activities may wind up split between floors, purely by chance. Activities may be preassigned to specific floors and to specific locations.

PLANET, like CORELAP, prints a layout of facilities in an irregular format. The program attempts to keep a department shape somewhat square, in order to avoid elongated shapes. However, relatively small departments may not appear in a desirable shape. PLANET permits any two-character symbol to be used to represent units of area on the final layout. Hence, the shipping department could be designated as *SH* while receiving would become *RC*. Final layouts are somewhat easier to interpret with this notational convenience. If a layout exceeds the capacity of a single output sheet, PLANET will split the layout into strips and print the strips on successive output pages.

Figure 13-1 depicts the general concept of each algorithm, in terms of the computer programming and appearance of output. The actual computer printouts of the programs are in terms of blocks of letters or numbers representing departments.

It can be seen that some of the computer print-outs must be adjusted into an acceptable rectangular shape. While this could be done by the computer, it would take a much longer routine. Besides, the adjustment process permits the analyst to exercise his experienced judgment in making final alterations, prior to working on the detailed layout.

And again—the reminder that the computer has only developed an *Area Allocation Diagram.* All the work implied by the preceding 12 chapters should

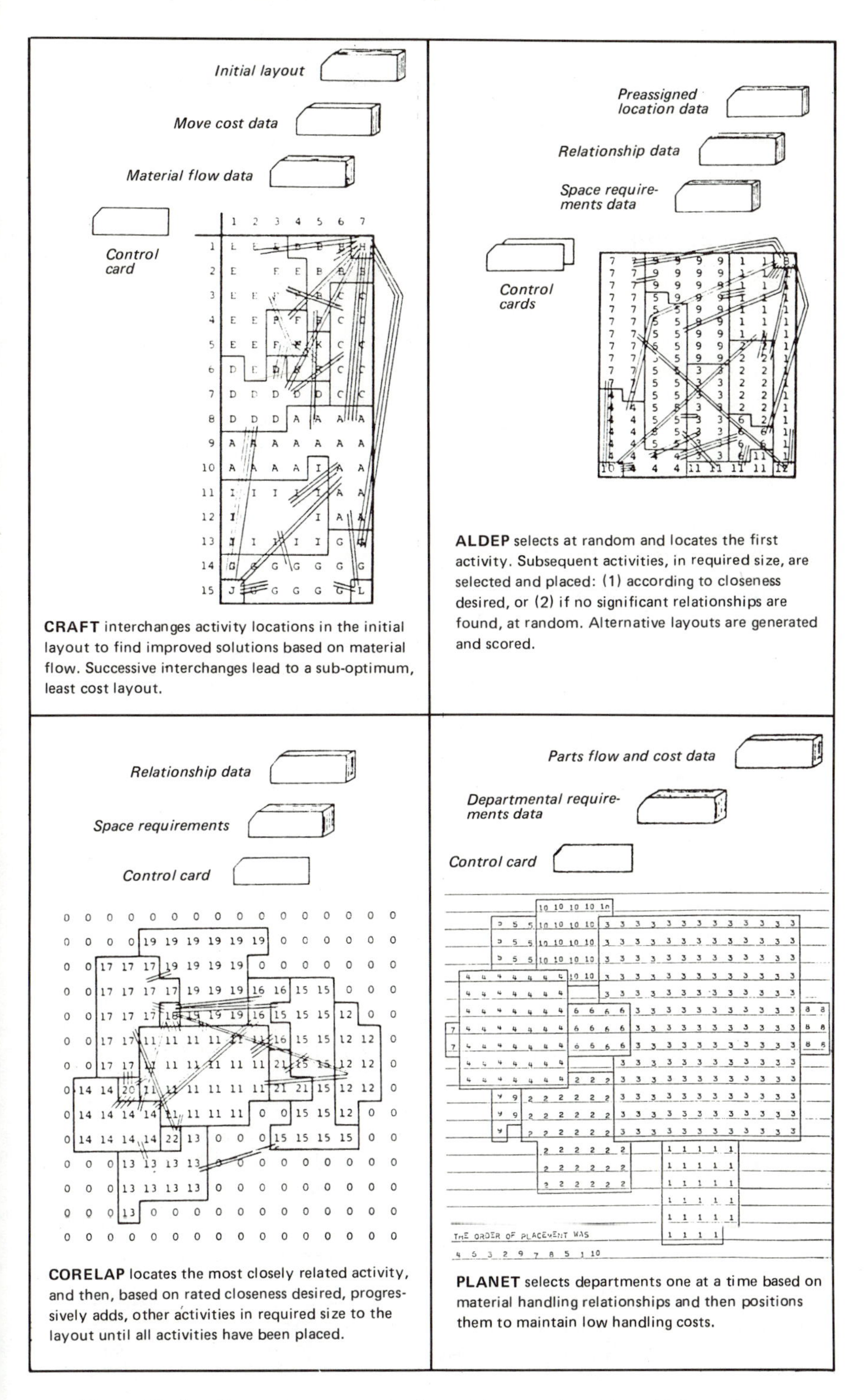

Figure 13-1—General concept of selected layout algorithms (Adapted from Muther and McPherson.)

have been done first. And, the further steps outlined in the following chapters still remain to be done.

Advantages and Limitations of Programs

Participants in the Seminar previously referred to listed the advantages and limitations of the four programs discussed here. The results, are given in Table 13-1 (these may not be completely up-to-date, as changes are continually being made in thc programs).

As suggested previously, there are several other layout algorithms, some proprietary, and some available for general use. Undoubtedly there will be more, as future experimenters try to develop more accurate and more easily applicable programs.

Comparison of Computerized Layout Techniques

With the several layout and design models available, it is only natural for the analyst to ask which is best. Several researchers have attacked this problem, and reported on comparisons of some of the more widely known programs. One of the studies is abbreviated below, and illustrated in Figures 13-2, 3, 4, and 5 showing variations of an original layout, as developed in the study. The authors concluded as follows:[18]

> Since CRAFT has the capability to evaluate a layout on a quantitative basis, we decided to appraise the output of CORELAP (after manual adjustment) and ALDEP through the use of the appraisal portion of CRAFT, and further, to let CRAFT try to improve on the best of CORELAP and ALDEP output. We, therefore, fed the output layouts to CRAFT. The basic materials handling data expressed in the CRAFT input remained invariant.
>
> CRAFT's evaluation of ALDEP output and an improved configuration is shown in Figure 13-5.
>
> This real life example uses CRAFT to measure the materials handling cost of an existing layout, and to measure these costs for CRAFT, CORELAP, and ALDEP output. A tabulation comparing the results is given below:

	Effectiveness Factor (Material handling cost per unit of time)	
	Input Layout	Best Generated Layout
CRAFT on original [Figure 13-2]	$102,781	$84,452
CRAFT on CORELAP's best [Figure 13-3]	115,592	90,805
CRAFT on ALDEP's best [Figure 13-4]	127,903	90,891

[18] Denholm and Brooks, "A Comparison of Three Computer Assisted Plant Layout Techniques," p. 77.

Table 13-1. Evaluation of the Computer Program

Advantages	Limitations
CRAFT	
1. Permits fixing specific location 2. Input shapes can vary 3. Short computer time 4. Mathematically sound 5. Can be used for office layouts 6. Can check previous iterations 7. Cost and savings printed out	1. Requires hand adjustment (output not directly usable) 2. Program tends to be "short sighted;" may not find best answer by switching only two or three departments at a time 3. Departments switched must be: 1) the same size, 2) adjacent to each other and 3) border on a common department 4. Input data needs careful structuring 5. Letter designation cumbersome 6. Does not generate an initial layout 7. Better adapted to rearrangements 8. Undesirable relationships not taken into account 9. Limited to 40 departments
CORELAP	
1. Easy to get going on computer 2. Generates a new layout 3. Input and output terms are the same 4. Based on the Relationship Chart 5. Each step visible during layout development 6. Most relationships properly honored	1. Cannot specify fixed activity locations 2. Does not calculate cost 3. Limited to 45 departments 4. Irregular shaped layout
ALDEP	
1. Can fix specific locations within confines of space available 2. Solution is within specified area 3. Many alternatives are developed 4. Honors most interrelationships 5. Has multi-level capability	1. Cost of movement not calculated 2. Undesirable (X) relationships not honored (i.e., not questioned) 3. Evaluation or scoring method questionable 4. Difficulty in evaluating production processes 5. Mandatory space configurations not taken into account 6. Limited to 53 departments
PLANET	
1. Based on From-To chart 2. Uses M.H. cost for a specified method of handling for each move in a predetermined operation sequence 3. *Requires* interaction between computer routine and engineer, to exercise his judgement 4. Applicable to any problem involving quanti fiable relationships between activities 5. Can fix specific activity locations and building features 6. No input layout required; i.e. generates an initial layout 7. Prints out handling cost per "activity relationship" plus total handling cost 8. Uses normal plant terminology and data as input 9. Allows choice of method of selecting and placing departments	1. Primarily useful for production layouts (vs. service depts.) 2. In need of actual application and experimentation 3. Input data needs structuring (but not much more than usually required for layout planning)

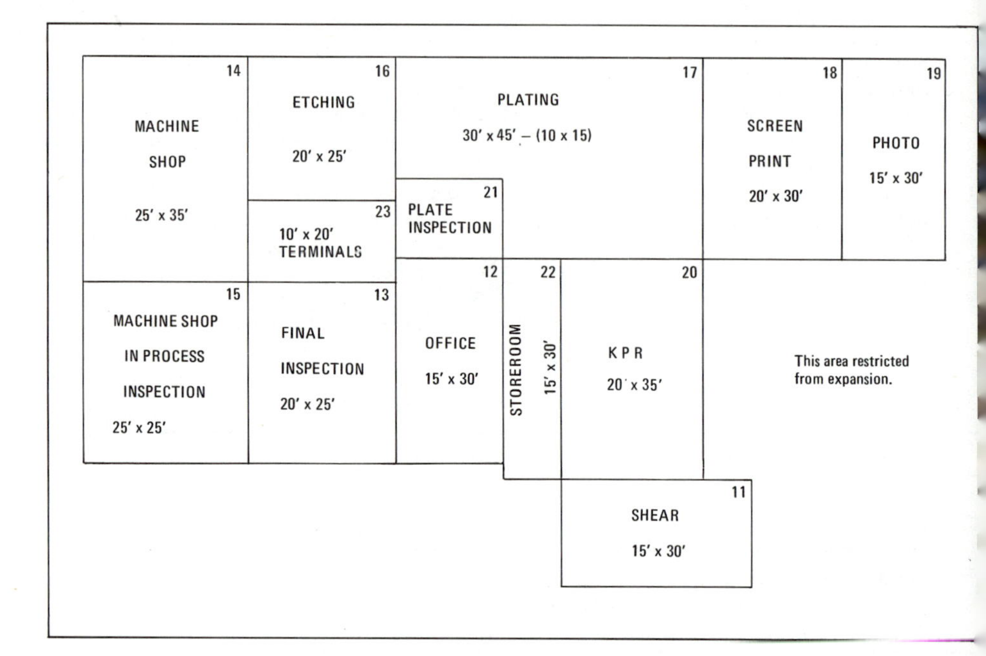

Figure 13–2—The original layout. (From Denholm and Brooks.)

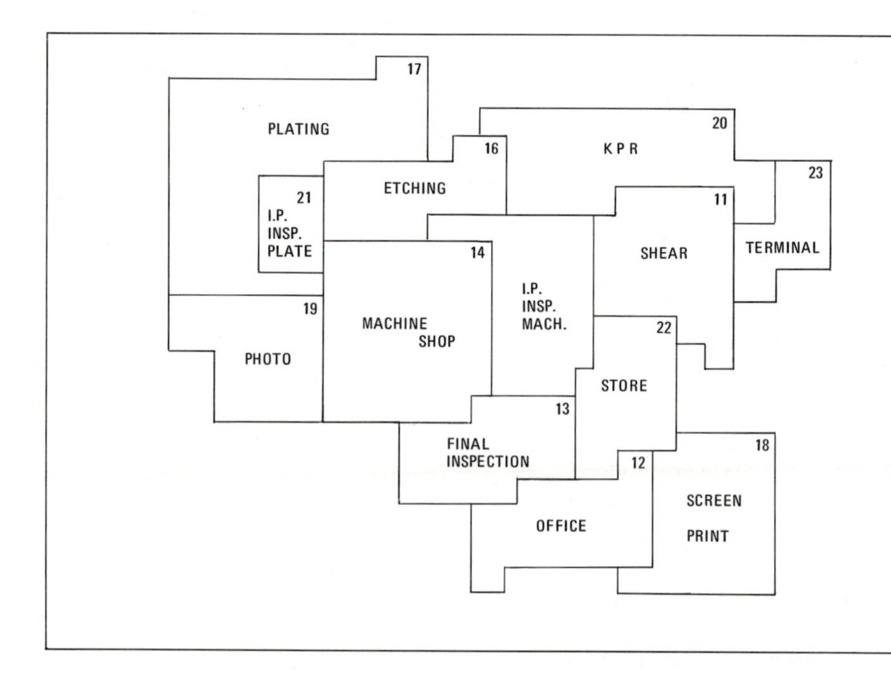

Figure 13–3—CORELAP output. (From Denholm and Brooks.)

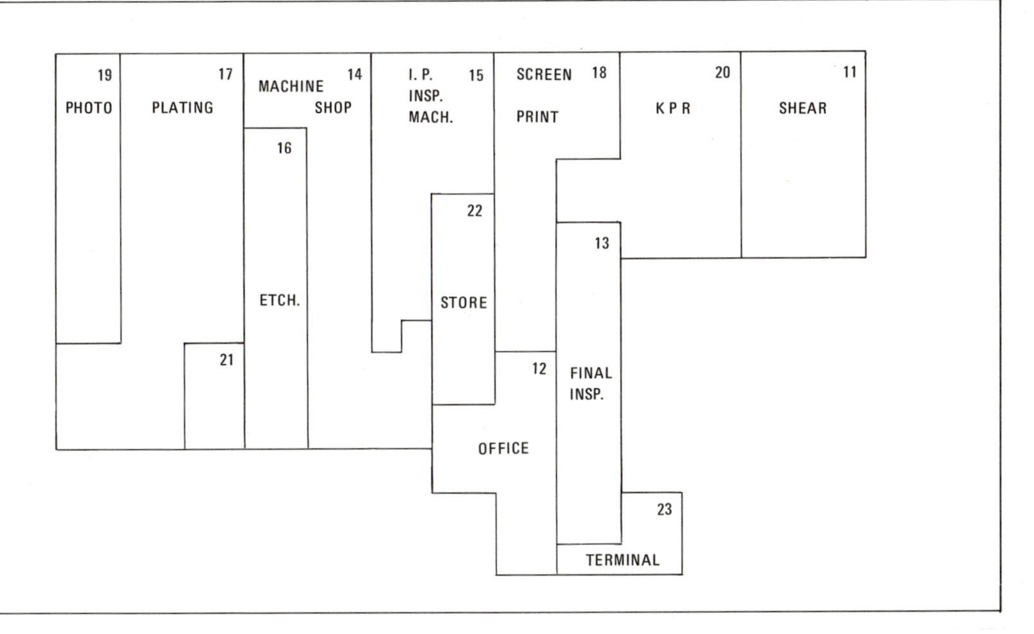

Figure 13–4—ALDEP output. (From Denholm and Brooks.)

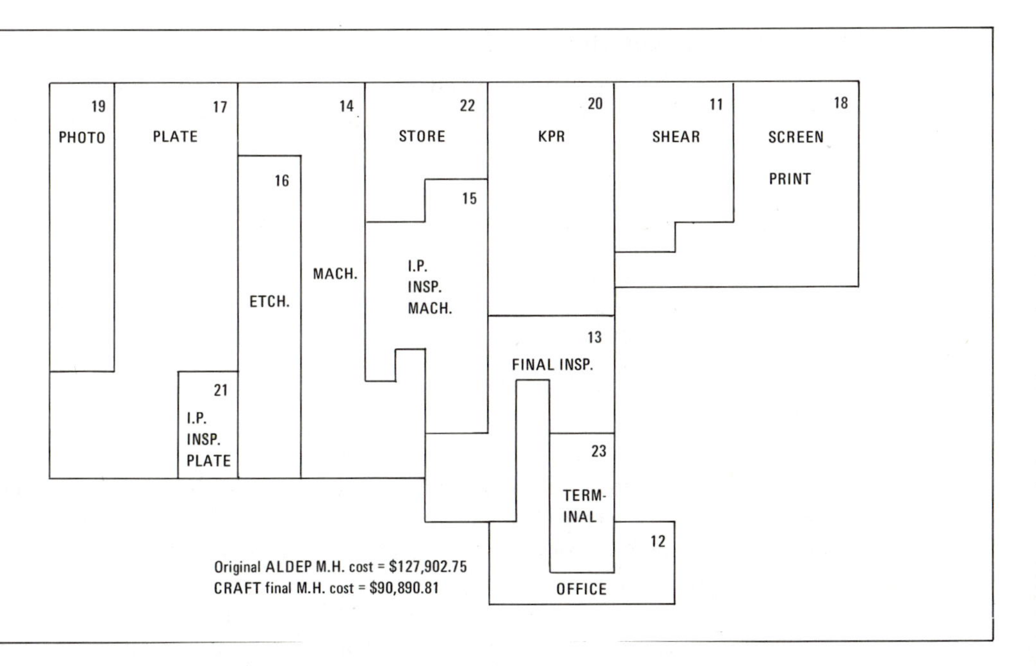

Figure 13–5—Craft's evaluation of ALDEP output. (From Denholm and Brooks.)

The improvement by CRAFT seems substantial in every case. In fact, CRAFT evaluated both ALDEP and CORELAP output as *worse* than the original. The fact that the best generated layout varied, is a peculiarity of CRAFT and is caused by the program's hueristic and suboptimal approach in order to avoid combinatorial difficulties.

We must acknowledge that material handling cost is not the only criterion against which to gauge an effective layout. However, CRAFT does have the potential to handle other quantifiable criteria, allowing the user to choose a criterion appropriate to his problem. Furthermore, it seems possible to use other criteria to "penalize" CRAFT's materials handling costs in some cases.

In summary, CRAFT output seems superior to the others. It improved the best output of each of the other two algorithms. Even the best output of ALDEP and CORELAP seemed more costly than the original layout in this example.

For further details. Those desiring additional information on the use of quantitative techniques making use of computer programs for facilities design should look in the bibliography for references, classified as:

A. General
B. Layout Program Comparison Studies
C. Fitting Machines into Existing Layouts
D. Computers in Warehouse Planning
E. Plant and Warehouse Location
F. Computers in Hospital Design
G. Computerized Material Handling Equipment Selection

The Future of Quantitative Techniques

As emphasized earlier, the computerized programs are extremely useful in facility planning for those who have the necessary information and access to a computer. As they become more refined and more accurate, they will gain broader acceptance and will probably become easily available to anyone who wants to use them in their facility design work—probably through consultants and universities. However, one last word of warning: *computerized layout is only one step in the facility design process.* Much work must precede it; much more must follow it. Though a very helpful and powerful tool, it is not a panacea.

Conclusion

This chapter has briefly reviewed computerized plant layout as a technique for use in the area allocation process. It should be used to assist, check, and reinforce the designer's experienced judgment. Those who desire further information will find plenty in the bibliography—in fact, probably enough for an entire course in computerized plant layout.

The following three chapters will deal with the material handling aspects of facility layout. Subsequent chapters will then continue the detailed development of the layout.

Questions

1. What is *computerized layout?*
2. What is commonly misunderstood about computerized layout?
3. What are the primary benefits from using computerized layout?
4. What were some of the early efforts in applying mathematical methodologies to layout problems?
5. Of what use are such ratios as developed by Gantz and Pettitt? (See also Chapter 18, on evaluating layouts.)
6. What were some of the misleading assumptions of the early researchers? Why were such assumptions made?
7. What are some criteria for a computerized layout program?
8. Distinguish between: (a) CRAFT, (b) CORELAP, (c) ALDEP, and (d) PLANET.
9. Comment on the differences in the approaches of CRAFT, CORELAP, and PLANET.
10. Should the program be designed to print out an optimum, rectangularized layout? Why? Why not?
11. Develop a list of the common limitations of the four programs. What might be involved in overcoming them, and still meet the criteria in *Question 7?*
12. What is the likely future of computerized layout programs? What can you visualize? How widely used? Discuss.

Exercises

A. Look up some of the early efforts in the first 13 footnotes for this chapter (or as assigned by instructor), and (1) report in class; or (2) write a critique or observation.

B. Specifically—in the following references—do as suggested:
   1. McClelland—analyze and comment. Why do you think the concept didn't catch on? Is it valid? Could it be improved upon?
   2. Gantz and Pettitt (article in *Modern Materials Handling,* Jan. 1953): Are the ratios valid? Can you think of, or develop, any more? Try some of them out in a local situation.
   3. Make up a *Flow Diagram* from an exercise in Chapter 6, or other data, as suggested by de Villeneuve.

C. Read the article in *Harvard Business Review,* Mar.–Apr. 1964, on CRAFT, and comment on (1) its strong points, (2) its weak points, and (3) its questionable points.

D. Look up other layout programs in the literature, and compare to those covered here.

E. Rectangularize the layouts in Figures 13-3, 13-4, and 13-5 to scale and compare. Comment on differences.

F. Look up selected (or assigned) articles under "Fitting Machines into existing Layouts" or "Computers in Warehouse Planning" in the Bibliography for this chapter. Report in class, or write a critique.

14

Introduction to Material Handling

In Chapter 5 it was suggested that the next step is to implement the flow. That is, the design was static; the implementation must make it dynamic—must cause material to flow. This chapter and the next two will introduce the concepts, procedures, and means for implementing the flow of material.

The primary function of material handling—the movement of material—is as old as man, and yet there is no universally acceptable definition. Many have been written, but they are primarily descriptive of the situations in which the handling takes place, what it moves, its objectives, and other qualifications as to what material handling is or is not. Rather than become involved in these, we will simply take it that *material handling is handling material.*

One of the few general conclusions we can draw about material handling is that its scope is expanding and its importance is becoming more widely recognized. And rightly so, since handling activity in the typical enterprise may easily account for 50 to 75% of the production activity, and not the 10 to 20% usually quoted. Too many people in business and industry are hiding their heads in the overhead and refusing to recognize the economic significance of improper handling methods.

In a somewhat chronological sense, material handling activity in an enterprise will go through three stages of development: (1) conventional, (2) contemporary, and (3) progressive, or system oriented. It might be said that in the conventional interpretation, material handling is primarily the movement of material from one point to another point within the confines of a facility—the problem is to move something from *point* A to *point B*. Too often the concern of the material handling analyst is simply for individual, isolated, independent problem situations, with little attention given to interrelationships between the separate situations.

More desirable would be the contemporary concern for the overall material flow. The analyst would become involved in all interrelated plant handling problems, establishing a general material handling plan, and tying each individual problem solution into all others—the common approach in many modern, well-managed plants today.

Chapters 14, 15, and 16 are condensed and adapted from Apple, *Material Handling Systems Design.*

In the more progressive interpretation, the analyst would visualize every material handling and physical distribution activity as part of one all-encompassing system, including:

1. The movement of materials from all sources of supply.
2. All intraplant handling.
3. The distribution of finished goods to all customers.

This all-inclusive point of view is called *the system approach,* with the goal of conceptualizing a solution or at least an approach, to the total handling problem in terms of a theoretical ideal system. The material handling analyst would then proceed to design and implement portions of the system that were feasible at the moment, while continuing to work on other phases, gradually implementing them as the means become technologically feasible or economically practical.

While this approach may seem too theoretical, it serves as a worthwhile goal for the design of an effective material handling system. The extent to which it should be carried out depends on the importance of material handling to the company, and the practical economics of attempting to extend the material handling system backward to its sources of material supply, and forward to the customers' locations. The least one might do would be to contact major suppliers and customers and work with them on the improvement of material handling related activity in which there is common interest.

These interrelationships are shown diagramatically by the sketch in Figure 14-1, *The Material Flow Cycle.* From the system point of view the concern would be for the entire flow of material as shown, from the sources of raw material to the delivery of goods to the customer—and, in many cases, with the flow of scrap and waste back through appropriate channels and into the system again.

Material Handling Interests and Activities

In carrying out his duties, the material handling analyst is concerned with a wide range of interest areas that have a bearing on the overall efficiency of production. Table 14-1 lists some of the areas and activities in which the handling analyst or engineer of a well-managed enterprise would be active—investigating, analyzing, writing specifications, setting standards, and conducting more formal studies and surveys. Of course, the number and extent of handling activities depends on the individual enterprise, the importance of material handling in its operations, and the interest of management in taking advantage of related cost improvement opportunities.

Objectives of Material Handling

In carrying out the above activities, the handling engineer is aiming at one overall goal—reduced production costs. This general objective is more easily understood

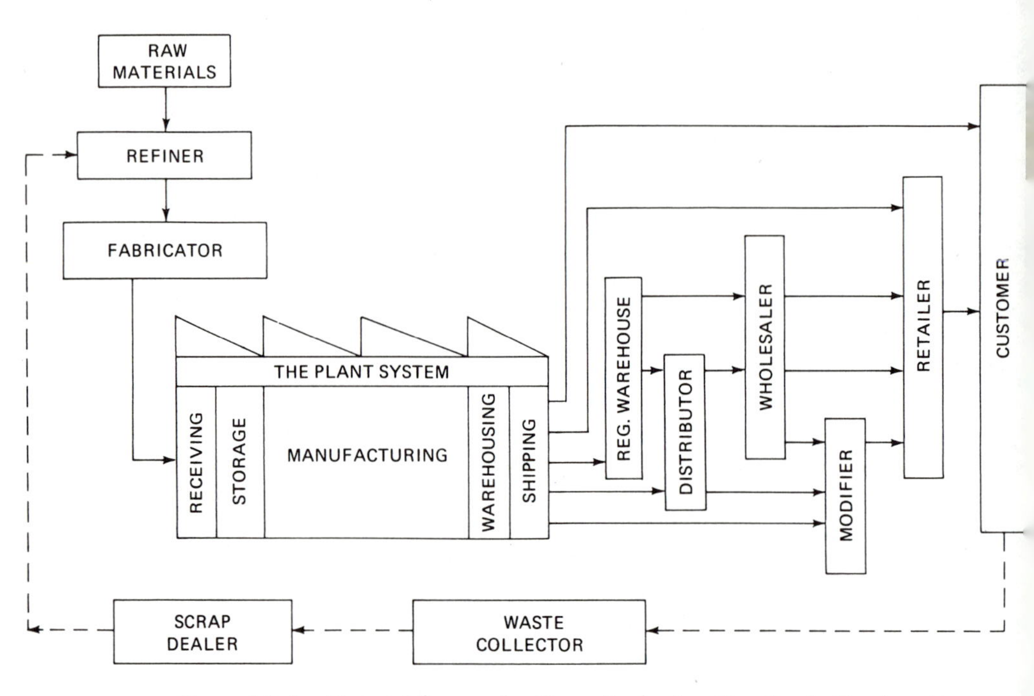

Figure 14–1—Material flow cycle. **(From Production Handbook, *The Ronald Press Co.*)**

if it is subdivided into more specific goals, such as:

1. Increased capacity
2. Improved working conditions
3. Improved customer service
4. Increased equipment and space utilization
5. Reduced costs

It can be seen that the handling engineer works in close contact with many other functions, which puts him in an outstanding position to reduce costs in many phases of plant operation.

The Material Handling Equation

If the handling problem is so important, so broad, and so all encompassing, there is a need for a concept or philosophy upon which to base the understanding and analysis of handling problems.

The *Material Handling Equation* has proven useful in visualizing the many aspects of a handling problem. Reference to Figure 14-2 will show that there are six major questions to be answered in searching for the solution to a material handling problem. They begin with the question "Why do this at all?" —which calls for a serious look at the problem to be sure that the problem has really been identified properly, and does, in fact, exist. Then, the analysis proceeds to the

Table 14-1. Material Handling Interests and Activities

Areas of Interest	Activities
1. Packaging and packing at the suppliers' plants	1. Handling methods
2. Loading at suppliers' docks	2. Storage methods
3. Transportation from suppliers	3. Loading and unloading techniques; methods
4. Unloading activities at "our" plant	4. Packaging (consumer)
5. Receiving operations	5. Packing (protective)
6. Storage of both material and supplies	6. Testing packaging, loading, and handling methods
7. Issuing material to production	7. Setting specifications and standards for handling, packing, and storage
8. In-process storage	8. Equipment feasibility studies
9. In-process handling	9. Handling and storage equipment selection
10. Work place handling	10. Auxiliary equipment evaluation and selection
11. Intra-departmental handling	11. Selection of containers (shop, packing, shipping)
12. Inter-departmental handling	12. Handling equipment repair and maintenance policy and procedure
13. Intra-plant handling	13. Damage prevention (material and product)
14. Handling related to auxiliary functions	14. Safety
15. Packaging (consumer)	15. Training
16. Warehousing of finished goods	16. Surveys to uncover savings opportunities
17. Packing (protective)	17. Handling costs and cost control methods
18. Loading and shipping	18. Keeping up to date on equipment, methods, procedures, etc.
19. Transportation to customers	19. Related paperwork, control and communication systems
20. Inter-plant transportation	
21. Related record keeping	

question of "What?"—with a concern for the *material* or item to be moved. Next, the analyst considers the questions of "Where?" and "When?"—which identify and specify the *move* to be made. And lastly, he considers the "How?" and the "Who?"—which pertain to *method.*

It might be suggested here, that one of the major difficulties in the alleged solutions to many material handling problems in the past, is that the analyst, in his haste to find a hardware-type answer to the problem, has jumped from the "*What?*" to the "*How?*"

But, having divided the problem into three major phases—the *material,* the *move,* and the *method*—it becomes necessary to examine the several factors and sub-factors involved in the proper analysis of each phase, listed in Figure 14-2, as related to the *material, move* and *method.* A detailed discussion of the factors and sub-factors is beyond the scope of this book, but it should be emphasized that no material handling problem can be properly solved unless adequate data pertaining to these factors are carefully accumulated, researched, and analyzed—and then given proper consideration in the design of any new handling method or system. This will be given attention in Chapter 16.

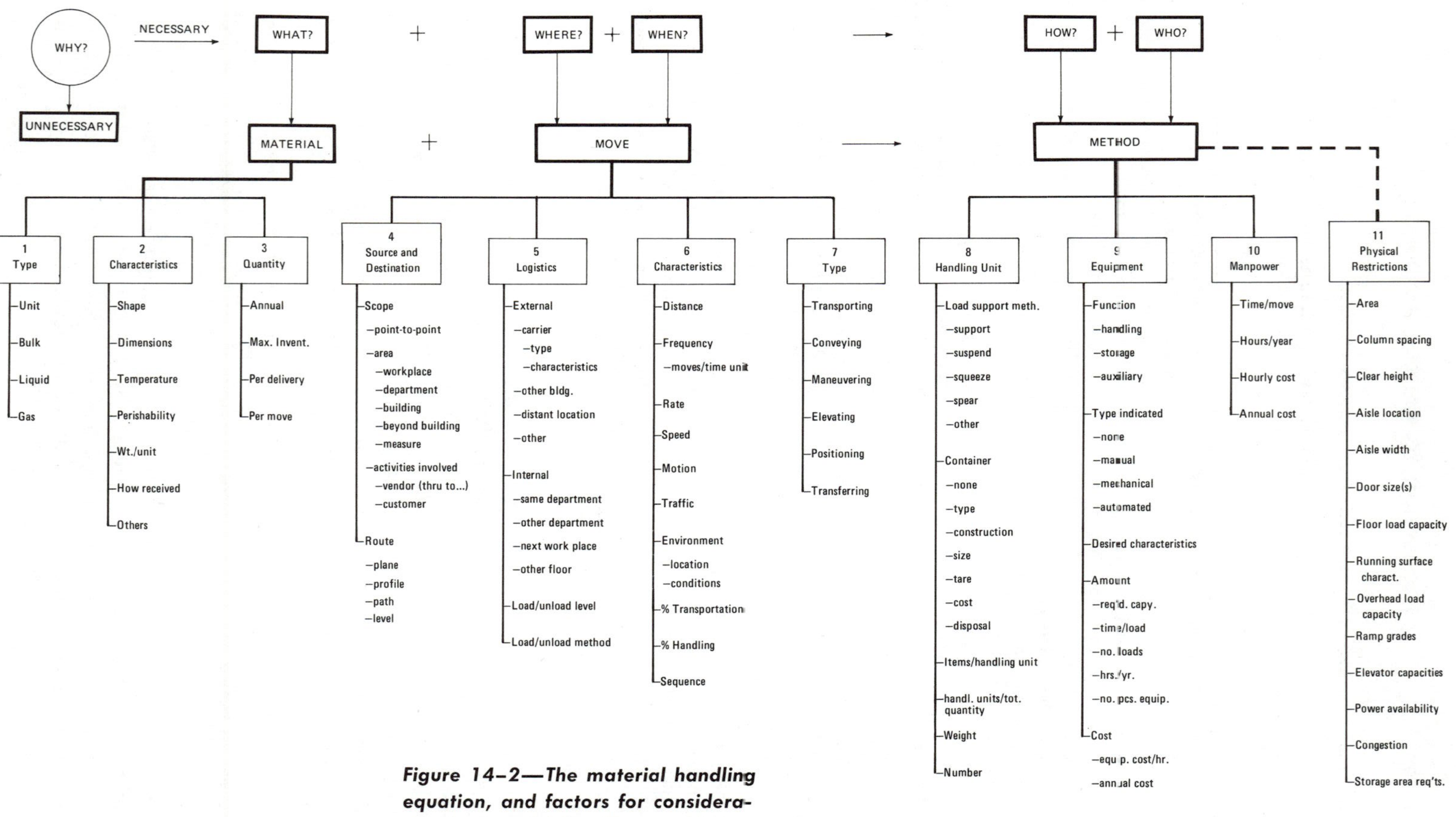

Figure 14-2—The material handling equation, and factors for consideration in material handling problem analysis

Table 14-2. Principles of Material Handling

1. All handling activities should be planned.
2. Plan a *System* integrating as many handling activities as is possible and coordinating the full scope of operations.
3. Plan an operation sequence and equipment arrangement to optimize material flow.
4. Reduce, combine, or eliminate unnecessary movements and/or equipment.
5. Utilize gravity to move material whenever practicable.
6. Make optimum utilization of building cube.
7. Increase quantity, size, weight of load handled.
8. Provide for safe handling methods and equipment.
9. Use mechanized or automated handling equipment when practicable.
10. In selecting handling equipment, consider all aspects of the *Material* to be handled, the *Move* to be made, and the *Method(s)* to be utilized.
11. Standardize methods as well as types and sizes of handling equipment.
12. Use methods and equipment that can perform a variety of tasks and applications.
13. Minimize the ratio of mobile equipment deadweight to pay load.
14. Equipment designed to transport materials should be kept in motion.
15. Reduce idle or unproductive time of both handling equipment and manpower.
16. Plan for preventive maintenance and scheduled repair of all handling equipment.
17. Replace obsolete handling methods and equipment when more efficient methods or equipment will improve operations.
18. Use material handling equipment to improve production control, inventory control, and order handling.
19. Use handling equipment to help achieve full production capacity.
20. Determine efficiency of handling performance in terms of expense per unit handled.

Source: See footnote 1.

Principles of Material Handling

One of the most important contributions to either analyzing or designing material handling systems is experience. But, it takes years of exposure to a wide variety of situations to accumulate this background. So, as is common in many fields, practitioners rely heavily on the experience of their predecessors for fundamental concepts. The concepts represent the experience of many people who have practiced in the field, which have been tried and tested and have been found worthwhile, passed on to others, and written down as guides for future practitioners (see Table 14-2[1]). However, they must be applied with extreme care and common sense. Often it will be found necessary to imply that they may only be true, *all other things being equal,* or *when practicable.*

Using the principles. Merely stating the principles of material handling and making some suggestions for their application does not assure that they will be

[1]The principles in Table 14-2 are adapted from those developed by the College-Industry Committee for Material Handling Education, which is sponsored by the Material Handling Institute, a trade association and promoter of education in material handling, and are discussed in detail in Chapter 3 of the author's *Material Handling Systems Design.*

Material Handling Check Sheet
Material Handling Methods

Plant ________________ Location ________
Dept. ________________ Operation ________
Observer ________________ Date ________

Indicators of Improvement Opportunities	Check Yes	Check No	Remarks
1. Moving one item at a time			
2. Not using gravity			
3. Insufficient storage at workplace			
4. Poor flow between work areas			
5. Stock control difficulties			
6. Production bottlenecks			
7. Scheduling difficulties			
8. Moving loose materials			
9. Production machines idle			
10. Materials delivered to wrong place			
11. Slow materials movement			
12. Haphazard handling methods			
13. Difficult manual handling tasks			
14. Unsafe handling methods			
15. Unsafe handling equipment			
16. Overmechanized handling methods			
17. Overloaded handling equipment			
18. Underloaded handling equipment			
19. Operators waiting for mat'ls.			
20. Heavy physical exertion			
21. Low production			
22. Excessive loading & unloading time			
23. Crowded work places			
24. Individual handling operations not coordinated			
25. Building restrictions impede handling operations			
26. Carrier restrictions impede handling operations			
27. Heavy cross-traffic			
28. Large, heavy items moving long distances			
29. Communications (paper work) delays mat'l. handling			
30. Manual coupling; uncoupling trailers, etc.			
31. Non-standardized handling equipment			
32. Unplanned handling methods			
33. Excess variety of handling equipment			
34. Non-standard in-process handling containers			
35. Zig zag flow paths			
36. Insufficient mechanization of handling			
37. No plans for scrap removal			

Figure 14–3—Typical checksheets, useful in identifying improvement opportunities and analyzing and solving problems. **(From Apple, *Material Handling Systems Design*.)**

Material Handling Check Sheet

General

Plant ______________________ Location ______________

Dept. ______________________ Operation ______________

Observer ______________________ Date ______________

Indicators of Improvement Opportunities	Check		Remarks
	Yes	No	
1. Crowded conditions			
2. Poor housekeeping			
3. Cluttered aisles			
4. Excess mat'ls. in process			
5. Empty floor space			
6. Materials piled *on* floor			
7. Unused space overhead			
8. Damaged materials			
9. Excessive scrap			
10. Long moves			
11. Complex flow patterns			
12. Backtracking			
13. Cross traffic			
14. Related operations far apart			
15. Obstructions to materials movement			
16. Traffic jams			
17. Floor poorly maintained			
18. Excess inter-floor handling			
19. Inadequate scrap removal			
20. Excessive mat'ls. at work place			
21. Unnecessary handling			
22. Re-handling			
23. Idle handling equipment			
24. Idle production equipment			
25. Operators waiting for mat'ls.			
26. High indirect labor cost			
27. High handling cost			
28. Unexplainable delays			
29. Poor flow pattern			
30. Crooked aisles			
31. Narrow aisles			
32. Aisles not marked			
33. Poor utilization of bay size			

properly used. The best way to utilize the principles effectively is by means of check sheets—dividing or sorting them or their coralaries into interest areas, then restating them more specifically. Figure 14-3 shows two examples of the use of the principles. Similar check sheets could be developed for other areas of activity or specific plants, to reflect their individual differences.

The Unit Load Concept

One of the principles of material handling, the *Unit Size Principle,* implies that the larger the load handled, the lower the cost per unit handled. This section will develop the unit load concept as one of the most universally applicable of the principles.

A *unit load* may be defined as: A number of items, or bulk material, so arranged or restrained that the mass can be picked up and moved as a single object, too large for manual handling, and upon being released will retain its initial arrangement for subsequent movement. It is implied that any single object too large for manual handling is a unit load. The many advantages of the unit load should be obvious—but there are also some disadvantages to be considered, such as:

1. Cost of unitizing and de-unitizing.
2. Equipment and space requirements.
3. Tare weight of unitizing medium.
4. Problem of returning empty pallets or containers.
5. Transfer equipment often not available on both ends of the move.

Basic ways to move a unit load. Since the unit load concept implies a fairly large quantity or volume, a primary consideration in its application should be how the load will be lifted and carried. This is usually accomplished in one of the following ways:

1. A lifting device under the mass.
2. Inserting a lifting element into the body of the load.
3. Squeezing the load between two lifting surfaces.
4. Suspending the load.

This leads directly to the various types of unit loads, and related devices:

1. On a platform
2. On a sheet
3. On a rack
4. In a container
5. Self-contained

The majority of unit loads are probably handled on pallets or skids. Some typical unit loads and methods of handling them are shown in Figure 14-4. Pallet or other unit load devices should be selected on the basis of such factors as:

1. Purpose for which unit load is intended.
2. Characteristics of item handled.
3. Handling system capabilities and limitations.
4. Carrier characteristics.

Figure 14–4—Typical methods of handling unit loads (above), and various pallets and skids. (From Modern Materials Handling Directory.)

5. Physical facilities—vendor, plant, customer.
6. Disposition of pallet, etc.
7. Common practice in the industry.
8. Pallet characteristics.
9. Building characteristics.

Significant among pallet characteristics are its dimensions. Standard sizes have been developed by A.N.S.I., as follows:

Rectangular (in inches)				*Square (in inches)*	
R–1	24 × 32	R–6	40 × 48	S–1	36 × 36
R–2	32 × 40	R–7	48 × 60	S–2	42 × 42
R–3	36 × 42	R–8	48 × 72	S–3	48 × 48
R–4	32 × 48	R–9	88 × 108		
R–5	36 × 48				

Unit load planning and design. In carrying out the design of the unit load, it should be evaluated against such criteria as:

1. Minimum tare weight
2. Low cost
3. Mechanical strength
4. Disposable (or expendable)
5. Universal in application
6. Optimum size
7. Low maintenance
8. Ease of de-unitizing
9. Ease of identification
10. Versatility
11. Transportable by conventional equipment
12. Interchangable
13. Optimum shape
14. Easy to store
15. Stackable
16. Meet customer requirements
17. Cost of unitizing

It is suggested here, as in many other situations throughout this book, that a logical procedure be followed in approaching such a complex task as designing a unit load and the unit load system. As an aid, the following procedure is suggested:

1. Determine applicability of the unit load concept.
2. Select the type of unit load.
3. Identify the most remote source of a potential unit load.
4. Establish the farthest practicable destination for the unit load.
5. Determine the size of the unit load.
6. Establish the configuration of the unit load.
7. Determine the method of building the unit load.

The above is only an introduction to the very important unit load concept. Considerably more detail can be found in the trade journals, or may be obtained from the literature of manufacturers furnishing unit load equipment and supplies.

Conclusion

This chapter has introduced the subject of material handling as an important factor in facilities planning and design. Since handling is such a large part of the productive activity in most enterprises, it should be very carefully thought through, and the handling methods designed into the layout—not superimposed upon it. The conscientious facilities designer should feel obligated to know far more about handling than is contained in this book. The author's *Material Handling Systems Design* contains an extensive coverage of handling, storage, warehousing, and physical distribution, as well as many references to even more information.

The next chapter will consider the problem of handling equipment as a means of implementing the material flow pattern. Then, Chapter 16 will discuss the problem of designing the handling system and selecting the equipment. The three chapters should be studied as a unit, before trying to resolve the handling aspects of facilities design.

Questions

1. What is understood by the term *material handling?*
2. What is the extent of material handling in industry? Explain.
3. What are the three so-called *chronological stages* in the development of material handling activity?
4. What is involved in the overall material flow cycle concept?
5. Name some of the areas of interest to the material handling analyst.
6. Name some of the activities of the material handling function in dealing with the areas of interest in *Question* 5.
7. Discuss each of the objectives of material handling.
8. What is the material handling equation?
9. Explain its significance to material handling problem analysis.
10. Name the eleven categories of factors involved in analyzing a material handling problem. Suggest several factors under each.
11. What *are* principles of material handling?
12. What is their significance?
13. State some of the principles and illustrate each.
14. How can the principles be used?
15. Describe the unit load concept.
16. What is the significance of standardizing pallet sizes?
17. Name some criteria for planning a unit load system.
18. List the steps in designing a unit load system.

Exercises

A. In a local situation, use the check sheets in Figure 14-3.

B. Recap the findings in *A* in the form of suggestions for improvement.*

C. Visit a local concern and observe the use (or lack of use) of the unit load principle. Report, or discuss in class.

D. Design a material handling related check sheet for a specific type of enterprise. Choose your own or use one of those suggested in Table 8-1.*

E. In a current material handling periodical, identify material handling principle applications, and list applications—or misapplications.

*For additional help, see Apple, ch. 3.

15

Material Handling Equipment

The material handling equation suggested that an analysis of the *material* to be handled and the *moves* to be made should indicate a *method*. However, before becoming involved in selecting equipment, a brief introduction to material handling equipment concepts may prove helpful in comprehending the overall objective of the handling analysis—to design the method. Since most handling methods include some type of handling equipment, a major task for the material handling engineer is to become acquainted with the ever growing field of handling equipment. Although it is next to impossible for an individual to become intimately acquainted with all types, the engineer's salvation lies in the fact that in a typical industry, business, or field of activity, only a fraction of the total number of types is commonly applicable. Probably less than 50 types (out of over 500), along with their attachments and accessories, will serve to implement the majority of an industry's handling activity.

It should be pointed out that relatively seldom can a single piece of equipment do a complete job. Each type and variety of equipment has specific uses and advantages, and often two or more are combined into the solution of a single problem. This is especially true if the problem is system-wide, since a handling operation large enough to be classified as a system will usually include many kinds of moves requiring a variety of equipment. In choosing the handling methods, the reader should constantly keep in mind that the successful solution to most handling problems involves, within the existing or contemplated physical facilities and environment, and against related costs, a proper matching of:

1. *Material characteristics*
2. *Move requirements*
3. *Method (equipment) capabilities*

The Place of Equipment in the Handling System

As implied previously, equipment is not always required in solving a handling problem—sometimes the simplest and most economical method will require no equipment at all. In fact, the *work simplification approach* suggests the following general procedure:

1. *Eliminate the move.*
2. *Combine the move* with some other function: such as processing, inspection, storage.
3. *Change the sequence* of activities, to shorten, eliminate, or alter the move requirements.
4. *Simplify the move,* to reduce the scope, extent, or distance or to improve on the method or the equipment selection.

Then, after having accomplished the above, the equipment should be selected—if necessary.

A factor frequently overlooked in the rush to mechanize or automate is that manual handling may in fact be the easiest, most efficient, and least expensive method of moving material. Only after it has been proven that manual handling is more costly, too dangerous, or too slow, should the analyst turn his attention to equipment.

Included among the several hundred types of handling equipment are many that are non-powered, or manually operated or controlled—that is, not mechanized.

The possibilities of both manual and non-powered handling methods should be checked before mechanized or powered equipment is seriously considered.

Basic Handling Equipment Types

There are four basic types of material handling equipment: (1) conveyors, (2) cranes and hoists, (3) trucks, and (4) auxiliary equipment. There are subclassifications under each, but the four basic types are a convenient framework for the treatment of handling equipment, and are briefly described in the following.

Conveyors—gravity or powered devices commonly used for moving uniform loads continuously from point-to-point over fixed paths, where the *primary function is conveying.* Common examples are:

1. Roller conveyor
2. Belt conveyor
3. Chute
4. Trolley conveyor
5. Bucket conveyor
6. Pneumatic conveyor

Cranes and hoists—overhead devices used for moving varying loads intermittently between points within an area, fixed by the supporting and guiding rails, where the *primary function is transferring.* Common examples are:

1. Overhead travelling crane
2. Gantry crane
3. Jib crane
4. Hoist
5. Stacker crane
6. Monorail

Industrial trucks—hand or powered vehicles (non-highway) used for the movement of mixed or uniform loads intermittently over various paths having suitable running surfaces and clearances, where the *primary function is manuvering or transporting*. Common examples are:

1. Lift truck
2. Platform truck
3. Two-wheel hand truck
4. Tractor-trailer train
5. Hand stacker
6. Walkie truck

Auxiliary equipment—devices or attachments used with handling equipment to make its use more effective. Common examples are:

1. Pallets; skids
2. Containers
3. Below-the-hook devices (for cranes)
4. Lift truck attachments
5. Dock boards and levelers
6. Pallet loaders and unloaders
7. Positioners
8. Ramps
9. Weighing equipment

Each of these basic types is further detailed in books, commercial literature, and periodicals. Attention is centered here on:

1. The capabilities of each basic type.
2. Their relationships to the factors to be considered in analyzing handling problems.

Some of these interrelationships are listed in Table 15-1.

Basic Handling Systems

In addition to the basic types of handling equipment outlined above, there are also several so-called *basic handling systems,* which provide an idea of the conventional way of looking at the handling equipment problem. And it should be pointed out that the word *system* in this context means a group of related handling devices, commonly used in combination with each other. This equipment system approach is sometimes helpful in conceptualizing problem solutions. It also serves as a method of classifying handling methods according to selected characteristics of the situation that may be helpful in solving the problem and selecting handling equipment. It should also be recognized that different companies will often use different systems for the same purpose. Differences in plant layout, processes, volume of production, type of goods, shipping and receiving procedures, and many other factors may suggest an entirely different system for one plant than for another. The systems described here are rather basic and may be used exclusively in a plant; or several may be used as components of an integrated plant handling system.

Equipment-oriented systems are described in terms of the three of the basic equipment groups: conveyors, cranes and hoists, and industrial vehicles.

Table 15-1. General Characteristics of Basic Material Handling Equipment Types

	TASK and EQUIPMENT CHARACTERISTICS / EQUIPMENT TYPE	CONVEYORS Moving uniform loads continuously from point to point over fixed paths where primary function is transporting	CRANES and HOISTS Moving varying loads intermittently to any point within a fixed area	INDUSTRIAL TRUCKS Moving mixed or uniform loads intermittently over various paths with suitable surfaces where primary function is maneuvering
MATERIAL	Volume	high	low, medium	low, medium, relatively high
	Type	individual item, unit load, bulk	indiv. item, unit load, variety	indiv. item, unit load, variety
	Shape	regular, uniform, irregular	irregular	regular, uniform
	Size	uniform	mixed, variable	mixed, or uniform
	Weight	low, medium, heavy, uniform	heavy	medium, heavy
MOVE	Distance	any, relatively unlimited	moderate, within area	moderate, 250–300 ft.
	Rate, Speed	uniform, variable	variable, irregular	variable
	Frequency	continuous	intermittent, irregular	intermittent
	Origin, Destination	fixed	may vary	may vary
	Area covered	point to point	confined to area within rails	variable
	Sequence	fixed	may vary	may vary
	Path	mechanical, fixed pt. to fixed pt.	may vary	may vary
	Route	fixed, area to area	variable, no path	variable, but over defined path
	Location	indoors, outdoors	indoors, outdoors	indoors, outdoors
	Cross traffic	problems in by-passing	can by-pass, no effect	can by-pass, maneuver, no effect
	Primary function	transport, process/store in move	lift & carry, position	stack, maneuver, carry, load, unload
	% Transport in operation	should be high	should be low	should be low
METHOD	Load support method	none, or in containers	suspension; pallet, skid, none	from beneath; pallet, skid, container
	Load/unload characteristics	automatic, manual, designated points	manual, self, any point	self; any point on available path
	Oper. accompany load	no	may or may not, usually does	usually does; may be remote
BUILDING CHARACT.	Cost of floor space	low, medium	high	medium, high
	Clear height	if enough, conv. can go overhead	high	low, medium, high
	Floor load capacity	depends on type conv. & mat'l.	depends on activity	medium, high
	Running surfaces	not applicable	not applicable	must be suitable
	Aisles	not applicable	not applicable	must be sufficient
	Congested areas	fair	good	poor

1. *Industrial truck systems.* One system is constituted of platform trucks and skids. The high-lift platform trucks and low-lift trucks pick up, transport, and set down skid-loaded materials. While the low-lift trucks are used for moving, the high-lift truck is used for stacking, maneuvering, and positioning.

2. *The fork truck and pallet system* is similar to the platform truck and skid system. The fork truck needs less clearance than the platform truck. The forks permit the use of pallets, usually shallower than skids, thereby saving space in tiering. Several types of powered hand trucks may be used with either the skid or pallet system. Economic travel distance is about 200 ft.

3. *The tractor–trailer system* is usually more economical for hauling larger quantities of material for distances over 200 to 300 feet. The cost per ton becomes much lower than with lift trucks, since one relatively inexpensive tractor can tow many loaded trailers at one time. Loading and unloading may have to be done by crane, hoist, platform lift, or fork lift truck—or sometimes by hand.

4. *Conveyors and conveyor systems* are commonly found in operations where items are of uniform size and shape and are transported over the same path repeatedly, or for long periods of time. Conveyor systems in general are more easily adaptable to higher speeds or to mass movement than the truck-based systems. The application of control devices, programming systems to conveyors, along with careful layout, can often reduce the cost of material handling to a negligible amount. However, conveyors become less economical when they must be loaded and unloaded frequently, especially if people must do the loading and unloading.

Conveyor systems are very useful in warehousing operations. For example, several packaging lines may feed into main-line accumulating conveyors, which in turn could feed into a battery of pallet loaders some distance from the production operation. The loaded pallets could even be delivered by conveyor and automatic pallet elevator to another floor or level, on which it was to be accumulated. From this point it might be picked up by a fork-lift truck and taken to the warehouse for storage.

5. *Overhead systems* using cranes and monorail equipment are common in operations where floor space utilization or product characteristics make the use of lift trucks or conveyors undesirable, and travel distances and paths are reasonably restricted.

Material (load)-oriented systems are commonly identified as: unit handling systems; bulk handling systems; and liquid handling systems. Only unit handling systems will be covered here. A *unit*—such as a box, bale, roll of material—can be either an individual item or a number of items grouped to form the unit load, as discussed in Chapter 14.

Generally speaking the unit load system is more flexible and requires less investment. It can usually be applied in any size organization with some degree of success, and is particularly applicable to operations involving non-repetitive handling, or where a variety of products and material are handled, although it is also applied in highly repetitive, large-volume operations.

In storage areas, unit loads can be arranged and stacked to permit direct access to every item, with identification and location of items based on a grid location system.

Recent developments have made it possible to eliminate pallets and skids in many applications, using such devices as carton clamps and slip sheet attachments.

Method (production)-oriented systems are commonly described in terms of the types of production in which they are used, such as: manual, mechanical, mass production, automated, or job shop.

1. *The manual system* implies the use of manual handling methods because of the nature of the operations, such as low volume, wide variety, and extreme fragility. Such situations would normally infer that anything other than manual handling would be undesirable.

2. *Mechanized or automated systems* suggest increased volume or standardization of product and therefore the use of more sophisticated, complex, or more mechanized equipment, including extensive use of conveyors, automatic controls, transfer devices, and other methods of mechanical and automatic handling between operations.

3. *Mass production handling systems* imply the presence of such high volume that it is possible to apply complex handling machinery. In mass production, the basic machinery is frequently no different from what is used in other applications. The advantages are derived from automatic control of repetitive handling operations by mechanical, electrical, electronic, photo-electric, or magnetic control devices. The mass production handling system demonstrates the value of basic handling techniques combined and integrated to simplify sequences of complex handling operations.

4. *Job shop handling systems* relate to the conventional concept of a job shop, as involving small-volume operations. The unit load system is frequently combined with both fixed and portable conveyors or sectional conveyor units. Establishing flexible or adaptable handling systems suitable to several similar products, requiring similar handling techniques, may make it possible for at least a part of the facility to approach the efficiencies of mass production.

The remaining activities must gain handling efficiency by judicious combinations of such equipment as lift trucks, unit load handling and storage devices, portable and adjustable conveyors, and good job shop production control. The handling efficiency of a job shop operation is more often dependent upon good management than on elaborate equipment applications.

Function-oriented systems classify equipment and activities by function, and appear useful largely from a problem-solving point of view. The categories are:

1. *Transportation systems*—horizontal motion over fixed or variable, level or nearly level routes, by pulling or pushing, on surface riding vehicles.
2. *Elevating systems*—vertical motion over fixed vertical or steeply inclined routes, with continuous or intermittent motion.
3. *Conveying systems*—horizontal, inclined or declined motions, over fixed routes, by gravity or power.
4. *Transferring systems*—horizontal, or vertical compound motions, through the air, over fixed routes or within limited areas, with intermittent motion.
5. *Self loading systems*—intermittent motion with machines that pick up,

Table 15-2. Classification of Selected Common Types of Material Handling Equipment

I. CONVEYORS

- **A. Belt** (I–A)
- **B. Chain**
  1. Power and free (I–B–1)
  2. Slat (I–B–2)
  3. Tow (I–B–3)
     a. Overhead (I–B–3a)
     b. Under-floor (I–B–3b)
  4. Trolley (I–B–4)
- **C. Chute** (I–C)
- **D. Roller**
  1. Gravity (I–D–1)
  2. Live (I–D–2)
- **E. Wheel** (I–E)

II. CRANES AND HOISTS

- **A. Jib** (II–A)
- **B. Bridge** (II–B)
- **C. Monorail** (II–C)
- **D. Stacker** (II–D)
- **E. Storage-retrieval unit** (II–E)

III. TRUCKS

- **A. Non-powered**
  1. 4-wheel (III–A–1)
  2. Jack lift (III–A–2)
  3. Trailer (III–B–8)
- **B. Powered**
  1. Fork lift (III–B–1)
  2. Load carrier (III–B–2)
  3. Narrow aisle (III–B–3)
  4. Order-picker (III–B–4)
  5. Reach (III–B–5)
  6. Side loader (III–B–6)
  7. Straddle (out-rigger) (III–B–7)
  8. Tractor-trailer train (III–B–8)
  9. Walkie (III–B–9)
- **C. Attachments** (III–C)

IV. AUXILIARY EQUIPMENT

- **A. Dock boards and levelers** (IV–A–1, –2)
- **B. Shipping containers** (IV–B)
- **C. Shop containers** (IV–C)
- **D. Supports**
  1. Pallets (IV–D–1)
  2. Skid (IV–D–2)
  3. Rack (IV–D–3)

Note: Numbers following equipment types refer to Figure numbers.

move horizontally, set down, and in some cases tier loads, without other handling. Also known as unit load systems.

Although the several methods of categorizing or classifying handling systems will not solve any handling problems, they do offer a method of thinking about handling problems, or characterizing them, which may be helpful in analyzing a problem or in conceptualizing problem solutions.

Common Material Handling Equipment

The pages that follow are devoted to definitions, characteristics, uses, and illustrations of about 35 types of handling equipment commonly used in mechanically oriented enterprises.[1] The illustrations are numbered to refer to the descriptive passages and to the outline of equipment types in Table 15-2.

[1] Abbreviated from Apple, *Material Handling Systems Design*, ch. 6, *q.v.* for additional details and for some 45 more types of equipment.

I. Conveyors

I–A. *Flat belt conveyor*—an endless fabric, rubber, plastic, leather, or metal belt operating over suitable drive, tail end, and bend terminals and over belt idlers or slider bed for handling material, packages, or objects placed directly upon the belt (shown carrying orders to packing area).

I–A

1. Top and return runs of belt may be utilized.
2. Will operate on level, incline up to 28 degrees, or downgrade.
3. Belt supported on flat surface is used as carrier of objects or as basis for an assembly line.
4. Belt supported by flat rollers will carry bags, bales, boxes, etc.
5. Metal mesh belts are used for applications subjected to heat, cold, or chemicals.
6. High capacity.
7. Capacity easily adjusted.
8. Versatile.
9. Can elevate or lower.
10. Provides continuous flow.
11. Relatively easy maintenance.
12. *Used for:*
    —Carrying objects (units,

cartons, bags, bulk material).
—Assembly lines.
—Moving people.

I–B–1. *Power and free conveyor*—a combination of powered trolley conveyors and unpowered monorail-type free conveyors. Two sets of tracks are used, usually suspended one above the other. The upper track carries the powered trolley conveyor, and the lower is the free monorail track. Load-carrying free trolleys are engaged by pushers attached to the powered trolley conveyors. Load trolleys can be switched to and from adjacent unpowered free tracks.

1. Free trolleys move by gravity, or by *pushers* supported from trolley conveyor on upper level.
2. Interconnections may be manually or automatically controlled.
3. Track switches may divert trolleys from power to free tracks.
4. Dispatching may be automatically controlled.
5. Free gravity tracks may be installed between two power tracks for storage.
6. Speeds may be varied from one power section to another.
7. Can include elevating and lowering units in free line.
8. Can recirculate loads on all or sections of system.
9. Can be computer controlled.
10. *Used for:*
    —Temporary storage of loads between points on machining, assembly, and test lines.
    —Routing loads to selected points.
    —Overhead storage for later delivery of loads to floor level.
    —Integrating production, assembly, and test equipment.
    —Provides for surge storage against a breakdown.

I–B–1

I–B–2. *Slat conveyor*—a conveyor whose carrying surface consists of spaced wood or metal slats, fastened at their ends to two strands of chain running in a suitable track or guide.

1. Slats made of wood, metal or combination.
2. Slats only $\frac{1}{4}$ to $\frac{1}{2}$ inch apart provide a relatively continuous surface.
3. Slats can serve as base for fixtures or be built as specially

designed supports for specific objects.
4. Slats can be mounted at work level or flush to floor—where slow speed permits foot traffic to cross over.
5. A variation uses rollers as slats.
6. Sturdy, heavy duty, low maintenance.
7. Inclines in a relatively short horizontal distance.
8. For inclines over 10°, requires cleats; then can operate to 30 to 40°.
9. *Used for:*
    —Heavy unit loads (crates, cartons, drums, rolls, bags, etc.).
    —Hot material (castings, forgings, molds).
    —Wet materials.
    —Warehousing (goods to and from storage).

I–B–3. *Tow conveyor*—an endless chain (a) supported by trolleys from an overhead track, (b) running in a track flush with or on the floor, or (c) running in a track under the floor.

1. Performs somewhat same function as tractor–trailer train (III–B–8).
2. Carts usually 3 by 5 ft; can be larger.
3. Cart can be specially designed for specific loads.
4. Track can be equipped with sidings or spurs.
5. Rugged; easy maintenance.
6. Automatic programmed pick-up and release of carts.
7. Carts removed from conveyor become free and portable to any point.
8. Can include moderate inclines and declines.
9. Can make use of carts required for other purposes.
10. Requires no operator.
11. Paces activity.
12. Relatively low cost per ton handled.
13. *Used for:*
    —Boxes, barrels, crates, cartons, freight
    —Warehousing:
        —Loads between receiving, storage, shipping
        —Order picking—operator attaches free cart to conveyor (or picks one up)
    —Intra-plant moves
    —Assembly lines
    —Continuous moving storage

I–B–3a. *Overhead tow* (shown moving stock-picking carts).

1. Track 8 or 9 ft from floor.
2. Frees floor; no interference with other traffic.
3. Track may dip to lower level for more convenient access to carts.
4. Unwheeled carts may be used for overhead transporting, with dips to working level as required.
5. Carts connected to conveyor by hook and chain—or link.
6. Cheaper to install than underfloor.

I–B–3b. *Underfloor tow* (shown moving picked orders to shipping).

1. Carts connected to conveyor by pin through slot in floor to pick up device on chain.
2. Commonly installed in new buildings.
3. Greater speed than overhead.
4. Pick-up action smoother than overhead.

I–B–4. *Trolley conveyor*—a series of trolleys supported from or within an overhead track and connected by an endless propelling medium such as chain, cable, or other linkage, with loads usually suspended from the trolleys.

1. Trolleys run (a) on flanges of structural tracks, or (b) inside rectangular or round tubes.
2. Multi-wheel trolleys or multiple-trolleys with load bar between used to distribute weight of large loads.
3. Load carriers suspended from trolleys and usually designed for optimum handling of object being moved.
4. Propelling medium can be chain, cable, or solid link.
5. May use sprocket wheel or caterpillar drive.
6. Functions in 3 dimensions (horizontal, vertical, incline).
7. Track 8 or 9 ft above floor.
8. Track may be elevated for move, then dip for access to operator or process.
9. Track easily routed around obstructions.

10. Frees floor space; no interference with other traffic.
11. Entire length can be used; i.e., no empty return run.
12. Relatively inexpensive to install and relocate.
13. Salvage value high.
14. Low operating and maintenance cost.
15. Relatively unlimited length and path.
16. Can follow complicated paths.
17. Easy to alter, shorten, lengthen path.
18. Paces activity.
19. Can be made automatic, or computer controlled.
20. Loads can be automatically switched to or from conveyor.
21. Can be hung from floor mounted supports.
22. *Used for:*
    —Moving nearly any material or load.
    —Overhead moving storage.
    —Intraplant movement.
    —Interplant movement.
    —Interfloor movement.
    —Recirculating materials.
    —Order picking, with goods on conveyor and picker selecting as items go past him.
    —Moving objects through continuous processes such as painting, baking, degreasing.

I–C. *Gravity chute*—a slide made of metal or other material and shaped so that it guides objects or materials as they are moved from one location to another. May be used on horizontal or declined planes, or as a spiral between extreme levels.

1. Usually slopes downward to utilize gravity.
2. Running surface may be wood, metal, composition.
3. May be straight, curved, or spiral; open or closed.
4. Spiral can be multiple flight.
5. Frequently custom designed and home made.
6. Low cost—no power, low maintenance.
7. Makes economical use of space (spiral).
8. Rate of descent determined by:
    —Contacting surface.
    —Atmospheric conditions.
    —Pitch.
    —Length.
9. Can be variable pitch; portable.
10. *Used for:*
    —Many kinds of materials and objects.

—Inter-floor moves.
—Inter-level moves.
—Warehouses, stores, terminals, industry.
—Loading and unloading carriers.
—Fire escapes.
—Between machines; storage ahead of a machine.
—From machines to container on floor.

I–D–1. ***Gravity roller conveyor***—a conveyor which supports the load on a series of rollers, turning on fixed bearings, and mounted between side rails at fixed intervals determined by the size of the object to be carried, which is usually moved manually or by gravity (shown with switch and lift sections). (See variations I–D–2 and I–E below.)

1. Rollers usually cylindrical tubing with a bearing on each end.
2. Rollers range from $\frac{3}{4}$ to $3\frac{1}{2}$ inches in diameter, with length governed by load.
3. Curved sections used for turns.
4. Rollers may be tapered for turns, or arranged differentially.
5. Tight corners may use ball table.
6. Requires 3 rollers under load at all times.
7. Rollers may be troughed or formed to conform to shape of load.
8. Standard roller spacing is 3, 4, and 6 inches.
9. Inexpensive, easy to install, minimum maintenance, long life.
10. Runs can include switches, spurs, gates, scales, deflectors, up-enders, processing and packaging equipment.
11. Can be arranged in spiral (chute) form.
12. Belt boosters used between levels.
13. In spite of apparent simplicity

often the basis for highly engineered installations.

14. *Used for:*
    —Almost any load with rigid riding surface that will contact 3 or more rollers.
    —Moves between areas, machines, buildings.
    —Storage between work stations.
    —Warehouses, docks, foundries, steel mills, canneries, manufacturing, assembly, packaging.
    —Loading and unloading carriers (portable sections and accordion type).
    —Bases for a handling system.
    —Integral segment of composite handling system.

I–D–2. *Live roller conveyor*—similar to gravity roller, except that power is applied to some or all of the rollers to propel the loads (note drive belt under rollers). Generally used for same purposes as gravity rollers (I–D–1) except for features noted below. Therefore, also similar to belt conveyor except better for heavy duty.

1. Power usually applied by:
    —Chain on sprockets.
    —Belting (underneath) held up against rollers at intervals by other rollers or other devices.
2. Can move objects on level runs, up slight grades, or restrain descent on down grades.
3. Permit controlled flow—articles are spaced.
4. Inclines possible to about 10°; declines to 17°.
5. Curves can be powered.
6. More rugged than belt conveyors.
7. More expensive than gravity or belt conveyors.

I–D–2

I–E. *Wheel conveyors*—a conveyor which supports the load on a series of skate-like wheels, mounted on common shafts in a frame or on parallel spaced rails, and with the wheels spaced to accommodate the size of the load to be carried. Also adapted to live, rack, and spiral versions as in the roller conveyor.

1. Very similar to the roller conveyor.
2. Objects usually moved by hand or gravity.
3. Wheels 2 inches in diameter and up, and staggered on shafts.
4. Lighter weight construction than roller conveyor.
5. Frequently made of aluminum with plastic wheels.

6. Easily portable.
7. Less expensive than roller.
8. Requires about 50% as much grade as rollers, except when used for storage.
9. Easy to set up and put away.
10. Low maintenance.
11. Comes in 5 and 10-ft sections.
12. Number of wheels per foot determines load capacity.
13. Must have 6 wheels under load.
14. Pitch of $1\frac{1}{2}$ to 3 in./10-ft section advisable.
15. *Used for:*
    —Warehousing.
    —Frequently carried in trucks for use in unloading and loading.
    —Ideal for curves, due to differential characteristics of construction.
    —Single wheeled rails useful in flow rack construction—on two sides of a lane.
    —Single rails useful as guides.

II. Cranes, Hoists, Monorails

II–A. *Jib crane*—a lifting device travelling on a horizontal boom that is mounted on a column or mast, which is *fastened to:* (a) *floor,* (b) *floor* and *a top support,* or (c) *wall bracket or rails.* (Shown: a heavy-duty floor-mounted jib crane lifting 15,000-lb steel coils.)

II–A

1. Can rotate to 360°.
2. Inexpensive and versatile.
3. Adapted to portable use by an outrigger-equipped wheeled stand.
4. Sometimes mounted on wheels and top and bottom rails along a wall or dock.
5. Heavy duty (hammer-head type) used for loads up to 350 tons.
6. *Used for:*
   —Serving individual work places in machine shops, etc., anywhere within its radius.
   —Loading and unloading carriers.
   —Handling molds in a foundry.
   —Supplementing an overhead travelling crane.

II–B. *Bridge crane*—a lifting device mounted on a bridge consisting of one or two horizontal girders, which are supported at each end by trucks riding on runways installed at right angles to the bridge. Runways are installed on building columns, overhead trusses, or frames. Lifting device moves along bridge while bridge moves along runway.

1. Covers any spot within the rectangular area over which the bridge travels, i.e., length of one bay.
2. Can be provided with cross-over to adjacent bay.
3. Provides 3-dimensional travel.
4. Designed as:
   —Top-running, where end trucks ride on top of runway tracks.
   —Bottom-running (underhung) where end trucks are suspended from lower flanges of runway tracks.
5. Hoist can also be top or bottom running.
6. Bottom-running usually limited to about 10 tons.
7. Bridge propelled by hand, chained gearing or power.
8. Two hoists (light and heavy duty) may be mounted on one crane.
9. Usually designed and built by specialist companies.
10. Does not interfere with work on floor.
11. Can reduce aisle space requirements.
12. Can reach areas otherwise not easily accessible.
13. Craneways can extend out of building.

14. Can be pendant or radio controlled from the floor.
15. *Used for:*
    —Low to medium volume.
    —Large, heavy and awkward objects.
    —Machine shops, foundries, steel mills, heavy assembly and repair shops.
    —Intermittent moves.
    —Warehousing and yard storage.
    —With attachments such as magnets, slings, grabs, and buckets, can handle an extremely wide range of loads.

II–C. *Monorail conveyor*—a handling system[2] on which loads are suspended from wheeled carriers or trolleys that usually roll along the top surface of the lower flange of the rail forming the overhead track, or in a similar fashion with other track shapes (shown with special carrier for moving items from warehouse to packing area).

1. Relatively low installation cost.
2. Low operating cost.
3. Little maintenance.
4. Track may be pipe, T, I, flat-bar or other formed structural shape.
5. Can be hand or motor propelled on both travel and lift.
6. Motor may be controlled by pendant switches, from integral cab, or automatically.
7. Removes traffic from floor.
8. Releases floor space.
9. Makes use of overhead space.
10. Easily extended.
11. Switches, spurs, transfer bridges, drop sections, swinging sections, cross-overs, turntables provide flexibility.

[2]Extracted from *Material Handling with Monorails* (Monorail Manufacturer's Association, 1967).

12. *Used for:*
    —Point-to-point moves.
    —Fixed path handling.
    —Low-volume moves.
    —Intermittent handling tasks.
    —Semi-live storage (on spur tracks).
    —Loading and unloading carriers.
    —Handling through processes (paint, bake, dry, plate, test).
    —Connecting buildings.
    —Pouring metal (From ladles suspended from monorail carrier).

II–D. *Stacker crane* —a device with a rigid upright mast or supports, suspended from a carriage, mounted on an overhead travelling (bridge) crane—or equivalent—and fitted with forks or a platform to permit it to place in or retrieve items from racks on either side of the aisle it traverses (see also II–E, Storage–retrieval unit).

1. Requires aisles only 4 to 6 in. wider than load.
2. Little if any obstruction of aisle when in raised position or out of aisle.
3. Serves both sides of aisle.
4. Saves both square feet and cubic feet.
5. Permits high selectivity.
6. Reduces order selection time.
7. Can be manned, pendant, electronic, card, or even computer controlled.
8. Can be transferred from aisle to aisle by transfer bridge.
9. Operator may ride in cab with load-carrying device.
10. Helps assure orderly storage operations.

11. Minimizes inventory control problems.
12. Usually requires one operator.
13. *Used for:*
    —Handling unit or containerized bulk loads.
    —Storage and warehousing operations (1 cu ft/ton in steel storage operations).
    —Adaptable to automatic warehousing operations.
    —With attachments can handle a wide variety of loads.
    —Excellent for long loads (metal bars, shapes, sheets, pipes, tubes, etc.).

II–E. *Storage–retrieval unit*—an outgrowth of the stacker-crane concept, commonly consisting of a mast or upright supports (a) suspended from a crane bridge, (b) fastened to rack-mounted rails, (c) suspended from a top-mounted monorail, (d) supported from the floor, on a wheeled truck, or (e) supported between top and bottom rails or tracks. Integral in the uprights are forks or a platen device that moves up and down the support, permitting it to place items in, or retrieve items from, the racks on either side of the relatively narrow aisle it traverses. It may be (a) captive, within the aisle it serves; (b) portable, by means of a transfer car at the end of the aisles, to permit it to be moved from one aisle to another; or (c) mobile, on its own wheels, with power to move to any racks located on a suitable running surface. (As shown, device operates between rows of racks by card, tape, or computer control.)

1. Permits random storage.
2. Provides high selectivity.
3. Requires minimum building, heat, light—only as required by product.
4. Can be manual, electronic, punched card, or computer controlled.
5. Permits automatic perpetual inventory.
6. Can guarantee first-in, first-out stock rotation.
7. Some can stack loads 2 deep, therefore increasing output by 25%.
8. Cost breakdown, approximately:
    —Racks, 40 to 50%.
    —Stacker and controls, 15 to 35%.
    —Conveyor and related equipment, 15 to 20%.
    (all, plus building cost)

9. About 5% are computer controlled (1975).
10. *Used for:*
    —Storage of materials and supplies.
    —Finished goods warehousing.
    —In-process storage.
    —Nearly any load on pallet or in container.

III. Industrial Trucks

III–A–1. *Four-wheel hand truck* —a rectangular load-carrying platform with 4 to 6 wheels, for manual pushing, usually by means of a rack or handle at one or both ends. Some have 2 larger wheels at center of platform for easy maneuverability.

1. May be fitted with box or other special body for variety of handling tasks.
2. Inexpensive.
3. Versatile.
4. *Used for:*
    —Manual handling of large loads.
    —Supplementing mechanical handling.
    —Low-frequency moves.
    —Low-volume movement.
    —Short distances.
    —Relatively light loads.
    —Temporary storage; in-process storage.
    —Handling awkward shapes.
    —Weak floors.
    —Small elevators.
    —Narrow aisles.
    —Crowded areas.

III–A–2. *Hand-lift (jack) truck*—essentially a wheeled platform that can be rolled under a pallet or skid, and equipped with a lifting device designed to raise loads just high enough to clear the floor and permit moving the load. Propulsion is by hand and lift is by hydraulic or other mechanism. Platform type is used for handling skids, and fork type for handling pallets.

1. Low cost.
2. Durable, minimum maintenance.
3. Light weight.

4. Compact.
5. Simple to operate.
6. Versatile.
7. *Used for:*
   —Loading or unloading carriers.
   —Supplementing powered trucks, spotting loads.
   —Moderate distances (50 to 200 ft).
   —Intermittent, low-frequency use.
   —Low volume moves.
   —Increasing utilization of powered equipment.
   —Captive use in a local area (economical).
   —Loading and unloading elevators.
   —Tight quarters; narrow aisles.

III–B–1. *Fork lift truck*—a self-loading, counterbalanced, self-propelled, wheeled vehicle, carrying an operator, and designed to carry a load on a fork (or other attachment) fastened to telescoping mast which is mounted ahead of the vehicle to permit lifting and stacking of loads.

1. May be powered by gasoline, diesel, battery, or LP gas engine.
2. Mast may be tilted foward or backward to facilitate loading and unloading.
3. Operator may ride in center or at back end of truck—or, with special attachments, on the lifting mechanism, with the load.
4. Operator may sit or stand.
5. Used with a wide variety of attachments to provide an extremely flexible and adaptable handling device.
6. Carries own power source—therefore useful away from power lines.
7. Wheels and tires can be provided for a variety of floor conditions or operating locations—wood, concrete, highway, yard.
8. Wide range of capabilities.
9. Electric type especially useful where reduced noise or no fumes are desired.
10. *Used for:*
    —Lifting, lowering, stacking, unstacking, loading, unloading, maneuvering.
    —Variable and flexible paths.
    —Medium to large units loads.
    —Uniform shaped loads.
    —Low to medium volume of material.
    —Intermittent moves.

III–B–1

III–B–2

III–B–2. *Platform truck (powered)*—a fixed-level, non-elevating, load-carrying powered industrial truck supporting the load on a platform. Smaller-capacity models are referred to as load carriers or burden carriers, for handling lighter loads and pick up use.

1. Straight frame—carrying surface above wheels.
2. Drop frame—carrying surface closer to floor with smaller wheels at end opposite power source.
3. Operator normally stands.
4. May be gas, diesel or battery powered.
5. Versatile.
6. Adaptable—with special chassis or attachments.
7. *Used for:*
   —Heavy loads.
   —Occasional use.
   —Relatively long loads with offset driver's seat—wall board, pipe, wood, etc.
   —Bulky loads.
   —Maintenance work—carrying tools to work or work to shop.
   Where platform lift or fork lift is not warranted by handling volume.

III–B–3. *Narrow-aisle truck*—in general, any one of several types of powered trucks capable of operating in a narrow aisle (6 ft down to 30 in. wide), by virtue of one of the following features: (a) *outriggers* (straddle *truck*) (III–B–7); (b) *extendable forks* (reach truck), with pantograph (III–B–5), sliding fork, or moving mast (III–B–6); (c) *four-way travel;* (d) *side loader design* (III–B–6); (e) *rotating mast;* (f) *side-motion* fork attachment; (g) *narrow* construction (III–B–4).

1. Uses less aisle space.
2. Relatively maneuverable.
3. Indoor models usually electric.
4. *Used for:*
   —Order picking.
   —Congested areas.

III–B–4. *Order-picking truck*—a truck designed or adapted to facilitate the order-picking process by making it easier for the operator to control the truck lift and travel while selecting orders. Variations include: (a) *walkie tractor,* with trailer; (b) *walkie tractor,* with pallet, rack, or container; (c) *walkie rider,* to permit easy leading through aisles and riding between picking locations; (d) *straddle,* with operating controls on a platform between the mast and forks—the vehicle is sometimes only 26 to 30 in. wide, with guide rollers or wheels on chassis which engage rails on bottom of racks to eliminate need for steering, and with operator riding on a platform so that he can pick from both sides of the narrow 36 in. (±) aisle (as shown).

III–B–5. *Reach truck*—a variation of the straddle truck in which the fork reaches out for the load on a pantograph-type device which permits the fork to travel forward to engage the load, lift it, and then retract it to the mast for travelling (the narrow-aisle truck shown can turn a right angle in a 6-ft aisle).

1. Uses less aisle space.
2. Maneuverable.
3. Weighs about 2,000 lb less than counter-balanced equivalent.
4. Some models can stack loads 2 deep on racks.
5. *Used for:*
   - —Warehousing.
   - —Narrow aisles.
   - —Tight quarters.
   - —Low floor load areas.
   - —Loading/unloading vehicles.

III–B–6. *Side-loading lift truck*—a powered, 4-wheel truck that picks up the load from the side by means of a mast, with forks centrally mounted on a bay at the center of the truck chassis. This arrangement permits the mast to travel transversely across the chassis and, in the extreme outboard position, lift loads within reach of the forks. The load is then retracted and placed on the chassis or deck for carrying.

1. No need to turn into the load.
2. May be gas, diesel, or battery powered.
3. Fast load pick up.

III–B–4

III–B–5

4. Very maneuverable, for size of load.
5. Quick, safe transport and stacking.
6. Can climb 15 to 20% grade.
7. One-man operation, even for most large loads.
8. Some have pneumatic tires for outdoor use.
9. Can travel on highway—about 25 mph.
10. Truck width equals load plus about 3 ft.
11. Heavy duty models have jacks for stabilizing while loading and unloading.
12. Can have guide rollers for use in narrow aisles.
13. *Used for:*
    —Narrow aisles.
    —Long loads, 40 ft or more.
    —Storing long loads (pipe, lumber, steel shapes, sheet metal, bar stock, etc.) on racks or in piles.
    —With attachment can tandem store 2 loads deep, therefore eliminating 2 aisles out of 5.
    —Yard storage work.

III–B–6

III–B–7. *Straddle truck (outrigger)*—a variation of the lift truck where vehicle is equipped with wheeled outrigger arms extending forward on the floor along either side of the load; arms perform the function of a counterbalance and keep the truck from overturning (shown with extension or reach fork).

1. Uses less aisle space.
2. Can be equipped with reach attachment.
3. Generally battery powered, therefore quiet, no fumes, low operating cost.
4. Operator rides, usually in standup position at rear of truck.

III–B–7

5. Maneuverable.
6. Relatively lightweight—about 2,000 lb less than counter-balanced truck of equal capacity.
7. *Used for:*
   —Narrow aisles ($6\frac{1}{2}$ ft).
   —Tight quarters.
   —Low capacity floors.
   —Loading elevators, trucks, etc.
   —Warehousing (stacking, picking, etc.)

III–B–8

III–B–8. *Tractor–trailer train*—a handling system consisting of a 3- or 4-wheeled, self-propelled vehicle designed for pulling loaded carts or trailers. Common versions are: (a) *rider type,* (b) *walkie type,* and (c) *electronically guided type.* (Shown: pulling two 4-wheel hand trucks; driver can also walk ahead and lead the truck by the control arm.)

1. Motive power is not tied up while trailers are being loaded or unloaded.
2. One power unit pulls several load units.
3. One tractor can keep three sets of trailers in use—one loading, one unloading, and one in transit, if loading and unloading labor is available at terminal points.
4. Low cost movement of large quantities.
5. Flexible route.
6. Three-wheeled-type tractor extremely maneuverable.
7. Electronically guided type requires no operator and follows a path described by a wire embedded in the floor (or a line, or tape) and can be programmed for automatic dropping of trailers and sounding horn or signalling arrival at intersections.
8. *Used for:*
   —Greater volume than fork-lift truck.
   —Distances of 200 to 300 ft and above.
   —Warehousing operations to haul loaded trailers from order picking area to order assembly area, and from receiving to storage.
   —Receiving and shipping, in conjunction with fork-lift truck for loading and unloading carriers.
   —Specific loads, with specially designed trailers, such as platforms, boxes, and racks.
   —Order picking where driver can also pick orders.
   —Collecting or delivering loads to a number of locations, as on a route.

III–B–9. *Walkie truck*—a term applied to many of the basic truck

types previously described, when designed to be power-propelled, and usually power-operated, but with the operator walking and operating the truck by means of controls on the handle. Designed to fill the gap between the "hand" trucks and the rider trucks, although some are designed as rider–walkies. Common types are: (a) *fork-lift,* (b) *narrow-aisle,* (c) *order-picker,* (d) *pallet,* (e) *platform,* (f) *reach,* (g) *skid,* (h) *stacker* (portable elevator), (i) *straddle* (outrigger), and (j) *tractor.* (Shown: jack-lift pallet type)

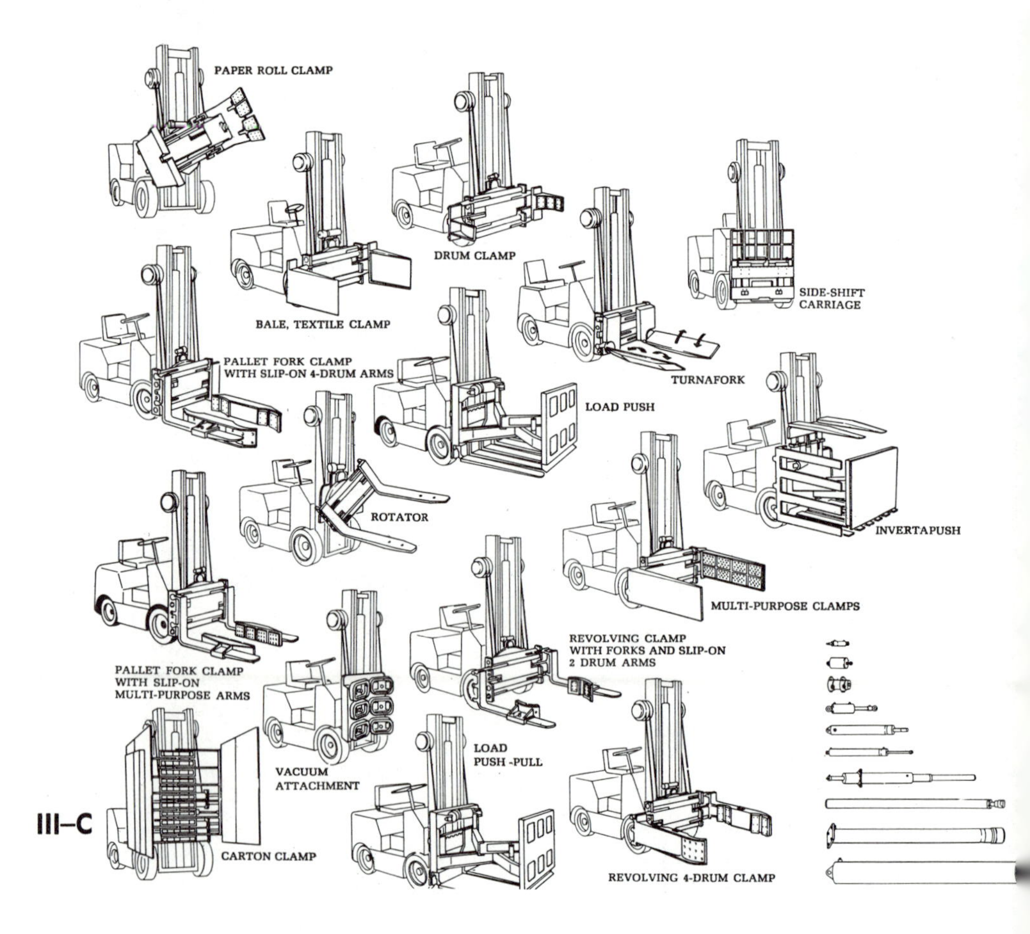

1. Smaller than rider-types.
2. Lighter than rider-types.
3. Slower than rider-types.
4. Usually battery powered.
5. Lower cost.
6. Adaptable.
7. Dependable.
8. *Used for* (see also rider types):
   —Lighter loads.
   —Hauls (up to 250 ft).
   —Congested areas.
   —Occasional use.
   —Servicing elevators.
   —Low floor load areas.
   —Supplement rider trucks.

III–C. ***Lift truck attachments***—any one of a large number (over 50) and wide variety of devices designed for attaching to lift trucks—to permit their adaptation to many tasks. Some of the more common types are listed below, and several are shown in the accompanying illustration.

IV. Auxiliary Equipment

IV–A–1. ***Dock board***—a specially designed platform device to bridge the gap between the edge of the dock and the carrier floor. Sometimes known as bridge plates. Carrier floors vary from 44 in. for rail cars, to 48 in. for pick-up trucks, to 52 in. for highway trucks, plus special bodies of even lower design.

1. Made in formed shape to provide strength and side guards.
2. Usually lightweight metal.
3. Often designed with loops to permit moving by fork truck.
4. Can be fastened to dock edge.
5. Some can be slid along a rail from one location to another.
6. Often have pins to lock lateral position.
7. Have non-skid surfaces.
8. May be flared for narrow (shallow) docks.
9. Should be carefully selected for intended use.

IV–A–1

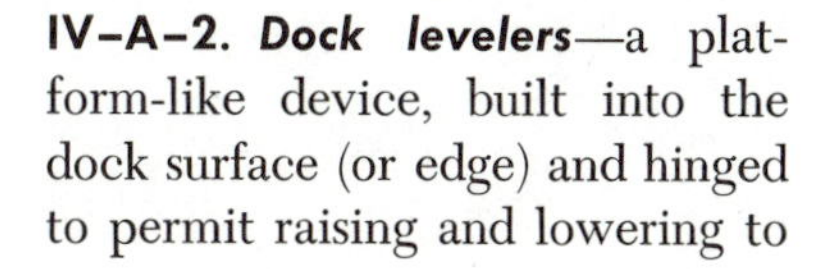

IV–A–2. ***Dock levelers***—a platform-like device, built into the dock surface (or edge) and hinged to permit raising and lowering to

IV–A–2

accommodate truck height when bridging the gap between dock and truck floor.

1. Permits extension of dock floor into carrier.
2. Adjusts up and down, left and right, or for vehicle tilt.
3. May be counterbalanced or hydraulically operated.
4. May be automatic; i.e., adjustment to truck initiated upon bumping by vehicle.
5. Has lip, to level out vehicle end of platform.

IV–B. *Shipping container*—a large container designed for consolidating material or goods to facilitate shipment by common carrier—usually 500 to 2500 cu ft; sometimes classified as pallet, cargo, and van containers (shown being unloaded from truck for transfer to flat car).

1. Common sizes: 8 by 8-ft cross section, and 10, 20, 30, or 40 ft long; pallet sizes: 50 to 100 cu ft.
2. Sealed by shipper.
3. Handled as a unit—usually direct to customer.
4. Reduced pilferage, damage, contamination, etc.
5. Reduced handling time (vs. individual items).
6. Reduced packaging and packing costs.
7. Lower insurance rates.
8. May be made of metal, wood, plastic, rubber, etc.
9. Some are collapsible.
10. Some have drop bottoms.
11. Many are designed for attachment to carriers.
12. *Used for:*
    —Over the road, rail air shipment.
    —Bulk, liquid, or unit materials.

IV–C. *Shop container*—any one of a number of varieties of relatively small containers for handling

material through the shop. (Shown: steel tote boxes.)

1. May be made of sheet metal, wire, wood, fiber board, corrugated board, or plastic.
2. Permit easy handling of small parts.
3. May be consolidated into unit loads.
4. May be easily stacked or placed in racks.
5. Custom liners or inserts permit safe handling of odd shaped or fragile items.
6. Some are stackable.
7. *Used for:*
   —Consolidating items.
   —Elimination of smaller containers such as bags, with attendant handling.
   —Increasing use of storage space.
   —Reducing freight cost.

IV–D–1. *Pallet*—a horizontal platform device used as a base for assembling storing and handling materials as a unit load. Usually consists of two flat surfaces, separated by three stringers (III–B–5 and 9). (Shown: pallets stored at the far left, and in use; note stacking frames that can be mounted for use with crushable loads.)

IV–D–1

1. May be expendable, general purpose, or special purpose.
2. May be single or double faced.
3. May be flush stringer, single or double wing.
4. May be one-way, two-way, or four-way entry.
5. Made of wood, plywood, metals, corrugated, plastic, etc.
6. Protects goods being moved from damage, pilferage, etc.
7. Facilitates inventorying.
8. Promotes cleanliness and good housekeeping.
9. Keeps material off floor, therefore easier to handle.
10. *Used for:*
    —Fork-truck-based systems.
    —Unitizing items.
    —Utilizing building cube.
    —Increasing load size.
    —Reducing handling of individual items.
    —Minimizing packaging of individual items.

IV–D–2. *Skid*—a load-carrying platform supported from the floor by two parallel stringers or supports. (See also Figures III–A–2 and B–7.)

1. Similar to the pallet (*q.v.*).

IV–D–2

2. Usually heavier and stronger than pallets.
3. Usually for use with heavier loads.
4. May be wood or metal.
5. May be built into a shipping container for ease of handling.
6. Used for same general purposes as pallets (*q.v.*).
7. Semi-live skids have 2 wheels and 2 legs for use with skid jack.

IV–D–3

IV–D–3. *Rack*—a framework designed to facilitate the storage of loads, usually consisting of upright columns and horizontal members for supporting the loads, and diagonal bracing for stability. (Shown: 8-high pallet racks of structural steel, being served by a side-loading truck.)

1. May be classified as *Selective:*
   —Bolted
   —Lock-fit
   —Palletless
   —Cantilever
   —Bar stock
   —A-frame
   —Custom
   or *Bulk:*
   —Drive-in
   —Drive-through
   —Live
   or *Portable:*
   —Integral unit
   —Rigid
   —Knock-down
   —Collapsible
   —Pallet stacking frame
   —Bolt-on (special or adapted pallet)
   —Snap fit (standard pallet)
   —Independent of pallet
2. Made of metal, wood, pipe, etc.
3. May be fixed or adjustable in shelf height.
4. Usually built for pallets, but may be used or adapted for skids, rolls, drums, reels, bars, boxes, etc.

5. May have shelves for storage of loads, but may be designed for drive-in or drive-through applications.
6. Facilitates inventory taking.
7. Rugged; minimum maintenance.
8. Live racks are designed for loads to flow to the unloading position.
9. Cantilever racks best for long items.
10. *Used for:*
    —Increasing utilization of storage space.
    —Increasing selectivity of goods stored.
    —Protecting goods.
    —Control of inventory.
    —Improving housekeeping.

Conclusion

This chapter has presented material handling equipment concepts, including the place of equipment in the handling situation, basic equipment types and basic handling systems. This is followed by a selection of more common types of equipment used in the typical manufacturing enterprise—as well as in many other types of businesses. Further information can be found in the sources listed in the bibliography. The next chapter will deal with the problem of designing the handling system into the layout, and then selecting the equipment.

Questions

1. Why doesn't the typical analyst have to be concerned with all the types of material handling equipment?
2. Why can a single type of equipment seldom solve a material handling problem?
3. In applying the material handling equation to equipment selection, the analyst is concerned with the (a) material ______, (b) move ______, and (c) method ______.
4. How does the work simplification approach apply to the material handling equipment problem?
5. Briefly identify the four basic types of handling equipment. Indicate several examples of each.
6. Identify the several basic handling systems, and give an example or two of each.
7. Suggest several handling tasks demanding a specific type of equipment.
8. Suggest tasks which could be equitably solved by several types of equipment. Name them and describe.
9. Suggest tasks requiring specially designed equipment, that is, not off-the-shelf.
10. Suggest handling situations requiring: (a) no equipment; (b) non-powered equipment; (c) manual equipment.
11. Discuss situations you may have seen, where a system other than the one being used might be appropriate.

Exercises

Visit a local concern and:

A. Identify equipment types in use.

B. Make a process chart of an item from delivery vehicle to its point of use in a product, and suggest method(s) for implementing each move.

16

Designing the Handling System

Based on the background acquired in the first 15 chapters, it is now time to design the handling system for implementing the material flow. The emphasis is to point out again the need for a great amount of preliminary work before enough information has been developed to design a handling system into the layout. Often, unfortunately, this is not done—except for some of the larger and more production-oriented enterprises. In most instances, management personnel are just not well enough informed on the significance of planned handling methods and their important contribution to efficient production and high productivity. Managements of progressive enterprises should insist on designed handling systems—not just a collection of equipment, hastily obtained, and thrown in at the last minute.

In terms of the overall facility design process, it should be pointed out that potential handling systems and equipment alternatives have very likely been given serious consideration during the design stages previously carried out. It is at this point in the planning process that ideas must be crystallized, and preparations made for designing or procuring any handling equipment required.

Designing the Handling System into the Layout

Efficient material handling does not just happen. It comes about only through a careful analysis and evaluation of the entire operation, with the objective of implementing a well-planned material flow pattern by means of appropriate methods and equipment. If there is such a thing as a procedure for doing this, it would be somewhat as reflected in the headings that outline the rest of this chapter.

1. *Understand the system concept.*[1]

2. *Review the system design criteria*—and try to keep them in mind while designing the handling system:

[1]For details on the systems concept and procedure, see Apple, *Material Handling Systems Design*, pp. 278–89.

a. Increased production; productivity
b. Cost reduction
c. Improved safety
d. Storage capacity
e. Expandability
f. Minimum product damage
g. Optimum control
h. Improved working conditions
i. Improved quality
j. Reduced dependency on labor
k. Dependability
l. Easy, low maintenance
m. Back-up capability
n. Continuous Flow
o. Flexibility
p. Information handling capability
q. Optimum space requirements
r. Standardized components
s. Adaptability
t. Combine handling with other functions
u. Optimize material flow
v. Handle as large a load as possible
w. Judicious use of mechanization
x. Minimum down-time
y. Planned for progressive mechanization
z. Make full use of equipment
aa. Meet system objectives
bb. Improved customer service
cc. Comply with present (and future) laws, codes, regulations
dd. Be compatible with balance of plant

3. *Establish objective of the handling system*—to be sure it is properly stated and compatible with the overall facility objectives and plans.

4. *Obtain data required*—as discussed in Chapter 3.

5. *Develop preliminary flow patterns*—as discussed in Chapter 5.

a. Check vs. flow pattern criteria on p. 108.
b. Check with Material Flow Design Evaluation Sheet on p. 118.

6. *Identify activities and plan activity interrelationships*—as in Chapter 8.

7. *Determine space requirements and make area allocation*—as in Chapter 11.

8. *Establish material flow pattern*—check as in 5a and 5b above, and pp. 113–20.

9. *Identify and document move requirements*

a. Path: (1) *Process Chart* p. 136; (2) *Flow Process Chart* p. 140; (3) *Flow Diagram* p. 138.
b. Characteristics: (1) M. H. equation—Chapter 15; (2) Primary characteristics (scope, source and destination, distance, path, frequency, speed, rate, etc.)

10. *Analyze material characteristics*

a. Quantity
b. Unit volume
c. Unit weight
d. Type, form
e. Uniformity
f. Properties

11. *Establish desired or existing building characteristics.*

a. Loading level
b. Unloading level
c. Column spacing
d. Clear height
e. Running surface

Table 16-1. Equipment Selection Guidelines

No Equipment	Equipment		
1. Low volume 2. Low rate of flow 3. Non-uniform flow 4. Small items 5. Short distances 6. Limited area 7. Infrequent handling 8. Occasional handling 9. Varying paths 10. Small percentage of time spent in handling 11. Little cost attributable to handling 12. Complex flow pattern 13. Obstacles in flow path 14. No alternative	A. *General* 1. Loads over 50 lb. (or other predetermined limit) 2. Two-man handling tasks 3. Travel time exceeds lifting and placing time 4. Unused space above floor B. *Manual* 1. Relatively light loads 2. Limited volume 3. Physical restrictions 4. Equipment equally useful for storage 5. Limited capital 6. Minimum maintenance facilities 7. Stand-by use 8. Wide variety of small or infrequent handling tasks (requiring flexibility of manual equipment) 9. Efficiency of manual methods relatively high 10. Low cost operation 11. Complex flow pattern	C. *Mechanized* 1. High volume 2. Continuous movement necessary 3. Much handling required 4. Direct labor performing handling tasks 5. Need for controlled rate of flow 6. Increased capacity 7. Hazardous materials 8. Operators waiting for materials 9. Manual handling undesirable 10. Production bottlenecks 11. Unit loads practicable 12. Dependable handling necessary 13. Limited space 14. Wasted cube 15. Flexible types of equipment adaptable to handling tasks	D. *Automated* 1. High volume 2. High percentage of han- dling in operation 3. Uniform product; material 4. Stable product 5. Practicable to combin movement with produc tion or other operation 6. Maintain process contro 7. Reduce cost 8. Limited number of path 9. Moves relatively fixed 10. Relatively fixed materia flow pattern

12. *Study basic handling systems*—as in Chapter 15.

a. *Equipment* oriented
b. *Material* (load) oriented
c. *Method* oriented
d. *Production* oriented
e. *Function* oriented

13. *Determine feasibility and desirability of mechanization.*

Some of the many factors determining the feasibility and level of mechanization are shown in Table 16-1, as general guidelines.

Before a decision can be made on equipment to be used for a specific situation, consideration must be given to determining a more specific degree of mechanization. However, it may not necessarily be the same for every handling operation in the process. Only by means of a complete analysis can the optimum level be reached for each handling task in the system.

The concept of *level of mechanization* has been adapted for application to material handling, and is shown in Table 16-2. Although the chart may not resolve the problem, it does show the thinking process involved in making the decision on the appropriate level of handling mechanization.

The selection of the level of mechanization becomes a problem of economic

Table 16-2. A Summary of the Levels of Mechanization of Material Handling Equipment

POWER	CONTROL	LEVEL OF MECH.	DESCRIPTION OF LEVELS OF MECHANIZATION OF MATERIAL HANDLING
ELECTRIC OR INTERNAL COMBUSTION ENGINE	COMPUTER	10	Automated-on-line, self regulating, instructions & control by "central" computer
		9	Mechanized-with instructions & control by integrated mini-computer
	ELEC-TRON.	8	Mechanized-with control by manual insertion of tape, cards, etc.
	PUSH BUTT.	7	Power propulsion and/or lift-remote control by push-button, switch, etc.
	MANUAL	6	Power propulsion and/or lift, rider control at site
		5	Power propulsion, walkie, conveyor, etc.
GRAV-ITY		4	Level or gravity conveyor, chute, etc. equipment indicated below
MANUAL		3	Hand trucks with shelving, racks, etc. to permit manual access to height
		2	Hand trucks, carts, etc.-equipment carries load, man propels equipment
		1	Hand—man carries load

feasibility. After the total handling system has been conceptualized, it is necessary to make engineering economic analyses of each move.

14. *Relate the material characteristics and move requirements to systems and equipment capabilities.*

At this point, the *Material Handling Analysis Re-cap Sheet,* Figure 16-1, may prove helpful in documenting *Steps 9* to *13*, at least partially. It, too, is based on the three major aspects of a material handling problem—the *material,* the *move,* and the *method*—and selected sub-factors under each phase, many of which must be taken into consideration in solving the problem. The *Material Handling*

			Item	1	2	3	4
MATERIAL	1. Type		UNIT, BULK, LIQUID, GAS	U			
	2. Characteristics		HOW RECEIVED	Trk			
			DIMENSIONS	10×16 ×12			
			WEIGHT/UNIT	55			
	3. Quantity		ANNUAL USAGE	20 M			
			MAXIMUM INVENT.	5 M			
			QUANT./MOVE	1600			
MOVE	4. Source and Destination		EXTERNAL–INTERNAL	E			
			LOAD/UNLOAD LEVEL	Trk.			
			LOAD/UNLOAD METHOD	F.T.			
	5. Logistics	Scope	POINT TO POINT	√			
			AREA	30 × 100			
		Route	PLANE	Str.			
			PROFILE	Hor.			
			PATH	VP-VP			
			LEVEL	FL.			
	6. Characteristics		DISTANCE	80'			
			FREQUENCY	1/Mo.			
			RATE	360/Hr			
			SPEED	—			
			MOTION TYPE	INT.			
			TRAFFIC	—			
			ENVIRONMENT	—			
			% TRANSP.	80			
			% HANDLING	20			
			SEQUENCE	—			
	7. Type		TRANSP., CONVEY, MANEUVER, ELEVATE, LOWER, POSITION, TRANSFER	T/M			
METHOD	8. Handling Unit		LOAD SUPPORT METHOD	Spt.			
			CONTAINER	Pall			
			WEIGHT	2M			
			ITEMS/HANDL. UNIT	36			
			HANDL. UNITS PER TOT. QUANTITY	44			
	9. Equipment		TYPE	F.T.			
			CAPACITY	4M			
			NO. OF UNITS	—			
	10. Manpower		HOURS PER YEAR	54			
			NUMBER OF MEN	—			
11. Physical Restrictions			COLUMN SPACING	—			
			CLEAR HEIGHT	24			
			AISLE WIDTH	9'6"			
			FLOOR LOAD CAPACITY	—			
			OVERHEAD LOAD CAPACITY	—			
			FLOOR CONDITION	Gd.			
			ELEVATOR CAPACITY	—			

(Column headings 1–4: MOVE NUMBER (from PROCESS CHART))

Figure 16–1—Material handling analysis re-cap sheet.

Analysis Re-cap Sheet, is an attempt to do just this. It will not select the equipment, but if the material handling analyst has the proper background, the *Re-cap Sheet—while* it is being filled in—will guide his thinking toward the equipment type by focusing his attention on the kinds of equipment (with which he is familiar) that come closest to matching the characteristics of the *material* and the requirements of the *move.*

15. *Make preliminary selection of basic handling systems and equipment types.*

In addition to the above consideration of material handling problem factors and levels of mechanization, the equipment selection process still depends largely on the analyst's knowledge of handling equipment. Table 15-1 showed some of the interrelationships between the factors and the general capabilities of the basic types of handling equipment. It represents the thinking process the analyst uses in applying his experience to the task of equipment selection, to determine the type best suited to each individual move or to the process as a whole.

16. *Narrow the choice.*

Such charts as Tables 15-1 and 16-1 cannot accurately depict relationships between the 500-plus equipment types and the 50 to 60 problem characteristics, but they will serve as guides to the selection of general types of equipment. In addition to the charts, there are other sources of help for narrowing down the alternatives to the most appropriate type or types, listed in the bibliography. As the reader views the equipment selection process, he will gain an insight into the complexity of the problem. He should not become discouraged, however, since a reasonable amount of practical experience will fairly well qualify him in his particular enterprise.

In narrowing the choice, the analyst might review each *move* in terms of equipment capabilities required, and review the layout in terms of equipment alternatives (and vice versa). Where the handling problem is more complex, or more of a system problem, he may want to:

a. Conceptualize system possibilities.
b. Structure alternative systems.
c. Simulate potential systems.
d. Select feasible system.

17. *Evaluate the alternatives.*

Having narrowed the choice to a relatively few types or pieces of equipment, it is now necessary to carefully evaluate the alternatives. This consists of several approaches to evaluation. First, there is the qualitative evaluation, by reviewing the objective of a method or system, and such equipment selection criteria as the following—does the method, equipment, or system:

a. Fit into the handling system?
b. Combine handling with other functions (production, storage, inspection, packing, etc.)?

c. Optimize material flow?
d. Seem as simple as practicable?
e. Utilize gravity wherever possible?
f. Require a minimum of space?
g. Handle as large a load as practical?
h. Make the move safely, in terms of both manpower and material?
i. Use mechanization judiciously?
j. Have flexibility, adaptability?
k. Have a low dead-weight to pay-load ratio?
l. Utilize a minimum of operator time?
m. Require a minimum of loading, unloading, and rehandling?
n. Call for as little maintenance, repair, power, and fuel as possible?
o. Have a long, useful life?
p. Have the capability of capacity utilization?
q. Perform the handling operation efficiently and economically?

Each alternative under consideration can be informally rated against each of the above criteria.

Other methods of evaluation[2] would include cost comparisons and a more formal examination of the intangible aspects, which may outweigh the cost factors.

18. *Arrive at a decision*—based on the above analysis, and involving:

a. Move requirements
b. Material characteristics
c. Equipment capabilities
d. Cost and economic factors
e. Intangible factors
f. System objectives
g. Selection criteria
h. Experience
i. Judgment
j. Common sense

Rather obviously, the above is not as easy as it sounds. Nevertheless, the problem must be resolved; and of course, the analyst finds it much easier as he gains in experience. In fact, some experts can merely look at a problem and almost instantaneously resolve the many factors—or at least the important ones—in their heads, and come up with the answer.

19. *Check the selection for compatibility.*

One of the criteria on the preceding list was concerned with fitting the equipment under consideration into the overall system. This suggests that, during the selection process, constant attention be given to the compatibility of prospective equipment types with other equipment in use, or to be used. It must be remembered that the handling system is often a composite solution to a number of interrelated handling problems. Each segment of the solution must be carefully integrated with the other segments—and a major aspect of the overall solution is the handling equipment.

20. *Prepare Performance Specifications.*

[2] For further details, see Apple, pp. 366–67, and chs. 13, 14, 15.

Having selected the equipment most appropriate to the problem solution, the next step is detailing the performance specifications to insure that competitive bids will be on the same or equivalent pieces of equipment.

A careful study of the type of equipment selected will be necessary to familiarize the analyst with what details should be specified, and the permissible variations he can allow. If the equipment being specified requires design work, it should be done at this point.

In complex systems, where several suppliers may be working together, and much design work is necessary, extra care must be taken in stating the first broad system specifications. If vendors are not allowed enough freedom in designing their portions of the system, it is possible that the analyst may have limited the flexibility and imagination of the vendor's designers.

21. *Procure the equipment.*

Once the specifications have been established, the procurement process can begin. If the analyst does not have a knowledge of sources of supply, he should consult the trade magazines and directories. After prospective suppliers have been selected, formal requests for quotations are issued, accompanied by specifications and drawings. Quotations received should be in writing, and should show evidence of including all items covered by the specifications. Upon their receipt, the engineer should evaluate them carefully. Figure 16-2 shows one format commonly used in comparing specifications and technical details of alternative pieces of equipment.

22. *Actual implementation.*

Rather obviously, the actual implementation process requires long and careful planning, in order to insure that all details are taken into consideration. Although some may have been considered earlier, they will include:

a. Developing a tentative budget.
b. Preparing a justification report or presentation.
c. Obtaining final approval.
d. Vendor follow-up.
e. Problems in delivery of equipment.
f. Scheduling the installation—involving the time frame (PERT), manpower, trades, etc.
g. Supervising installation.
h. Orientation and training of personnel involved.
i. Human relations considerations.
j. Start-up and de-bugging of the installation.
k. Auditing the performance—at installation, as well as periodically in the future.
l. Formal acceptance—for payment approval.

The above list embodies many important details that should be carefully reviewed and thought through to make sure nothing important is overlooked.

	BIDDER A		BIDDER B	
GENERAL:				
1. Capacity	20/5 Ton		20/5 Ton	
2. Span	90'0"		90'0"	
3. Lift	35'0"		35'0"	
4. Total Net Weight	88,000 lb		77,000 lb	
HOISTS:	**Main Hoist**	**Aux. Hoist**	**Main Hoist**	**Aux. Hoist**
1. Hoist Speed	25 FPM	54 FPM	25 FPM	54 FPM
2. HP of Hoist Motor and Rating	40 HP 30 M.-55° C.	20 HP	40 HP	20 HP
3. Computed HP Required	39 HP	19.7 HP	39 HP	19.7 HP
4. Number and Parts of Rope	8 parts 5/8"	4 - 1/2"	8 - 9/16"	4 - 3/8"
5. Type of Wire Rope	6/19 Improved Plow Steel	6/19 Improved Plow Steel	6/19 Improved Plow Steel	6/19 Improved Plow Steel
6. Diameter of Hoist Drum	20"	15"	15"	10"
7. Material in Drum	Welded Steel	Cast Iron	Cast Iron	Cast Iron
8. Type of Bearing	Hyatt Roller	Hyatt Roller	Hyatt Roller	Hyatt Roller
9. Make & Type of Gears & Pinions	Spur Gears Welded Steel Forged Steel Pinions	Spur Gears Welded Steel Forged Steel Pinions	Spur Gears Welded Steel Forged Steel Pinions	Spur Gears Welded Steel Forged Steel Pinions
10 Material of Gears and Pinions	Gears SAE 8630 Pin. SAE 8742 Heat Treated Hardened	Gears SAE 8630 Pi. SAE 8742 Heat Treated Hardened	Gears SAE 1040 Pinions SAE 1045	Gears SAE 1040 Pinions SAE 1045
TROLLEY:				
1. Trolley Speed	200 FPM		125 FPM	
2. HP of Trolley Motor and Rating	7½ HP		3 HP	
3. Computed Running HP Required	3.5 HP		1.96 HP	
4. Service Factor Used	2.14		1.53 HP	
5. Diameter of Trolley Wheels	13½"		12"	
6. Spread of Trolley	8'0"		7'0"	
7. Type of Bearings	Hyatt 5214		Hyatt 5212	
8. Make & Type of Gears & Pinions	Welded Spur Gears Forged Steel Pinions		Welded Spur Gears Forged Steel Pinions	
9. Material of Gears and Pinions	Gears SAE 8630 Pinions SAE 8742 Heat Treated Hardened		Gears SAE 1040 Pinions SAE 1045	
10. Weight of Trolley complete	18,400 lbs.		13,000 lbs.	
BRIDGE:				
1. Bridge Speed	300 FPM		300 FPM	
2. HP of Bridge Motor and Rating	25 HP		20 HP	
3. Computed Running HP Required	8.85		8.75	
4. Service Factor Used	2.83		2.29	
5. Girder Section at Center of Span	Top-28 x ¾" Bot.-28 x ½" Webs-60 x ¼"		Top-28 x 5/8" Bot. - 28 x 3/8" Webs - 54 x 5/16"	
6. Number & Dia. of Bridge Wheels	4-24"		4-21"	
7. Maximum Wheel Load	50,600 lb		44,000 lb	
8. Wheel Base	13'8"		12'3"	
ELECTRICAL:				
Control –				
Make & Type of Hoist Control	Semi-Magnetic Drum		Drum Control	
Make & Type of Aux. Hoist Control	Semi-Magnetic Drum		Drum Control	
Make & Type of Hoist Limit Switches	Motor Circuit		Control Circuit	
Make & Type of Trolley Control	Drum		Drum	
Make & Type of Bridge Control	Drum		Drum	
Make & Type of Overload Protection	Overload Relays		Fuses	

Figure 16-2—Form for comparative analysis of overhead crane bids. (From Frank M. Blum, "How to Evaluate Crane Bids," Courtesy Harnischfeger Corp., Milwaukee.)

Conclusion

This chapter has presented an orderly approach to one of industry's most perplexing problems—the design and implementation of a handling system to convert a static material flow pattern into a dynamic flow of material, in an efficient and economical manner. As stated previously this process is often a nebulous and complex task. In order to avoid the mistakes of the snap judgment or off-the-cuff decisions, a rather detailed procedure has been described. Although it may appear to be a forbidding process, the only proper approach to such complex situations is an orderly one. The procedure itself, however, will neither solve any problems, nor select any equipment. And since each problem differs from others, it will be found necessary to vary the procedure to better fit the problem at hand. Constant use of the procedure outlined here will make it easier for the analyst to cope with such problems. (Further details on the topics covered in this chapter will be found in *Material Handling Systems Design,* Chapters 5, 7, 9, 11, 13, 14, and 15.) Now that the material handling system has been designed, the following chapter will continue with the layout process, and deal with the problems of actually constructing a scale layout of the facility.

Questions

1. Why is it important to design a handling method or system?
2. Name some criteria to keep in mind in designing a method or system.
3. Briefly outline the overall procedure in designing a handling system.
4. What problem characteristics suggest (a) no equipment, (b) manual equipment, (c) mechanized equipment, (d) automated equipment.
5. Where can the analyst obtain additional information on handling equipment?
6. Name some of the criteria for selecting handling equipment.
7. After selecting the equipment, what is involved in implementing the solution?

Exercises

Copy (or revise) Figure 16-1, and fill it out for a sequence of moves in a local situation (or use the one you documented for Chapter 15, *Exercise B*).

17

Constructing the Layout

The time has finally arrived, after sixteen chapters describing preliminary planning and related details, for the actual construction of the layout. Because of the careful preplanning done in previous steps, the actual making of the layout should be a relatively easy task. The layout planner should be warned, however, that there still remain many important decisions to be made and details to be worked out.

The general objective, at this point, is the coordination of all previous planning activities and the melding of preliminary plans into a master layout. This consists primarily of transferring the *Area Allocation Diagram* to a master layout grid sheet, and then working out the details of the individual production departments, auxiliary services, and work places—along with the material handling plans. Normally, several rough layouts may be partially developed, as a basis for analysis and evaluation, before detailed work is started on the final layout. Preliminary layouts may have been done in the *Area Allocation Diagram* phase, as shown in Figure 17-1. Once the area allocation has been agreed upon, work can be started on the master layout—the original—which will be kept as a record of the layout of the area under consideration. Prints or copies will be made and issued to those who require them for installation, plant operation, or re-evaluation.

Methods of Constructing the Layout

There are four common methods of constructing or representing the layout. They are:

1. *Drawn in the conventional drafting manner* on drawing or tracing paper.
2. *Constructed with two-dimensional templates* (cut-outs, to scale, representing the shape and size of each piece of equipment), mounted on a suitable base or grid depicting the building drawing.
3. *Constructed with three-dimensional scale models* in place of templates.
4. *Constructed with a combination of scale models and templates,* for ease of reproduction.

Unfortunately, entirely too many layouts are still being made by Methods 1 and 2, using construction paper and paste. Such methods are extremely inefficient, as the

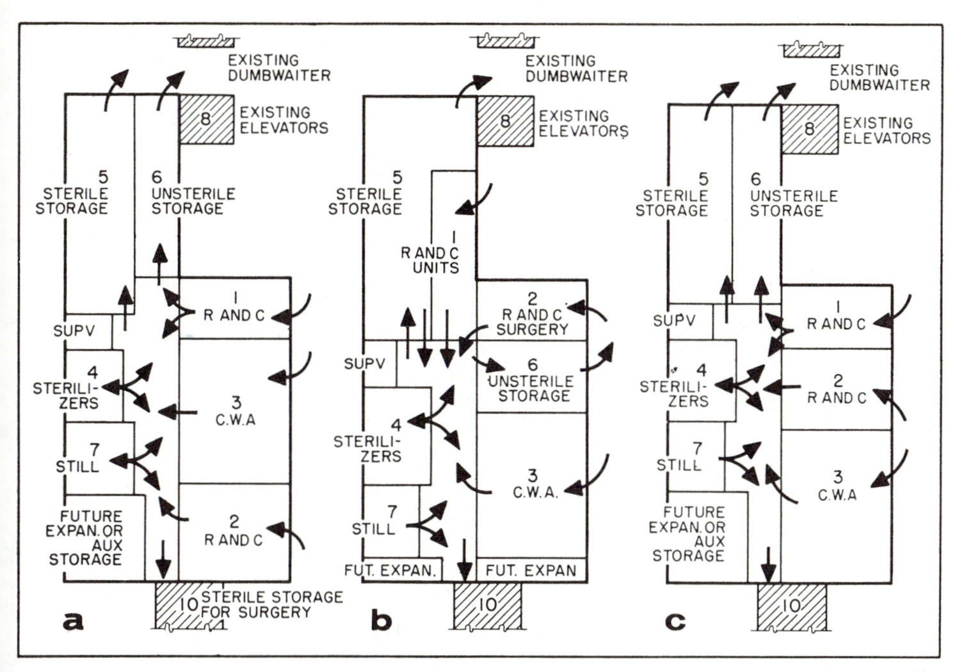

Figure 17–1—Preliminary alternatives for evaluation prior to developing master layout. (Courtesy of Plant Engineering Magazine.)

former makes it difficult to make (and discourages) adjustments and alterations, without using more eraser than pencil. Both make poor use of the analyst's time, in draftsman's work.

Figures 17-2, 17-3, 17-4, and 17-5 illustrate the basic techniques listed above. It will be noted that Figures 17-4 and 17-5 are for the same area and complement each other. In general, the method selected for a specific project should be based on the characteristics tabulated in Table 17-1, which compares four methods. The probable conclusion to be drawn is that *Method IV* is preferable when its cost can be justified. This may not be as difficult as it may seem, when one calculates that the cost of a complete scale model and matching template layout amounts to about $0.07/sq ft of actual plant area for all necessary materials. This is less than 1 per cent of the cost of construction. The architect's fee will range from 5 to 7 per cent. For example, on a 100,000-sq-ft building, the comparison might be somewhat as follows:

Building—100,000 sq ft at $10/sq ft—	$1,000,000
Architect's fee—$1,000,000 × 6%—	$60,000
Scale model—100,000 sq ft at 6¢/sq ft—	$6,000
Template layout—100,000 sq ft at 1¢/sq ft—	$1,000

It should be recognized that the models and templates are as much a tool of the plant layout designer as a screwdriver is to a production operator. Would management ask the operator to do without, or to make his own screwdriver?

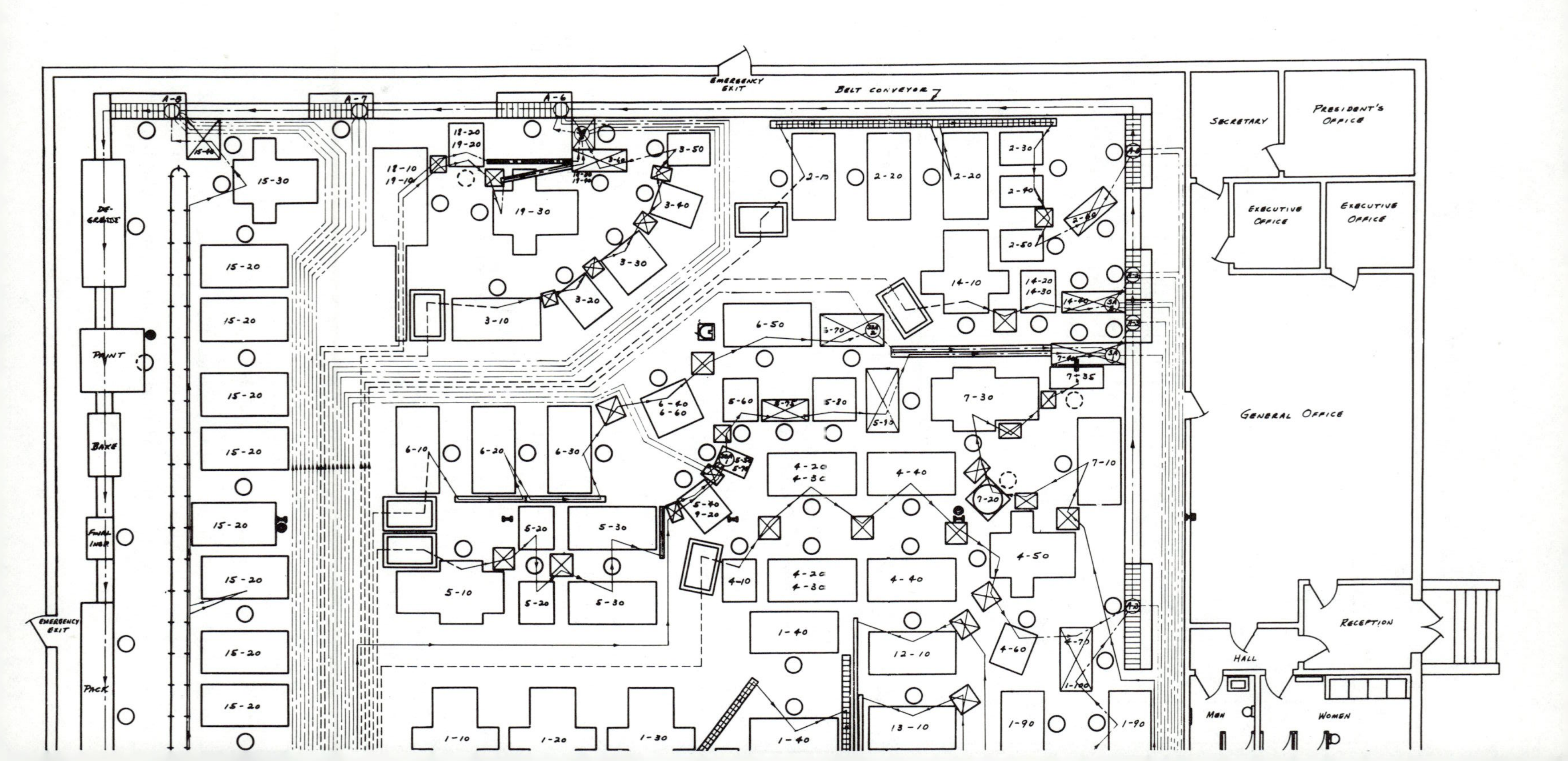

PRESIDENT'S OFFICE
SECRETARY
EXECUTIVE OFFICE
EXECUTIVE OFFICE
GENERAL OFFICE
RECEPTION
HALL
MEN
WOMEN
BELT CONVEYOR
EMERGENCY EXIT
EMERGENCY EXIT
DE-GREASE
PAINT
BAKE
FINAL INSP
PACK

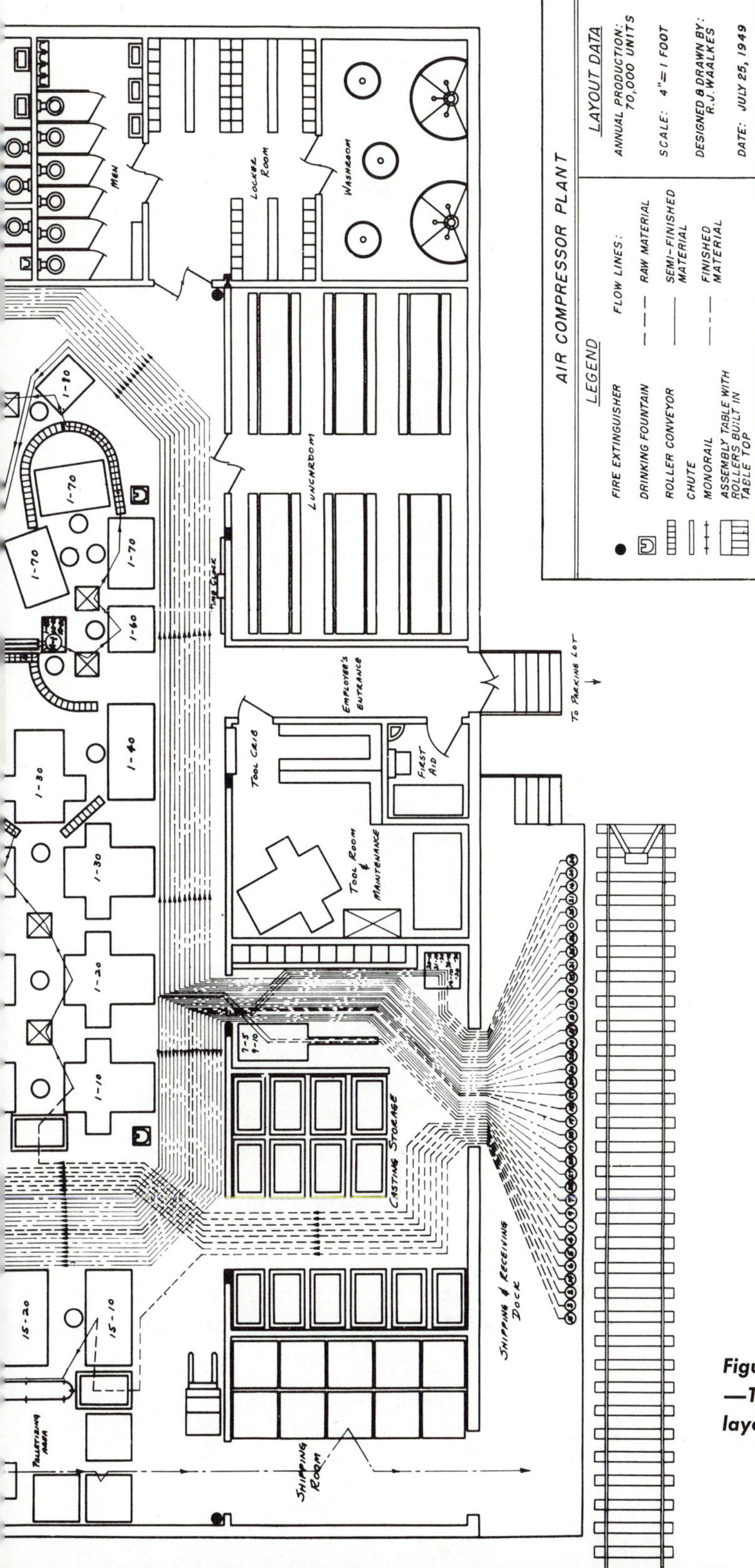

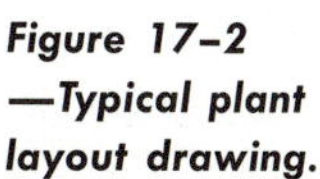

Figure 17-2 —Typical plant layout drawing.

The "plain" drawing can be properly used:

1. For simple projects.
2. For preliminary sketching.
3. When "in a hurry."
4. When no reproduction equipment is available or necessary.

Plant Layout Tools and Techniques

The basic materials used in plant layout work are as follows.

Base material upon which the layout is constructed. Common materials for this are:

1. Transparent plastic sheet (0.0075 to 0.0100 inch), with grid lines.
2. Transparent plastic board ($\frac{1}{4}$ in.), with grid lines scribed on back surface.
3. Metal faced plywood with appropriate finish on metal surface.
4. Sheet metal, plywood, or composition board with appropriate finish.
5. Drawing or tracing paper.

Tapes, etc., to represent architectural and other details, commercially available as follows:

1. Lines—solid, broken, dotted, in various colors.
2. Walls—in accurate architectural symbols.
3. Aisles—cross-hatched and labeled, in various widths.
4. Arrows—continuous, to represent flow lines.
5. Building columns, posts, stairways.
6. Operators—2-ft circles or sketches.
7. Conveyors—roller, wheel, slat, belt, overhead, drag-chain, etc.
8. Railroad tracks.
9. Symbols—numbers, letters, power and telephone outlets, telephones, hydraulic and pneumatic connections.
10. Pallets, gondolas, drums, skids, tables, benches, etc.

Templates to represent physical facilities. These can be hand drawn and reproduced on plastic film or sheet, but are often those commercially available on sheet acetate, Mylar, etc., in:

1. Black and white outline style.
2. Black and white negative style (also in color).

Those of film or sheet can be fastened to the grid sheet with double-coated pressure sensitive tape, rubber cement, or have pressure sensitive adhesive preapplied and protected with backing paper.

3. Black and white, color sheet plastic—with magnet attached.
4. $\frac{1}{8}$-inch plastic—with magnet inserted for use on metal surfaces.

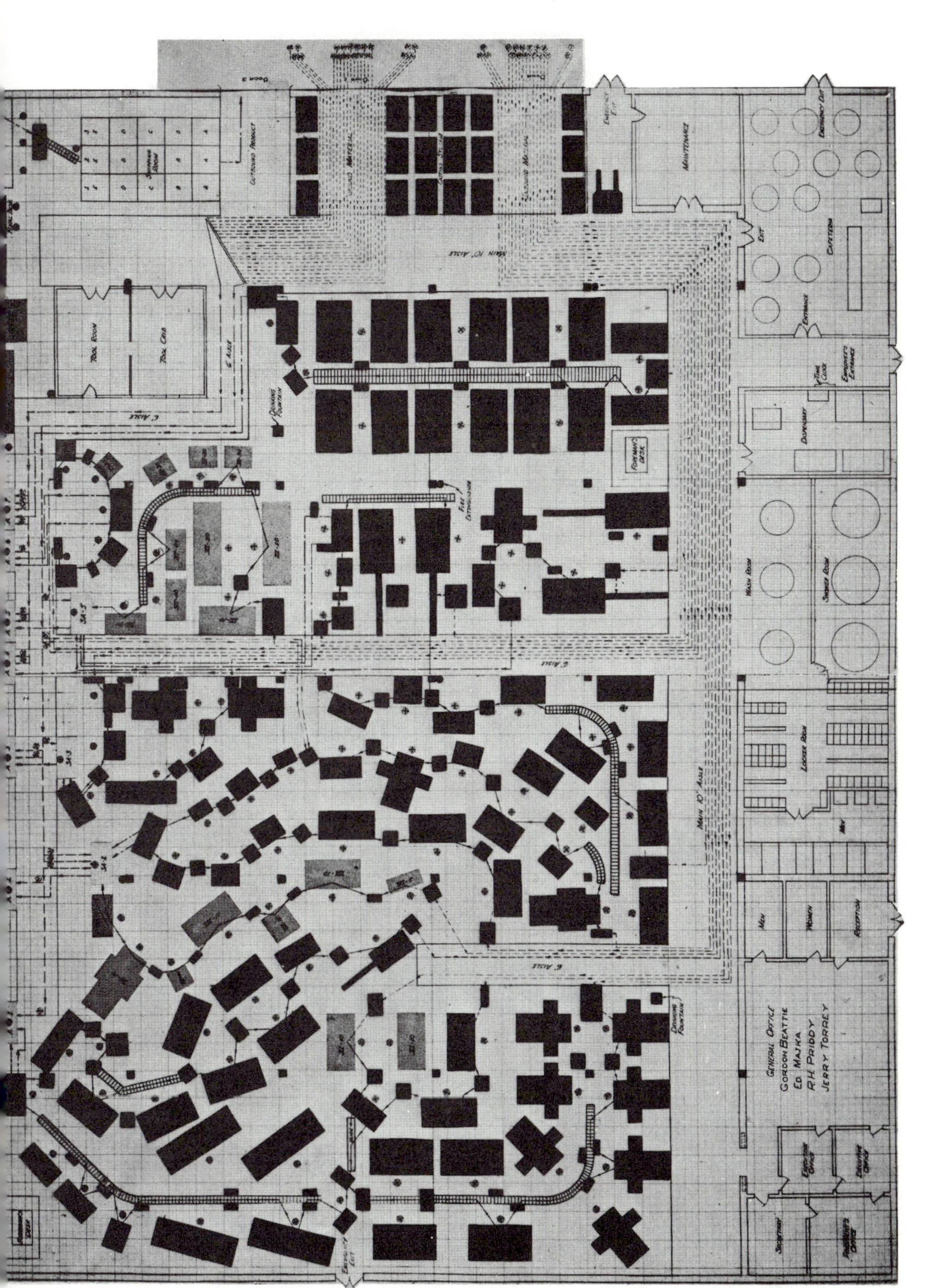

Figure 17–3—Plant layout using "home-made" templates.

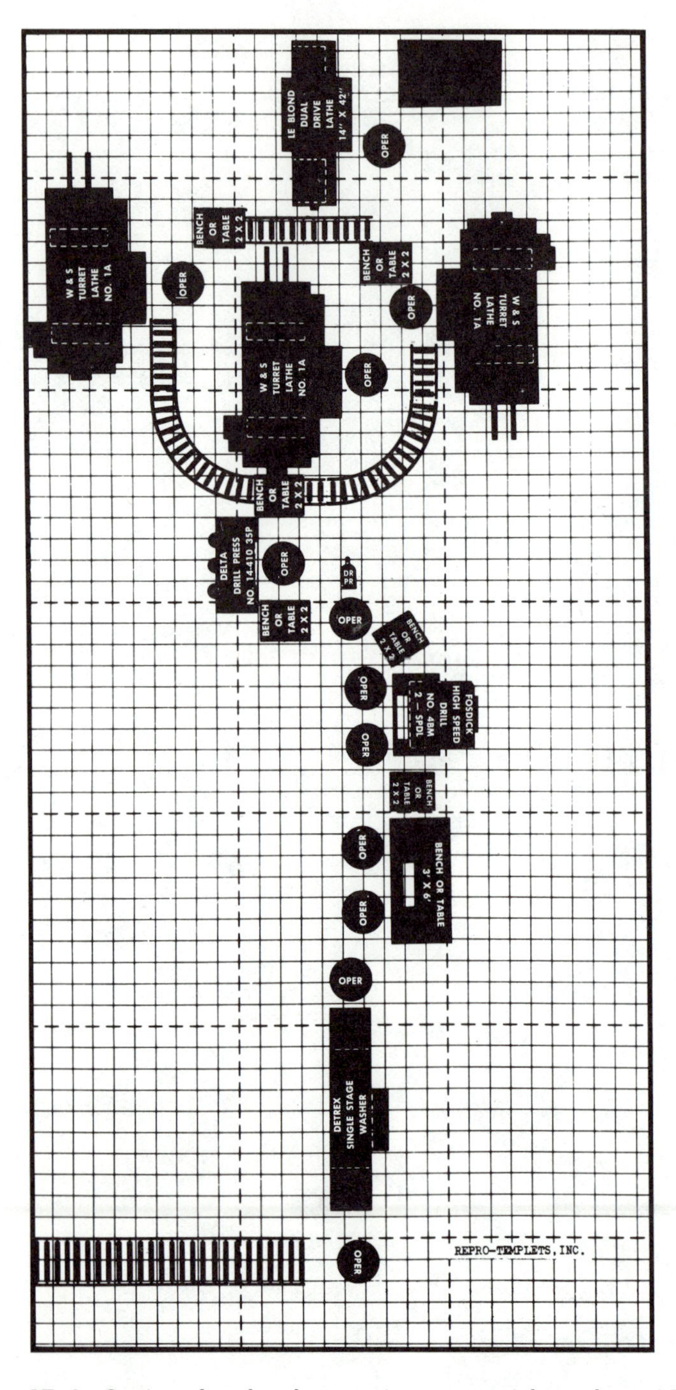

Figure 17-4—Section of a plant layout using commercial templates. (Courtesy of Repro-Templates, Inc.)

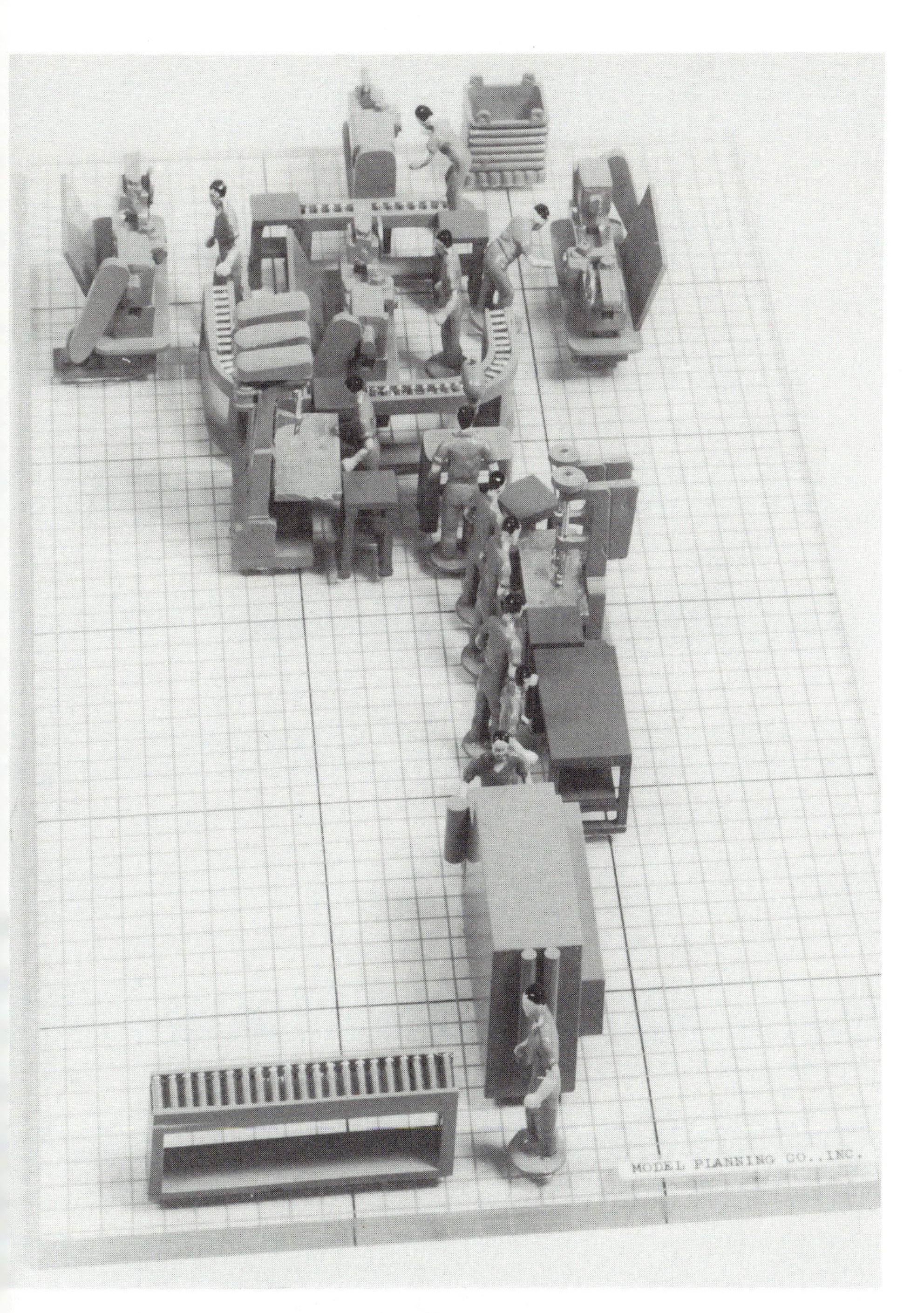

Figure 17–5—Section of the layout shown in Figure 17–4, using commercial models. (Courtesy of Model Planning Co.)

Table 17.1. Comparison of Basic Layout Methods

Characteristic	Conventional I. 2-Dimensional Templates	Reproducible II. 2-Dimensional Templates	3-Dimensional III. Scale Models	Combination of IV. Templates & Scale Models
Materials	Heavy paper or cardboard, pinned or glued to base material	Transparent grid sheets, templates, tapes, etc.	Exact scale models of all plant equipment and buildings	Combination of II and III
Design and Construction Characteristics	Roughly drawn and cut out from paper or cardboard	Carefully drawn to show exact physical dimensions, travel of machine components, and service area. Reproduced on transparent plastic.	Constructed to exact scale in three dimensions, and to actual height, to represent specific make and model accurately. Assembled from Lucite, cast in metal, or a combination of both.	Combination of II and III
Engineering Value	Practically none – due to probable inaccuracies	Good, for trained personnel and engineers: accurate and detailed.	Better, since third dimension adds note of realism and enables easy interpretation by persons inexperienced in plant layout work.	Best, because it combines the characteristics of both two and three dimensional methods.
Relative Cost	Engineer's time to measure and sketch: Draftsman's time to draw up and reproduce. Time required to cut out. Total estimated at one hour per template or about \$5-\$6 each. Probably most expensive of all four methods.	Commercially available at 2¢–18¢ each; average cost 9¢ each. Special items might go to \$1.00 each. Would amount to about 1¢/actual square foot of building, on 40-50,000 sq. ft. order	Average cost about \$2-\$4/model for typical machine tool. Small items less, unusual or special items may be considerably more. Would amount to about 6¢ per square foot of actual building, including all plant equipment and building model on 40-50,000 sq. ft. order	Combination of II and III, or about 7¢ per square foot of actual building.
Advantages	None	–Accurate detailed layout –Less time required to produce layout than no. I –Clearly indicates actual relationships in terms of	–Enables non-technical personnel to study and evaluate the layout –Quickly rearranged to study alternative layout plans –Photos of alternate proposals can be studied –Shows overhead details and clearances –Assures more correct location of each piece of equipment	

Table 17.1 (continued)

		floor space requirements –Serves as permanent record –Minimum storage space required –Economical upkeep –Prints easily produced –Special templates easily made –Flexibility permits easy changes –Overlays and underlays easily made and used –Eliminates drafting –Trial layout can be printed; alternate arrangements worked out; compared with original; and remade into original form, using trial layout print as a record of the original layout.	–Saves expensive moving of equipment after installation. –Models last indefinitely –Invites participation in planning –Aids in selling the project. –Exposes design errors –Shows up potential danger spots –Permits accurate study of manpower requirements –Speeds actual plant construction work, using model as a guide –Facilitates study of congested areas, especially where overhead conveyors, piping, etc., are involved –Easier to interpret than blueprints –More accurate visualizing of space utilization, thereby making it easier to plan proper use of building cube –Makes entire organization layout conscious –Useful in employee orientation and training –More nearly approaches the actual situation –Shows details unobtainable on drawings –Better depicts mass, depth, height and clearances –Leaves less planning to the imagination –Shows plans in all planes –Subject to less mis-interpretation –Reduces executive time required to study and approve –Shows up errors prior to installation –Aids cooperation from all concerned –Provides a better understanding of the problem	
Disadvantages	–Difficult to visualize –Alternate plans must be drawn and redrawn –Time consuming –Expensive –Requires tracing for reproduction –Inaccurate –Invites errors and discrepancies	–Vertical dimensions impossible to visualize –Planners must carry all vertical details "in their heads."	–Difficult to obtain copies unless template reproduction is made –Does not include detailed information shown on templates –Costs more than template method, no. II	–Highest initial cost of the three modern methods (II, III, IV), but probably LESS than no. I, in the long run, if accurate costs are kept.

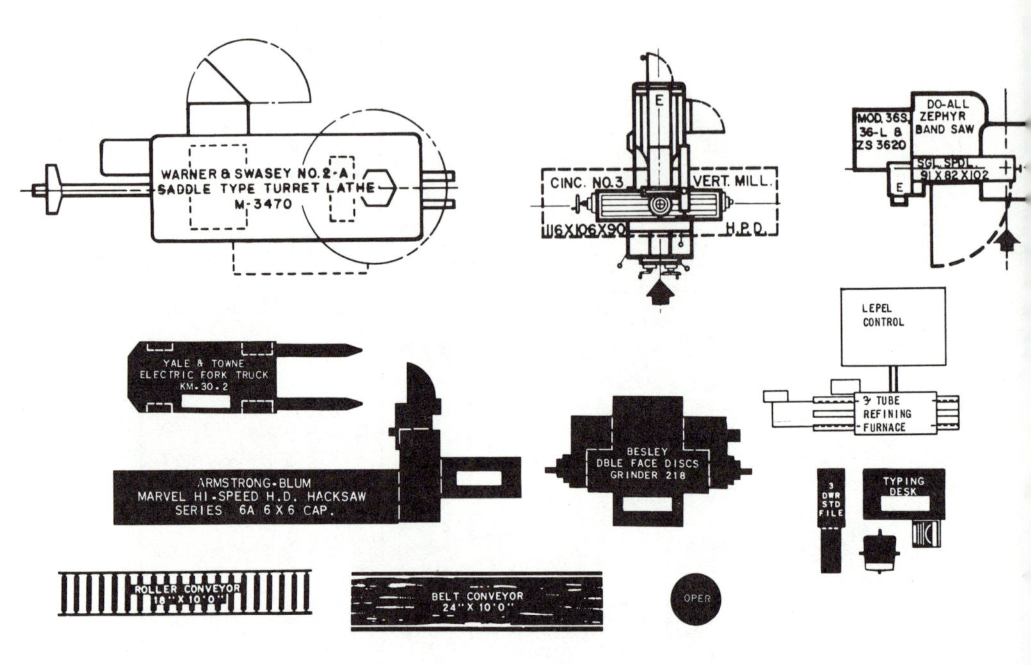

Figure 17–6—Common templates used in plant layout work.

Scale models commercially available or homemade, to represent physical facilities, etc., can be of:

1. Plastic—built up or molded—some with cast components, etc.
2. Cast metal
3. Wood

Figures 17-6 and 7 illustrate some of the many types of templates and models in common use.

Model and Template Standards

In 1949, a subcommittee of the American Society of Mechanical Engineers drafted a standard for plant layout tools, which was subsequently adopted. Basically, the standard outlines the characteristics of templates and models, which have been used as guides by both commercial suppliers of such materials and by companies that choose to construct their own. The scale recommended is $\frac{1}{4}$ in. = 1 ft, which has become the most commonly accepted scale. Occasionally, a larger scale is used to show greater detail; or a smaller scale is used for large areas, when little detail is required, such as on plot plans or building outlines used for planning purposes.

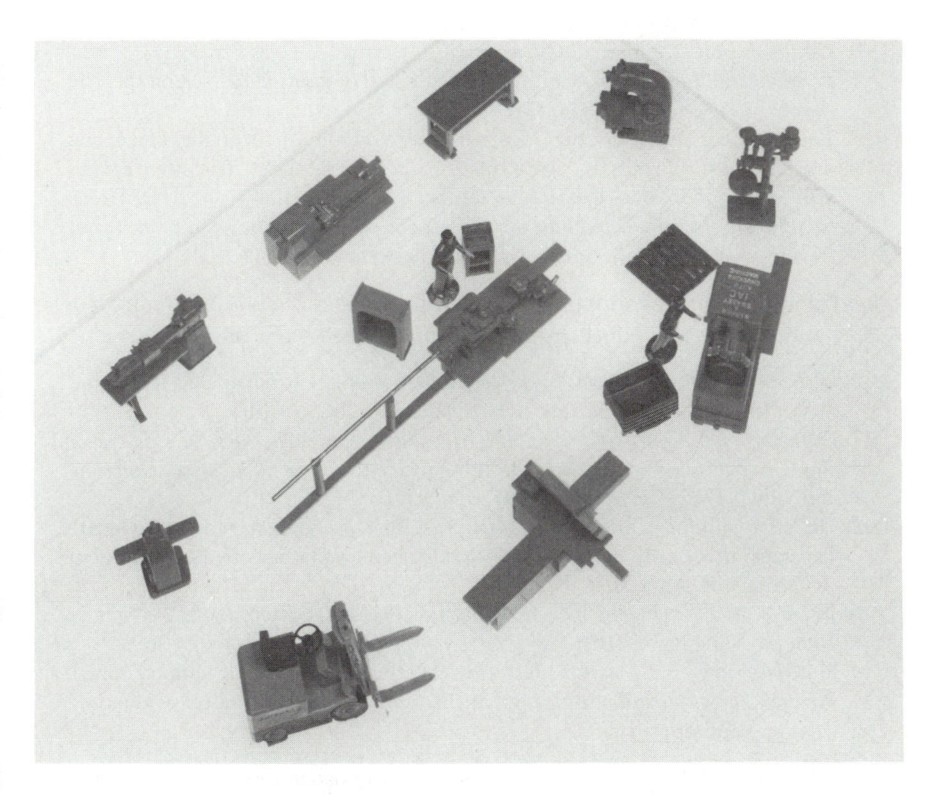

Figure 17–7—Common models used in plant layout work.

Layout Construction Procedure

The actual construction of the final layout requires that the many detailed plans and decisions previously worked out be resolved and integrated into one master plan or representation of the ideas developed. As a guide to the integration process, the following step-by-step procedure is suggested:

1. *Gather all planning data* for easy reference. This will include:
   - a. *Parts Lists*
   - b. *Assembly Charts*
   - c. *Operation Process Charts*
   - d. *Production Routings*
   - e. *From–To Charts*
   - f. *Activity Relationship Diagrams*
   - g. *Area Allocation Diagrams*
   - h. *Operation Charts*
   - i. *Layout Planning Charts*
   - j. *Flow Diagrams*
   - k. *Space Calculations*
2. *Determine rough overall size of layout* from *Area Allocation Diagram*
3. *Decide on appropriate scale* to be used. Usually $\frac{1}{4}$ in. = 1 ft; for larger layouts, $\frac{1}{8}$ in., or $\frac{1}{16}$ in.
4. *Obtain grid sheet* for base of layout. Work on opposite side from printed or scribed cross section lines.

5. *Obtain templates or models* and related supplies.
6. *Establish bay size and column spacing.*
7. *Estimate number of bays* (width and length) and determine their fit on the grid sheet.
8. *Locate a fixed corner* or other location (such as shipping, receiving, office) and apply tapes for 2 walls, to locate this area on the grid sheet. The other 2 walls will "seek" their own location as the layout develops.
9. *Tentatively position columns* on grid sheet. Columns may shift as layout develops—to maintain uniform spacing between walls.

Note: If building or area in which layout is included already exists omit *steps 6, 7, 8, 9* and merely reproduce present walls, columns, etc., on grid sheet.

10. *Indicate probable locations of aisles* on the floor plan, if possible at this point. (Use aisle tape or $\frac{1}{16}$-in. line tape to show borders of aisles.)
11. *Transfer preliminary plans to the grid sheet,* gradually, with templates or models, necessary tapes, etc., working from the *Area Allocation Diagram,* individual *Flow Diagrams,* and *Work-place Layouts.*
12. *Adjust preliminary plans* continually as they are transferred, to re-locate columns, aisles, or to make other desirable changes that may become apparent as the layout progresses.
13. *Work out final arrangements and details* for auxiliary, service, and activity areas indicated in Figure 11-18.
14. *Add lettering*—with transfer lettering sheets (available at art supply stores).
15. *Arrange for duplication,* if desired at this time; or wait until after evaluation, critique, and approvals.

It will be noted that the entire layout has been constructed without the use of pencil, paper, or eraser!

Some construction suggestions. One use of transparent sheets (acetate, etc.) is for overlays of the master layout to show, without cluttering up the original, such features as:

1. Material flow
2. Utilities
3. Millwright or contractor instructions
4. Consecutive changes

Colored tapes, or colored (adhesive) art film, can be used in overlays to identify different items, variations, or stages. Also, a variation, the *layout book* (a multi-layered layout) may be considered.[1]

An easy way of making custom transparent templates, is as follows:

1. Make outline drawings, to scale, of equipment items (it is more economical to do a whole sheet of templates or labels at one time)
2. Outline them with a fairly wide ($\frac{1}{16}$-in.) dark black felt-tip pen. Typing (dark) can be done on the drawing for letters, machine numbers, etc.

[1]Described by S. H. Isaacs, in "Transparent Layout Books Aid Long Range Planning," Modern Materials Handling, Oct. 1954:116.

3. Make a xerographic transparency of the sheet; or (a) make a xerographic *copy* of the sheet, and then (b) make a heat-process transparency[2] of that.
4. Cut out individual templates and apply transparent, double-sided tape to the back of the template.

Master Plans

In many cases, a building-outline master drawing is kept on file[3]—showing only the various activity areas within the facility. Another master contains the equipment only—no other information or data. Then, translucent copies of either are ordered whenever a particular area is being studied or altered. Changes can then be made, and detail plus labels and even instructions added for millwrights or contractors working on the facility.

A variation of the master plan is the grid concept used by an aircraft manufacturer.[4] The entire plant site was laid out on a grid of 200-ft squares, with permanent markers set at intersections throughout the site. On drawings, the grid was divided into 100-ft squares, then of 25-ft sub-grids (for columns) and 8-ft–4-in. modules (for lighting circuit feeders, air conditioning ducts, interior partitions and corridors). Each utility was placed on a particular grid line. A primary advantage is the location of future additions or buildings on the original, within the grid pattern, and maintaining an organized overall approach to long-range facilities planning.

The Key Plan

For the larger facility, covering several acres or with a number of buildings, a *Key Plan* may be established. This may be based on an accurate scale drawing of the entire property, similar or equal to the *Plot Plan.* An overlay is then prepared, by ruling off the entire area into rectangular plant segments, probably corresponding to a module that will fit onto the largest sheet of drafting material in common use in the organization for layout prints (say 36 by 48 in.)—at the scale used in layout work, that is, $\frac{1}{4}$, or $\frac{1}{8}$ in. (etc.) = 1 ft. Then, each rectangle or panel, is given a key number. This procedure facilitates the coordination of the individual layouts of sections of a facility with the whole, or with each other. Careful attention

[2]The 3M Company has an adhesive-backed, translucent polyester film that combines (or takes the place of) *steps 3(a)* and *(b)*, and applying the adhesive; and Deitzgen has an adhesive-backed Mylar film, which is diazo (ammonia) sensitive, for similar use on a blue-print (or diazo) machine. Both are more expensive, however, than the transparency (material) plus the double-sided tape.

[3]For details, see B. L. Roberts, "Master Layout Drawings Help Control Plant Design," *Plant Engineering,* May 1966:134; see also R. O. Riccetti, "Layout Drawing Package Makes Up-dating Easy," *Plant Engineering,* Aug. 21, 1969:56; and for a similar progressive concept, although not on overlays, see W. A. Schmucker, "Mechanical Arrangement Drawings—Final Steps to Good Plant Design," *Plant Engineering,* July 8, 1971.

[4]"Grid Concept Unifies Plan for 'Today-Tomorrow' Building." Plant Engineering, July 1964:118.

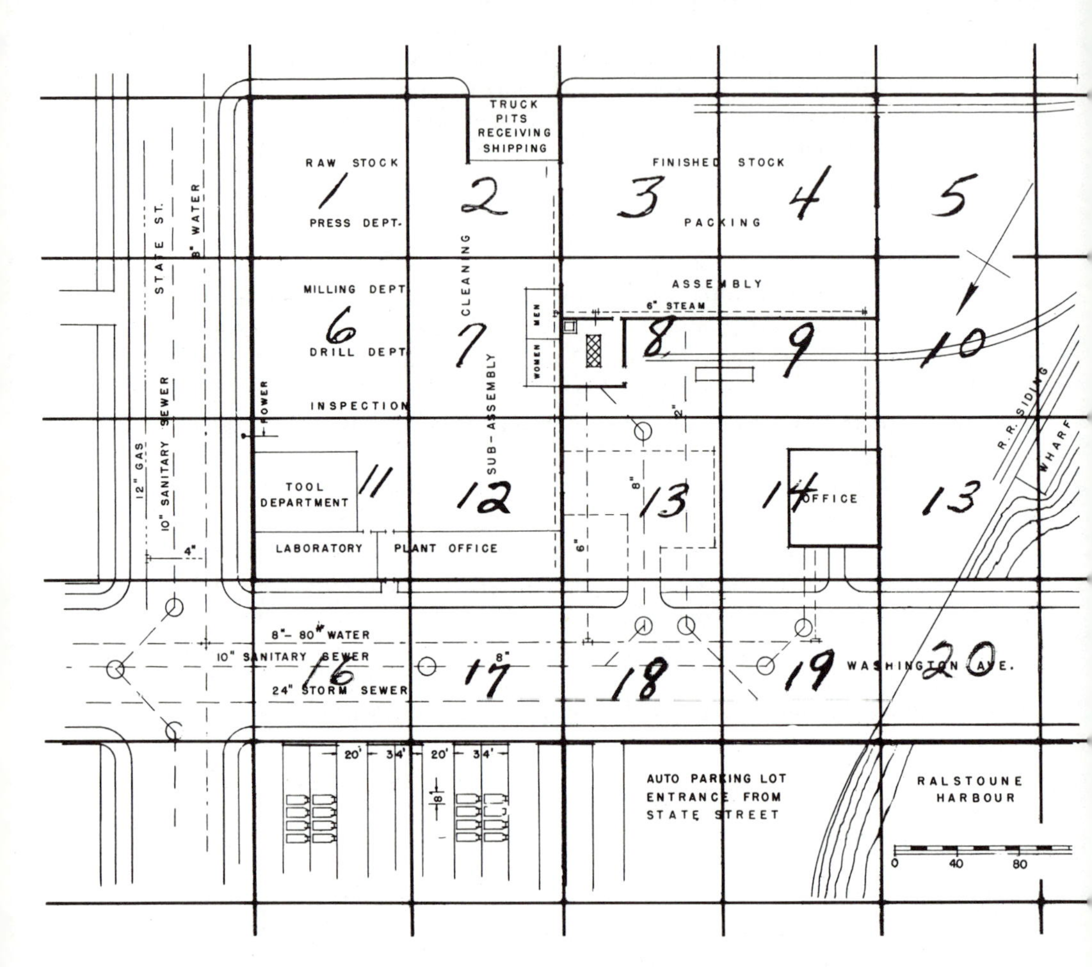

Figure 17–8—Key plan for relating smaller segments of the layout to the whole plant site.

should be paid to the natural divisions of the property, such as roads, walls, and columns, so they can be fitted to the *Key Plan,* if practicable. That is, the rectangles need not all be the same size—if adhering to natural boundaries will yield any benefit in the division process. This concept is shown in Figure 17-8.

Finishing Up

In spite of all the planning done so far, the layout problem still resolves itself into an intricate process of moving, shifting, and rearranging of areas, templates, or models until a satisfactory solution has been achieved. While it is true that preliminary sketches and plans have been made, they must all be resolved into a workable and economical solution. The reason for the detailed planning has been to eliminate as much time and as many mistakes as possible in the preparation of the final layout.

Extreme care was suggested throughout the layout process, because it is far easier to correct mistakes on paper than to correct them after the machinery and equipment are installed. And this certainly will happen if the layout is hurriedly thrown together without adequate planning. The work outlined should be continued until a satisfactory solution has been reached and the layout is complete. It will probably be a compromise between what had been planned in previous steps, and what would be an ideal situation—if one were possible.

Figure 17-9 is the final layout for the Powrarm plant, which has been followed throughout the text. A reexamination of Figures 6-1, 6-4, 8-4, 11-12, 11-13, and 12-21 will show that the basic planned relationships have been carried through the various steps. As mentioned, compromises have been made as necessary along the way. Nevertheless, the orderly procedure followed has greatly aided the development of the final layout shown.

Flow Lines

As the layout progresses, or when it is complete, it is highly desirable to add flow lines to represent the actual paths to be traveled by major parts being processed. The flow lines should indicate the path until it becomes part of an assembly or leaves the building as a finished product. The lines should follow as nearly as possible the exact path the part will follow. If several parts flow through a single machine, work place, or other piece of equipment, a separate line should be shown for each, if practical.

These flow lines are an important part of the layout, as they help to indicate how well the principles of good layout have been followed, especially in regard to straight-line flow, backtracking, and congestion in aisles or specific areas. A large number of flow lines in a given aisle or area obviously represents heavy traffic and possible operating difficulties.

The flow lines may be added to the layout by several different methods. On the plastic grid sheets, they may be applied with narrow tapes or by means of arrow templates commercially available. Or it may be desirable to make a print of the layout and draw the flow line with colored pencil, or felt-tip pen. Another alternative would be to make an intermediate (brown-line or sepia) copy, apply the tape or drawn lines, and then make a print of this. A fourth method is to mount a print of the layout on a composition board, insert pins or tacks at appropriate locations, and represent the flow lines with colored strings, as in Figure 6-9; the strings can even be removed and measured to determine travel distance for a specific part. Or, an overlay of clear plastic sheet could be used, with tapes applied to denote material flow. Figure 17-10 shows a layout with taped flow lines added.

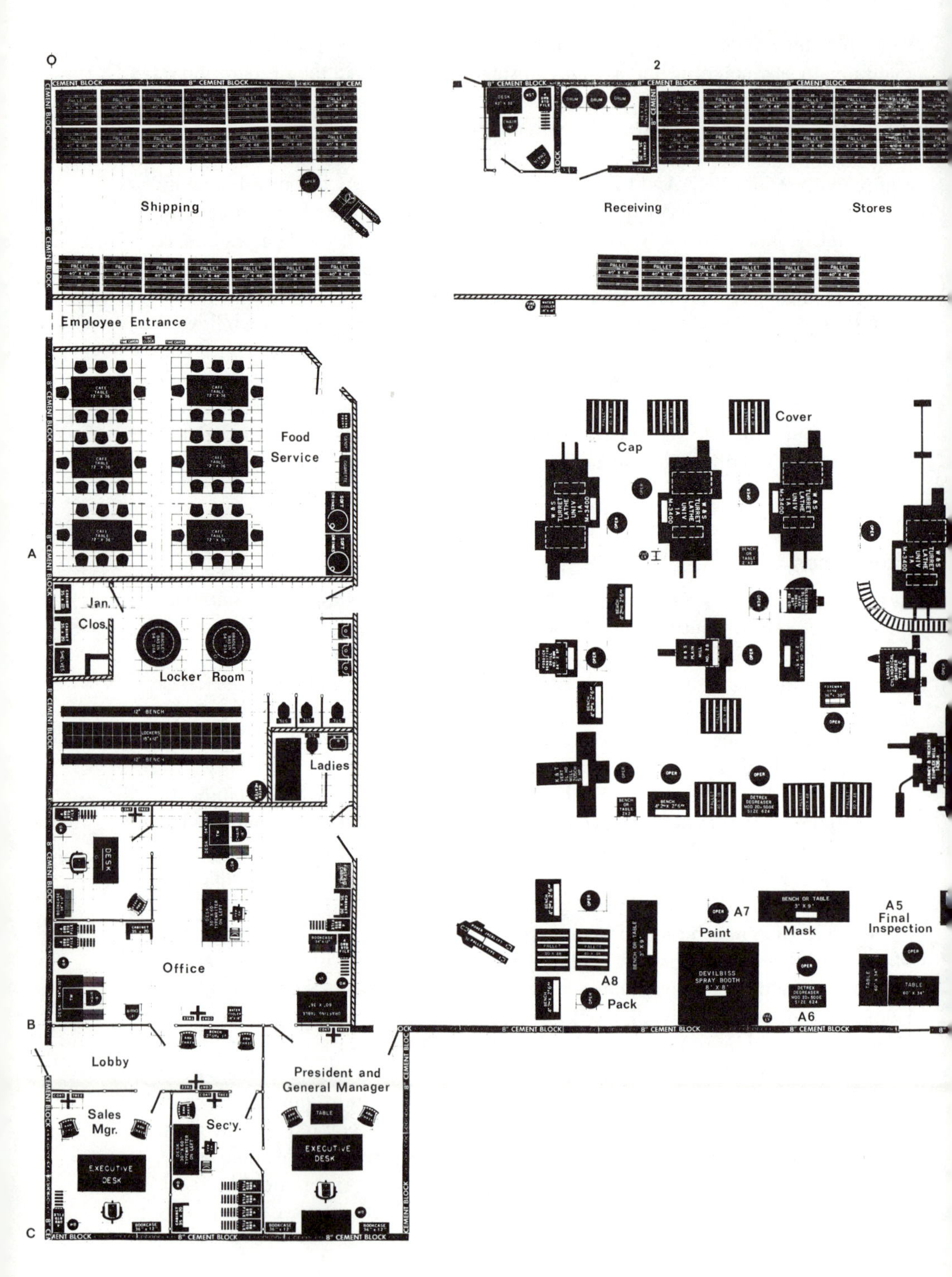
2
Shipping
Receiving
Stores
Employee Entrance
Food Service
Cap
Cover
A
Jan. Clos.
Locker Room
Ladies
Office
A7
Paint
Mask
A5 Final Inspection
A8 Pack
A6
DEVILBISS SPRAY BOOTH 8' X 8'
B
Lobby
President and General Manager
Sales Mgr.
Sec'y.
EXECUTIVE DESK
C
8" CEMENT BLOCK

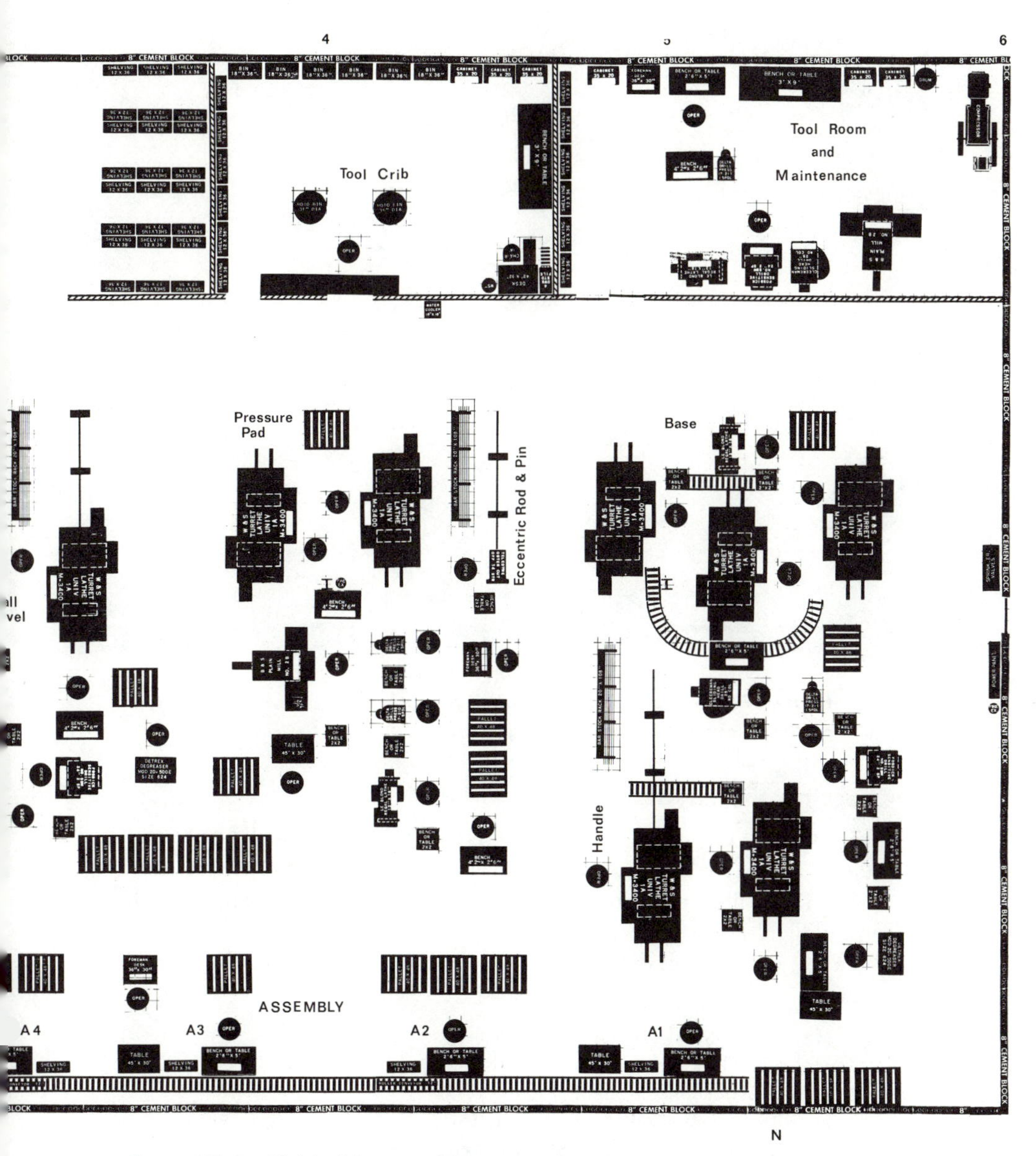

Figure 17–9—Finished layout of Powrarm plant based on plans made throughout text.

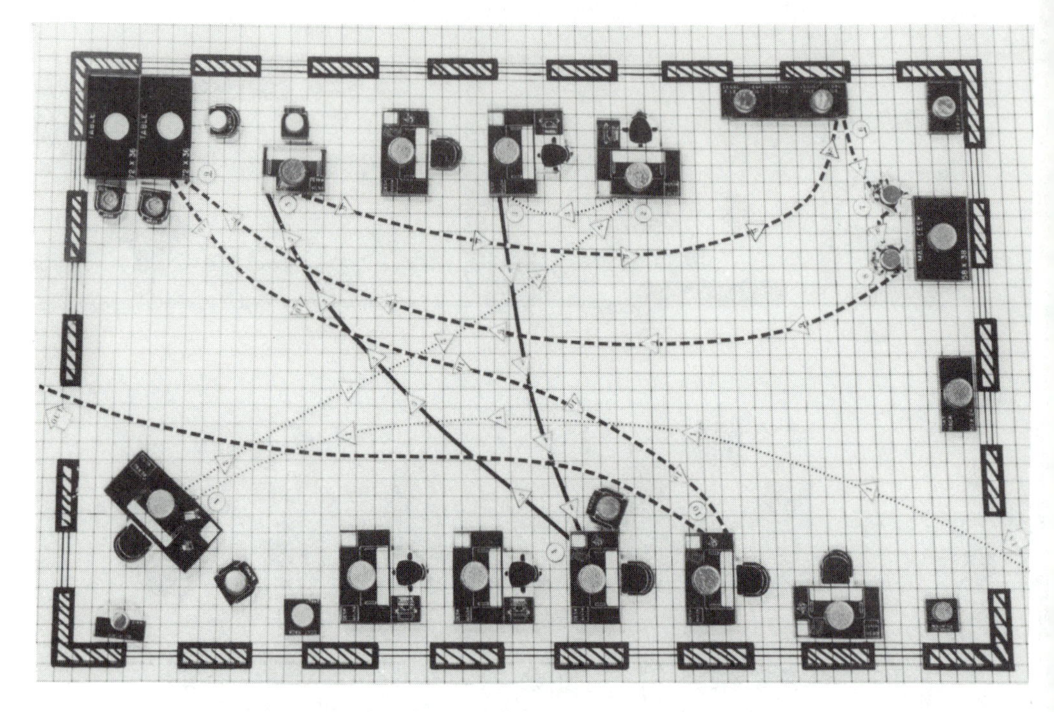

Figure 17-10—Flow lines on finished layout—note use of magnetic templates. (Courtesy of Magne-Plastic Corp.)

Checking the Layout

After the plant layout has been completed (and before it is presented to others for review), the layout engineer should carefully check every detail (see Figure 17-11). This will assure that he has not overlooked any important items that may cause difficulties or embarrassment later on.

In the checking or evaluation process, the following may be useful:

1. Layout objectives (pp. 7–10)
2. Marks of a Good Layout (pp. 18–19)
3. Material Flow Design Evaluation Sheet (pp. 118–19)
4. Plant Layout Check Sheet (Figure 17-11)
5. Facility Layout Evaluation (Figure 18-1)
6. Facility Evaluation Check Sheet (Figure 18-6)
7. Layout Evaluation Sheet (Figure 18-7)

After the layout engineer has carefully and thoroughly checked the entire layout, he should ask others to check for specific items of interest to them and their functions. Many persons will have to live with a layout once it is put into effect. For this reason, as many interested people as practicable should be given a chance to participate in checking the layout, before final approval. Chapter 18 details the evaluation process.

Conclusion

This chapter has brought together all the concepts and procedures of the preceding sixteen—and has hopefully resulted in a efficient layout of the proposed facility. The material flow, the production function, the service and auxiliary activities, the individual work places, and the material handling plans will have been melded into a single overall plan for the arrangement of the facilities within the structure that is to house them. The layout is now almost ready to be turned over to the architect or contractor, for a proper building to be designed around it.

But, some important steps remain—the evaluation of the layout, before it can be presented for final approval, and its implementation. These problems are the subject of the next chapter.

Questions

1. Why is so much preliminary work necessary before the final layout is constructed?
2. What are the principal methods of making the master layout? Give some advantages and disadvantages of each.
3. When are pencil-and-paper layouts acceptable?
4. What are some of the tools and techniques used in constructing the layout?
5. Why are templates used along with a scale model of a layout?
6. Why are template and model standards advisable?
7. Briefly describe the layout construction process.
8. What is an overlay? For what purposes can it be used?
9. What is a *Master Plan*?
10. Describe the *Key Plan* concept.
11. What are some uses or advantages of keeping master building drawings on file?
12. What are the advantages of adding flow lines to the layout? How may they be represented?
13. Who might be consulted in checking the layout—prior to presentation for approval? Why would each one be selected?

LAYOUT CHECK SHEET

Company ______________________ **Plant** ______________

Checked By ______________________ **Date** ______________

Before submitting the layout for official approvals, check against the following items

INDUSTRIAL ENGINEERING	not applic.	OK	remarks
1. Machinery and equipment arranged to make full use of capacity?			
2. Machinery and equipment accessible for material supply and removal?			
3. Machinery and equipment located for maximum operator efficiency?			
4. Line production used where practical?			
5. Proper use made of mechanical handling?			
6. Processing combined with transportation?			
7. Minimum walking required of operators?			
8. Finished work of one operator easily accessible to next?			
9. Machinery and equipment "block in" any operators?			
10. Machine overtravel extend into aisles or interfere with operator?			
11. Adequate storage space at work stations?			
12. Efficient work place layouts?			
13. Service areas conveniently located: (tool room, tool crib, maintenance, etc.)			
14. Easy for supervisor to oversee his area?			
15. Machine arrangement permit maximum flexibility in case of product change?			
16. Space allocated for foremen and production control records?			
17. Related activities located near each other?			
18. All required equipment included in layout?			
19. Floor area fully utilized?			
20. Provisions made for expansion?			
21. Provisions for scrap removal?			
22. Crowded conditions anywhere?			
MATERIAL HANDLING			
1. Incoming materials move directly to work area?			
2. Processes involving heavy or bulky materials close to receiving area?			
3. Sub-assemblies flow into final assembly?			
4. Plans for auxiliary material flow in case of tie-up?			
5. Minimum of backtracking?			
6. Obstacles in material flow?			

MATERIAL HANDLING (cont'd)	not applic.	OK	remarks
7. Excessively long moves?			
8. Conveyors "box in" anyone?			
9. Material handling equipment bring materials to operators?			
10. Material handling equipment remove materials from operators?			
11. Each material handling method integrated into overall system?			
12. Conveyors from receiving to processing areas?			
13. Conveyors from assembly to shipping area?			
14. Minimum of manual handling?			
15. Material handling equipment carry materials in a position to conserve space?			
16. Widely separated areas connected by suitable mechanical handling system?			
17. Aisles and doors wide enough for maximum loads and fire equipment?			
18. Ramps at lowest possible grade?			
19. Proper use of conveyors for efficient handling?			
20. Adequate storage space for containers and material handling equipment?			
21. Use of conveyors for floats and banks of material?			
22. Adequate storage for materials in process?			
23. Shipping and receiving docks covered?			
24. Proper dock heights and/or levelling devices?			
25. Maximum use made of building cube?			
26. Aisles straight and clearly marked?			
PERSONNEL AND SAFETY			
1. Exits, fire doors and fire escapes adequate and properly located? Free of obstruction?			
2. Easy exit from any location in building?			
3. Plans checked by local fire, police and safety officials?			
4. Plans approved by insurance companies?			
5. Lifts and hoists provided for loads over 50 lbs.?			
6. Adequate work area for each operator?			
7. Easy access to all safety devices?			
8. Hazardous and unpleasant operations isolated?			
9. Adequate storage for inflammable materials?			
10. Drinking fountains in enough locations?			
11. Maximum use made of natural light?			
12. Adequate artificial light provided for?			
13. Proper ventilation provided where required?			

Figure 17–11—Plant layout check sheet.

PERSONNEL AND SAFETY (cont'd)	not applic.	OK	remarks
14. Employee service areas provided for and conveniently located:			
a. first aid			
b. toilets			
c. locker rooms			
d. smoking areas			
e. coat racks			
f. food service			
g. time clocks			
h. stretchers			
15. Parking space provided for employees?			
PLANT ENGINEERING			
1. Floor loads within allowable limits?			
2. Overhead clearances adequate?			
3. Proposed overhead loading within limits?			
4. Special foundations or equipment mountings required?			
5. Floor and roof drains provided for?			
6. Elevators required, located, specified?			
7. All possible use made of under-floor, overhead, and on-the-roof space?			
8. Layout make efficient use of building shape, size, cube?			
9. Machinery and equipment accessible for maintenance?			
10. Layout conducive to good housekeeping?			
11. Provisions for rubbish collection, storage, and removal?			
12. Utilities provided for and properly located:			
a. air			
b. gas			
c. water			
d. electricity			
e. steam			
f. telephone			
g. sprinklers			
h. sewers			
i. air conditioning			
j. heating			
k. ventilation			
l. lighting			
13. Detailed drawings of special equipment or installations?			
14. Fire extinguishers and fire protection equipment provided for?			
15. Layout satisfy all safety codes?			
PRODUCTION CONTROL			
1. Inventory checking easy?			
2. Adequate space and facilities for salvage operations?			
3. Proper protection for material and finished goods storage?			
4. Valuable materials protected from pilferage?			
5. Adequate space for shipping and receiving?			

PRODUCTION CONTROL (cont'd)	not applic.	OK	remarks
6. Adequate area for trucks waiting at shipping and receiving?			
7. Layout permit operations to be performed in logical sequence?			
8. Easy adjustment to changing schedules?			
9. Layout permit paced production?			
10. Will equipment breakdown shut down line or plant?			
11. Inspection points located at strategic points?			
12. Straight flow lines?			
OTHER COMMENTS OR SUGGESTIONS			

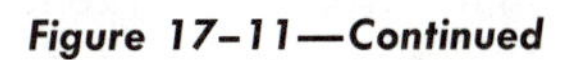

Figure 17–11—Continued

18

Evaluating and Implementing the Layout

After the layout has been completed, it must be evaluated by the designers and other interested parties; then it must be approved; and finally, it must be installed, or implemented. This chapter will deal with these last three steps in the facility planning and design process. In spite of the considerable effort that will have gone into designing the layout, there is no way to guarantee that it is the one best, or that it includes all the desirable objectives, criteria, and ideas. So the completed layout should be evaluated by one or more of the approaches outlined below.

Evaluating the Layout

The need for layout evaluation may arise from either of two possibilities:

1. *Evaluation of an existing layout* for the purpose of discovering improvement opportunities.
2. *Evaluation of an alternative layout* under consideration for a single problem or project area.

But before any evaluation can take place, there must be some basis for the evaluation, which might include:

1. *The objectives* developed at the beginning of the layout process.
2. *Layout criteria* or marks of a good layout.
3. *Cost comparison* with other alternatives.
4. *Return on investment* in the new facility.
5. *Indeterminate or imponderable factors,* of the type normally not given consideration because of the difficulty in quantifying them.
6. *Intangible factors,* with little or no real basis for conversion to numerical values for comparison purposes—which must, therefore, be evaluated by judgment.

Most of these will be considered in one way or another in the following discussion of techniques and procedures, which may be either qualitative or quantitative. That is, they can consist of a relatively simple balancing of advantages vs. disadvantages, or some quantitative means of measuring the value of the layout or layouts. If the layout is a relatively simple one, or there is no alternative at this point, then the evaluation might involve no more than the use of a check sheet. For a more complex project, or for comparing several alternatives, the evaluation should be more formal or sophisticated.

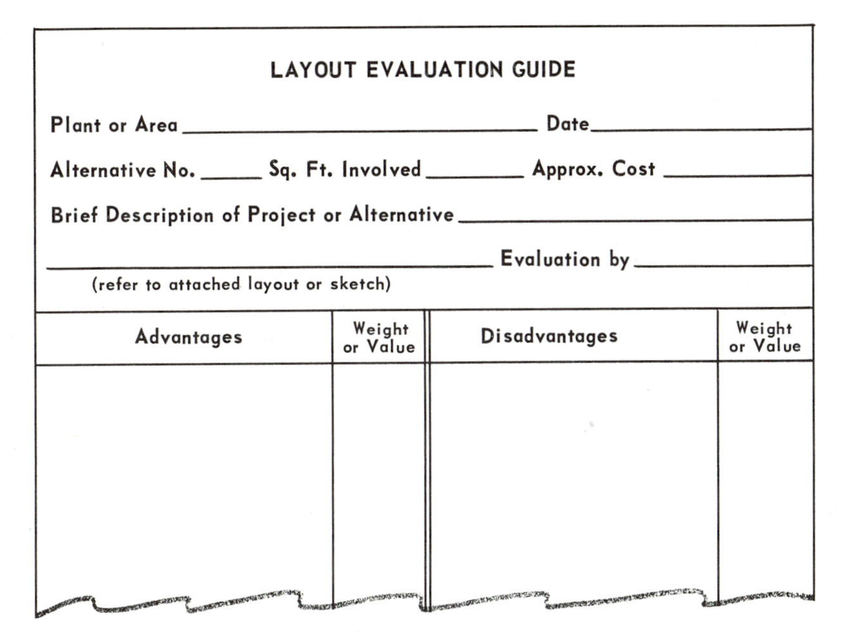

LAYOUT EVALUATION GUIDE

Plant or Area ______________________ Date ______________

Alternative No. _____ Sq. Ft. Involved ________ Approx. Cost __________

Brief Description of Project or Alternative ______________________

______________________ Evaluation by ______________

(refer to attached layout or sketch)

Advantages	Weight or Value	Disadvantages	Weight or Value

Figure 18-1—Facility layout evaluation.

Qualitative Evaluation Techniques

The most common and simplest approach to layout evaluation is the listing of advantages and disadvantages of the layout at hand or the alternatives being considered. Such a tabulation might appear as in Figure 18-1. Numerical values may be placed in the appropriate columns on the form shown in the figure, possibly assigned as follows:

Perfect	10	Good	6	Poor	3
Ideal	9	Fair	5	Unsatisfactory	2
Excellent	8	Average	4	Unacceptable	1
Very Good	7				

Each advantage or disadvantage should be assigned a value relative to the importance of the item in the overall results of the proposed solution.

A more complete qualitative evaluation, useful as an audit of an existing operational layout, can be made with the aid of the charts[1] shown in Figures 18-2 and 3. These charts have been designed to simplify the tasks of:

1. *Spotting indicators* of layout problems.
2. *Identifying causes* of the problems indicated.
3. *Finding areas for possible elimination of the problems.*

On the two charts, there are:

[1]J. M. Apple, "How to Spot Your Plant Layout Problems," *Factory*, Jan. 1959.

CHART 1 – How to Recognize Plant Layout and

Indicators of Layout and Related Problems		1. Poor housekeeping	2. Crowded conditions	3. No over-all plan	4. No alternatives	5. Excess mtls. on hand	6. Poor schedule, dispatch	7. Bldg. cube unused	8. Delayed mtl. movement	9. Unsuitable containers	10. Non-use of mech. equip.	11. Manual hndlg., rehndlg.	12. Loads not unitized	13. Long hauls	14. Wrong move, first time	15. Mtls. piled on floor	16. Poor storage equip.	17. Poor flow pattern	18. Unbalanced oprns. seq.	19. No flexibility	20. No space	21. Inadequate equip., fac.	22. Bad space allocation	23. Poor plant layout	24. Poor process methods	25. No aisle markings	26. Poor placing, spacing	27. Poor loc., source or dest.	28. Poor loc. related activ.	29. Employees waiting	30. Material-flow blocks	31. Scattered buildings	32. Rel. areas partitioned	33. Crooked aisles
		GENERAL						MATERIALS HANDLING										PLANT LAYOUT																
GENERAL	a. Crowded aisles	✓	✓		✓	✓	✓	✓	✓	✓			✓		✓	✓	✓	✓			✓	✓	✓	✓		✓					✓			✓
	b. Hazardous areas	✓	✓							✓	✓	✓	✓								✓	✓		✓		✓	✓				✓			✓
	c. High accident rate	✓	✓							✓	✓	✓									✓	✓			✓	✓	✓				✓			✓
	d. High overhead	✓	✓			✓	✓	✓	✓	✓	✓	✓	✓	✓	✓	✓	✓	✓	✓	✓	✓	✓	✓	✓	✓		✓			✓	✓	✓	✓	✓
	e. Crowded conditions	✓			✓	✓	✓	✓	✓	✓			✓			✓	✓	✓			✓		✓	✓		✓	✓				✓			✓
	f. Poor housekeeping		✓			✓		✓	✓	✓			✓			✓	✓				✓	✓	✓			✓	✓				✓		✓	✓
PRODUCTION	a. High indir. labor				✓		✓				✓	✓	✓	✓	✓			✓							✓							✓	✓	✓
	b. Unexplainable delays	✓	✓	✓		✓	✓		✓		✓	✓	✓	✓		✓		✓	✓	✓	✓	✓		✓	✓				✓	✓	✓	✓	✓	✓
	c. Scheduling trouble	✓	✓	✓	✓	✓	✓		✓		✓	✓		✓	✓			✓	✓	✓		✓		✓	✓				✓	✓	✓	✓	✓	
	d. High in-process mtl.			✓	✓	✓	✓		✓									✓	✓	✓		✓		✓	✓					✓				
	e. Too many men collecting scrap, waste									✓	✓		✓	✓		✓		✓				✓		✓	✓			✓						
	f. Material-flow blocks	✓	✓		✓	✓		✓	✓	✓	✓			✓		✓		✓			✓		✓	✓			✓	✓				✓	✓	✓
	g. Equipment moved often			✓	✓													✓				✓	✓	✓	✓		✓							
	h. Related work scattered			✓	✓			✓										✓		✓			✓	✓	✓			✓	✓		✓	✓	✓	
	i. Uneven flow rate		✓	✓			✓		✓		✓	✓		✓		✓		✓	✓	✓		✓		✓	✓			✓	✓	✓	✓	✓	✓	✓
	j. Scattered buildings			✓	✓			✓															✓	✓				✓	✓					
	k. Vanishing aisles	✓	✓	✓		✓		✓	✓				✓			✓	✓				✓		✓	✓		✓	✓				✓			✓
	l. Walls, partitions, betw. related areas				✓													✓			✓			✓				✓	✓			✓		
	m. Blocked doorways	✓	✓		✓	✓			✓	✓						✓	✓				✓			✓					✓					
	n. No flexibility	✓	✓	✓	✓			✓	✓	✓								✓			✓	✓	✓	✓	✓						✓	✓	✓	
	o. Makeshifts perpetuated			✓	✓															✓	✓	✓		✓						✓	✓			
	p. Working in aisles		✓	✓	✓	✓		✓										✓			✓		✓	✓		✓	✓				✓		✓	✓
	q. Mtnce. hard to perform	✓	✓	✓			✓									✓						✓			✓		✓					✓		
	r. Mfg. cycle too long	✓	✓			✓	✓		✓			✓				✓		✓	✓			✓		✓	✓					✓	✓	✓	✓	✓
	s. Idle production equip.						✓		✓										✓	✓					✓				✓	✓	✓	✓		
	t. Overloaded prod. equip.				✓	✓	✓											✓	✓		✓				✓									
	u. Low production density	✓	✓	✓	✓	✓	✓		✓		✓	✓						✓	✓	✓	✓	✓		✓	✓					✓	✓	✓		✓
	v. Vacant floor space			✓	✓																	✓	✓	✓			✓							
	w. Poor quality	✓	✓	✓							✓	✓		✓		✓	✓					✓			✓									
	x. Line-machine breakdowns			✓	✓		✓											✓		✓		✓			✓									
	y. Inaccessible machines	✓	✓	✓		✓		✓	✓	✓						✓		✓			✓		✓	✓	✓		✓		✓			✓	✓	
	z. Bottleneck operations			✓	✓	✓	✓		✓		✓							✓	✓	✓		✓		✓	✓				✓	✓				
	aa. Crooked aisles		✓	✓	✓			✓								✓		✓			✓		✓	✓		✓	✓		✓				✓	
MANPOWER UTILIZATION	a. Too many men moving materials			✓	✓	✓	✓		✓	✓	✓	✓	✓	✓	✓	✓	✓	✓	✓			✓		✓	✓		✓	✓	✓	✓	✓	✓	✓	
	b. Handling by skilled men				✓	✓	✓		✓	✓	✓	✓			✓	✓			✓			✓			✓		✓	✓	✓	✓				
	c. Excess manual handling				✓	✓	✓		✓		✓	✓	✓		✓	✓		✓				✓			✓					✓				
	d. High labor turnover	✓	✓	✓																					✓									
	e. Walking for tools, mtls.				✓		✓		✓	✓	✓	✓	✓		✓	✓	✓	✓	✓			✓		✓	✓		✓	✓	✓	✓	✓	✓	✓	

Figure 18–2—Indicators and causes of plant layout

76 *indicators*—or symptoms of the basic difficulties of inefficient layouts
33 *causes*—typically found, which produce the indicators or symptoms observed
73 "solutions"—or areas for investigation in hopes of removing the causes

Observing the chart in Figure 18-2, it will be seen that there are check marks *opposite* each indicator and *under* probable or possible causes of the basic problems or difficulties. the chart attempts to relate the appropriate causes to the

Related Problems – and How to Spot the Causes

Indicators of Layout and Related Problems / Possible Causes of Problems	1. Poor housekeeping	2. Crowded conditions	3. No over-all plan	4. No alternatives	5. Excess mtls. on hand	6. Poor schedule, dispatch	7. Bldg. cube unused	8. Delayed mtl. movement	9. Unsuitable containers	10. Non-use of mech. equip.	11. Manual hndlg., rehndlg.	12. Loads not unitized	13. Long hauls	14. Wrong move, first time	15. Mtls. piled on floor	16. Poor storage equip.	17. Poor flow pattern	18. Unbalanced oprns. seq.	19. No flexibility	20. No space	21. Inadequate equip., fac.	22. Bad space allocation	23. Poor plant layout	24. Poor process methods	25. No aisle markings	26. Poor placing, spacing	27. Poor loc., source or dest.	28. Poor loc. related activ.	29. Employees waiting	30. Material-flow blocks	31. Scattered buildings	32. Rel. areas partitioned	33. Crooked aisles
	GENERAL						MATERIALS HANDLING										PLANT LAYOUT																
MANPOWER UTILIZATION																																	
f. High load, unload time				✓							✓				✓						✓			✓				✓					
g. Cramped work quarters	✓	✓		✓	✓	✓	✓	✓	✓			✓					✓			✓		✓	✓	✓		✓				✓		✓	✓
h. Idle personnel				✓		✓		✓		✓							✓	✓			✓		✓	✓					✓				
i. Awkward work cycle	✓	✓		✓	✓				✓	✓	✓						✓	✓		✓	✓	✓	✓	✓		✓		✓		✓		✓	
j. Difficult handling	✓	✓							✓	✓	✓	✓	✓	✓	✓		✓				✓		✓	✓		✓	✓	✓	✓	✓	✓	✓	✓
k. Waiting for mach. cycle				✓														✓			✓			✓				✓					
l. Waiting for materials				✓		✓		✓	✓	✓		✓	✓	✓			✓	✓						✓			✓	✓		✓	✓		
m. Waiting for help				✓				✓		✓					✓			✓						✓							✓		
n. Unsafe conditions	✓	✓		✓							✓									✓	✓			✓	✓	✓					✓		
o. Operator away from work				✓		✓		✓			✓							✓	✓										✓				
MATERIALS HANDLING																																	
a. Backtracking		✓	✓	✓		✓				✓	✓			✓			✓	✓		✓		✓	✓	✓			✓	✓		✓	✓	✓	✓
b. Zig-zag flow lines		✓	✓	✓		✓				✓	✓			✓	✓		✓	✓		✓		✓	✓	✓		✓	✓	✓		✓	✓	✓	✓
c. Traffic jams	✓	✓			✓	✓	✓			✓	✓	✓	✓	✓	✓		✓			✓		✓	✓		✓	✓	✓	✓		✓			✓
d. Long hauls			✓	✓		✓				✓				✓			✓					✓	✓	✓			✓	✓		✓	✓	✓	
e. Overloaded hndlg. equip.				✓	✓	✓		✓		✓	✓	✓	✓	✓			✓				✓			✓				✓					
f. Underloaded hndlg. equip.						✓			✓			✓		✓			✓							✓				✓			✓		✓
g. Rehndlg. and transfer		✓		✓	✓	✓	✓		✓	✓		✓	✓	✓	✓	✓	✓			✓	✓	✓	✓	✓			✓	✓		✓	✓		
h. Two-man hndlg. tasks				✓					✓	✓											✓			✓						✓			
i. Idle hndlg. equipment	✓	✓				✓	✓			✓					✓	✓			✓				✓					✓					
j. Manual handling		✓		✓				✓	✓	✓	✓	✓		✓	✓	✓	✓				✓		✓	✓		✓	✓			✓			
k. Piece-at-a-time hndlg.				✓					✓	✓	✓	✓		✓	✓						✓			✓		✓	✓		✓	✓			✓
l. Mtls. damaged, pilfered	✓	✓			✓		✓		✓	✓	✓	✓			✓						✓			✓							✓		
m. Slow mtls. movement	✓	✓		✓	✓	✓		✓	✓	✓	✓	✓	✓	✓	✓	✓	✓				✓			✓		✓	✓	✓		✓	✓	✓	✓
n. Much hndlg. betw. oprns.		✓			✓						✓			✓	✓		✓				✓		✓	✓		✓	✓			✓	✓	✓	
o. High equip. mtnce.																																	
p. Obsolete hndlg. equip.			✓							✓																							
RECEIVING, SHIPPING, STORAGE																																	
a. Transfer betw. containers				✓		✓			✓			✓		✓		✓					✓			✓				✓					
b. Excess temp. storage					✓	✓	✓	✓	✓			✓				✓					✓							✓					✓
c. Manual loading, hndlg.				✓	✓				✓	✓		✓			✓						✓			✓					✓				
d. Mtls. piled on floor				✓	✓	✓	✓		✓	✓		✓									✓			✓					✓				
e. Disorderly storage area		✓	✓		✓	✓	✓	✓	✓			✓			✓	✓				✓	✓	✓			✓					✓			✓
f. Unused overhead space									✓	✓		✓			✓	✓					✓												
g. Poor storage equip.																					✓												
h. Misplaced materials	✓	✓			✓	✓		✓	✓		✓	✓		✓	✓	✓	✓			✓	✓									✓	✓	✓	✓
i. Waiting carriers						✓		✓	✓	✓	✓	✓	✓							✓	✓			✓			✓	✓		✓	✓		✓
j. Shipping delays					✓	✓		✓	✓	✓	✓	✓									✓			✓						✓	✓	✓	✓
k. Demurrage charges					✓	✓	✓	✓	✓	✓		✓				✓				✓	✓			✓									
l. Unstandardized containers			✓									✓				✓					✓												

***and related problems.* (Courtesy of Factory Magazine.)**

indicators spotted. Figure 18-3 uses the same format to relate suggestions for improvement to the problems identified in Figure 18-2. The *same* causes are listed on the left of Figure 18-3 as were found at the top of Figure 18-2. In the case of the second chart, the "remedies" are listed along the top. In using the charts:

1. *Systematically review activities* in each activity area with the aid of the chart in Figure 18-2.

CHART 2 – What Causes Plant Layout and

Possible Problem Causes	Suggested Problem Answers	GENERAL															SHIPPING, RECEIVING, STORAGE															
		a. Eliminate	b. Change oprns. seq.	c. Improve housekeeping	d. Reduce inventory	e. Cut goods-in-process	f. Use smaller equip.	g. Use larger equip.	h. Use proper equip.	i. Respect rated capacity	j. Mark aisles	k. Subcontract some work	l. Devise master plan	m. Review over-all plan	n. Improve communications	o. Consult other staffs	a. Cut pickup, dlvry. spots	b. Cut mtls. in storage	c. Make storage accessible	d. Minimize storage time	e. Use outdoor storage	f. Use point-of-use strg.	g. Use unit loads	h. Use same cont'r. through oprns.	i. Minimize containers	j. Receive in usable cont'rs.	k. Use proper cont'rs., racks	l. Design cont'rs. for easy mtl. preposition & disposal	a. Relate hndlg. to layout	b. Construct master layout	c. Change via master layout	d. Plan mtls. flow
GENERAL	1. Poor housekeeping	✓		✓	✓	✓					✓		✓	✓		✓	✓	✓			✓		✓	✓	✓	✓	✓					
	2. Crowded conditions	✓		✓	✓	✓	✓				✓	✓						✓		✓	✓		✓	✓	✓	✓	✓		✓	✓		
	3. No over-all plan												✓	✓		✓													✓	✓		
	4. No alternatives														✓	✓													✓		✓	
	5. Excess mtls. on hand	✓		✓	✓	✓		✓		✓				✓	✓	✓		✓		✓	✓		✓	✓	✓	✓	✓					
	6. Poor schedule, dispatch	✓	✓	✓			✓						✓	✓	✓	✓	✓	✓	✓	✓			✓	✓	✓	✓	✓	✓				✓
MATERIALS HANDLING	7. Bldg. cube unused			✓					✓				✓	✓		✓		✓	✓	✓			✓	✓			✓		✓	✓	✓	
	8. Delayed mtl. movement	✓							✓				✓	✓	✓	✓		✓	✓	✓		✓	✓	✓	✓	✓	✓	✓	✓			✓
	9. Unsuitable containers								✓				✓	✓		✓							✓	✓	✓	✓	✓	✓				
	10. Non-use of mech. equip.	✓	✓	✓					✓					✓		✓													✓			
	11. Manual hndlg., rehndlg.	✓	✓	✓					✓							✓			✓				✓	✓		✓	✓		✓			
	12. Loads not unitized								✓				✓	✓		✓							✓	✓		✓	✓					
	13. Long hauls	✓							✓							✓	✓												✓	✓	✓	✓
	14. Wrong move, first time		✓										✓	✓	✓	✓	✓		✓			✓				✓			✓		✓	✓
	15. Mtls. piled on floor	✓		✓		✓								✓	✓	✓	✓	✓	✓	✓	✓		✓	✓		✓	✓				✓	
	16. Poor storage equip.	✓							✓				✓	✓	✓	✓		✓	✓	✓		✓	✓	✓							✓	
PLANT LAYOUT	17. Poor flow pattern		✓	✓									✓	✓		✓	✓												✓	✓	✓	✓
	18. Unbalanced oprns. seq.	✓	✓											✓		✓													✓	✓	✓	✓
	19. No flexibility						✓							✓		✓								✓		✓	✓		✓	✓	✓	✓
	20. No space			✓	✓	✓	✓						✓	✓		✓		✓		✓	✓		✓	✓			✓		✓	✓		
	21. Inadequate equip., fac.								✓	✓		✓		✓		✓				✓	✓						✓				✓	✓
	22. Bad space allocation												✓	✓		✓		✓		✓			✓				✓		✓	✓	✓	✓
	23. Poor plant layout												✓	✓		✓		✓											✓	✓		✓
	24. Poor process methods	✓	✓					✓	✓	✓		✓		✓		✓												✓	✓	✓	✓	✓
	25. No aisle markings			✓							✓		✓	✓		✓													✓	✓		
	26. Poor placing, spacing		✓										✓	✓		✓													✓	✓	✓	✓
	27. Poor loc. source or dest.		✓											✓		✓	✓												✓	✓	✓	✓
	28. Poor loc. related activ.		✓										✓	✓	✓	✓	✓												✓	✓	✓	✓
	29. Employees waiting	✓	✓											✓	✓	✓	✓		✓			✓	✓	✓		✓	✓		✓			✓
	30. Material-flow blocks	✓	✓	✓	✓	✓					✓					✓	✓	✓		✓	✓		✓	✓	✓		✓		✓	✓	✓	✓
	31. Scattered buildings		✓										✓				✓												✓	✓	✓	✓
	32. Rel. areas partitioned	✓	✓											✓		✓													✓	✓	✓	✓
	33. Crooked aisles	✓		✓							✓			✓		✓											✓			✓	✓	✓

Figure 18–3—Causes and suggested answers to plant layout

2. *List indicators* in the first column of the worksheet in Figure 18-4.
3. *Select causes of indicators* from top of chart in Figure 18-2, and list (by code *number*) in column 2 of worksheet.
4. *List suggested* solutions from the top of the chart in Figure 18-3 (by code *letter*) opposite each cause.
5. *Eliminate duplication* by circling each cause the *first* time it appears in a sub-column or column of the worksheet.
6. *Enter all appropriate solutions* in the proper column, using the *Layout Audit Assignment Sheet* (Figure 18-5), for each major indicator.
7. *Indicate those to whom responsibility is to be assigned*—departments or individuals—for further investigation and analysis of possible causes of each problem indicated.

Related Problems – and What to Do About Them

	PLANT LAYOUT											MATERIALS HANDLING																WORKPLACE LAYOUT														
	e. Plan direct mtls. flow	f. Plan for best flow rate	g. Minimize backtracking	h. Plan for flexibility	i. Plan controlled flow	j. Plan for line prodn.	k. Bring related jobs close	l. Use heavy mtl. near rcvg.	m. Provide aux. flow lines	n. Plan growth, shrinkage	o. Make aisles adequate	a. Larger unitized lds.	b. Don't pile on floor	c. Use mech. hndlg. equip.	d. Design built-in. hndlg.	e. Take direct to use point	f. Schedule, dispatch for min. hndlg.	g. Use gravity	h. Combine proc. & transp.	i. Utilize bldg. cube	j. Mechanize scrap removal	k. Use flexible hndlg. equip.	l. Use low-cost hndlg. equip.	m. Use space-saving equip.	n. Use movable equipment	o. Use var.-spd. cnvrs.	p. Floats, banks on cnvrs.	a. Supply and remove mtls. at proper rate	b. Automate processing	c. Plan opms. sequentially	d. Keep workers on prodn.	e. Plan nec. manual hndlg. meth.	f. Make work space adequate	g. Scan man-machine rel'ships	h. Integrate hndlg. & process	i. Search for improvements	j. Store min. mtl. at place	k. Use multi-oprn. equip.	l. Minimize walking	m. Reduce manual hndlg.	n. Plan equip. arrang't.	o. L-O for motion econ.
			✓							✓	✓		✓			✓				✓	✓			✓	✓		✓						✓			✓	✓				✓	
		✓				✓				✓	✓		✓	✓						✓	✓			✓	✓		✓			✓			✓			✓	✓				✓	
				✓						✓			✓				✓																			✓					✓	
				✓					✓	✓																										✓						
		✓		✓	✓	✓	✓			✓						✓	✓		✓	✓				✓	✓	✓	✓	✓	✓	✓	✓		✓		✓	✓	✓	✓				
	✓	✓	✓	✓	✓	✓	✓	✓	✓			✓	✓	✓			✓					✓				✓	✓	✓	✓	✓	✓	✓		✓	✓	✓		✓		✓	✓	
										✓			✓	✓	✓					✓			✓	✓	✓		✓									✓						
	✓	✓	✓	✓	✓	✓	✓	✓	✓		✓	✓	✓	✓	✓	✓	✓	✓	✓			✓	✓				✓	✓	✓	✓	✓	✓		✓	✓	✓			✓	✓	✓	✓
				✓								✓		✓	✓							✓		✓											✓	✓						
		✓			✓								✓	✓	✓						✓	✓	✓		✓			✓	✓		✓			✓		✓						
		✓	✓		✓		✓					✓	✓	✓	✓	✓	✓	✓	✓		✓	✓	✓	✓			✓	✓	✓		✓	✓		✓	✓	✓		✓	✓	✓	✓	✓
				✓								✓	✓	✓																						✓						
	✓		✓		✓		✓		✓			✓		✓		✓	✓	✓	✓			✓	✓	✓	✓		✓	✓				✓				✓	✓		✓	✓	✓	
	✓	✓	✓	✓	✓	✓	✓	✓								✓	✓											✓								✓						
													✓	✓			✓			✓	✓	✓	✓	✓	✓		✓	✓				✓	✓			✓	✓			✓		
										✓		✓	✓		✓					✓			✓	✓			✓						✓		✓	✓	✓					
	✓	✓	✓	✓	✓	✓	✓	✓	✓	✓	✓					✓	✓		✓										✓	✓	✓					✓		✓	✓		✓	✓
		✓		✓	✓	✓																				✓		✓	✓	✓	✓			✓	✓	✓		✓			✓	✓
				✓					✓	✓	✓															✓				✓				✓		✓					✓	
										✓	✓									✓				✓	✓		✓	✓					✓			✓	✓	✓			✓	
										✓								✓		✓			✓	✓	✓		✓	✓							✓	✓	✓					
								✓		✓	✓									✓				✓	✓								✓			✓	✓				✓	
	✓	✓	✓	✓		✓	✓	✓	✓	✓	✓								✓											✓						✓					✓	
		✓	✓	✓	✓	✓	✓							✓					✓			✓				✓	✓	✓	✓	✓	✓	✓		✓	✓	✓		✓	✓	✓	✓	✓
										✓	✓																									✓						
		✓	✓	✓		✓	✓			✓													✓							✓	✓					✓			✓	✓	✓	
		✓	✓	✓			✓	✓		✓						✓	✓						✓					✓		✓						✓			✓	✓	✓	✓
		✓	✓	✓		✓	✓			✓													✓					✓	✓	✓					✓	✓			✓	✓	✓	✓
	✓	✓	✓	✓	✓	✓	✓		✓			✓		✓	✓	✓	✓	✓	✓		✓	✓	✓	✓		✓	✓	✓	✓	✓	✓	✓	✓	✓	✓	✓		✓		✓	✓	✓
	✓	✓	✓		✓	✓			✓	✓	✓						✓	✓		✓		✓	✓				✓	✓		✓		✓	✓			✓	✓	✓	✓	✓	✓	
	✓	✓	✓		✓	✓	✓		✓	✓								✓				✓					✓	✓		✓						✓			✓	✓	✓	
	✓	✓	✓		✓	✓	✓		✓	✓																	✓	✓		✓						✓			✓	✓	✓	
	✓		✓								✓																									✓	✓					

and related problems. (Courtesy of Factory Magazine.)

8. *Plan for a follow-up* to assure that:
   —Proper causes have been identified.
   —Best possible solutions have been devised.
   —Solutions have been installed.
   —Solutions are solving the indicated problem(s).
   —Solutions are *not* creating additional problems.

While such a procedure may appear rather complicated, it will be seen that it *does* provide an orderly method of suggesting one, or a combination, of 73 possible solutions or areas for investigation in improving the layout difficulties. The alert analyst would be aware of many of the suggestions indicated, but the proper solution may lie in one of the areas of which he was not cognizant.

WORKSHEET FOR SPOTTING AND ANALYZING LAYOUT PROBLEMS

PLANT AREA Acme Mfg Co **OBSERVER** T. L. Foster **DATE** Oct. 24

Indicators from Audit Check Sheet	Possible Causes of Problems Indicated	Preliminary Tabulation of Suggested Solutions or Areas for Investigation				
		General	Shipping Receiving, Storage	Plant Layout	Materials Handling	Workplace Layout
j. scattered buildings	3 lack of planning	ⓛ, ⓜ, ⓞ	—	ⓐ, ⓑ, ⓗ, ⓝ	ⓑ, ⓕ	ⓘ, ⓝ
	4 lack of alternatives	ⓝ, o	—	a, ⓒ, h, ⓜ, n	—	i
	7 failure to use bldg. cube	ⓒ, ⓗ, l m, o	ⓑ, ⓒ, ⓓ, ⓖ, ⓗ, ⓚ	a, b, c, n	b, ⓒ, ⓓ, ⓘ, ⓛ, ⓜ, ⓝ	i
	22 improper space allocation	l, m, o	b, d, g, k	a, b, c, ⓓ, ⓛ, n, ⓞ	i, m, n	ⓕ, i, ⓙ, n
	23 poor plant layout	l, m, o	b	a, b, d, ⓔ, ⓕ, ⓖ, h, ⓙ, ⓚ, l, m, n, o	h	c, i, n
	27 poor location of source or destination	b, m, o	ⓐ	a, b, c, d, f, g, h, k, l, n	ⓔ, f, l	ⓐ, c, i, ⓛ, ⓜ, n, ⓞ
	28 poor location of related activities	ⓑ, l, m, n, o	a	a, b, c, d, f, g, h, j, k, n	l	a, ⓑ, ⓒ, ⓗ, i, l, m, n, o

Figure 18–4—Worksheets for spotting and analyzing layout problems.

Check sheets can also be used as a means of evaluating or comparing alternative layouts. Those shown in Figures 18-6 and 7 may be used, or more specific check sheets may be designed. Figure 18-6 was developed from a combination of the left-hand and top lists of indicators and causes from Figure 18-2. Somewhat more accurate would be the actual weighting of criteria, plus a rating of each of the proposed alternatives. The Layout Evaluation Sheet (Figure 18-7) contains selected items from previous check sheets, principles, objectives, etc., and contains those which it is felt there is a reasonable chance of observing, and evaluating "on paper". For a specific layout, the form might be adapted or rewritten to match the project at hand.

In using the *Layout Evaluation Sheet,* it is necessary to:

1. *Identify the project.*
2. *Establish the relative weight* of each major criterion, i.e., items 1–10. This should be based on company policies, aims, or objectives and should be done by appropriate management personnel. The weights should be allocated to equal 1, 10, 100, etc. If desired, the sub-items might even be weighted within each major group. If this much detail is not desired, the sub-items can be used as guides in observing and rating the major factors. Enter weights in column two.

LAYOUT AUDIT ASSIGNMENT SHEET

Indicator j. Scattered Buildings

Causes	Suggested Solutions or Areas for Investigation	Assigned for Investigation to:
3. lack of planning 4. lack of alternatives	Change oprns. seq. Improve housekeeping Use proper equipment Devise master plan Review overall plan Improve communications Consult other staffs	R.T.S.
7. Failure to use building cube 22. improper space allocation	Cut pick-up delivery spots Cut materials in storage Make storage accessible Minimize storage time Use unit loads Use same cont'r, through oprns. Use proper cont'rs., racks	N.F.B
23. poor plant layout 27. poor location of source of destination	Relate hndlg. to layout Construct master layout Change via master layout Plan mtls. flow Plan direct mtls. flow Plan for best flow rate Minimize backtracking Plan for flexibility Plan for line prodn. Bring related jobs close Use heavy mtl. near rcvg. Provide aux. flow lines Plan growth, shrinkage Make aisles adequate	J.T.J.
28. poor location of related activities	Don't pile on floor Use mech. hndlg. equip. Design built-in hndlg. Take direct to use point Schedule, dispatch for min. hndlg. Utilize bldg. cube Use low-cost hndlg. equip. Use space-saving equip. Use movable equipment	D.R.G.
	Supply and remove mtls. at proper rate Automate processing Plan oprns. sequentially Make work space adequate Integrate hndlg. and process Search for improvements Store min. mtl. at place Minimize walking Reduce manual hndlg. Plan equip. arrang't. L-O for motion econ.	G.L.K.

Figure 18-5—Plant layout audit assignment sheet.

FACILITY EVALUATION CHECK SHEET							
Indicators and Causes of Layout Problems	Alternative Layouts						Suggestions, Comments Remarks on Items To Be Checked Out
	1		2		3		
	OK	Check Into	OK	Check Into	OK	Check Into	
I. General							
1. Crowded aisles							
2. Hazardous areas							
3. High accident rate							
4. Crowded conditions							
5. Poor housekeeping							
6. No overall plan							
7. Existing physical cond's.							
8. Excess equipment							
9. No alternative							
10. Excessive mat'l. in process							
II. Production							
1. High indirect labor							
2. Obstacles in material flow							
3. Related work scattered							
4. Uneven flow rate							
5. Scattered buildings							
6. "Disappearing" aisles							
7. Partitions separating related areas							
8. Blocked doorways							
9. No flexibility							
10. Continued use of temp. arrangement							
11. Working in aisles							
12. Maintenance hard to perform							
13. Manufacturing cycle too long							
14. Idle production equipment							
15. Overloaded production equipment							
16. Low production density							
17. Vacant floor space							
18. Poor quality							
19. Machine breakdown stops line							
20. Inaccessible equipment							
21. Bottleneck operators							
22. Crooked aisles							

FACILITY EVALUATION CHECK SHEET—Cont'd.							
Indicators and Causes of Layout Problems	Alternative Layouts						Suggestions, Comments Remarks on Items To Be Checked Out
	1		2		3		
	OK	Check Into	OK	Check Into	OK	Check Into	
III. Manpower Utilization							
1. Too many men moving materials							
2. Handling by skilled employees							
3. Excess manual handling							
4. Walking for tools, equipment, etc.							
5. High load/unload time							
6. Cramped work quarters							
7. Idle personnel							
8. Awkward work cycle							
9. Difficult handling							
10. Waiting for machine cycle							
11. Waiting for material							
12. Waiting for help							
13. Unsafe conditions							
14. Operator away from work							
IV. Material Handling							
1. Backtracking							
2. Zig-zag flow lines							
3. Long hauls							
4. Traffic jams							
5. Overloaded handling equipment							
6. Underloaded handling equipment							
7. Rehandling							
8. Two-man handling task							
9. Idle handling equipment							
10. Manual handling							
11. One-at-a-time handling							
12. Material damaged, pilfered							
13. Slow material movement							
14. Excess handling between operations							
15. High MH equipment maintenance							
16. Obsolete MH equipment							
17. Failure to use building cube							
18. Delays in material movement							

Figure 18–6—Facilities evaluation checksheet.

FACILITY EVALUATION CHECK SHEET—Cont'd.

Indicators and Causes of Layout Problems	Alternative Layouts 1		2		3		Suggestions, Comments Remarks on Item To Be Checked Out
	OK	Check Into	OK	Check Into	OK	Check Into	
19. Lack of suitable containers							
20. Unstandardized containers							
21. Not using mechanical equipment							
22. Loads not unitized							
23. Wrong move, 1st time							
V. Plant Layout							
1. Poor flow patterns							
2. Unbalanced operation sequence							
3. No flexibility							
4. No space							
5. Inadequate equipment							
6. Poor space allocation							
7. Poor layout							
8. Ineffective processing methods							
9. No aisle markings							
10. Poor machine spacing							
11. Poor machine placement							
12. Poor location-source or destination							
13. Poor location of related activities							
14. Material flow blocked							
VI. Receiving, Storage, Shipping							
1. Excessive temporary storage							
2. Transfers between containers							
3. Manual unloading, loading							
4. Material piled on floors							
5. Disorderly storage area							
6. Unused overhead space							
7. Poor storage equipment							
8. Misplaced material							
9. Waiting carriers							
10. Shipping delays							
11. Demurrage charges							
12. Unstandardized containers							

Figure 18-6—Continued.

3. *Evaluate each factor or sub-factor* on a basis of 1 to 10, as indicated in the previous tabulation. If more than one layout is being evaluated, rate all layouts on a specific item at one time to aid in comparison. Enter ratings in appropriate columns.
4. *Multiply factor weight by rating* to obtain weighted rating for each, and enter result in appropriate column.
5. *Add each "weighted-rating" column* to determine the score for each layout.

In using any such evaluation technique, it is necessary to apply quantitative measures to factors that are usually considered qualitative. Nevertheless, such a technique will force better thinking from those involved in the evaluation process than could be obtained by purely qualitative or subjective means. These forms can be used alone, or in conjunction with the procedure accompanying Figure 18-2, etc. They can also be used as a preliminary survey of a facility, in a search for improvement opportunities.

Many of the "graphical" techniques explained in Chapter 6 may be used in the evaluation process, especially if used in the *before and after* context. Such techniques as:

1. *Assembly chart*
2. *Operation Process Chart*
3. *Multi-Product Process Chart*
4. *String Diagram*
5. *Process Chart*
6. *Flow Diagram*
7. *Flow Process Chart*
8. *From–To (Travel) Chart*
9. *Procedure Chart*

could be applied for each alternative, or *before* situation, to compare with (1) each other, or (2) an *after* version.

Efficiency Indices

A still more mathematical approach involves a set of ten indices,[2] reproduced here for consideration or adaptation to a specific problem.

1. *Index of Indirect Material Handling* $= a/b$

a = The sum of the distances that a part moves automatically from machine to machine, without external materials handling. (*External material handling* means manual movement of production material from one location to another, in boxes, tote pans, and the like.)

b = The total actual distance that a part travels on the production route from raw stores to finished stores. For dealing with smaller organizational units, this can be rephrased as: the distance from the layout area entrance to the exit from the layout area.

This index has been found to be consistent and accurate and is recommended as a good measure of the efficiency of the production route with respect to the mechanized handling of material.

[2]S. P. Gantz and R. B. Pettit, "Plant Layout Efficiency," *Modern Materials Handling*, Jan. 1953.

LAYOUT EVALUATION SHEET

Plant or Project ______________________________ Date ________

Brief Drescription of Project ______________________________

______________________________ Evaluated by: ________

Criteria, Factor, Characteristics	Weight	Alternate		Alternate		Alternate		Comments and Notes
		Rating	Weight. Rating	Rating	Weight. Rating	Rating	Weight. Rating	
1. GENERAL a. overall appearance b. crowded conditions c. excess or duplicate equipment d. ease of supervision e. ease of production control f. provisions for inspection g. access for repairs h. adequate exits								
2. FLOW OF MATERIALS a. planned b. "straight" line c. good equipment arrangement d. good equipment utilization e. adequate aisle space f. straight aisles g. straight flow lines h. minimum back-tracking i. related operations close together j. obstacles in material flow k. no apparent bottlenecks l. line production where practical								
3. FLEXIBILITY TO MEET CHANGING CONDITIONS								
4. EXPANDABILITY WITHOUT MAJOR DISRUPTION								
5. SPACE UTILIZATION a. fully utilized b. effective use of space available c. effective use of "cube"								
6. MATERIALS HANDLING a. materials handling planned for —to production —through production —to assembly —through assembly —to shipping b. minimum handling c. short hauls d. mechanized where practical e. integrated system f. use of unit loads g. provisions for scrap handling								

LAYOUT EVALUATION SHEET (Continued)

Criteria, Factor, Characteristics	Weight	Alternate		Alternate		Alternate		Comments and Notes
		Rating	Weight. Rating	Rating	Weight. Rating	Rating	Weight. Rating	
6. MATERIALS HANDLING (Continued) h. operations during transit i. materials used from vendor's cont'rs. j. effective use of gravity								
7. STORAGE ARRANGEMENTS a. rough or raw materials b. in-process materials c. finished parts or components d. finished products or assemblies								
8. SHIPPING AND RECEIVING a. provisions for common carriers b. docks covered c. receiving close to first operations d. provisions for receiving inspection e. mechanical handling where practical f. shipping close to last operations g. provisions for packing								
9. SERVICE ACTIVITIES a. convenience of location b. adequate coverage —first aid —toilet & wash facilities —smoking areas —drinking fountains —lockers or coat racks —food service —parking —tool room —tool crib —maintenance —offices adequate and convenient —rubbish collection —fire extinguishers, sprinklers								
10. BUILDING AND UTILITIES a. size adequate, reasonable b. shape practical c. bay size reasonable d. clear height sufficient e. entrances, exits adequate & convenient f. necessary utilities provided for —light —heat —water —sewage —drains —gas —air —electricity —telephone								

Figure 18-7—Layout evaluation sheet.

2. *Index of Direct Material Handling* $= b$. This value represents the exact distance a part or piece is required to travel during production. It is not an index, properly speaking, but simply a number of feet. It is a good measure of the efficiency with which the production route is laid out and can be used to compare plants or areas manufacturing the same type of product. It was found to be more accurate than any ratio investigated.

3. *Index of Gravity Utilization* $= d/e$

d = The sum of the vertical distance that gravity feed is used in a multi-story plant (it gives peculiar and unreasonable results when applied to single-story operations).

e = The total vertical distance up or down that a part moves, involving either machine or human effort, from the layout area entrance to the layout area exit of a multi-story plant.

Although this index has not been fully evaluated, it can be considered a good indication of the extent to which gravity is used in moving parts up and down.

4. *Prime Index of Automatic Machinery Loading* $= f/100g$

f = The sum of the percentages of machine down-time from all cases where the individual percentages of down time are equal to or less than 50% of the individual work cycles. (*Down time* is the portion of the work cycle in which the machine is loaded and unloaded.)

g = Total number of operators on these machines.

This index is an accurate indicator of the efficiency obtained by grouping machines for multimachine operation. It should be noted that it is used only when the machine time portion of the over-all work cycle is automatic and machines may be left unattended while in operation.

5. *Secondary Index of Automatic Machinery Loading* $= h/100g$

h = The sum of the percentages of machine down-time from all cases where the individual percentages of down-time are greater than 50% of the individual work cycles.

This criterion is similar to *Index 4* above, except that it is used only for odd groupings of machines that might not be adapted to *Index 4*.

6a. *Index of Production Line Flexibility* $= j_1/k_1$

j_1 = The number of machines or work stations performing operations on the part under consideration, so designed that they can be moved to a new location in the same production line in one working shift.

k_1 = The total number of machines or work stations performing operations on the part under consideration, in the production line.

Machine—a nonportable device with a separate or individual power source.

Work station—the area covered by the tools, equipment, machines, and in-process material necessary to the performance of a given operation.

6b. *Index of Work Station Flexibility* $= j_2/k_2$

j_2 = The number of machines or work stations within the area under consideration so designed that they can be moved to any other location in one working shift.

k_2 = The total number of machines or work stations within the area under consideration.

Index 6a is satisfactorily used as a measure of machine flexibility in the production line in relation to the flow of the part. *Index 6b* is a successful measure of machine and work station arrangements in terms of utilization of men and machinery.

7. *Index of Floor Area Loading Density* $= \dfrac{(m + 2)(n + 2) + p}{q - (r + u)}$

m = Extreme machine length.

n = Extreme machine width.

p = The total work area normally required by an operator in the performance of his job.

q = Total layout floor area.

r = Total aisle area.

u = Total floor area occupied by temporary or controlled storage of material, or tools and equipment required to modify this material.

This index is an accurate indication of the efficiency with which plant floor space is utilized. *Machine* here means all production machinery, including conveyors resting on or near the floor, but excluding overhead conveyors that pass above and clear of other machinery. It should be emphasized that the areas occupied by the machines, work stations, and operators may be totally independent of each other; that is, a productive work area may consist of machines or work stations operated by workers, or of machines operating independently, or of workers operating independently.

8. *Index of Aisle Space* $= r/q$

r = Total aisle area.

q = Total layout floor area.

This index gives a true indication of the overall utilization of layout floor area for aisles. An increase or decrease in aisle area is readily reflected by an increase or decrease in this index value—the particular manufacturing conditions encountered will determine whether a high or low value is desirable.

9. *Index of Storage Space* $= (q - u)/q$

q = Total layout floor arca.

u = Total floor area occupied by temporary or controlled storage of material, or tools and equipment required to modify this material.

This index is an adjunct to the *Index 7* (Floor Area Loading Density), where u retains the same meaning. It gives a true indication of the overall utilization of layout floor area for storage of in-process material or the tools and equipment required to modify this material. It readily reflects an increase or decrease in storage area.

10. *Index of Storage Volume Utilization* $= v/w$

v = Volume occupied by raw material or finished goods at the normal maximum level of storage.

w = Total volume available for storage of raw material or finished goods.

This criterion has not been thoroughly tested but shows promise of being an excellent measurement of the cubic utilization of storage or warehouse spaces, such as receiving and shipping. It is also a good potential measure of proper packaging, palletizing, or material handling, as applied to storage systems.

These indices might be calculated for several alternative layouts and tabulated to determine which layout is best.

Another set of indices (as ratios) was developed later,[3] borrowing from and building upon the above, to make the index concept more practicable; these are:

A. *Material Handling Labor Ratio*

$$= \frac{\text{Personnel assigned to material handling duties}}{\text{Total operating work force}}$$

B. *Direct Labor Handling Loss Ratio*

$$= \frac{\text{Material handling time lost by direct labor}}{\text{Total direct labor time}}$$

C. *Movement/Operation Ratio*

$$= \frac{\text{Total moves}}{\text{Total productive operations}}$$

D. *Manufacturing Cycle Efficiency*

$$= \frac{\text{Sum of all production operation cycle times}}{\text{Elapsed time in the production cycle}}$$

E. *Space Utilization Efficiency*

$$= \frac{\text{Cubic feet usefully occupied}}{\text{Net usable cube}}$$

F. *Equipment Utilization Ratio*

$$= \frac{\text{Actual output}}{\text{Theoretical capacity}}$$

G. *Aisle Space Potential*

$$= \frac{\text{Current aisle floor space minus theoretical optimum aisle floor space}}{\text{Current aisle floor space}}$$

Although the original source is now out of print, the basic data can be found in the article referenced.

[3] J. R. Bright, *A Management Guide to Productivity* (Eaton Corp., Industrial Truck Division, 1961); also in *Mill and Factory,* July 1961:77.

Cost Evaluation of Layouts

Probably the ultimate and most desirable evaluation of a layout is in terms of dollars. While this is more difficult to accomplish, certain costs can be determined and compared. The costs should be worked up into a form to suit the project or projects being evaluated and compared. The project showing the lowest total or the greatest savings would then have to be studied in the light of other factors—advantages, disadvantages, objectives, intangibles, etc.—in coming to the final decision.

One of the more common methods of comparing alternative projects calculates the *Return on Investment* (ROI) for each. The details are beyond the scope of this book, but can be found in any engineering economy textbook.

Quantitative Evaluation Techniques

Although some of the above may be considered mathematically quantitative in the modern sense, *quantitative* usually implies more sophisticated mathematical, statistical, modelling or computer-oriented approaches. These were discussed in Chapter 7, and some may be found useful in the evaluation or comparison process.

Still another approach is in the use of the computerized layout models discussed in Chapter 13. These techniques actually score or rate alternatives, in terms of numerical values representing closeness of related activities, or cost of handling. If time permits, it would be useful to extract the necessary data from the new layout, and put it into the PLANET, the CRAFT, or the CORELAP program. Although this test should probably have been made at the *Area Allocation Diagram* stage of the layout procedure, it could well be repeated here, after the layout has been completed. But, it should be remembered that these techniques consider neither the amount of work subsequently done on the layout, nor the many judgment or intangible factors that play such an important role in an efficient layout design. Neither do they consider the experience, nor even desires, of the experts or management personnel.

Any or several of the above evaluation techniques might be used on a given layout project. The analyst must study each individual problem and determine which techniques would be most useful and applicable to the situation at hand.

Evaluation Procedures

In general, the evaluation will probably be made by those who developed the layout, or their supervisors. However, other concerned persons may be brought into the process, such as:

1. Supervisory personnel
2. Personnel manager
3. Safety director
4. Production control supervisor
5. Material handling supervisor
6. Plant engineer
7. Local fire and police department officials
8. Security experts
9. Consultants

One important aspect of the evaluation process is the recording of comments, suggestions, and criticisms as they are made. Only in this way can the analyst be sure to consider them all, after the evaluation process is concluded.

Making the Alterations

The last step in the evaluation process is making the necessary alterations to the master layout—as agreed to by those in a position to authorize such changes. In general, the approach would be:

1. *Review and evaluate* suggestions.
2. *Select* those for implementation.
3. *Actually make* the alterations.
4. *Re-evaluate.*

When the necessary alterations have been made, the layout must be formally presented to those who will approve it.

Presenting the Layout to Management

One of the last steps in the layout project is the presentation and selling of the layout to management. The importance of this step[4] can not be over-emphasized. Whereas company officials must be concerned down to the lower levels of installations and facilities, some are too seldom exposed to such concerns to appreciate the particulars without a rather extensive supporting commentary detailing how reduction of expenditures and increase in production can be expected from the arrangements.

In the light of the exhaustive specialized work that goes into the design of a layout, it is not surprising that clarification of the layout, with "correlated facts, sound calculations, and satisfactory interpretations," is necessary before foremen and top executives will be able to proceed with the evaluation that only they are qualified to pursue. And for the presentation itself, it is only the layout engineer who is qualified to deal with the inevitable queries, either as the one who actually presents the layout or at least as a consultant during the proceeding.

Making up the total presentation can be:

1. *The Visual Presentation:* (a) the layout itself; (b) supplementary details and facts; and (c) supplementary charts and displays.
2. *An Oral Report.*
3. *A Written Report.*

The visual presentation. Needless to say, the first consideration is the accumulation of all pertinent data and exhibits for supporting the presentation and for examination by management. These could include:

1. *Final layout.*

[4] Air Material Command, *Plant Layout Engineering Manual,* No. 66-9 (Wright-Patterson Air Force Base, Ohio).

2. *Flow diagrams* (or other simplified versions to represent overall flow pattern in simpler terms).
3. *Evaluation forms* or tabulations.
4. *Cost comparison sheets.*
5. *Summary of intangible benefits.*

The actual presentation itself should be well planned and rehearsed, and the above data should be at hand for reference or display if called for.

The formal written report, where desirable or required, might include:[5]

A. Title Page
B. Letter of Transmittal
C. Table of Contents
D. Discussion of Project:
    1. *Introduction:*
        a. Purpose of project and presentation of the problem.
        b. Basic facts about facility—as necessary:
            —Product
            —Production methods
            —Annual sales
            —Number of employees
            —Plant size
            —Building construction
            —Plant location
    2. *Condensed statement of proposal.*
    3. *Discussion of major activities:*
        a. Receiving and shipping
        b. Storage
        c. Maintenance, tool crib, etc.
        d. Employee facilities
        e. Office
        f. Production—discussion of each activity, to include flow of material, special features, material handling problems, etc.
        g. Others as appropriate
    4. A *statement of costs* for installing the layout.
    5. A *statement of expected savings.*
    6. A *statement of the expected increases in operating expenses.*
    7. A *cost analysis and engineering economy study* summarizing the effects of the proposals and comparing the proposals with present conditions.
    8. *Recommendations for improvements* in related areas.
    9. *Schedule of installation.*
    10. *Procedure for re-layout* or re-arrangement.
    11. *Conclusions:*
        a. General comments and observations
        b. Recommendations
    12. *Appendix:*

[5]*Manual,* No. 66-9.

a. Production data and related information
—Sales Forecast
—Drawings
—Bills of Material
—Production calculations
—Routings
b. Material flow design
—Assembly Chart
—Operation Process Chart
—Flow Diagram
c. Other items, as necessary
—Activity Relationship Chart and Diagram
—Production area determination
—Storage and warehousing area determination
—Service area space determination
—Area Allocation Diagram
—Plot Plan
—Final layout
—Indices of layout efficiency, or other evidence of layout evaluation

It should be remembered that the more complete the picture presented to top management, the better the chance of gaining approval.

In this connection it may be well to re-emphasize the value of scale models in selling the layout. Even though the models may not be too fancy or exact, they add immeasurably to the ability of a person not familiar with the details to evaluate the planned layout.

A Michigan automobile manufacturer spent $200,000 making a full-size mock-up of a portion of a new assembly line that was to introduce a radical idea in progressive assembly. This full size model helped in many ways. It sold the management on the general idea, and also pointed out to methods engineers that they had a few kinks to iron out.

No step is more important to the plant layout engineer than the presentation of his final work to management for approval. No effort should be spared in the preparation of any necessary visual aids or supplementary data. Certainly it would be disastrous to lose the battle at this crucial point in a long and arduous project. The presentation deserves top effort in preparation—as well as whatever time must be spent in practice presentations or rehearsals.

Displaying the Layout

Good layout practice demands that the master layout not only be kept up-to-date, but that it be conveniently stored or displayed for easy reference or use. Both models and 2-D layouts are frequently referred to by builders, architects, and management personnel and must therefore be mounted, both for protection and for ease of examination or up-dating.

Some mount the boards on swinging panels fastened to the wall, so that they can be folded flat against the wall or swung out as desired.

A technique used by organizations having multi-story buildings is to build a frame, representing in scale and shape the building itself. Layout boards are then inserted as floors in their respective slots or positions.

By far the two most common methods are mounting (1) on a table or tables at a convenient work height, or (2) on the wall. If the layout is of a very large area (some represent over 1,000,000 sq ft of plant area), either method may become a space problem, and many ingenious ways have been devised to overcome this.

One method is to mount the layout in sections, on say 4 by 8-ft panels, and hang them vertically, in layers, like window sashes—one in front of the other—with counterweights or sash springs.[6]

When a large master layout is placed on the wall, it is usually made up in dismountable sections that can be removed for updating, etc.

Another ingenious method was used, with slight variations, by two companies. The layouts were mounted in one case on a triangular[7] table, and in the other on a pentagonal[8] table. Actually, the tables were three and five surfaced prisms mounted on trunions for easy rotation. The 3 and 5 sides allow ample foot or knee room under the down sides, when a side is in working position.

One company put layouts representing two floors on separate boards, mounted on frames, and supported the frames by auto bumper jacks built into the frames. For working purposes, the upper floor was jacked up sufficiently to permit work on the lower one.

And one of the most elaborate,[9] was the model for a five-story building. Each floor was suspended between supports in much the same manner as the moving span of a vertical lift bridge, with each floor individually counterweighted.

Many variations and combinations of these methods of presentation are in use, and only the ingenuity of the layout engineer limits the extent to which one may go. Figures 18-8 through 13 illustrate some of the points discussed above.

In all cases it is wise to protect the layout from dust, light, and tampering by unauthorized people, by a plastic cover, or curtains.

The Follow-up

Even though the layout project may be assumed to be complete, with the filing of the prints or the display of the model, this is not entirely true. Like almost everything else in industry and business, the layout must be kept up-to-date and must be continually studied for further improvement opportunities.

A procedure should be planned and established to insure that any and all changes made in the building are also made on the layout, so it is current at all times. This is especially important immediately after completion of the installa-

[6]C. H. Parks, "Disappearing Boards Show Plant Layout in Sections," *Factory*, Dec. 1947:81.
[7]C. C. Williams, "Plant at a Glance," *Factory*, July 1938:46.
[8]A. M. Bowers, "Pentagonal Layout Table Saves Office Space," *American Machinist*, Nov. 18, 1948:49.
[9]"Scale Model Helps Plan Multi-story Handling," *Factory*, June 1947:74.

Figure 18-8—A plant layout model serves as an excellent basis for study and discussion. (Courtesy of Fisher Body Div., General Motors Corp.)

tion, as many minor changes may have been made while the project progressed. All layouts, drawings, and models should be corrected to reflect as-installed conditions before they are filed away.

Needless to say, even the final, revised layout is seldom—if ever—perfect. Continuous study of plant operations will reveal the need for periodic changes—as will the several situations described in Chapter 1, under Plant Layout Problems. The alert layout engineer should always be on the lookout for improvement opportunities and the adaptation of the layout to keep abreast of the latest developments in methods and processing.

Approval of the Layout

After the layout has been made and checked as indicated above, there remains the problem of having the layout approved. Persons who should approve the layout will vary from organization to organization depending on procedures, and size of

Figure 18–9—Plant scale model showing use of a roof to protect it from dust. (Courtesy of Borg-Warner Corp.)

enterprise. Some of those who probably will either want to, or will be required to, approve the layout before putting it into effect are:

1. Department foreman
2. General superintendent
3. Production manager
4. Production engineer
5. Methods engineer
6. Plant layout engineer
7. Plant manager
8. General manager or top executive

Each of these will, of course, have been consulted during the construction of the layout, as will many others. Some companies make it a practice to gain the acceptance of the employees who will be working in the area, as pointed out previously. One plant placed a three-dimensional scale model of a proposed new layout of its manufacturing department in the cafeteria and solicited suggestions on the spot from the employees. Several suggestions were received on points that might have been overlooked and thus have caused difficulties after the layout was installed.

Frequently, the official approval or signatures will come as a result of the meeting at which the layout is presented in its final form and discussed by interested persons. When the organization is large, and the layout consists of a great many individual areas, it may be wise to maintain an actual log of approvals,

Figure 18–10—Filing of completed layouts on overhead trolleys. (Courtesy of Industrial Photo Products, Inc., and Pontiac Motor Div., General Motors Corp.)

by layout section. A form for this purpose is shown in Figure 18-14. One company used a large rubber stamp with spaces provided for representatives of indicated departments to sign-off.

Reproducing the Layout

Another necessary step is making a reproduction of the master layout, since the layout is made to be used. Some of the persons who will want to use, or at least refer to, the layout are the millwrights, the plant engineer, electricians, building contractor, architects, sub-contractors, methods engineers, and plant executives.

As it is usually desired to file and preserve the original layout, some method must be devised to make reproductions, which at this stage might consist of:

1. Grid sheet and templates
2. Prints of grid sheets
3. Scale models
4. Drawings or tracings
5. Any combination of these

Figure 18–11—Cabinets for storing template layouts. (Courtesy of Marathon Corp.)

All of these forms are commonly used by people who must refer to them. Probably the blue-print or white-print is the most common type of reproduction. When scale models are used, photographs can be made for reference. One point should be mentioned here about photographing scale model layouts. Photograph only a small section at one time. The small sections can then be fastened together to reconstruct the layout. If too large a section is taken there will be distortion near the edge of the area. This distortion makes the print difficult to use in determining the precise location of facilities. It is for this reason that many plants using models for layout planning also make template layouts of the final arrangement.

Installing the Layout

Last, but by no means least, comes the actual installation of the completed layout. All the preceding detailed planning is of little avail if the installation is not in exact accordance with the carefully made plans. It is of the utmost importance that the layout engineer supervise the work in connection with the culmination of his project.

When the final layout is turned over to the architects or contractors, the

Figure 18–12—Unusual arrangement for storing scale model of multi-story plant; each "floor" is supported by counterweights. (Courtesy of Alderman & Alderman, Architects.)

layout engineer must carefully observe all subsequent planning to insure that his work is properly incorporated into the building drawings. Often, much of the value of the layout work can be nullified by small changes in such details as building shape, column spacing, truss height, orientation on the property, interior partitions, window and door location, and floor drains. Each step in the building design should be followed closely to catch any changes that may be inadvertently made by the building designer or contractor—or even purposely made to suit *their* objectives, or cost relationships.

Before actual construction and installation can begin, much additional work must be carried out to complete the detail plans for equipment procurement and installation, or the actual move itself. Among the more important items of work yet to be done are:

1. *Detail drawings* of certain phases.
2. *Precise specification* of production equipment.
3. *Precise specification* of material handling equipment.
4. *Detailed listing* of all equipment involved, and the utility requirements.
5. *Actual plans and schedule* for the building construction or the move into the proposed space.

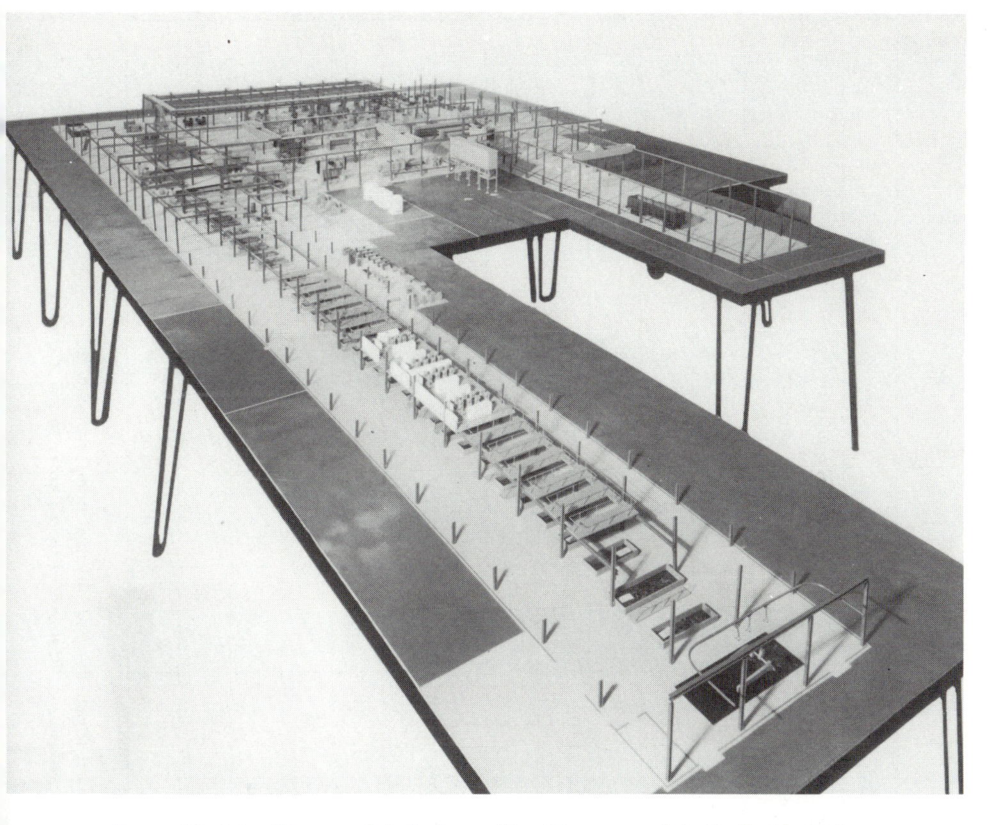

Figure 18–13—Plant model display table. (Courtesy of A. O. Smith Co.)

Detailed drawings should be made (possibly by plant engineering, the architect, or the contractor) of any unusual features of the layout to insure compliance with the original plans. This might include such features as mezzanines, balconies, docks, pits, equipment overhead or on the roof, and foundations for certain equipment.

Exact production equipment specifications should be on hand now, from the planning work. If not, complete details must be worked up for use as purchase specifications or as an aid to installation. For actual use during installation, the necessary data are recapped on a form similar to that in Figure 18-15, which is followed closely to insure proper utility connections at each machine location.

The entire construction and installation procedure is best planned and coordinated with either a Gantt chart (Figure 18-16) or a critical path network (Figure 6-20)—or possibly both, since the Gantt chart may offer a better visual picture of progress vs. time.

CPM and PERT

One of the most helpful techniques for both planning and implementing a layout and construction project is critical path analysis and scheduling. The two most

LAYOUT APPROVAL FORM											
AREA DESIGNATION	FUNCTION	OPERATING APPROV.		STAFF APPROVALS							
		SECTION HEAD	FOREMAN	PROCESS ENG'G.	PROD'N. CONTROL	INDUSTR. ENG'G.	INSPECTION	PLANT ENG'G.	PERSONNEL	SECURITY	OTHER
A-1	Receiving	A. Brown	I. Jones		M. Noel	P. Quick	S. Tile	U. Venn	W. Young	X. Zole	
B-1	Stores	B. Clay									
C-1/5	Machine Shop	C. Doe	J. Kopp	L. Moore	N. Moon	Q. Rowe	T. Wick				
B-2/4	Assembly	D. Evans									
B-5	Finishing	E. Fort	K. Long								
A-5	Packaging	F. Guy									
A-3/4	Warehousing	G. Ham	I. Jones		O. Pike	R. Stone		V. Will			
A-2	Shipping	H. Ivey									

Figure 18-14—Layout approval form; the person approving an area to initial and date the appropriate square, inappropriate squares to be X'd out.

common versions are: (1) CPM—*Critical Path Method*, and (2) PERT—*Program Evaluation and Review Technique.*

All systems based on critical path techniques use a network that is developed to provide a pictorial description of a plan, by showing the interrelationships of all required activities as to when they can start or finish. Others have no flexibility, and if delayed will delay the whole project. These are critical and form the critical path. The *critical path* of any project is the connected sequence of jobs that forms the *longest time duration path* through the project. There is always at least one critical path in every project and there may be, and frequently is, more than one (see Figure 6-20).

One main advantage of CPM/PERT over Gantt or bar charts is that one can see the interrelationships of the project jobs at a glance. In addition, the diagram shows the scope of the project and helps pinpoint responsibility for each job. These two techniques are covered in detail in a number of books. It is strongly recommened that anyone involved in facilities planning become well acquainted with them.

Making the Move

The general procedure. The one most important aspect of the move process is *adequate planning.* To overlook anything, or leave anything to chance, will seriously jeopardize the entire operation. An executive of one company that made a successful move stressed[10] the importance of planning, which "starts before the architects begin and ends on the day the movers come in," and that failure to recognize the value of detailed planning can lead to "higher moving charges, delayed startup, an alienated work force, and a recovery time that could stretch out as long as three years." Of great help was a "precedence network," which showed "sequential relationship and the critical path" for all projects, with overdue projects to be given concentrated attention. A lunch time slide program, and frequent bus trips to the new site, helped to pre-familiarize employees with the new building.

Table 18-1 contains a list of most of the major steps necessary in making a successful move. While it is beyond the scope of this book to discuss all of these steps, a few require a bit of explanation.

Move committee. The first, and one of the most important steps, is the appointment of a committee to plan the actual procedure and methods. The committee must be carefully constituted, to be sure the proper people are on it—that is, people who (1) can and will do the necessary work, in addition to their normal work-loads, and (2) properly represent all groups in the organization involved in the move or greatly affected by the move. Likely candidates would be people involved in the evaluation and approval process (see pp. 433–39), plus some

[10]"Any Questions? Move Out!" *Industry Week*, June 15, 1970:36.

PRESS EQUIPMENT	ELECTRICAL MOTORS	ELECTRIC DRIERS/HEATERS	NATURAL GAS	WATER	SANITARY SEWER	COMPRESSED AIR	PROCESS STEAM	EXHAUST
Web Offset Multi-Press	50 HP 440V 3φ Main Drive 25 HP 440V 3φ Auxiliaries 75 HP	-----	1,200 cu ft/hr	19,200 gals/24 hrs at 50° F	2 floor drains & Traps	20 CFM	-----	5,000 CFM
22¾" Web Offset	50 HP 440V 3φ Main Drive 25 HP 440V 3φ Auxiliaries 75 HP	-----	1,200 cu ft/hr	19,200 gals/24 hrs at 50° F	2 floor drains & Traps	20 CFM	-----	5,000 CFM
Variable Multi-Press (Web Offset)	50 HP 440V 3φ Main Drive 25 HP 440V 3φ Auxiliaries 75 HP	-----	1,200 cu ft/hr	19,200 gals/24 hrs at 30° F	2 floor drains & Traps	20 CFM	-----	5,000 CFM
Press No. 28	60 HP 440V 3φ Main Drive 40 HP 440V 3φ Auxiliaries 100 HP Main Drive has Thymatrol DC Converter (450V DC)	1-40 kw luker temp. control	1'4" Line Pressure 6" WC average 900 cu ft/hr Max Demand 2000 cu ft/hr (1000 BTU/CF)	19,200 gals/24 hrs at 30° F	2 floor drains & Traps	10-2 CFM Note 4 For supplying entire Lord St. Plant: Presses 1,2,3, 4,5,7,8,9,10, 15,16,19,20, 21,22,23,24, 25,26,27,28 All Bindery Eq. except Binders 1 & 2 Foundry Composing Maintenance Laboratory Misc. 2 Air Compressors are used: 1) 12x11 75 HP 220V 3φ - 304 CFM 2) 9x9 30 HP 220V 3φ 25C - 153 CFM Total 457 CFM (90 P.S.I.)	Provide Steam for fire Prof. in Exh. hoods. Note 5: For supplying entire Lord St. Plant: Presses 21, 28, 27 Foundry Misc. 4,500 gals/month No. 6 oil, for process steam (Press 21 uses 40%, 27 & 28 use 10%, Foundry uses 50%	Note 6: For entire Lord St. Plant: Presses 1,2,3,4, 5,7,8,9,10,19, 20,21,22,23,24, 25,26,27,28 All Bindery Equipment except Binders 1 & 2 Foundry Composing Maintenance Lab - Misc Process Exhaust is: Pressroom Exh. 50,000 CFM Foundry Exh. 15,000 CFM Balance of Bldg. 25,000 CFM Total 90,000 CFM
Press No. 27	50 HP 440V 3φ Main Drive 28 HP 220/440V 3φ Auxiliary Main Drive has Thymatrol DC Converter (450V DC)	-----	1'4" Line 6" w.c. average 900 cu ft/hr. Max Demand 2000 cu ft/hr (1000 BTU/CF)	19,200 gals/24 hrs	Required	10-20 CFM		
Press No. 25	25 HP 115V DC Main Drive 14 HP 440V 3φ Auxiliaries	-----	½" Line 6" w.c. Pressure Requirement is small	-----	-----	10 CFM	-----	
Press No. 26	25 HP 115V DC Main Drive 6.5 HP 440V 3φ Auxiliaries	-----	3/8" Line, 6" w.c. Pressure Requirement is small	-----	-----	10 CFM	See Note 5 above	

Figure 18–15—Typical equipment and facilities requirement schedule. (Courtesy of The Austin Co.)

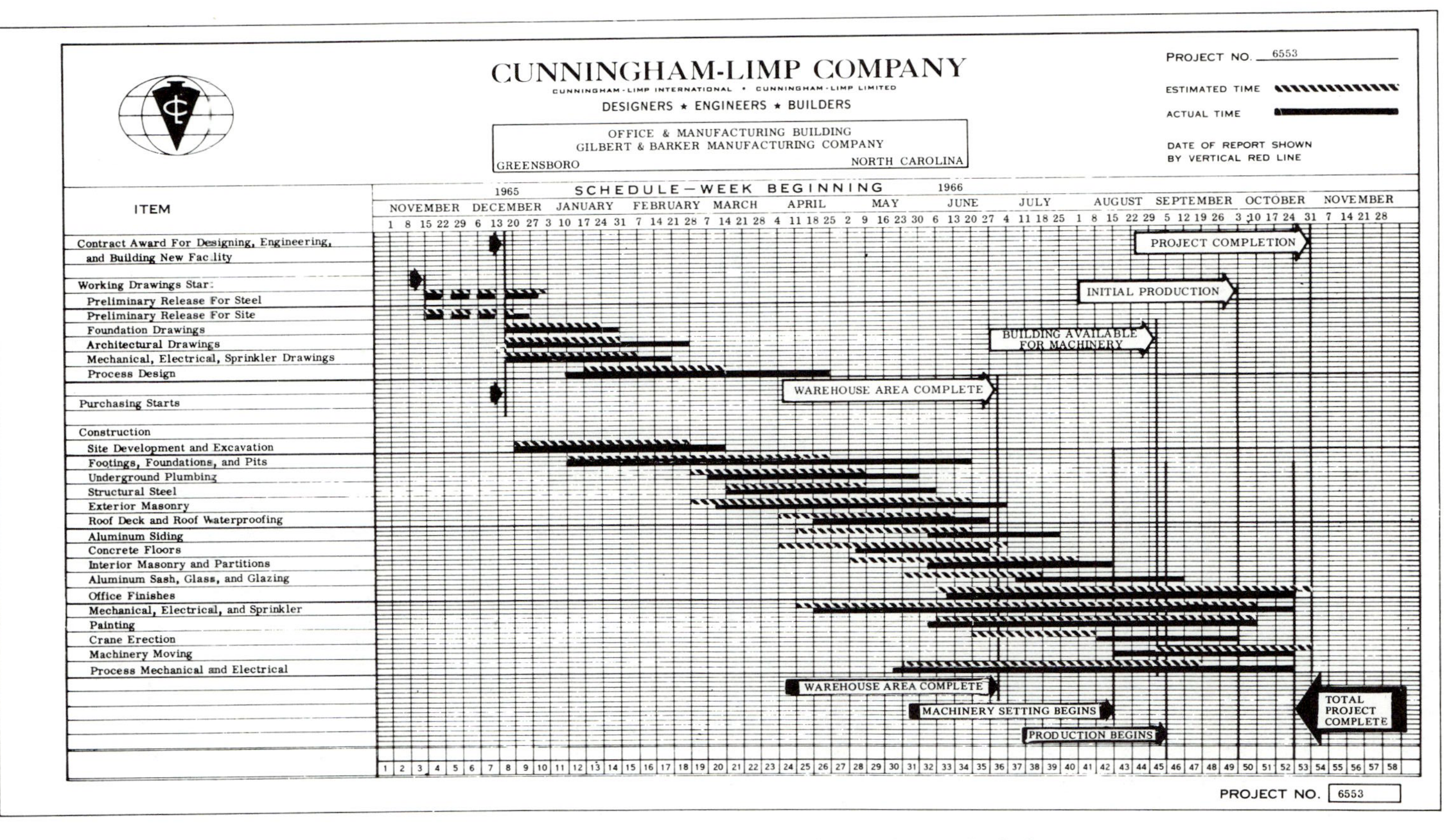

Figure 18-16—Gantt chart for planning, scheduling and control of plant construction and installation of equipment. (Courtesy of Cunningham-Limp.)

Table 18-1. Procedure for Making a Move

1. Select a *move committee.*
2. Determine information and data needed.
   a. Equipment inventory—production, auxiliary
   b. Work-in-process
   c. Material in Stores; Finished goods in Warehouse
   d. Area identification map for new plant (etc.)
3. Plan for accumulation of information and data.
4. Establish overall time table and progress schedule.
5. Design necessary forms and procedures for recording information and data.
6. Determine sequence of moving specific departments.
7. Check plans with each supervisor.
8. Perform housecleaning.
9. Determine methods of moving.
10. Place asset tags on all equipment, tooling, etc.
11. Obtain layout with asset numbers on all equipment.
12. Mark off floor area in new plant to designate exact location of each item, and mark with asset (or other) number.
13. Number stock bins, storage locations, etc., in new plant.
14. Establish department moving schedule by (1) departments, (2) machines, etc.
15. Tag all items to be moved.
16. Select move coordinators for each area, for (1) old building, (2) new plant.
17. Design forms and procedures for controlling move.
18. Prepare *Plant Move Master Data Sheets* for all areas.
19. Determine manpower requirements.
20. Determine container needs; obtain; assign.
21. Notify suppliers of new address.
22. Determine transportation requirements.
23. Prepare specifications for movers.
24. Obtain bids from movers.
25. Brief move coordinators.
26. Conduct employee move briefing and orientation—including visit to new location.
27. Each employee fill out *Individual Work Station Equipment List.*
28. Award move contract.
29. Make dry run (if deemed necessary).
30. Move
31. Install equipment.
32. Resume production.

members of the Industrial Engineering or Plant Engineering departments to carry on the actual day-to-day work of planning and implementation.

Housecleaning is another necessary task—a massive housecleaning. There is no better way to clean house than to move; and, there is absolutely no use moving "junk." Housecleaning, however, is not an easy task. People are extremely reluctant to admit that an item is not worth moving—from the janitor to the president. One plant engineer found it necessary to paint a big red X on all items he felt should be discarded. By a certain date, they were to (1) be out, or (2) have a *Move Tag* attached, with a justification written on the back.

Scheduling and control. Some of the data accumulated can be tabulated on a form similar to that shown in Figure 18-15. Other data can be entered on the *Plant Move Master Data Sheet,* shown in Figure 18-17.

Scheduling can be done with the aid of a Gantt chart, or a CPM network diagram, such as are shown in Figure 6-20. Probably the data should be tabulated in time-table format, for easier understanding by those involved.

The *Move Tag* (Figure 18-18) is a key document in the move process. It (1) identifies an item to be moved, (2) tells its new location, (3) provides records of:

1. *Leaving old location*—bottom torn off and entry made on *Plant Move Master Date Sheet* at old location.
2. *Arrival at new location*—check-in at new plant and enter data on *Data Sheet* at new location.
3. *Ready for use*—remainder of tag removed and returned to old location for check-off—as moved.

Departmental and individual work station instructions should be spelled out as shown in Figures 18-19 and 20. And the overall control may be maintained by the use of the procedure shown in Figure 18-21, with the *Plant Move Master Data Sheet* (Figure 18-17) maintained at the several necessary locations, such as:

1. Old department
2. New department
3. Plant engineering department

Briefing and orientation of employees is another important step. A series of meetings should be held until all have been given a clear explanation of all aspects of the move, and especially their own roles in the move process. If possible, all should be given an opportunity to visit the new location—possibly at an open house, so families can also see the building, the department, and employee's work area.

When to move is an important decision to be made. Some possible times for moving are:

1. During a normally low production period.
2. At model change-over time.
3. During a vacation period—if many plant personnel are not involved.
4. On a number of week-ends.
5. At night, when traffic is at an ebb.
6. When the movers are available.
7. When utilities are available.
8. When all (new) equipment is on hand and installed.
9. At a low inventory period.

The above procedure and related forms were developed by the author for use in a move of all production facilities from an original multi-story building in a downtown location to a new single-story building several miles away. Over 10,000 items were moved from one location to the other, including all work-in-process, which had to be sealed in double plastic film covers to prevent the accumulation of dust in the precision assemblies.

PLANT MOVE MASTER DATA SHEET

MASTER SHEET NO. 22 OF 90
DEPT. SHEET NO. 1 OF 4
DEPT. NAME Sheet Metal Fabrication DEPT NO. H 500

ITEM NO.	ASSET TOOL PART OR JOB NO.	DESCRIPTION NAME	UTILITIES: REQUIRED SERVICE	UTILITIES: READY AT NEW LOCATION	TRADE REQUIRED: DISCONNECT OLD LOCATION	TRADE REQUIRED: CONNECT AT NEW LOCATION	NOTIFIED	LOCATION: PRESENT FL./BAY	LOCATION: NEW	LOCATION: FL. MARKED	LOCATION: LINE/STAT	MOVE DATE: SCHED	MOVE DATE: LEFT	MOVE DATE: CPLTD	REMARKS &/OR ADDITION DATA OR INFORMATION
760	19876	SPOT WELDER	440V 3PH 60 Cyc 150 KVA	Yes	Plumber Elect'	Plumber Elect'	White	4/6	14A	X	A/1	8-4	8-4	8-5	
761	7959	SPOT WELDER	440 V 3 PH 60 Cyc 100 KVA	Yes	Plumber Elect'	Plumber Elect'	White	4/6	14A	X	A/2	8-4	8-4	8-5	Power Supply Interlock Required
762	20101	SPOT WELDER	440 V 3 PH 60 Cyc 100 KVA	Yes	Plumber Elect'	Plumber Elect'	White	4/6	14A	X	A/3	8-4	8-4	8-5	
781	6954	RIVETER	None	--	None	None	White	4/7	14B	X	C/1	8-5	8-5	8-5	
782	7458	RIVETER	None	--	None	None	White	4/7	14B	X	C/2	8-5	8-5	8-5	
740	3828	100-Ton PUNCH PRESS	440 V 3 PH 60 Cyc	Yes	Plumber Elect'	Plumber Elect'	White	4/8	15C	X	A/1	8-4	8-5	8-7	
741	3925	100-Ton PUNCH PRESS	"	"	"	"	"	"	"	"	A/2	"	"	"	
742	22012	20-Ton PUNCH PRESS	440 V 3 PH 60 Cyc	Yes	New Equip.	Plumber Elect'	"	--	15C	X	A/3	8-4	--	8-4	Vendor delivers to new site 8-4
743	22013	20-Ton PUNCH PRESS	"	"	"	"	"	--	15C	X	A/4	8-4	--	8-4	" "
744	7389	BUFFING JACK	440V 3 PH 60 Cyc	Yes	Elect'	Elect'	"	4/2	15A	X	B/6	8-4	8-4	8-4	
745	9656	BELT SANDER	440 V 3 PH 60 Cyc	Yes	Elect'	Elect'	"	4/2	15A	X	B/7	8-4	8-4	8-4	
746	10202	POWER BRAKE	440 V 3 PH 60 Cyc	Yes	Elect'	Elect'	"	4/3	15C	X	B/1	8-4	8-4	8-4	
747	10301	HAND BRAKE	None	Yes	None	None	"	4/2	15A	X	B/5	8-4	8-4	8-4	
748	10703	CIRCULAR SHEAR	440 V 3 PH 60 Cyc	Yes	Elect'	Elect'	"	4/3	15C	X	B/2	8-4	8-4	8-4	
749	16780	FOOT SHEAR	None	Yes	None	None	"	4/3	15A	X	B/5	8-4	8-4	8-4	
750	10760	BAND SAW	440 V 3 PH 60 Cyc	Yes	Elect'	Elect'	"	4/3	15C	X	B/3	8-4	8-4	8-4	
LEAR 62.1-21															

Figure 18–17—Plant move master data sheet for accumulation of all data pertinent to a plant move. (Courtesy of Lear, Inc.)

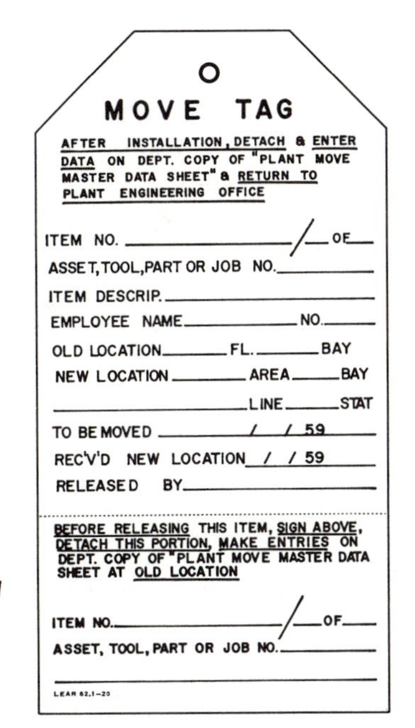

MOVE TAG

AFTER INSTALLATION, DETACH & ENTER DATA ON DEPT. COPY OF "PLANT MOVE MASTER DATA SHEET" & RETURN TO PLANT ENGINEERING OFFICE

ITEM NO. / OF

ASSET, TOOL, PART OR JOB NO.

ITEM DESCRIP.

EMPLOYEE NAME NO.

OLD LOCATION FL. BAY

NEW LOCATION AREA BAY

LINE STAT

TO BE MOVED / / 59

REC'V'D NEW LOCATION / / 59

RELEASED BY

BEFORE RELEASING THIS ITEM, SIGN ABOVE, DETACH THIS PORTION, MAKE ENTRIES ON DEPT. COPY OF "PLANT MOVE MASTER DATA SHEET AT OLD LOCATION

ITEM NO. / OF

ASSET, TOOL, PART OR JOB NO.

LEAR 62.1-20

Figure 18–18—Move tag, attached to every item in a plant move. (Courtesy of Lear, Inc.)

Conclusion

This chapter brings to an end the treatment of the layout procedure—a step-by-step process which began with the blue prints and specifications of a product or service, and has ended with a carefully planned flow pattern and arrangement of equipment and related facilities.

It has been the objective of this book to show how to design an effective layout for any kind of facility. As pointed out in the first chapter, the productive process is carried out in many kinds of enterprises besides manufacturing. The planning procedure described can be adapted to any type of enterprise in which the productive effort consists of a number of interrelated activities, with the flow of work moving among them. So the process is applicable to a warehouse, office, retail store, garage, post office, bank, campus, school building—or even to the many interrelated activity centers in a home.

The terminology should be translated into the vernacular of the particular facility under consideration, and some of the forms should also be adapted. It is hoped that the many forms, lists of factors, check sheets, and procedures will be found helpful in the design process. Most of them have been used many times, in the design of various kinds of facility. Unfortunately, too many facilities are still being built without such careful planning, which can only result in some of the problem situations indicated on Figure 18-6. It is felt that following the methodology described here can eliminate or minimize such difficulties.

The next step is up to the architect or builder, to design a suitable structure to enclose the facility, and give it a pleasing appearance. Then, of course there is the problem of where to locate the facility. The latter problem will be covered in the next chapter.

DEPARTMENTAL MOVE PROCEDURE AND CHECK SHEET	
1. Complete housecleaning	
2. Attach "Move Tag" to each item and container to be moved. Make sure each tag has item number and that it is entered on Plant Move Master Data Sheet.	
3. Attach "Do Not Move" labels to all items to remain in Plant 1.	
4. Remove all other "portable" items and materials.	
5. Assign 2 persons to coordinate the actual move (one for the Ionia Plant, one for the Eastern Ave. Plant).	
6. Coordinators check floor markings for location of equipment in Plant #6. Layout prints will be available.	
7. Be sure enough containers are on hand for packing all auxiliary items and materials.	
8. Plan for collection and packing of auxiliary items and materials.	
9. Plan to "disconnect" each piece of equipment.	
10. Prepare all equipment for moving	
a. Remove and pack "extensions"	
b. Secure all loose parts (handles, levers, cords, etc.)	
c. Fasten all drawers, doors, typewriter carriages, etc. (use glass reinforced tape where applicable).	
d. Pack all auxiliary equipment and related items.	
11. Brief all employees on move	
a. Move in general	
b. His part in preparation	
c. Making out Work Station Equipment list	
d. His part in setting up new area	
e. The date of his move	
f. Where to report.	
12. Check your move schedule.	
13. Coordinators personally supervise removing, loading, and unloading of items.	
14. Remove proper portion of move tag and enter appropriate data on copy of Plant Move Master Data Sheet.	
a. At Ionia Plant—remove bottom portion, enter data on Master Data Sheet.	
b. At Eastern Ave. Plant—Remove balance of tag, enter data on Master Data Sheet, and return tag to Plant Engineering.	

Figure 18–19—Departmental procedure and check list for use in a plant move. (Courtesy of Lear, Inc.)

LEAR, INCORPORATED
GRAND RAPIDS, MICHIGAN

Sheet No.__ of ______

WORK STATION EQUIPMENT LIST

Department __________ Scheduled Moving Date __________

Work Station Designation __________
(line no., station no., operation name and/or no., etc.)

Work Station Location __________
(Dept., area, bay, line, station, etc.)

Operator(s) Name(s) __________

In order that you can start work at your new location, you must have on hand everything you need. To help you in this task, it is requested that you:

1. List all items in your work area on this sheet (a day or two before your scheduled move).
2. See that all items are on hand the day your work area is scheduled to move.
3. Check off each item as you pack it for moving.
4. Attach Move Tag to each item and/or container to be moved. Be sure tag has item number and that it is entered on Plant Move Master Data Sheet.
5. Attach "Do Not Move" labels to all items NOT to be moved.
6. Return this Work Station Equipment list to your Coordinator when you are through packing. It will be returned to you (or your counterpart on the next shift) at your new location.
7. Check off each item as you unpack it in Plant #6.

ITEM NO. (to be assigned)	DESCRIPTION	QUANTITY	ON HAND BEFORE MOVE	PACKED	ON HAND AFTER MOVE
			Enter check mark in space at proper time		
	EQUIPMENT – jigs, fixtures, meters, test equipment, etc.				
	AUXILIARY ITEMS – hand tools, etc.				
	MATERIALS, WORK-IN-PROCESS, & STOCK – for first day at new location				

Figure 18-20—Work station equipment list for recording all items in a work station at time of plant move. (Courtesy of Lear, Inc.)

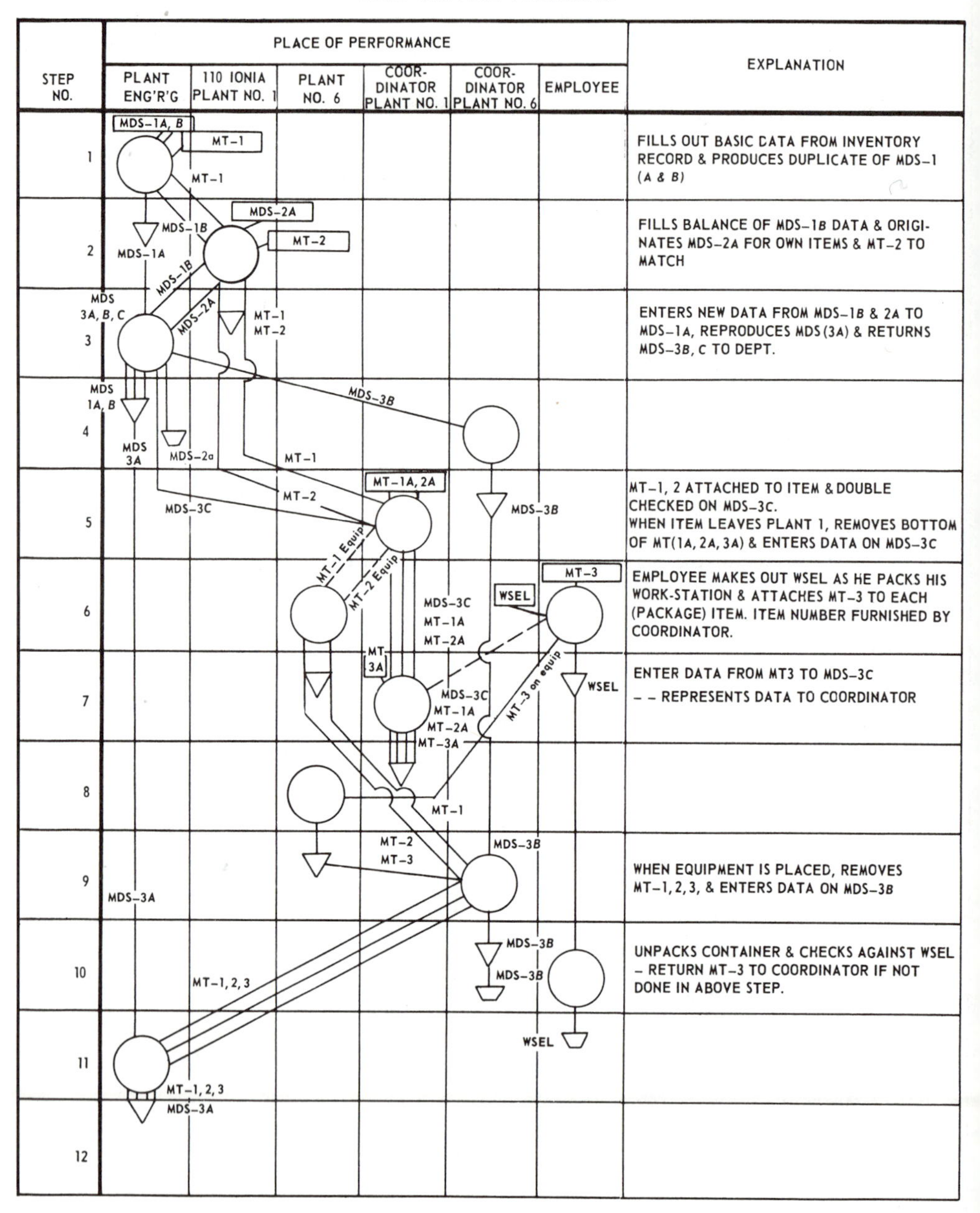

Figure 18-21—Plant move control procedure. (Courtesy of Lear, Inc.)

Questions

1. What are the purposes of evaluating a layout?
2. What are some bases for evaluating the layout.
3. Differentiate between qualitative and quantitative evaluations. Which is preferable? Why?
4. What is meant by an *efficiency index?* Discuss the concept.
5. Discuss the importance of presenting the layout.
6. Who should make the presentation? Why?
7. In addition to the final layout, what else is necessary for the presentation to management?
8. Why must the plant layout engineer become involved in the installation of the layout?
9. What further work is necessary after the final layout is made—to supplement the layout print or model? —to aid in the installation?
10. What are some ways the layout might be sold to employees? Why is this necessary?
11. What are some of the ways in which the master layout can be stored or displayed and protected for future use and reference? (Prints? Grid master? Models?)
12. What is involved in the follow-up activity?
13. Why, and by whom, should the final layout be approved?
14. What is meant by *implementing the layout?*
15. What are some common ways of reproducing the layout?
16. Even after the layout is completed, what kinds of things must be done before it can be implemented?
17. What techniques could one use to keep track of the layout installation—to make sure everything is done, and done on time?
18. Briefly discuss or outline the overall procedures in making the move into a new facility.
19. What are some of the suggested times for a plant move? Why?

Exercises

A. Try out the plant layout audit concept (on pp. 417–23) in a local situation—and develop an *assignment sheet,* as in Figure 18-5.

B. Develop a facility layout *check sheet* (as in Figure 18-6) for a non-manufacturing facility (possibly based on one of your previous exercises).

C. Same for a layout *evaluation sheet* (as in Figure 18-7).

D. If available, evaluate several alternative layouts for the same facility. Or, for several house plans in a home plan book (newsstand or library).

E. Try to develop efficiency indices for other than manufacturing facilities. What are the measurable characteristics?

F. Develop a procedure or list of steps for moving into a new home (or other non-industrial facility).

19 Facility Location

The two final stages in the facility design are the selection of a location and the design and construction of a building. Which comes first, is pretty much a chicken-and-egg problem—as some insist that the site should be selected to fit the building, while others claim that the building should be designed to fit the site. Many factors may contribute to the decision making process, some of which will be covered later.

Selecting a location for an enterprise is of vital importance to an entrepreneur, as many cost factors are determined by conditions over which management can exercise little or no control after the facility is located. Research, based on cost analysis, is essential in determining where a company can obtain the largest economic benefits in production and distribution expenses.

The primary reason for the existence of any enterprise is to supply a need or service to a demand as economically as possible. In order to supply this need or service, the enterprise must satisfy three basic requirements: (1) accumulation of required material at a location, (2) processing the material into a product, (3) distributing the product to the market. These basic requirements represent totals of various individual costs, and the relative importance of these costs will be different at different locations. Therefore, the ideal site is that in which the total cost of the three basic requirements is the least, provided that the enterprise is socially and economically acceptable to the community.

However, in spite of the serious economic and engineering implications of location, many executives locate their buildings with little or no thought to an imposing list of important factors. It may be honestly said that many commerical and industrial buildings are located where they are because the entrepreneur has lived there all his life, because he likes the climate, because the building was available, because founders of the business split off from a similar local concern, because a chamber of commerce offered a free plant site, or any one of a number of other equally unscientific reasons.

As will be seen later, there are many, many factors to be taken into consideration—some lists contain hundreds—many of them intangible.

Pitfalls in Site Selection

In spite of the fact that the economic consequences of a poor location may result in unnecessary handicaps, mistakes are commonplace, and consequences costly or even disastrous. Interestingly enough, errors in plant site selection seem to fall into a pattern. Some of the most common errors are:[1]

1. *Lack of thorough investigation* and consideration of factors involved.
2. *Personal likes and prejudices* of key executives or owners overriding impartially established facts.
3. *Reluctance of key executives* to move from traditional, established home ground to new and better locations.
4. *Moving to congested areas* already or about to be over-industrialized.
5. *Preference for acquiring an existing structure* (usually at an imagined bargain) that is improperly located or not designed for the most efficient production.
6. *Choice of a community with low cultural and educational standards,* so that key administrative and technical personnel eventually accept employment elsewhere.

Many of these can be avoided by a careful, open-minded analysis of the entire problem—with personal whims and apparently overwhelming factors giving way to reason.

The Need for Guidelines

A pre-determined list of criteria may prove extremely helpful in steering the analysis clear of the shoals of misconception, prejudice, oversight, and ignorance. Some guidelines might be:

1. *The most economical construction and operation* of the activity—considering costs of labor, transportation, utilities, taxes, and other pertinent factors—should be the objective of site selection.
2. *Site evaluation to be kept confidential,* to prevent premature publicity and embarrassment that might result from untimely disclosure of company plans.
3. *Ultimate employment* to be no more than ______ (considering future expansions in the next ______ years).
4. *At maximum employment,* plant should not require more than ______ % of the people in the labor area or ______ % of the eligible work force.
5. *Dispersal of employment* into more areas where customers are concentrated.
6. *Employment in different labor market* from other company operations.
7. *Minimum land area* should be five times eventual floor area, or ___ acres—whichever is larger.
8. *Provide facilities* of __________ capacity. (This capacity may be expressed in dollars, or other units of measurement)
9. *Future expansion* of ______ % at minimum expense, should be provided for.
10. *Complete on schedule,* by __________ .
11. *Complete within the appropriation* of $ __________ .

As can be seen, criteria may be very general, and all-encompassing, or they may be rather specific, such as completion date—but they should be stated, and checked periodically during the site selection and plant design process.

[1]Condensed from an article by A. K. Ferguson, in *Pacific Factory,* April 1954:70.

Basic Data Required

One of the early tasks is the collection of preliminary information and data with which to begin the decision making for the site selection process. Probably the first item of importance is the size of the piece of property to be obtained—which is, of course, dependent on the layout. One authority[2] suggested that care should be taken not to plan for too large an area but still give full consideration to the common land use requirements of "vehicle circulation and parking, buildings, utilities areas, landscaping, setback restrictions and special facilities," with particular consideration for local zoning ordinances—which may allow building coverage of gross land area of from 25% for open areas up to 100% for urban industrial sites. A complete statement of intended land use was strongly recommended to help in site selection; and then for determining the appropriate land area, the following sequence was suggested:

1. *Establish the total building floor area,* including any upper floor levels, to determine probable ground building area. Project the building area in accordance with potential expansion of facilities.
2. *Check codes for allowable ground coverage,* or estimate the total land as a total of all usage factors. Compare this total with legally specified setbacks from property lines to determine that net remaining land will be sufficient. Consider these site usage factors:

Use Factor	Percentage of Total Site Area	
Building structures	20–40	50–80
Parking,[a] including interal circulation	30–40	
Primary onsite roads, landscaped areas	25–30	20–50
Setback, natural,[b] or unusable area[c]	15–25	

[a] Conservatively, 100 cars per acre. [b] Land qualifying as open space excludes buildings, roads and parking areas. [c] Nonbuildable ground.

3. *Establish parking areas.* Parking lots often require an area equalling the net ground area occupied by buildings. Obviously, this depends on the number of car spaces needed. Suburban plants frequently require parking spaces equal in number to 75 per cent of their employees. Urban locations, with public transit, may require parking spaces for only 35 per cent.

Contingency planning, always wise where variables have yet to be resolved, should leave it open whether to acquire more land than minimum conditions would call for. Additional land area can allow for unexpected opportunities for expansion; and an open site appearance can help make it attractive to desired employees. Also suggested to be taken into account is basic information on such items as:

[2] R. Hansen, "Determining Land Requirements," *Plant Engineering,* Aug. 5, 1971:64

1. *Available capital.*
2. *Relationship of product* to an area of the country—such as snowmobiles, and peanut butter.
3. *Necessary support facilities*—such as outbuildings, water treatment, waste treatment, and large storage piles.
4. *Business trends*—industry; company.
5. *Competition.*

Many other items and factors are mentioned in lists below—some of which also belong in a later data-gathering phase.

It should be recognized that there are a number of factors to be considered for each of the above items. Some factors will be considered only once, and be disposed of; others must be reconsidered a second or even a third time. Table 19-1, Factors for Consideration in Facility Location, and other similar lists, should be studied carefully to extract the factors of greatest significance to the individual enterprise. They may be entered on a *Preliminary Selection of Specific Factors Sheet* (Figure 19-1), as a start in the decision making process.

The General Procedure

As in so many of the previous stages of facility planning and design, an orderly method of procedure is highly desirable to avoid such pitfalls as previously listed. The following overall facility location process was suggested in a government publication as follows.[3]

Start with a base map, as large-scale as practicable, and:

1. Pinpoint all suitable sites, and such major features as (a) industrial zoning boundaries; (b) road, rail, air, and water transportation networks; and (c) major utility lines—water, sewage disposal, gas, power.
2. Then depict each site on a separate map, showing (a) the above items in detail, (b) topographic features, (c) flooding possibilities, (d) prevailing winds, (e) adjacent land use, and any similar data helpful in site selection.

Make preliminary identifications on the base map:

1. Outline land areas having a slope gradient of 10% or less—referring to topographic maps.
2. Outline vacant tracts available for industrial use—referring to recent aerial photographs, land use maps, and zoning maps. (Note that re-zoning of non-industrial tracts may be possible if it can be demonstrated that a tract is suitable for industry.)
3. Add lines parallel to, and about 1 mile from, paved highways and railroads—the maximum distance from these facilities that plants will usually be located.
4. Plot electric transmission lines, gas lines, water supply lines, and sewage disposal systems.

[3]"How to Make an Industrial Site Survey," U.S. Department of Commerce, 0–713–509, 1963.

Table 19-1. Selected Factors for Consideration in Facility Location

A. Market

1. Location
2. Population trends
3. Income trends
4. Consumer characteristics
5. Retail sales trends
6. Industrial markets
7. Competition

B. Raw materials

1. Type
2. Location of sources
3. Characteristics
4. Price
5. Terms of sale
6. Availability
7. Storage facilities

C. Labor

1. Cost
2. Attitudes
3. Union situation
4. Supply
5. Location
6. Personnel policies in area
7. Legislation
8. Recruiting
9. Community pattern
10. Relocation

D. Transportation

1. Modes of transportation
2. Methods of operation
3. Handling methods
4. Facilities
5. Costs
6. Freight rates

E. Utilities

1. Power
2. Fuel
3. Water
4. Waste
5. Communications
6. Other

F. Climate

1. Maximum and minimum temperatures
2. Direction of winds
3. Weather changes
4. Humidity
5. Elevation
6. Climatic effects
7. Special weather hazards

G. Federal activities in area

H. Representation in congress

I. Financing

1. Financial aid available
2. Requirements
3. Credit factors
4. Factors effecting loan terms
5. Special inducements—taxes, loans, buildings

J. State government and taxes

1. Structure
2. Financial assistance
3. Government attitudes
4. State regulations and legislation
5. State taxes
6. Future tax prospects

K. Local government and taxes

1. Structure of government
2. Financial conditions
3. Civic attitudes
4. Local taxes

L. Community facilities

M. Community appearance

N. The size of the community

O. Population trend

P. Community planning and zoning

1. Planning commission
2. Industrial zoning
3. Building codes
4. Traffic and parking
5. Streets

Q. Others

1. Cost of living
2. Insurance
3. Construction costs
4. Location relative to competition
5. Size and character of other local industries
6. Buying habits

R. Individual sites

1. Requirements
2. Size and shape of site
3. Geologic considerations
4. Cost of development
5. Location within the community
6. Availability to transportation
7. Topography
8. Price of land
9. Availability of building
10. Personal factors
11. Intangible considerations

For on-site inspection and selection:

1. Tentatively selected site areas can be visited by a site-survey team.
2. Eliminate from consideration all sites having serious drawbacks for plant operation and service.
3. Determine what improvements would be required for sites still under consideration.
4. Determine availability of sites.
5. Obtain definite price agreements for comparison.

PRELIMINARY SELECTION OF SPECIFIC FACTORS FOR CONSIDERATION IN THE LOCATION OF—					
A.B.C. Manufacturing Co.					
ANALYST(s): K. E. Holmes				DATE: Nov. 17	
Tangible Factors					
Item No.	**Factor**	**Priority**	**Basis for Measurement**	**Value**	**Comments**
A 1	Market location	1	Market survey	100	A must!
J 5	State taxes	3	Published rates	80	Among lowest considerations
Intangible Factors			**Criteria for Evaluation**		
C 2	Labor attitudes	1	Investigate local situation	100	Appears ok
M	Community appearance	3	Personal investigation	80	Looks good
Q 1	Cost of living	2	Local figures	90	10% less probable

Figure 19–1—Form for preliminary selection of factors for consideration in site analysis.

The above outline of overall approach to facility location can be divided into a step-by-step procedure somewhat as follows:

1. *Choose a site selection committee*—which should be composed of people knowledgeable on the facts and factors—as well as those responsible for operating the facility. They should be charged with such responsibilities as:[4]

a. *Determining scope and best method* of survey.
b. *Identifying and giving consideration* to all pertinent requirements.
c. *Conducting the selection process* in a thorough, competent manner.

[4]"Site Selection Logbook," Factory, May 1966:A–142.

d. *Providing an economic analysis* of likely locations.
e. *Making the selection objectively.*
f. *Scheduling site acquisition steps* in coordination with all company functions.
g. *Reaching accord* on final choice of site.

2. *Determine company needs*—primarily by (1) establishing the goals, objectives and criteria, as discussed above, and (2) obtaining enough data on specific physical needs—such as size of building (based on layout) and size of site—as described earlier.

3. *Obtain knowledge of selection process and techniques*—first of all, an understanding of the three-phase concept of location:

a. Choosing the *region.*
b. Picking the *community.*
c. Selecting the *site.*

In this way, the normal selection process will narrow down the alternatives from a potentially endless number to a single specific site.

In addition to the overall process, there is the consideration of the several techniques available for assisting in the analysis and decision making process, some of the more commonly used techniques being:

Checklists. The comparison of potential plant locations can often be facilitated by using a checklist as a guide in collecting and evaluating data. By means of a checklist, the investigator is able to save time, record the data in a uniform way, avoid omitting necessary information. Each factor can be given a weighted importance if desired.

Questionnaires. A questionnaire is simply a list of questions as a basis for obtaining opinions or answers. It can be very similar to a checklist, and also facilitates collecting data.

Punched cards. The digital computer has proven itself to be a most useful tool for selecting the desired data from a vast amount of information. However, the data must be (1) collected and (2) entered on cards—or tape. Then it must be retrieved in the format programmed in the computer. Some governmental units use computerized site selection techniques. They consist primarily of a matching of company needs with the characteristics of available sites.

Spread sheets. A large single sheet of paper containing all information tabulated from the checklists and questionnaires for ready observation and evaluation is probably the most common technique for comparing sites (a sample spread sheet is shown in Figure 19-2).

Mathematical models. There are various mathematical approaches to the selection of plant location, all of which derive from or are variations of the least-cost theory of location—in which major location factors are expressed in an equation. Equations may be combined into a model. The least cost total is then selected as the site location. (Further details are given below.)

Charts. Photographs, specialized maps, sketches, etc., can be classified under the chart technique. These relatively simple methods can be utilized in the rapid elimination of undesirable geographic locations. In the hands of a trained observer aerial photographs and topographic maps can yield a surprising amount of information. Use is also being made of infra-red aerial photography to identify features and resources otherwise not readily indentifiable.

Cost comparisons. For the cost-oriented factors, a spread sheet can be designed to provide for the accumulation of comparative costs on selected factors. Columns can be added to obtain totals; rows can be scanned to identify favorable values for a particular factor. (A typical cost comparison is shown later in this chapter.)

Ranking and rating. A technique commonly used, especially with intangible factors, is the ranking or rating process in Table 19-2, for the market factor only.

Combinations. The investigator will probably use a combination of the above methods to facilitate site selection. Any method or combination of methods is satisfactory as long as it leads to correct answers to questions being considered.

4. *Decide on the overall approach*—this decision has been described[5] as an "empirical" and "trial and error" process, whereby "alternatives are selected, evaluated, compared, and the best chosen." This centers attention upon the best alternative "for the criteria that have been used," so successful site selection as much depends upon choice of criteria to be applied as upon choice of alternatives to be compared. The initial selection of alternatives being crucial to the final selection of site, it is safe to say that while individual acumen plays an important role, only in special circumstances can an intuitive approach by itself be even reasonably effective. Within rather rigid limitations a mathematical approach will enable "extremely rapid comparison of the merits of numerous sites," while remaining of "quite limited scope;" for example, one mathematical procedure very quickly works latitude and longitude of proposed sites against certain basic data on the proposed facility to determine an optimum site—reliable only when the criteria are "geographically consistent." Transportation cost based largely on distance "are especially conducive to use of this approach," and labor cost, where there is a distinct geographical differential, can also be used to good effect. But the mathematical method of selection of alternatives can offer only approximations—in cities, real estate costs and taxes may vary from block to block.

Another aspect of the choice of method of approach is the decision whether the actual selection process will be carried out by:

1. The committee itself
2. A public agency, utility, etc.
3. A consultant
4. A plant location service
5. Some combination of these

5. *Accumulate the data*—this step involves, first, determining what data are needed. In general, the data will be such as will quantify or otherwise support the questions implied by the factors listed in the tabulation in Figure 19-1.

Then, it will be necessary to make plans for and actually collect the information and data. This involves first the locating of sources of data; then the actual

[5]M. J. Newbourne and C. Barrett, "Guide to Industrial Site Selection," *Transportation and Distribution Management*, Aug., 1971:30.

Table 19-2. Evaluation and Comparison of Selection Factors

	Region			
	A	B	C	C
Location	10	2	8	4
Population trends	7	1	5	7
Income trends	9	3	6	8
Consumer characteristics	7	6	8	3
Retail sales trends	6	6	5	7
Industrial markets	8	7	6	5
Competition	7	6	3	5
Totals	54	31	41	39

collection; and finally the tabulation or other arrangement into usable formats. A tremendous amount of work will be necessary in the collection process, but it must be done, to insure that the best location is selected.

MAIN FUNCTION												LAND COSTS					ELEVATION					
Key							Proposed Layout															
OLD	NEW	General Location	Distance (in miles) From City Hall	Section	Front on Street	Side Street	Symbols: Slough, R. R., Spur, Street	Width (a and c)	Projected Length (b and d)	Square Feet of Ground (approximately)	Acres	Asking Price per Acre	Major Ground Improvement Cost per Acre	Total Cost per Acre	Server Railroad	Property Owner	Street Elevation	Main Railroad Elevation	Mean Elevation of Property	Defense Housing on Property? Available? Other Features	Type of Plant Layout	Spur Entrance

					F 1		F 2			FREIGHT TRANS				UTILITIES (AVAILABILITY AND BASE RATE)		
														WATER	SEWER	GAS
Section	Area 50 Mile Radius	Miles to Big City	City and State	1940 Population 1947 if Possible	Freight Out (LCL)	Freight Out (LTL)	Yearly Incoming Freight Originating in City	Raw Material from Outside City	All Freight In and Out F1 + F2	Belt Line Interlock	No. of Railroads Serving City. Names of Railroads	Switching Charge per Car	Truck Highways Serving City – Direction	Cost per 1000 CF	Storm, Sanitary, Restricted Waste?	BTU Available Cost per 1000 CF

6. *Analyze the data*—after the data have been collected, it becomes necessary to analyze all the facts, figures and information accumulated. In general, this involves:

a. *Sort and classify.*
b. *Check for completeness.*
c. *Summarize* the various aspects.
d. *Determine practicability* of data.
e. *Average, weight, or otherwise treat data* to yield useful information.
f. *Develop charts, graphs, maps, and tables,* to properly present data.
g. *Check carefully* for inconsistencies, omissions, errors, and unnecessary or irrelevant data.

When it comes to organizing the data, it is probably best to arrange it in spread sheet format, as shown in Figures 19-2 and 3. Figure 19-2 includes charts for comparing site characteristics (the chart on the right for sites within one area), and Figure 19-3 is a chart for summarizing pertinent costs for each site surveyed.

UTILITIES										
WATER		SEWER			GAS		ELECTRICITY			
Demand Av. CF/Mo C. 100,000 CF/Mo	Av. Cost 1000 CF. Location and Size of Main. Pressure	Demand Max. Flow Hr. Gal. at 4 to 5 p.m.	Location and Size		Demand CF/Mo	Av. Cost 1000 CF. Kind (BTU/CF). Size of Main. Location, Pressure	Demand Av. KW/H	Av. Cost 1000 Kwh. Elec. Power Co. HT. Voltage. Substation Location	Bus and Streetcar Transportation	Truck Roads	Adjacent Plants
			STORM	SAN.							

	LABOR				UNOCCUPIED ACREAGE	
ELECTRICITY						
Available ELP Co. Cost per 1000 Kwh.	Labor Supply – Type Transient?	Base Rates Mill Machinists	State Labor Legislation	Industry Tax Condition	Range in Cost Improvement per Acre	Parcel Number

Figure 19–2—Charts for comparing site characteristics—the chart on the right for sites within one area. (From Shaffer & Lormer, "Preplanning for Plant Locations," Factory Management & Maintenance, Oct. 1949:100.)

ESTIMATED COST COMPARISON CHART FOR TYPICAL MANUFACTURING COMPANY
Present Location Versus Recommended Communities

OPERATING EXPENSES	Present Location	Community A	Community B	Community C	Community D
Transportation					
Inbound	$ 202,942	$ 212,209	$ 207,467	$ 220,009	$ 216,778
Outbound	480,605	361,268	393,402	365,198	410,637
Labor					
Hourly Direct and Indirect	$1,520,943	$1,339,790	$1,146,087	$1,223,416	$1,178,809
Fringe Benefits	304,189	187,571	126,070	159,044	141,457
Plant Overhead					
Rent or Carrying Costs	$ 271,436	$ 290,000	$ 280,000	$ 295,000	$ 275,000
Real Estate Taxes	43,345	39,000	34,000	39,000	36,000
Personal Property and Other Locally Assessed Taxes	16,899	–	–	8,500	–
Fuel for Heating	19,260	11,000	9,500	13,000	16,000
Utilities					
Power	$ 56,580	$ 61,304	$ 41,712	$ 49,007	$ 43,112
Gas	18,460	19,812	13,767	16,633	14,986
Water	12,474	8,200	4,500	4,500	7,000
Treatment of Effluent	6,376	–	2,300	–	3,400
State Factors					
State Taxes	$ 67,811	$ 73,400	$ 44,920	$ 71,000	$ 58,250
Workmen's Compensation Insurance	30,499	24,000	14,000	17,000	21,000
TOTAL	**$3,051,819**	**$2,627,554**	**$2,317,725**	**$2,481,307**	**$2,422,429**
Savings through construction of new plant					
New Plant Layout		($ 210,000)	($ 210,000)	($ 210,000)	($ 210,000)
Reduced Materials Handling		(38,000)	(38,000)	(38,000)	(38,000)
Elimination of Present Local Interplant Movements		(60,000)	(60,000)	(60,000)	(60,000)
Reduced Public Warehousing		(30,000)	(30,000)	(30,000)	(30,000)
Reduced Supervisory Personnel		(27,000)	(27,000)	(27,000)	(27,000)
Savings through new construction		($ 365,000)	($ 365,000)	($ 365,000)	($ 365,000)
Annual operating costs	**$3,051,819**	**$2,262,554**	**$1,952,725**	**$2,116,307**	**$2,057,429**
Potential annual savings over present location		**$ 789,265**	**$1,099,094**	**$ 935,512**	**$ 994,390**
Percentage of saving		**25.9%**	**36.0%**	**30.7%**	**32.6%**

Figure 19–3—Typical estimated cost comparison chart. **(From A. S. Damiani, *"Selecting a Plant Site?"* Plant Engineering, April 1, 1971:41.)**

7. ***Evaluate the alternatives***—the actual evaluation process eventually requires the quantification of as many tangible factors as possible, with a judgment evaluation process applied to the intangible factors. (Some of the techniques useful in such work were discussed in Chapter 18.)

The most common method of making the comparison of factors is by giving selected factors a value or weight. Table 19-3 indicates how one company

Table 19-3. Example of Site Factor Weighting

Factor	Wt. in Units	Factor	Wt. in Units
Labor	250	Transportation	50
Fuel—for processing	330	Water supply	10
Power—from utility	100	Taxes and laws	20
Living conditions	100	Selection of site—(specific site)	10
Climate	50		
Supplies	60	Construction costs	20
			1,000

weighted its selection of factors; a basis for weighting might be established from the results of a survey of community attributes, conducted by a trade journal, as indicated by Table 19-4; and a tabulation or spread sheet of selected factors for comparing 6 sites in 3 states is given in Table 19-5.[6]

In any event, it is necessary to (1) select the factors of most specific interest to the enterprise, (2) rank them in order of importance, and (3) evaluate each region, community, or site for each location alternative—in cost or other quantitative terms, when possible. By one or several of these methods, it should be possible to accomplish the next step.

8. *Reduce possible sites to a workable minimum*—somewhere along the line it is necessary to reduce the sites under consideration to a half-dozen, more or less. This will have been done by comparing, first *regions;* next *communities;* and

[6] J. B. Dower, "Choosing an Industrial Site," *Distribution Worldwide,* Feb. 1974:24.

Table 19-4. Evaluation of Site Factor Importance by Survey

	% of Respondents			
Factor	Critical	Significant	Average	Minimum
Highway access	37	40	17	3
Scheduled air service	12	25	31	28
Water transportation	3	5	9	80
Scheduled rail service	23	17	22	34
Piggyback facilities	5	12	25	54
Industrial water supply—processed	23	22	29	22
Industrial water supply—raw	16	17	27	35
Natural gas service	31	27	25	13
Industrial sewage processing	20	26	32	19
Solid waste disposal	17	25	35	20
Soil load bearing capability	14	22	36	24
Plant site size	23	39	30	4

Source: J. B. Dower, "Choosing an Industrial Site," *Distribution Worldwide*, Feb. 1974:24.

Table 19-5. Comparative Community

State	Community	POPULATION		COUNTY				TRANSPORTATION			
	County	Community	County	% Change 10 Years	Native Born	Percent Foreign Born	Negro	Rail	Air	Truck	Bus
OHIO	A	503,000	864,000	+19	98	2	10.4	B&O ERIE L&N	AA DAL EAL	128	5
OHIO	B	472,000	683,000	36	97	3	7.6	NYC C&O N&W	LCA PA TWA	100	7
INDIANA	C	480,000	700,000	26	95	5	9.5	PA MONON IC	UAL OZA AA	130	15
INDIANA	D	150,000	166,000	3	90	10	5.5	C&EI SOU NYC	DAL EAL LCA	30	7
KENTUCKY	E	400,000	611,000	26	96	4	10.2	B&O C&O L&N	AA DAL OZA	78	6
KENTUCKY	F	63,000	132,000	31	94	6	13.3	SOU CND&TP C&O	DAL EAL PA	24	2

finally *sites*. Some factors appropriate to each phase are listed in Table 19-6. By the consideration of such facts, factors, and techniques as suggested above, the analyst should be ready to proceed in greater detail.

9. *Investigate the alternative sites in detail*—each potential site should be personally visited by selected committee members, along with such other personnel as deemed advisable. This could mean inviting consultants, appraisers, architects, and builders, to join in the final evaluation process. It is often necessary, however, to keep such visits extremely confidential, to avoid any leaks that might result in increasing prices of land suspected to be under consideration. In actual cases, company officials have even registered under assumed names at local hotels. At the least, such investigations should be undertaken very quietly. Most top local officials will recognize the need for secrecy and will respect such requests.

10. *Collect and analyze further data*—if it is found necessary or helpful in the final decision making process, additional information and data should be obtained. Especially needed at this time will be information on the specific sites; the local governments; living accomodations for personnel to be transferred; school, shopping and recreational facilities, etc.

When it is felt that sufficient information is at hand, all data should be carefully evaluated. A final review of the situation should permit the committee to proceed with the final step.

Data—Typical Analysis

HOTELS		MOTELS		HOSPITALS		BANKS		Churches	SCHOOLS				DEPARTMENT	
									Public & Parochial				Police	Fire
No.	Rooms	No.	Rooms	No.	Beds	No.	Deposit (Million)		Grade	Jr.	High	College	Employee Eqpt.	Class Employee Eqpt.
14	6000	18	700	18	4000	12	$1,360	470	192 48	12 Comb. 16	20	9	800 260	1 330 66
12	3400	20	900	15	3600	6	800	400	170 40	Comb. 26 Comb. 15		7	620 200	1 300 58
20	4500	23	875	14	3200	6	1,200	440	185 42	10 2	21 8	5	700 200	2 310 60
3	700	8	250	5	1100	5	187	150	58 14 Comb	4	8 2	4	220 62	2 102 24
12	2700	12	880	10	2400	9	680	375	155 21	8 Comb. 10	14	3	580 180	1 260 55
8	900	8	420	3	400	6	40	110	34 2 Comb.	3	6 1	2	95 30	3 40 12

11. *Select and obtain the site*—no words can adequately describe the actual decision process. Suffice it to say that the facts, data, and information collected to this point should have been adequate to lead to the selection of the site. And it should be remembered that many of the intangible, judgment, and even sentimental factors will play a big role in the final decision—but they must not be allowed to override solid cost factors, which could lose hundreds of thousands of dollars if a poor selection were to be made.

Once the site has been selected, it must be procured. This final step involves many details of a technical and legal nature. For instance, a survey and related data must be obtained, and checked carefully to be sure of such items as:

a. Corner markers
b. Accuracy of legal description
c. Easements and their grantees
d. Records of conveyances
e. Rights of way
f. Covenants, conditions, restrictions
g. Oil or mineral rights
h. Soil conditions
i. Contour map details

One authority suggests the *Survey Check List* of Table 19-7[7] for making sure the survey contains all pertinent data. There are also such possibilities as the state taking over for access to a potential new expressway, or unannounced plans for public improvements in the area.

Next, is probably a meeting with owners, and realtors, to negotiate price and

[7] M. Schwartz, "Civil Engineering for the Plant Engineer–Land Acquisition and Survey," *Plant Engineering*, July 1967:131.

Table 19-6. Factors for Consideration in Choosing Sites

A. Choosing the region

1. Material
2. Market
3. Transportation
   - a. Relative to material
   - b. Relative to market
   - c. Types available
   - d. Relative costs
4. State laws—corporation, labor, etc.
5. State taxes—corporation, sales, income, etc.
6. Climate

B. Selecting the community

1. Transportation available
2. Local laws
3. Local taxes
4. Labor
   - a. Supply
   - b. Skills
   - c. Wage rates
   - d. History
5. Utilities
   - a. Types available
   - b. Costs
   - c. Future trends
6. Community attitude and government
7. Service facilities
8. Social, cultural, educational institutions
9. Taxes—types, rates
10. Zoning laws
11. Building codes
12. Building costs
13. Population—present and trend
14. Cost of living

C. Picking the site

1. Size
2. Cost
3. Traffic network
4. Utilities—proximity to site
5. Orientation in relation to community—wind direction, residential areas
6. Topography
7. Soil conditions
8. Zoning
9. Available buildings

terms. Then, a meeting with local government officials, on tax adjustments; and with the local utilities, on extending their services.

Industrial Parks

No treatment of facility location would be complete without some mention of the industrial park—a development of the 1960's. Technically, an industrial park is a large tract of land, sub-divided into parcels for individual tenants, which:[8]

1. Provides industrial zoning and protective deed covenants.
2. Has full utility development.
3. Has a controlled harmonious construction and land program.
4. Has good access to various forms of transportation.
5. Is limited, with specific exceptions, to wholesale, distribution, or manufacturing activities.
6. Has a continuous management relationship with the tenants.

Other than the advantages inherent in the above definition, industrial parks greatly simplify the acquisition process and the related zoning, permits, and utilities problems. They can also offer such additional advantages as:[9]

1. A recognized, effective spokesman to represent plant owners to municipal, county, and state officials.
2. Well organized, dependable, and economical service—snow removal, catering, taxis, etc.
3. Compatible, appropriate neighbors.

[8]"Industrial Evolution," *Atlanta,* Dec. 1970:74.
[9]P. J. Marotta, "Have Industrial Parks Lost Their Lustre?" *Factory,* July 1968:15.

Table 19-7. Site Survey Checklist

A. Legal

1. Lot number
2. Block number
3. Street number
4. Tract number
5. Recording agency
6. Vicinity map
7. Easements
8. Metes and bounds

B. Dimensions

1. Compass
2. Plot
3. Lot lines
4. Side yards
5. Rear yards
6. Set-back
7. Loading space
8. Angles
9. Scale
10. Monuments

C. Grades

1. Datum
2. Natural grade
3. Finish grade
4. Stakes
5. Floor line
6. Elevation
7. Front
8. Rear

D. Utilities

1. Water
2. Gas
3. Electric
4. Telephone
5. Sewer
6. Storm drain
7. Steam
8. Sprinklers

E. Existing buildings

1. Location
2. Type
3. Occupancy
4. Stories
5. Dimensions
6. Materials of construction
7. Encroachments

F. Improvements

1. Streets
2. Alleys
3. Center lines
4. Sidewalks
5. Curbs
6. Drive aprons
7. Parkways
8. Trees
9. Catch basins
10. Fire hydrants
11. Power poles
12. Telephone poles
13. Lamp posts
14. Storm drain
15. Sewers
16. Railroad spur
17. Future expansion

From M. Schwartz, "Civil Engineering for the Plant Engineer-Land Acquisition and Survey," *Plant Engineering*, July 1967:131.

4. The possibility of a packaged deal—architect, engineer to assist in construction, lease-back arrangements, and complete mortgage financing.
5. Top architectural and engineering consultants for expansion, for tough installation or production problems, and for advice on technical innovations.
6. Built-in markets (similar industries attract each other and nearby markets develop accordingly).
7. Competent legal and public relations counsel, retained by the park; possibly an advisory council of plant managers to discuss mutual problems.
8. Other miscellaneous services may include police assistance in directing traffic, a central treatment plant for sewage, or for heating and cooling, a substation for standby power generation, full-time maintenance people, group protection for certain types of insurance, on-premises fire protection and medical facilities.

While these seem to distribute many service costs, provide a plant ready for a quick move, and purport to offer an attractive return on capital investment, industrial parks do, of course, have some disadvantages. Some have been stated as:

1. More expensive.
2. Too much open acreage required.
3. Higher building costs.
4. Building designed and constructed for convertibility in case a tenant moves out.
5. Monotonous architecture.
6. Traffic problems.
7. Higher labor turnover and labor unrest, due to proximity of other job opportunities.

As with all other aspects of the facilities design process, the pro's and con's must be evaluated for an individual situation before a decision is reached on an industrial park site vs. a private site.

Conclusion

The selection of a facility location is not a problem to be taken lightly. It is one of the most important decisions a firm is called upon to make, and the wisdom of a choice can influence the profits of a company for years to come.

This chapter has presented the problem of facility location in the framework of a three-phase project: *region, community,* and *site.* A procedure has outlined the step-by-step process involved, as well as a variety of tools, techniques and aids.

It can be concluded that site selection remains a problem whose solution will require the services of a variety of talents and the combined judgment of a number of officers. Approach to the problem from the standpoint of cost analysis will most readily insure determining the site where the sum of all costs of operation and distribution will be at a minimum.

Questions

1. Why is the location of a plant of great concern to management?
2. What are some common pitfalls in site selection?
3. What are some worthwhile guidelines in locating a facility?
4. What factors affect the size of property that should be purchased?
5. What are the categories of factors to be considered in facility location? Give some examples of each.
6. Briefly describe, or list the steps in, the overall approach to plant location.
7. What are some common techniques used in the selection process?
8. What are some of the many sources of information and data useful in the selection process? What can be obtained from each?
9. What are the three major aspects of the facility location problem?
10. What are some factors to be considered in each aspect?
11. When studying individual sites, what particular details, technical and legal factors, should be considered or obtained?
12. Discuss the concept of industrial parks. Advantages? Disadvantages?

Exercises

A. List some worthwhile guidelines in selecting a site for an enterprise of your (or instructor's) choice. (Home, store, gas station, motel, restaurant, park, camp, etc.)

B. Develop a list of factors for consideration in a non-industrial location selection process (for example, those suggested in Table 8-1, or above).

C. Develop a location selection guide or check sheet, by which alternative locations could be evaluated quantitatively for your choice of facilities (see *Exercise* A).

D. Review copies of *Industrial Development* magazine, and report on interesting aspects of the location problem covered in selected (or assigned) articles.

E. Visit a representative of a local—
   1. Chamber of Commerce
   2. Utility company
   3. Railroad or airline
   4. City government
   5. Industrial realtor
   6. State industrial development agency
   7. Port authority

   and report on their (a) interests, and (b) activities in regard to plant location.

F. Invite a representative of one or more of the above to visit class and present his case.

G. Obtain a local zoning ordinance or building code and search out location aspects (or invite local representatives to discuss in class).

Bibliography

General References

Bolz, H. A., and G. E. Hageman, eds. *Materials Handling Handbook,* New York: The Ronald Press Company, 1958.

Carson, G. B., H. A. Bolz, and H. H. Young, eds. *Production Handbook,* 3rd ed. New York: The Ronald Press Company, 1972.

Footlik, I. M., and J. F. Carle. *Materials Handling and Plant Layout.* Cleveland: Lincoln Extension Institute, 1968.

Immer, J. R. *Layout Planning Techniques.* New York: McGraw-Hill Book Co., 1950.

Ireson, W. G. *Factory Planning and Plant Layout.* Englewood Cliffs, N.J.: Prentice-Hall, Inc., 1952.

Lewis, B. T., and J. P. Marron, eds. *Facilities and Plant Engineering Handbook.* New York: McGraw-Hill Book Co., 1973.

Mallick, R. W., and A. T. Gaudreau. *Plant Layout Planning and Practice.* New York: John Wiley & Sons, Inc., 1951.

Maynard, H. B., ed. *Industrial Engineering Handbook.* New York: McGraw-Hill Book Co., 1971.

Moore, J. M. *Plant Layout Design.* New York: The Macmillan Co., 1962.

Muther, R. *Practical Plant Layout.* New York: McGraw-Hill Book Co., 1955.

———. *Systematic Layout Planning.* Boston: Cahners Books, 1973.

Reed, R. *Plant Layout.* Homewood, Ill.: Richard D. Irwin, Inc., 1961.

———. *Plant Location, Layout, and Maintenance.* Homewood, Ill.: Richard D. Irwin, Inc., 1967.

Periodicals

Automation, Cleveland, Ohio.
Factory Management, New York, N.Y.
Industrial Engineering, Norcross, Ga.
Material Handling Engineering, Cleveland, Ohio.
Modern Materials Handling, Boston, Mass.
Plant Engineering, Barrington, Ill.
Plant Operating Management, New York, N.Y.
Production, Bloomfield Hills, Mich.

Chapter 3: Preliminary Enterprise Design Activities

Dible, D. M. *Up Your Own Organization.* New York: Hawthorn Books, Inc., 1973.

Goslin, L. W. *The Product Planning System.* Homewood, Ill.: Richard D. Irwin, Inc., 1967.

Karger, D. W. *The New Product.* New York: Industrial Press, 1960.

———. *New Product Venture Management.* New York: Gordon & Breach, 1972.

Maynard, H. B., ed. *Top Management Handbook.* New York: McGraw-Hill Book Co., 1960.

Morrison, R. S. *Handbook for Manufacturing Entrepreneurs,* 2nd ed. Cleveland: Western Reserve University Press, 1974.

Chapter 4: Designing the Process

Bright, J. R. *Automation and Management.* Cambridge, Mass.: Harvard University Press, Division of Research, 1958.

Burbidge, J. L. *The Introduction of Group Technology.* New York: John Wiley & Sons, Inc., 1975.

Doyle, L. E. *Tool Engineering.* Englewood Cliffs, N.J.: Prentice-Hall, Inc., 1950.

Eary, D. F., and G. E. Johnson. *Process Engineering for Manufacturing.* Englewood Cliffs, N.J.: Prentice-Hall, Inc., 1962.

———, and E. A. Reed. *Techniques of Pressworking Sheet Metal.* Englewood Cliffs, N.J.: Prentice-Hall, Inc., 1958.

Martino, R. L. *Integrated Manufacturing Systems.* New York: McGraw-Hill Book Co., 1972.

Wild, R. *Mass-Production Management.* New York: John Wiley & Sons, Inc., 1972.

Chapter 5: Designing Material Flow

Apple, J. M. *Material Handling Systems Design.* New York: The Ronald Press Company, 1972.

Muther, R. *Production Line Technique.* New York: McGraw-Hill Book Co., 1944.

Chapter 6: Conventional Techniques for Analyzing Material Flow

Apple. *Material Handling Systems Design.*

Burbidge. *Group Technology.*

Chapter 7: Quantitative Techniques for Analyzing Material Flow

Francis, R. L., and J. A. White. *Facility Layout and Location—An Analytical Approach.* Englewood Cliffs, N.J.: Prentice-Hall, Inc., 1974.

Morris, W. T. *Analysis for Materials Handling Management.* Homewood, Ill.: Richard D. Irwin, Inc., 1962.

See also footnotes in the chapter.

Chapter 8: Planning Activity Relationships

Muther. *Systematic Plant Layout.*

Chapter 9: Production and Physical Plant Services

Apple. *Material Handling Systems Design.*

Army, Department of the. *Storage and Material Handling,* TM 743–200. Washington, D.C.: U.S. Government Printing Office, 1955.

Briggs, A. J. *Warehouse Operations, Planning and Management.* New York: John Wiley & Sons, 1960.

General Services Administration. *Warehouse Operations.* Washington, D.C.: U.S. Government Printing Office, 1969.

Jenkins, C. H. *Modern Warehouse Management.* New York: McGraw-Hill Book Co., 1968.

Chapter 10: Administrative and Personnel Services

"Action Office 2." Zeeland, Mich.: Herman Miller Co., 1969.

Lorenzen, H. J., and D. Jaeger. "A Systems Concept." *Contract Magazine,* Jan. 1968.

"Making Office Walls Come Tumbling Down." *Business Week,* May 11, 1968.

Pile, J. F. "The Nature of Office Landscaping." *AIA Journal* 52(July 1969).

———. "Clearing the Mystery of Office Landscaping." In chapter on Future Directions, *Interiors' Second Book of Office Design;* Whitney Library of Design. New York: Watson-Guptill Publications, 1969.

Quickborner Team. "America's First Office Landscape." *Office Design,* 1969.

Planas, R. E. "Office Landscape Layout—Pro and Con." *The Office* 70(July 1969).

Propst, R. *The Office—A Facility Based on Change.* Elmhurst, Ill.: The Business Press, 1968.

Chapter 11: Space Determination

See footnotes in the chapter.

Chapter 12: Area Allocation

See footnotes in the chapter.

Chapter 13: Computerized Facilities Layout

See also footnotes in the chapter.

General

Apple, J. M., and M. P. Deisenroth. "A Computerized Plant Layout and Evaluation Technique—PLANET." *Proceedings of AIIE Annual Conference,* May 1972:121.

Buffa, E. S., G. C. Armour, and T. E. Vollman. "Allocating Facilities with CRAFT." *Harvard Business Review* 42(Mar.–Apr. 1964):130.

Denholm, D. H., and T. E. Morgan. "School Facilities Design Aided by Computer Assisted Layout." *Proceedings of AIIE Annual Conference,* May 1972:137.

Francis, and White. *Facility Layout and Location.* The only book on quantitative approaches to layout problems; includes a rather detailed chapter on CRAFT, ALDEP, and CORELAP.

Lee, K. "Computer Programs for Architects and Layout Planners." *Proceedings of AIIE Annual Conference*, May 1971:139.

Lee, R. C., and J. M. Moore. "CORELAP—*CO*mputerized *RE*lationship *LA*yout *P*lanning." *Journal of Industrial Engineering* 18(Mar. 1967):195.

Matto, R. "LAYOPT—General Purpose Layout Optimizing Program." Unpublished working paper, Technical University, Helsinki, 1969.

Moore, J. M. "Computer Program Evaluates Plant Layout Alternatives." *Industrial Engineering* 3(1971):19–25.

———. "Computer Aided Facilities Design: An International Survey." *2nd International Conference on Production Research* (Copenhagen). London: Taylar and Francis Ltd., 1973.

Ritzman, L. P. "The Efficiency of Computer Algorithm for Plant Layout." *Management Science* 18(Jan. 1972):240.

Seehof, J. M., and W. O. Evans. "Automated Layout Design Program (ALDEP)." *Journal of Industrial Engineering* 18(Dec. 1967):690.

Vollman, T. B., C. E. Nugent, and R. L. Zartler. "A Computerized Model for Office Layout." *Journal of Industrial Engineering* 19(July 1968):321.

Vollman, T. E., and E. S. Buffa. "The Facilities Layout Problem in Perspective." *Management Science* 12(1966):450–68.

Whitehead, B., and M. Z. Edlars. "The Planning of Single Story Layouts." *Building Science* 1(1965):116–25.

Zoller, K., and K. Adendorff. "Layout Planning by Computer Simulation." *AIIE Transactions* 4(1972):116–25.

For Program and User's Manual

ALDEP—IBM Reference Library for Systems 360: No. 360–D–15.0.004.

CORELAP—J. M. Moore, Dept. of Industrial Engineering, Virginia Polytechnic Institute and State University, Blacksburg, Va.

CRAFT—SHARE Library No. SDA 3391 (IBM).

PLANET—M. P. Deisenroth, School of Industrial Engineering, Purdue University, Lafayette, Ind.

Layout Program Comparison Studies

Denholm, D. H., and G. H. Brooks. "A comparison of Three Computer Assisted Plant Layout Techniques." *Proceedings of AIIE Conference,* 1970.

Eldin, H. K., and J. W. Wood. "A Survey of Basic Industrial Engineering Computer Applications," *Proceedings of AIIE Conference,* 1971:55. One portion of this study is devoted to 5 plant layout techniques: CORELAP, CRAFT, ALDEP, RMA COMP 1, and CDUPL. Their relative efficiency is not compared, but a table compares input, output, concept, features, and the limitations of each.

Muther, R., and K. McPherson. "Four Approaches to Computerized Layout Planning." *Industrial Engineering* 2(Feb. 1970):39. CRAFT, ALDEP, CORELAP, and RMA COMP I are presented and compared as to approach and methodology.

Nugent, C. E., T. E. Vollman, and J. Ruml. "An Experimental Comparison of Techniques for the Assignment of Facilities to Locations." *Operations Research* 16(1968):150–73. This research compares CRAFT with a procedure developed by Hillier, one by Hillier and Connors, and one called *biased sampling*—a variation of CRAFT. Several cross-comparisons are made and results presented in tables comparing average final costs, average time required on a GE 165, and level of significance and count of optima.

Roberts, S. D. "Computerized Facilities Design—An Evaluation." The University of Florida, Department of Industrial and Systems Engineering, Gainesville, Fla., TR30, Project THEMIS, ARO-D Contract No.DAH/CO4/68C/0002, Nov. 1969.

Fitting Machines into Existing Layouts

Eyster, J. W., J. A. White, and W. W. Wierwille. "On Solving Multi-facility Location Problems Using a Hyperboloid Approximation Procedure." *AIIE Transactions*, Mar. 1973.

Hintzman, F. H. "A Computer-Assisted Method for Optimizing Floorspace Layouts." *Western Electric Engineer* 12(Apr. 1969).

Jagadish, J., and I. Gupta. "A Program for Plotting Plant Layout." *Industrial Engineering* 1(Mar. 1969):35.

Moore, J. M. "Optimal Locations for Multiple Machines," *Journal of Industrial Engineering* 12(Sept.-Oct. 1961):307.

———, and M. R. Mariner. "Layout Planning: New Role for Computers." *Modern Materials Handling*, Mar. 1963:38.

Computers in Warehouse Planning

Francis, R. L. "On Some Problems of Rectangular Warehouse Design and Layout." *The Journal of Industrial Engineering* 18(1967):595–604.

Kooy, E. D., and D. L. Peterson. "Use of the Computer in Warehouse Layout and Space Planning. *Proceedings of AIIE Conference*, 1971:131.

Roberts, S. D., and R. Reed. "On the Problem of Optimal Warehouse Bay Configurations." *AIIE Transactions* 4(1972):178–85.

Spaziano, R. T. "Applications of Advanced Industrial Engineering Techniques in Warehouse Sizing." *Proceedings of AIIE Conference*, 1971:417.

Way, H. "SPACE I Progress Report." Washington, D.C.: Physical Distribution Services Div., Marketing Publications, Inc. This brief report describes 3 "SPACE" programs, which: (a) "construct" loads of packages in optimum configurations to fit pallets, racks, etc., (b) determine combinations of alternatives resulting in least-cost warehousing, and (c) design warehouse layouts based on forecasts of volumes and product mixes.

White, J. A. "On the Optimum Design of Warehouses Having Radial Aisles." *AIIE Transactions* 4(1973):333–36.

Plant Location

Koopmans, J. C., and M. Beckmann. "Assignment Problems and the Location of Economic Activities." *Econometrica* 25(1957):53–76.

Spielberg, K. "Algorithm for the Simple Plant-Location Problem with Some Side Conditions." *Operations Research* 17(1969):85–111.

———. "Plant Location with Generalized Search Origin." *Management Science* 16(1969):165–78.

Whybark, D. C., and B. M. Khumawala. "A Survey of Facility Location Methods." Institute Paper 350, Herman C. Krannert Graduate School of Industrial Administration, Purdue University, Lafayette, Ind., 1972.

Warehouse Location

Ballou, R. H. "Dynamic Warehouse Location Analysis." *Journal of Marketing Research* 15(1969):271–76.

———. "Locating Warehouses in a Logistics System," *The Logistics Review* 4(1968):23–40.

Baumol, W. J., and P. Wolfe. "A Warehouse Location Problem." *Operations Research* 16(1958):252–63.

Feldman, E., F. A. Lehrer, and T. L. Ray. "Warehouse Locations Under Continuous Economies of Scale." *Management Science* 2(1966): 670–84.

Khumawala, B. M. "An Efficient Branch and Bound Algorithm for the Warehouse Location Problem." *Management Science* 18(1972):718–31.

———, and D. L. Kelly. "Warehouse Location with Concave Costs." Institute Paper 360, Herman C. Krannert Graduate School of Industrial Administration, Purdue University, Lafayette, Ind., 1972.

———, and D. C. Whybark. "A Comparison of Some Recent Warehouse Location Techniques." *The Logistics Review* 7(1971):3–9.

Shannon, R. E., and J. P. Ignizio. "A Heuristic Programming Algorithm for Warehouse Location." *AIIE Transactions* 2(1970):334–39.

Watson-Gandy, C. D. T., and S. Eilon. "The Depot Siting Problem with Discontinuous Delivery Cost." *Operational Research Quarterly* 23(1972):277–88.

Computers in Hospital Design

Delon, G. L., and H. E. Smalley. "Quantitative Methods for Evaluating Hospital Designs." *Health Services Research* 5(Fall, 1970):187.

Freeman, J. R. "Quantitative Criteria for Hospital In-Patient Nursing Unit Design." Doctoral Dissertation, Georgia Institute of Technology, 1968.

Jelinek, R. C. "Nursing: the Development of an Activity Model." Doctoral Dissertation, University of Michigan, 1964.

Sendler, R. N. "Quantitative Evaluation of In-Patient Nursing Unit Design." Master's Thesis, Georgia Institute of Technology, 1968.

Souder, J. J., *et al.* "Planning for Hospitals: A Systems Approach Using Computer Aided Techniques." *American Hospital Association* 38(1964).

Computerized Material Handling Equipment Selection

"Can You Computerize Equipment Selection?" *Modern Materials Handling,* Nov. 1966:46.

Jones, P. S. "A Least-Cost Equipment Selection Technique for Distribution Warehouses." Unpublished Dissertation, Stanford University, 1971; available through University Microfilm, Ann Arbor, Mich.

Chapter 14: Introduction to Material Handling

Apple, J. M. *Lesson Guide Outline on Material Handling Education.* Pittsburgh: The Material Handling Institute, Inc., 1975.

———. *Material Handling Systems Design.*

Basics of Material Handling. Pittsburgh: Material Handling Institute, Inc. 1973.

Bolz and Hageman. *Materials Handling Handbook.*

Footlik, I. M. *Industrial Materials Handling.* Cleveland: Lincoln Extension Institute, 1968.

Hardie, C. *Materials Handling in the Machine Shop.* London: Machinery Publishing Co., Ltd., 1970.

Sims, E. R. *Planning and Managing Materials Flow.* Lancaster, Ohio: E. R. Sims Associates, 1968.

Stocker, H. E. *Materials Handling,* 2nd ed. Englewood Cliffs, N.J.: Prentice-Hall, Inc., 1951.

Woodley, D. R. *Encyclopedia of Materials Handling,* 2 vols. London: Pergamon Press, 1964.

Chapter 15: Material Handling Equipment

Apple. *Material Handling Systems Design.*

Belt Conveyors for Bulk Materials. Washington, D.C.: Conveyor Equipment Manufacturers Association. 1966.

Bowman, D. *Lift Trucks—A Practical Guide for Buyers and Users.* Boston: Cahners Books, 1972.

Conveyor Terms & Definitions. Washington, D.C.: Conveyor Equipment Manufacturers Association, 1966.

Haynes, D. O. *Material Handling Equipment.* Philadelphia: Chilton Book Company, 1957.

———. *Material Handling Applications.* Philadelphia: Chilton Book Company, 1958.

Hetzel, F. V., and R. K. Albright. *Belt Conveyors and Belt Elevators.* New York: John Wiley & Sons, Inc., 1940.

Keller, H. C. *Unit Load and Package Conveyors.* New York: The Ronald Press Company, 1967.

Kraus, M. W. *Pneumatic Handling of Bulk Materials.* New York: The Ronald Press Company, 1968.

Material Handling Engineering Directory and Handbook. Cleveland: Industrial Publishing Co. Biennial.

Smith, D. K. *Package Conveyors.* London: Charles Griffin & Co., Ltd., 1972.

Stoess, H. A. *Pneumatic Conveying.* New York: John Wiley & Sons, Inc., 1970.

Chapter 16: Designing the Handling System

Apple. *Material Handling System Design.*

Morris. *Materials Handling Management.*

Muther, R. *Systematic Handling Analysis.* Kansas City, Mo.: Management and Industrial Publications, 1969.

Chapter 17: Constructing the Layout

Moore. *Layout Planning and Design,* Ch. 7.

Muther. *Systematic Plant Layout,* Ch. 12.

Reed. *Plant Layout,* Ch. 21.

Chapter 18: Evaluating and Implementing the Layout

Moore, *Layout Planning and Design,* Ch. 8.

Muther. *Systematic Plant Layout,* Ch. 14, 15.

Reed. *Plant Layout,* Ch. 22.

Chapter 19: Facility Location

Greenhut, M. L. *Plant Location in Theory and Practice.* Raleigh, N.C.: University of North Carolina Press, 1956.

Isard, W., *Location and Space Economy.* Cambridge, Mass.: The M.I.T. Press, 1956.

Karaska, G. J., and D. F. Bramhall, eds. *Locational Analysis for Manufacturing.* Cambridge, Mass.: The M.I.T. Press, 1969.

Weber, *Theory of the Location of Industries.* Chicago: The University of Chicago Press, 1929.

Yaseen, L. C. *Plant Location.* Larchmont, N.Y.: American Research Council, 1960.

See also under Chapter 13, regarding computers and plant and warehouse location.

See also footnotes for periodical references.

Index